JOH·ALPH·BORELLUS
DE
MOTU ANIMALIUM.
Editio Novissima,
Auctior
et
emendatior.
HAGÆ COMITUM.
Excudit PETRUM GOSSE, Bibliop.
MDCCXXXXIII.

JOH. ALPHONSI BORELLI

Neapolitani Matheseos Professoris,

DE

MOTU ANIMALIUM,

PARS PRIMA.

Editio Nova, a plurimis mendis repurgata,

AC

DISSERTATIONIBUS PHYSICO-MECHANICIS

DE

MOTU MUSCULORUM, ET DE EFFERVESCENTIA, ET FERMENTATIONE,

CLARISSIMI VIRI

JOH. BERNOULLII

MATHESEOS PROFESSORIS BASILEENSIS,

Aucta, & ornata.

HAGÆ COMITUM,

Apud PETRUM GOSSE

MDCCXLIII.

Giovanni Alfonso Borelli

On the Movement of Animals

Translated by
Paul Maquet

With 18 Tables Inside Back Cover

Springer-Verlag
Berlin Heidelberg New York
London Paris Tokyo

Giovanni Alfonso Borelli (1608–1679)
formerly: Professor of Mathematics in Naples

Dr. Paul Maquet
25, Thier Bosset, B-4070 Aywaille, Belgium

The Portrait of Giovanni Alfonso Borelli on the cover was reproduced by kind permission of the Director of the Vatican Library.

ISBN-13: 978-3-642-73814-2 e-ISBN-13: 978-3-642-73812-8
DOI: 10.1007/978-3-642-73812-8

Library of Congress Cataloging-in-Publication Data
Borelli, Giovanni Alfonso, 1608–1679. [De motu animalium. English] On the movement of animals / Giovanni Alfonso Borelli; translated by P. Maquet. Translation of: De motu animalium. Bibliography: p. Includes index. 1. Animal locomotion—Early works to 1800. I. Title. QP301.B613 1989 591.1'852—dc19 88-21841.

Softcover reprint of the hardcover 1st edition 1989

2124/3140-543210

Translator's Preface

Acknowledgement

De Motu Animalium of Borelli is sometimes referred to in papers on biomechanics. Having acquired the book, I was amazed by the concepts which the figures illustrating the work suggested. These figures reminded me of diagrams published recently to support ideas considered as new or even revolutionary. My curiosity thus aroused, I attempted to find somebody willing to translate the book, but in vain. Therefore, I returned to my Latin grammar and dictionary which I had stored away more than forty years ago. After some difficulty in the first pages, I finally understood and could proceed more quickly. I intended to limit myself to the first chapters but, stumbling from curiosity to curiosity, I arrived at the end of the book after more than a year. Fortunately, Professor Charles Fontinoy (Liège) helped me for the understanding of the Introduction and Prefaces which are particularly literary and difficult. Professor Jean Lecomte (Liège) read the translation of the second part and kindly commented on it. Professor Robert Halleux (Liège) spent much time helping my endeavour. He gave me the meaning of some words and expressions the understanding of which requires a good knowledge of the history of sciences in the seventeenth century, and he patiently answered my many questions. His help was invaluable. Professor Halleux and his team also collected information on the author and his work. I thus learned that no complete translation of *De Motu Animalium* was known so far. Mr Ronald Furlong, F.R.C.S. helped me in the editing the book. As usual, also, my wife indefatigably typed and corrected my writing, suggesting improvements here and there. After I submitted a first draft to Springer-Verlag, Dr Götze, enthusiastically agreed to join in the venture and publish the *De Motu Animalium* in its English version.

The Author

But who was Borelli? Settle (in Gillispie's *Dictionary of Scientific Biographies*, vol 2, 1970, New York, pp 306–314) has published a thorough study of his life on which most of the following notes are based.

Born in Naples on 28 January 1608 Giovanni Alfonso Borelli was the son of a Spanish infantryman and his Italian wife. He became a student of Castelli along with Torricelli. He must have been in Rome when Galileo published his

Dialogo and during his subsequent trial. In 1635 or shortly thereafter Castelli's recommendation obtained for him the public lectureship in mathematics in Messina, Sicily. There he published several works among which one was on fevers which he had observed during an epidemic, *On the cause of the malignant fevers of Sicily in the years 1647 and 1648.* In 1658 he was given the chair of mathematics at Pisa. Malpighi who was the professor of theoretical medicine recalled: "What progress I have made in philosophizing stems from Borelli. On the other hand, dissecting living animals at his home and observing their parts, I worked hard to satisfy his very keen curiosity". Malpighi in turn focused Borelli's interest on the movements of living creatures (Boorstin D., *The Discoverers,* New York 1983, Random House).

Borelli was one of the most distinguished members and seems to have been the principal animus of the Accademia del Cimento, organized formally for purely experimental research by Prince Leopold, brother of the Grand Duke of Tuscany, Ferdinand II. While in Pisa Borelli wrote *De Vi Percussionis* published in 1667, and *De motionibus naturalibus a gravitate pendentibus* published in 1670. In 1670 he also published *Historia et meteorologica incendii Aetnaei anni 1669* (Regio Julio). He was one of the founders of "iatrophysics", the application of physics to medicine, corresponding in a wider sense to what we call today biomechanics. In 1675 he joined the Accademia Reale founded in Rome by Queen Christina of Sweden converted to Catholicism and living in this city. Hoping to be received as a member of the Académie Royale des Sciences, recently created by Louis XIV in Paris, he presented as credentials two voluminous manuscripts, *De Motu Animalium*. But, as he had only one copy, he would not trust it to the unreliable mail between Rome and Paris.

He ended his life poorly in Rome where he accepted the hospitality of the fathers of the Casa di S. Pantaleo. For his last two years he taught mathematics at its Scuole Pie. Suffering from a pleurisy, he died in the night of 31 December 1679.

Finally in 1679 Queen Christina had agreed to bear the costs of the publication of the main work of Borelli's life, *De Motu Animalium*. This was published after his death, the first part in 1680, the second in 1681.

The Work

De Motu Animalium thus comprises two parts. The first one is divided into 23 chapters and 224 propositions and deals with the movements of the limbs and displacements of man and animals.

Borelli appears as a genial precursor. He raises problems some of which stimulated the curiosity and endeavours of many generations of researchers and found their solution only recently. For example, Borelli showed that the muscles act on the limbs with short lever arms, whereas the part of the body and load thus carried act with much longer lever arms. Consequently, the joints transmit forces which are several times the weight of the supported part of the body. Pauwels in 1935 surprised the community of orthopaedic surgeons when he compared the hip with scales provided with unequal arms and asserted that during normal walking the hip joint thus transmits a force more than four times the body weight. Borelli already knew this.

Borelli calculated the force exerted by an arm carrying a load in different positions. Only in 1954 did Pauwels solve this problem accurately by criticizing and correcting a previous work of Braune and Fischer.

Borelli experimentally determined the position of the centre of gravity in the human body. Only in 1889 did Braune and Fischer improve on this research by using frozen cadavers and parts of cadavers which they balanced on steel rods in different planes.

Borelli considered the gait of man and other animals in different conditions. Between 1895 and 1904 Braune and Fischer and then Fischer alone, using a photographic method, published a thorough study of normal human gait providing the mathematical basis for further research.

In his calculation of forces, however, Borelli considers that the force equals the resistance only if they are directly opposed. In most instances the acting force is equal to twice the resistance. In other words, for him, the action is not always equal to the reaction. The equality of action and reaction was to be asserted by Newton some years later, in 1687.

In part one the author also analyses running, jumping, moving on ice with skates, the flight of birds, and the displacements of fishes and other animals in water. He ends by describing a diving-bell, a diving-suit, and a submarine.

The second part of *De Motu Animalium* is divided into 22 chapters and 233 propositions. In this part Borelli deals with physiology and analyses the working of the viscera considered as machines.

Borelli is a mechanist. For him the operations thus described are carried out by displacements of particles which are either simply in contact or fitting into each other. These particles are provided with proper shapes. They are absorbed by the organs and tissues through canals the orifices of which are shaped like the appropriate particles and which act as a sieve. Chemical reactions, called fermentations, require huge spaces and stasis of the fermentable substances. Such spaces and delay in the circulation are not found in the body except in the digestive tract and in the testicles. There are thus probably no fermentations outside the digestive tract and the testicles. Biochemistry is ignored and was to develop on a scientific basis only much later.

In all his work Borelli relies on an axiom which he does not question: Nature always acts using the simplest and most economical means. The differences which are observed are due to mechanical necessities. Conversely, when Nature carries out an operation, it must be concluded that this operation is the simplest possible, that it is carried out according to the laws of mechanics and that it is impossible to do otherwise or better. Such was also the opinion of Descartes.

The author keeps comparing the anatomy and physiology of man, animals terrestrial as well as aquatic and flying, and plants. From this comparison he draws general conclusions.

Borelli develops his subject by logical reasoning based on examples taken in everyday life and on experiments in the laboratory. Common sense prevails.

He writes: if blood-letting was actually useful, more febrile patients would heal in France and Spain where they are all subjected to blood-letting than in Italy and elsewhere where they are not. If blood-letting was actually harmful more patients would die in France and Spain. Since no difference is observed, he concludes that blood-letting is neither very useful nor very harmful.

To test whether the heart contains a vital flame he opens the chest of a live stag, incises a ventricle and sets a finger in the heart cavity. His finger is not burned. But it is squeezed transversely. This entails deductions on the pulsation of the heart. Introducing a thermometer in the heart shows him that the temperature there is 40°C and is the same as in the other viscera.

Borelli thus not only observes but also measures. He measures the weight which a muscle or a group of muscles can raise and hence calculates the force exerted by the muscles or the group of muscles.

He also designs geometric models. Using such a model he calculates the increase in volume of the chest due to inspired air. He builds a spirometer to measure the expired air and the reserve air. With a geometric model he demonstrates that the intercostal muscles are involved in inspiration only whereas they cannot provoke expiration which occurs by resilience.

Borelli, however, asserts that there is no attraction in Nature. He claims that almost everybody finds the concept of attraction ridiculous (part II, prop. CXXXIX). This was a few years before Newton!

Checking the calculations of Borelli does not always yield his results (for example in part II, proposition CXLVIII).

The Translation

I could not obtain the first edition of the work and translated the edition published in 1743 by Peter Gosse in The Hague. This edition comprises additionally two dissertations by J. Bernoulli which are quite distinct from the work of Borelli. Their translation will be published separately. The division of the work in short propositions provides continuous and easy reference to the original text without the need of overloading the margins of the translation with indications of pages. I reproduced the references of Borelli as he did. At the end of the book I listed the authors cited by Borelli and their main works, when I could trace them. Some researchers may investigate the works thus mentioned and complete the references in a modern way. I had neither the time nor the means to do it. In translating I attempted to keep very close to the original text and to convey the way of thinking of the author. This may sometimes make some sentences look cumbersome or their meaning questionable. For instance, when Borelli uses the word "moment of a force" he means either the magnitude of the force, or its moment in its modern sense (product of the force and its lever arm), or the product of the force and the velocity of the object moved by this force, or simply the importance of the force. I also used the chemical designations of the time. For example, "spirit of vitriol" could have been translated as sulphuric acid. But the modern term would not have conveyed the exact definition meant by the author. To compensate for this inconvenience a glossary at the end of the present book explains the obsolete meaning of some terms.

Despite its imperfections, may this translation stimulate more research and discussion on the extraordinarily imaginative work of Borelli.

Aywaille, Christmas 1988 P. Maquet

Contents

Part I

On the External Motions of Animals and the Involved Forces

To the august Queen Christina

Giovanni Alfonso Borelli, academician of the Queen S and F

It is well known, my Lady, that people have an insatiable wish for learning. They are indeed instinctively attracted by and forced to wonder at the works and contrivances devised by Nature with utmost cunning. As a consequence of this commendable tendency, people have always honoured the authors of similarly admirable works, have praised them and loved them to the point of giving them divine honours.

Through these stimuli, these seeds or rather these sparks of His inaccessible and ethereal splendour rooted into the human souls, the supreme Goodness has condescended to show Himself to us, to call us to Him and to charm us. He drew our attention by exposing publicly the treasures of His infinite Wisdom which appear in the open book constituted by the heavens, the earth and all the visible creatures and, in the first place, in the anatomy of animals and man. Through them as by degrees the invisible intelligence of God appears. The created structures sing a perpetual hymn to the glory of God and claim the supreme prominence, the unspeakable excellence and the goodness of the Creator. These wonders are described in this small work on the structure of animals which we examine.

All men are called to examine and read this divine book and none is excluded from its vision. However, it is not given to everyone to enter its sanctuary. It is not allowed to everyone to read and understand the secret sentences which are written in the living characters of this book, since this has required untiring study by the past and present most learned men only to find and describe the component parts of the animals and their multiple use. But obviously that which is most secret and divine in them has neither been noticed nor perceived so far. This occurred because simple anatomists and ordinary philosophers are not very keen on understanding the language which the Author of Nature uses to describe his concepts in this sensible book. The language which the Creator speaks in His works is geometry. This was expressed most clearly by divine Plato. When asked what God did, he answered: "God exerts geometry". This famous sentence remarkably commented by learned men seems to apply to our subject.

Animals are bodies and their vital operations are either movements or actions which require movements. But bodies and movements are the subject of Mathematics. Such a scientific approach is exactly Geometry. Similarly, the operations of animals are carried out using instruments and mechanical means such as scales, levers, pulleys, winding-drums, nails, spirals, etc. The scientific knowledge of these means is the province of geometry. Thus it is true that, in building the organs of animals, God exerts geometry. To understand them we need geometry which is the unique and appropriate science to enable one to read and understand the divine book written on animals.

You, my Lady, after having enriched your mind with the treasures of the sciences, through highest intelligence and assiduous study, you understand "the pure works built on the theory of wise men and you are able to judge others and to discover their errors." (Lucr. I. 2)

You have wished to look at the sanctuary of the Divine Book exposed to our eyes while, although unworthy, I was showing the light and teaching. You have wanted to discuss the geometry which the divine hand drew in the sensible and animal world. This geometry publishes the existence, the prominence and the goodness of the Creator. If I am allowed to perceive some small part of these geometrical reasonings which are expressed by the divine Wisdom in the structure and the operations of the animals, the literary community must thank your beneficent and heroic hand. With my endeavours I am your devoted client and slave. Greetings. In Rome, from the pious school of S. Pantaleo. On the Calenders of December 1679.

Carlo Giovanni from the regular clerics of Jesus, general superior of the arm and pious scholars of the Mother of God.

Benevolent Reader, greetings

The book on the Movement of Animals by Giovanni Alfonso Borelli expected for so many years is finally published. Unfortunately this edition could not rejoice the eyes of the author and the author could not carry out himself the ultimate correction which confers to the work perfection and polish. Actually this work is so pleasant and so valuable as to silence envy. It was published posthumously by the most fortunate protection of Queen Christina, tutelary divinity of literature. Her generosity provided the poor Author with considerable funds to improve his destiny and for his work. By her Royal Munificence She behaved as if She were the heir of the genius. Our specialists in mathematics for whose information the author elaborated, rewriting, modifying, adding bricks and sand to the work, as people say, fully understand the meaning of the Teacher. So that which is not yet explained or which is not illustrated in the figures or which is added or changed here and there on different pages, does not disappoint new readers, Queen Christina accepted our help as had been the desire of the Author.

We aim at satisfying three groups of persons: the Author by preparing the result of such a great work which will immortalize his name; the good Queen by displaying the testimony of her generosity to the Author and her love of Virtue; the literary community by exposing a bright treasure of science dealing with the mechanics of the formation and survival of animals, which was unknown so far. This theory should open a large field for further study and new observations.

Be satisfied, good Reader, and congratulate us for this glory which is bestowed upon us. Several Academies from the remotest parts of the world have desired this glory. Several universities in Holland, France and Italy indeed have asked the Author insistently for his work, to publish it at their expense. Borelli, however, has preferred the august name of Christina and has felt honoured by her most human Majesty. She ordered the present work, more lasting than bronze, to be carried out as a remarkable monument to his intelligence.

Something must be said here on the Author and his way of thinking. He sufficiently expresses his way of thinking in his Introduction. I shall thus be brief. What shall I say of the Author since he is known all over the world for his famous intelligence? Naples, the most prosperous city of Italy, very fecund mother of heroes who shine in arms, literature and all sorts of virtues, rejoices to have added to their list Giovanni Alfonso on January the 28th 1608. His parents, Michaele Alonso and Laura, served Philip III, unvanquished king of Spain, in the garrison of a fortified citadel called Castel Nuovo. As a result of his studies in philosophy and mathematics, he was offered the most famous Chairs in Florence and Pisa where he worked, invited by the Princes, and published in Italian in 1649 different works such as the Discourse on the Causes of the Malignant Fevers. In 1658 he published a third edition of Euclid, completed and improved, paid for by Alexandro Falconieri, a most distinguished adolescent who with all his noble family bestowed particular benevolence on his teacher, Borelli. He published the books V, VI and VII of Apollonius from Perga on conical sections in 1661, the Theories of the Medical Influence of Planets in 1666, On the Force of Percussion in 1677, History and Meteorology of the Eruption of Etna of 1669 in 1670. He also published a Treatise on the Natural Movements Resulting from Gravity. In each of these different works he was equal to himself in wisdom and clarity. During the last years he was chosen by Queen Christina and received in our School of Saint Pantaleo close to the Janiculum Hill in Rome to teach mathematics to the future clerics. At the same time, having worked on the elements of Apollonius and on the works of Archimedes he added them to the rest in 1679. He happily lived in Rome for almost two years, remembering the discussions which he used to have with our community of Florence and particularly with Father Francisco from St Joseph, lecturer of Mathematics in the Chair of Pisa, who having returned to secular life is called Famiano Michelini and published a work on the Direction of the Streams, and with Father Angelo from St Dominico who was a disciple of the famous Galileo and who by his scrupulous probity enhances the reputation of wise elders, and with many others. He was such an example of modesty, sobriety and equanimity that he looked like Socrates, Plato or some other of the ancient Founders of schools except that he adhered to the Catholic faith in which he always displayed the purest conviction. When he taught astronomy, a discussion arose about the systems. Whatever others had said, he thought their opinion had not to be taken into account: "The Holy Church teaches such, and such must be believed. We must submit to it." His piety was obvious. Every day he took part in the religious offices. He frequently received the Holy Penitence and the Sacrament of Eucharist. He met Father Master Giacomo Ricci from the Dominicans, secretary of the Holy Congregation of the Index and his germane brother Michael Angelo Ricci, secretary of the Holy Congregation for Indulgences and Relics whom he admired and considered as the two lights of the century and the City because of both their sciences and high principles. Before he went to bed he was often seen kneeling and praying. He would never allow that a small picture of the Holy Virgin be removed from his bed. He had for it the greatest veneration and confidence. A few examples of this kind show how happily he combined sciences and Catholic piety. Moved by these qualities I used to say to his disciples that they were lucky to have a Teacher who was such an example by his religious principles as well as by the display of his theories.

Suffering from pleurisy and knowing that his death was imminent, he piously and humbly asked for the Sacraments. He died with all the signs of Catholic piety on the eighteenth day of his disease between the last hour of 1679 and the first of the new year, assisted as is usual by our religious brothers reading the last prayers which he said were his best consolation. He was buried in our church to which he had entrusted his body. This body is precious for us and our followers.

These are some notes on the Author. You will probably want to know more either from us or from others in another part of this book which is set in press. But first do appreciate this first part.

Greetings. Rome. Ides of August 1680

Introduction

I tackle the difficult physiology of movements in animals. The study of this subject has been attempted by many ancient authors and by more recent ones. To my knowledge, however, nobody has described or even suspected the innumerable, remarkable and interesting problems which are involved and can be discussed. Nobody has succeeded so far in confirming or solving these problems by using demonstrations based on Mechanics.

I undertook this work, not only to illustrate and enrich the part devoted to Physics by mathematical demonstrations but also to enlist Anatomy into Physics and Mathematics not less than Astronomy. If my endeavour does not succeed completely, at least cleverer and more learned men, under my stimulus, will be able to improve and complete this science by firmer reasoning and better methods.

The books on the force of percussion and on the natural movements resulting from gravity were first published. They are now followed by the main work on the movements of animals. This work describes the causes which enable one to predict the motions, their range, the involvement of forces, the mechanical organs which make the movements possible, the contrivances and the reasons why the ordinary processes were ordered by wise Nature.

The present work is divided into two parts. In the first we shall discuss at length the apparent motions of animals, the movements of their limbs, such as flexion and extension, and finally walking, flying, swimming and related conditions.

In the second part we shall analyse the causes of the movements of the muscles, the internal motions, that of the fluids which flow in the vessels and viscera of the animal. We shall not proceed according to the ordinary order of the matter but according to the requirements of a clearer doctrine. We shall analyse the structure of the muscles and show which forces and which mechanical organs move the limbs of animals. Then we shall explain how the muscles work, discuss the motive force led by the nerves to the muscles. Finally we shall consider the internal motions which do not depend on the will, heart pulsations, blood circulation, breathing, its way of action, its organs and its function. We shall discuss the spirits or nervous fluids directing movements and sensations, and involved in nutrition, their motion, their action on displacement. We shall demonstrate the necessity of eating, the digestion of foods, the cleansing of the chyle, and the way nutrition is carried out and how the excrements are discharged by the intestines, glands and kidneys. We shall expose the circulation of bile in the abdomen and that of the genital semen. We shall consider sleeping and waking. Finally we shall discuss the disturbances of some internal motions, diseases, such as convulsions, tiredness and fevers.

In the meantime, learned Reader, know that I will often use the terms of will, order and such in a wide acceptance of these terms even when considering the brutes, by an analogy resulting from our way of speaking.

Chapter I

Enumeration of what must be assumed in dealing with movement of the animals.

Methodically to discuss the displacement of the animals one must consider all the types of movements which animals carry out. First of all one knows that an animal can go from one place to another by moving all its mass from one place to the other. If this displacement takes place on earth it is called walking. If it takes place in water it is called swimming. If it takes place in air it is called flying. One must also consider the different motions and displacements of the parts of the animal which are either external, such as carried out by the hands, thighs, head, etc., or internal such as those of the viscera, heart, arteries, veins, muscles, bones and other parts. There are also flowing and motions of liquids, such as blood and other fluids, through cavities and vessels. To study the ways, organs and mechanisms by which Nature carries out the external motions the obvious must not be neglected: everybody agrees that the principle and the effective cause of movement of animals is the soul. The animals live through their soul and keep moving as long as they live. When dead, i.e. when the soul stops working, the animal machine remains inert and immobile.

Many motions of the animal result from choice or natural tendency. This is obvious and admitted as such by everybody.

It is also evident that ideas and will by themselves are unable to move the parts of an animal but require the necessary instruments without which movements cannot be carried out.

The instruments of movement are either active or organic and purely passive. The active instrument of the soul is called will or motive facultly and is supposed to reside in the animal spirits.

The organ by which the motive faculty of the soul moves the parts of the animal, according to Aristotle, are the spirits which flow from the heart through the arteries to the nerve ends, up to the recess of the limbs. They move the bones, pulling them together thus flexing the limbs or repelling them from each other, thus extending the limbs. However, this theory is rejected by Galen and all the others, and by the evidence that muscles are instruments and machines by which the motive faculty of the soul moves the limbs and the parts of the animal.

This is confirmed by the fact that transverse division of the muscles stops any flexion of that limb in the vicinity of which the muscle was attached. However, movement of the limb which depends on other muscles persists.

Similarly, it is known that muscle by itself is a dead and inert machine in the absence of an external motive faculty. The latter orders, stimulates the muscle from its lethargy and forces it to move. Therefore, in sleep an intact muscle, for example the biceps, does not move the arm except if forced to act by the will.

So far it has been investigated as to by which ways the order of the soul and the motive faculty are transmitted to the muscle. Common sense and experience answer this question. Arteries, veins and nerves arrive to the muscle from outside. Neither arteries nor veins transmit this command since despite ligature or division of arteries and veins, the muscle continues to carry out its motions as previously, when the vessels were intact. Consequently, the motive faculty is not transmitted though the veins or the arteries to the muscles to be moved. Contrariwise, immediately after division or ligature of the nerve which ends in a muscle any movement of this muscle stops and the muscle remains inert and as immobile as in a cadaver. Therefore, the nerve is the path through which the motive faculty transmits from the soul to the muscle the order to be stimulated, to move or to carry. But what is actually transmitted to the muscle by the nerves? Is it an immaterial faculty or a gas or some liquid or some movement or an impulse or anything else? How does it overcome the resistance of considerable weights? We shall answer these questions in due time. Now it is enough to know that the nerves transmit the order of the motive faculty of the soul, without which no voluntary movement can be carried out.

Chapter II

Description and function of the muscle.

As usual in the other physico-mathematical sciences, we shall attempt to explain the science of the movement of the animals from observations and facts. Since the muscles are the principal organs of animal movement we shall at first consider their structure, their constitutive parts and their obvious actions.

Proposition I

Structure of the muscle.

A muscle is an organ consisting of tendon, membrane, flesh, veins, arteries and nerves. Tendon is observed mostly at the origin and insertion of the muscle. It seems to have a nervous consistency and to take part in the connection with the bone. The first tendinous part of the muscle is called the head, its end the tail and its intermediate part the body. The body is full of muscular flesh. Some authors believe that this flesh does not really constitute the muscle but presents the appropriate consistency to fill the interstices of the fibres and thus avoid injury of these fibres during movement. I think this view is erroneous. Inside the nervous or tendinous membrane, there are many bundles which have a prismatic shape, either hexagonal or quadratic or triangular. The individual prismatic fasciculi are constituted of many filaments or tendinous fibres. These fibres are themselves prismatic and parallel to each other. They are linked firmly together if they are not continued into tendons or membranes and sometimes they are connected directly to bones or fleshy fibres.

Moreover, in some places these fibres are bound together by countless transverse fibres as appears in a muscle boiled and then dessicated. These nervous fibres seem to constitute reticular membranes with afferent and efferent blood capillaries. Mixed with these fibres there are strong and hard nerves which resist tension and rupture when we attempt to dissociate the fibres with a needle.

Although the muscular fibres appear red and bloody, all are white. Their red colour results from the blood which fills them like sponges and impregnates them. This blood can be removed by washing. Then the fibres remain quite white very similar to tendinous fibres with which they are confused. They also display the same firm consistency as tendons and nerves. Actually they resist tension as shown by the gracilis muscle which supports more than 80 pounds without rupture of its fibres. Despite their softness, these fibres not only are not stretched but they contract spontaneously.

Individual fibres swell after being boiled. Under a microscope they appear as small cylinders similar to the twigs of a tree. They do not seem to be empty tubes as are reed stalks. They appear to be filled by a substance, some medulla which must be spongy as in elder. Since any soft twig when swollen by some afferent humour becomes turgescent and straightens, the fibres must be porous and filled by watery particles as if it were by wedges, as occurs in a wet rope.

This is also suggested by the fact that in the fibres of muscles impregnated by blood and dried, such as are seen in ham examined under a microscope, one observes some droplets of blood or fibrils longitudinal and transverse, separated from each other like in porphyry. Such a picture could not be seen if the internal substance of the fibres was not spongy.

On the other hand vessels and capillary nerves are prismatic connective bundles thinner than the columns or muscular fibrils although the latter are thinner than a female's hair.

Finally only loose transverse bindings are observed in the muscular bundles. Like tendons, nerves and membranes do not contract when the muscular fibres do. This appears in vivisection, mainly in the membrane of the diaphragm which wrinkles when the included muscular fibres contract.

Proposition II

Muscle does not differ from flesh.

An error of the ancient authors must be refuted. They distinguish between muscle and flesh and think that muscle is formed by tendinous fibres. Flesh would be an addition different from fibres, hairy stuff containing blood and surrounding the tendinous fibres. They invoke the following argument. In emaciated or starved animals the muscular fibres are very thin and deprived of flesh whereas they are thick in healthy animals. If the muscles of healthy animals are battered, squeezed and crushed there remain in them the same thin fibres as in animals dead from exhaustion or hunger.

But, if I am right, flesh cannot be different from muscular fibres. Whether they are thinned or not, if they are washed in water several times, they always appear white under the microscope and of tendinous consistency. No stuffing appears in them. Even in the fleshy muscles which are believed by these authors to be loaded by this fleshy stuff, these fibres seem to have the same consistency and the same columnar arrangement as in muscles washed by water. The same appears in muscles which have been boiled or are preserved in salt.

However, it is true that this thinness of the fibres in emaciated or compressed muscles may result from a lack of the nutritive juice which filled the pores of the fibres. This occurs also in dried sponge and in dessicated tree leaves. In these there is no lack of stuff but only of the juice which previously filled the cavities. Consequently, muscular fibres and flesh are the same.

However, muscular fibres are different from the fibres of the tendons and membranes. How different are these muscular fibres spongy and impregnated by blood, nervous and lymphatic juice from the tendinous and membranous fibres? How much do they differ in their structure, in their action, in their energy? These questions will be answered in due time.

Proposition III

Types of muscles.

Finally the different types of muscles must be mentioned as well as which parts of the muscles act, how they act and in what direction. There are two types of muscles. Some are made of bundles of fleshy fibres which constitute a rectangular prism (Fig. 1, Tab. I*). They can be called straight prismatic.

Others constitute a rhomboid (Fig. 2, Tab. I). They can be designated as rhomboidal.

Others are made of intersecting fibres, resulting from two rhomboids (Fig. 3, Tab. I). They can be called decussated.

* Tables I-XVIII can be found as loose cards in the pocket inside the back cover.

Others result from two rhomboidal bundles which do not intersect (Fig. 4, Tab. I). They are said to be multipennate.

Others are made of orbicular fibres and are called radial.

Others are circular and are called circular sphincters.

Others are constituted by circumvoluted fibres like balls of wool. They are called spiral, orbicular.

In the tail of a crab I also saw muscular structures not like a web but rather made of bundles of complicated fibres, as women's hairs are usually arranged.

Proposition IV

Muscle acts by contracting.

In muscle we see that only the fleshy threads AB, CD, EF and C (Figs. 1, 2, 3, 4, Tab. I) contract when the muscle acts. The tendons at the extremities BH to which the fleshy fibres are attached do not contract but retain their initial length. This is obvious in vivisection.

Consequently, only the fleshy fibres AB, CD, EF, GH and C exert a force by contracting when they carry considerable weights. The tendons BH are subjected to a force when moved by the contraction of the fleshy fibres. They undergo this force like a handle to which the fibres are attached.

Proposition V

Examination of the structure of muscle, as recently discovered, and of its mode of action.

New thoughts on the true shape of muscle and on its mode of action have appeared in the last few years. I shall express my opinion about these new thoughts (Tab. I, Figs. 5, 6, 7 and 8).

Some authors assume that simple rhomboidal muscles have been found in animals, such as the muscle ABCD (Fig. 5, Tab. I). Its tendon AC is firmly attached to the bone EAC or is fixed at point E. The opposite tendon BD is parallel to AC. Both tendons are separated from each other. There are two opposite forces. One is the weight R which pulls the tendon BD downwards from B towards F. The other is the contraction force of the fibres which pulls the weight R upwards obliquely from B towards A and from D towards C. The authors also assume that this action is carried out by the tension of the fibres without addition of a new body since the fibres do not appear to swell, to increase or to decrease in volume. Finally they claim that every time that, in the rhomboid ABCD in which the two opposite planes AC, BD remain unchanged in dimension and the mass of solid substance mentioned above is neither increased nor decreased, only the oblique fibres AB, CD are shortened, the obliquity of the parallelogram ABCD necessarily must diminish and tend to straightness AGHC. Henceforth, the acute angle BAC opens and becomes GAC. This pulls the weight R upwards.

All this reasoning is based on the third Euclidean proposition: two parallelograms ABDC and AGHC formed by two parallel lines with the same base AC are equal, and inversely. Consequently, the equal parallelograms thus mentioned have neither the same length nor the same width since the one ABDC which is more inclined is longer and narrower than the one which is less inclined AGHC, and since where the sides AB, CD are the more shortened, the width of the parallelogram is the more increased. Let us see whether the reasoning of these famous authors coincides with the principles and with the experiments. When the fibres AB, CD of the parallelogram are shortened and have become AG, CH, these fibres must become wider. Otherwise they would not fill the space. Therefore, the muscular fibres when contracted become thicker, which is against their hypothesis.

Secondly, all of the fibres in a straight muscle shorten parallel to each other. Therefore, since there is no interpenetration of bodies they must swell and thicken. Consequently, the thickness of the whole muscle must be increased, which the authors also denied.

Thirdly, when ribs are brought closer together all the fibres of the oblique intercostal muscles shorten. The interstices between these fibres cannot increase since the obliquity of the fibres rather increases. Consequently, the whole volume of the muscle would decrease, which the authors denied.

Finally, what is important in this matter is the mechanism by which, using the organ, the force of the muscle moves a resistance. However, the structure and arrangement of the muscle or of a rhomboidal instrument, appear to be very inappropriate to raise a weight R. This could be demonstrated easily as will be explained further. But, to avoid disturbing the didactic order, it is sufficient to close the debate by sensible experiments.

Two equal straight rods AC and BD (Fig. 6, Tab. I) are taken and connected by several threads AB, CD, etc., of equal length. An extremity A of a rod is attached to a nail at E and a weight R is suspended from the extremity D. You observe firstly that the shape of the parallelogram ABCD is destroyed, the rod BD moves to the contact of the rod AC, the two rods unite and form a straight vertical line AC, DR.

If the number or the thickness of the intermediate threads prevents contact between the rods, the parallelogram appears constricted and lengthened (Fig. 7, Tab. I). Its diagonal ADF carries out an oblique movement towards a vertical position. The same happens if the fibres AB, CD are solid but flexible as are the twigs of a tree. But in this instance, the parallelogram remains larger. We see that shortening the threads AB, CD, either by pulling them upwards or by wetting them, provokes raising of the ropes with the suspended weight R. We also observe (Fig. 5, Tab. I) that, to impede contact and union of the rods BD and AC and inclination of the whole parallelogram, it is necessary that the rod BD be retained either by transverse ties or by forces pulling transversely in X and Z. Then, when the threads contract, BD moves towards AC remaining parallel to itself. Never do the ropes rise towards AG about the centre A as long as the rod BD is pulled downwards by the weight R. Therefore, the moving power of the fibres cannot raise the resistance R by using a single rhomboidal muscle.

However, the proposition can be verified in a particular instance. If the fibres was attached to a fixed bone EAC (Fig. 8, Tab. I) and the side BD of the rhomboid were retained in a smooth and slippery groove LF carved in a column, then contraction of the fibres AB, GH, CD could pull a tendon BD obliquely upwards with the suspended weight R. But this hypothesis does not apply to animals in which there are no such single muscles having the shape of a rhomboid with a tendon or a mobile side BD moving in a smooth groove. In conclusion, such single muscles are not seen normally and do not act in the way these famous authors think they do. This mode of action can occur only with some muscles composed of several rhomboids as we shall see later. It cannot occur with these single muscles which constitute one rhomb alone about which these authors speak and which they give as examples.

Proposition VI

True picture of muscles.
Tab. I, Figs. 9, 10, 11.

Autopsy shows that, in animals, there are columnar muscles made of parallel fibres such as ABCD (Fig. 9, Tab. I). They pull the resistance R in the direction AB of their fibres. Such muscles are found particularly in the tongue of the woodpecker, in the tails of crabs and in many others.

There are also simple rhomboidal muscles AB, DC (Fig. 10, Tab. I). Their oblique fibres AB and CD are inserted in the bone TV and pull another bone or tendon RS in their direction. Such are the abdominal and intercostal muscles and others. The bone RS resisting tension is moved transversely and parallel to itself towards the fixed bone TV. It must be noted that the angle VCD does not become less obtuse but, on the contrary, more obtuse.

Finally, there are multipennate muscles (Fig. 11, Tab. I). They were first observed and described by Casserio of Piacenza. They are built most appropriately for the ends which will be explained further. Their oblique fibres act by contracting. A suspended weight R is subjected to their tension like a weight which is suspended by two or more oblique ropes, the force and way of action of which we shall describe.

Proposition VII

The muscles exert a double force, their own and an instrumental force.

Experience shows that muscles exert two forces. One is due to the fibres themselves and results from their normal structure. The other results from an external cause and is used by the muscles and by instruments to carry considerable weights. Indeed the fibres are far from being inert. They have some contraction force. In cadavers, shortly after death, divided muscles contract. In life, such contraction is carried out without painful exertion. It is the same contraction which is carried out by zither players. This contraction results from the structure of the little machines of which the fibres are made.

The degree of the motive faculty of these little machines is small. This appears in the short contraction of the muscle after division of one of the opposite ligaments or after division or removal of an antagonist. If no contracture occurs, the power of the contraction can be overcome by a weight of a few pounds.

When a joint is flexed as much as possible, the muscle located in the concavity remains loose and, therefore, cannot contract further. Then the muscle at the convexity can exert all its force without antagonistic opposition. However, we see that it is too weak to move a weight or to overcome the resistance of the limb itself since it does not succeed in straightening the joint spontaneously. This appears from the following. When the forearms, the lower legs or other limbs are completely flexed and pointing upwards perpendicularly, the weight of the erected bone and limb obviously does not oppose flexion and extension. Then the extensor muscles which are no longer opposed by their antagonists, should act spontaneously and develop their natural energy to flex the joint, either without notice or even against our will. But this does not happen and we, without any feeling of being tired, resist the natural action by which the machines of the muscular fibres attempt to contract. Moreover, we do not perceive any fatigue from the continuous action of the muscles against their antagonists to keep the joints in their normal position by their tone.

Besides the weak contraction of their fibres which the muscles exert against their antagonists, they also develop a much more powerful voluntary contraction by which they raise considerable weights.

Some people seem to confuse this latter action with the former. They are completely different. Therefore, I shall distinguish them by calling the former proper action of the fibre and the latter vital action of the muscle, carried out by will or by wish.

Chapter III

Magnitude of the vital motive power of the muscles
according to the ancients.

In this first part of the work it is investigated how great the vital motive force of the muscles is in relation to the resistance. First of all we must know what our predecessors transmitted to us on this subject.

Proposition VIII

It is commonly thought that Nature raises considerable weights by using the machines of the muscles with a weak moving force.

The magnitude of the vital force of the muscles must be measured. This force sustains, raises and moves not only an arm or a leg, but the whole animal machine, enabling it even to dance. Besides the mass of the animal, heavy enough by itself, this force carries, pulls and pushes considerable weights.

Aristotle above all dealt with this matter. He did not recognize the muscles but imagined spirits which pull and push the limbs. This perspicacious author sensibly remarked how difficult it would be for the huge mass of an elephant to be moved and pushed by tenuous spirit or wind. He met the difficulty by saying that Nature moves the joints and limbs of the animal by using very small force. He said that this results from the work of the machine by which motions initially small soon lead to large and multiple displacements. Similarly, small and easy motion of a pole or of the rudder provokes large displacements of the bows and of the boat. Then he considered the nature of a pole from a mechanical point of view and said that the operation is carried out by way of a lever. Therefore, it is not surprising that huge weights can be moved and displaced by a small force. Lucretius used the same example.

> "One hand steers the ship which moves forwards with the impetus which you want and one rudder turns in one direction."

And then mentioning other instruments he said:

> "Using pulleys and winding-drums it moves many heavy things and the machine raises them with little effort."

Gassendi also claims that using a machine the animal faculty can move heavy weights with little forces but he doubts that levers, pulleys and winding-drums can be found in animals since muscles alone which are rather like the ropes of pulleys are observed. However, he finds in muscle a mechanical contrivance made of pulleys and ropes. He says that the folds which appear in the fibres of muscles when the length of the muscle is shortened by contraction play the role of pulleys. A series of pulleys results from these folds like in tackle the force of which is almost boundless.

Galen also says that a tendon is like a lever. He thinks that, consequently, a small force of the animal faculty can pull and move heavy weights.

Others also, but for a different reason, claim that muscles raise heavy weights by way of a machine, despite a small force of the spirits.

This general opinion seems to be so likely that, to my knowledge, not surprisingly, it has been questioned by nobody. Who indeed would be stupid enough to look for a machine to move a very light weight with a great force, i.e. use a machine or a contrivance not to save forces but rather to spend forces? This is as if one who can move and carry a weight of one pound directly without any machine by exerting a force equal to one pound, with complete disregard for economy, was looking for levers, pulleys and other intruments requiring forces ten or a hundred times greater to raise this one pound. And, if this is rightly considered as stupid, how is it possible that wise Nature, everywhere looking for economy, simplicity and facility, builds with great industry in animals machines to move, not heavy weights with a small force but, on the contrary, light weights with almost boundless force? This seems strange and against common sense, I agree, but I can convincingly demonstrate that this is what happens and, given permission, that the upholders of the opposite opinion have been mistaken. I shall demonstrate that multiple and different machines actually are used in the motions of animals but that light weights are carried by large and strong force rather than heavy weights being supported by small force. This is such that often the motive force is a hundred or a thousand times greater than the weight of the bones and limbs which are raised and is never smaller. This is the main subject of this first part of my work.

Chapter IV

Theorems useful to show the hugeness of the motive power of the muscles.

Proposition IX

The flexion movement of limbs is spherical or circular or takes place over a conical surface about an imaginary centre.
Tab. II, Figs. 1 and 2.

During all flexions of limbs it must be seen at first how the bones move and what figure they describe when moving.

Firstly, there is no doubt that almost all the movements of the animal limbs are circular or made of circular components. This appears from the fact that any movement of the bone of a limb occurs about the extremity of another bone forming the joint. This bony extremity constitutes a fulcrum about which the first bone moves. If the extremity of the second bone is immobile then the movement of the first bone considered will be circular in the same plane. If the fulcrum is not fixed but is unsteady (Fig. 1, Tab. II) and moves, the movement of the articulated bone will no longer be circular, but straight or curved in variable ways. Illustrating the first assumption, it appears that, the humerus AE being immobile, the ulna AB moves about the joint or fulcrum A and this movement is circular.

But, although the movements of the limbs are circular, the position of the centre of rotation of the limbs and bones is not obvious and must be found. If the bones of the limbs were solid lines their contact would be a point which should be considered as the centre or fulcrum. However, since the bones have some thickness their extremities cannot easily contact and articulate at one point to rotate about this point of contact. This would be possible if the extremity of one of the bones was pointed, like the apex of a cone or of a pyramid, and if this tip was attached to the cavity of the opposite bone which would be immobile. Then the point of contact would be the fulcrum and the centre of rotation. But this would be very inconvenient and fragile. If indeed the ulna ended in a conical apex and if this conical extremity were set in a conical cavity with greater slope, carved in the extremity of the humerus, such a pointed protuberance could easily be crushed and broken and the joint could not be linked strongly enough to avoid unsteady contact at the apex and deviation. Actually, foreseeing and wise Nature, to avoid these inconveniences, created a joint which is easy, safe, stable, resistant, and minimally liable to dislocation. She shaped the ends of the bones round, one convex, the other concave so that contact does not occur at one point but over a wide area. This avoids crushing and fracture. Moreover, in this way the bony extremities can be attached more easily and more firmly without risk of dislocation during various and multiple movements. This does not give a centre of rotation but rather a cylindrical fulcrum over which the bone moves. Any point in which the moving bone rests over the immobile bone is not a fixed or stable point and, therefore, there can be no centre of rotation. The articular end of the bone AB (Fig. 2, Tab. II) is spherical or cylindrical ADEF.. That of the bone GD is concave. It contains and surrounds exactly the extremity of the other bone AB. No point can be considered as immobile and stable in the bone DG. Any point of this bone describes an arc of a circle during movement. The circles described by the points are unequal. They are the greater, the closer the points are to the extremity G of the bone. All of them necessarily describe predictable movements about a point I which is not in the bone DG but which is the centre of the sphere or of the cylinder ADF. It is as if a straight line were conceived drawn from the extremity G entering the epiphysis ADF and passing across the centre I of this epiphysis and all this line moved, except for one of its points. This point is the centre and fulcrum about which the bone rotates. This centre of the joint lies outside the moving bone DG, being at the centre I of the epiphysis of the immobile bone.

If, inversely, the bone DG is immobile and the bone BA moves, the centre and fulcrum is not at the contact DC but at point I in the epiphysis. Therefore, the centre of rotation of the humerus or of the femur is exactly in the middle of the head of these bones which is contained and attached in the immobile glenoid cavity of the scapula or socket of the pelvis. In these two joints the mobile hemispherical extremity is protruding. On the contrary, the centre of the cylindrical articular extremity of the ulna is outside the ulna, in the middle of the immobile epiphysis of the humerus which is surrounded by the articular extremity of the ulna to which it is attached. The same can be said of other similar joints.

It must also be noted that the movements of the limbs are spherical for some, circular in one plane for others and often they take place over the surface of a cone. As a rule, whenever a bone can move anywhere about one fixed point, then the movement is spherical and possible to the right, to the left, upwards, downwards, forwards or backwards. When the movement must take place about two poles or about an axis it must be a flexion-extension either circular in one plane or over the surface of a cone. An example of spherical movement is given by the humerus. Its proximal end is spherical and globulous. Thanks to this sphericity of its epiphysis the humerus can move in all directions since all the radii from the centre of the humeral head to its surface are equal. Therefore, contact is equally good everywhere over the humeral head which moves with the greatest facility in the concentric glenoid cavity. The movement of the ulna about the humerus and that of the tibia in the knee are different because in these joints there is no centre of rotation but an axis between the two poles of a cylinder. Actually the distal extremities of the humerus and of the femur are not spherical but cylindrical with furrows which are as so many pulleys which enhance stability and prevent dislocation during motions. Because of these cylinders, points on the limbs must move over circles with the same centre as that of the cylinder and deviate neither to the right nor to the left in the direction of the poles. This results from the nature of the cylindrical shape: in a cylinder there is no midpoint equidistant from all the points on the surface except if one takes points on a circumference at equal distances from the bases of the cylinder. All the circles of the cylinder are equal and parallel to the bases. Such a construction only allows movements of the ulna and tibia along circles parallel to the base of the cylinder. If the mobile bone is at right angles to the axis of the distal cylinder of the humerus or femur, then the ulna or tibia moves in the plane of one of the circles which are at right angles to the said axis of the cylinder. If the tibia is oblique to this axis, in moving it will necessarily describe the surface of a cone, inclined towards the side where it forms an acute angle with the axis.

It is also interesting to know the trajectories described by the extremities of the limbs when these are not fixed but move. These trajectories can be straight, conical or irregular. Others have written on this subject with much erudition and elegance. We shall refer to them.

Chapter V

Although contracted with a maximum effort by a determined magnitude of motive power, a muscle sometimes may exert very little or no force.

Beside the positions of the muscles and their insertions their motive forces must also be considered. Although they are thought to have one and the same magnitude as a result of their nature it may happen that the muscles exert only a smaller force or no force at all. I do not speak of their moment which depends on other circumstances as will appear later. I speak of the motive power itself which, though exerting its forces with the greatest effort, may induce no force as if it were totally absent. This assertion may seem to be absurd. But it is easy to demonstrate its truth.

Proposition X

Muscles inserted in the concavity of flexed limbs become loose.
Tab. II, Figs. 3 and 4.

As appears at autopsy, muscles seldom span one joint alone. More often they bridge two, three, four or more. Let us assume that three bones AB, BC and CD form two joints. The two extreme bones articulate similarly with the middle one about the joints B and C. The bones being aligned, a muscle AED is attached at A and D. The same muscle is tightly connected to the midbone by ligaments and membranes. Let us assume that the extreme bones AB and CD are flexed, not by the force of the muscle AED but by some other external cause in such a way that the muscle exerts no force at all. The muscle does not contract and takes a concave shape FHEIG (Fig. 4, Tab. II). This results from the following. The joints are made of three mobile parts articulated similarly at B and C and the muscle FEG is attached at the extremities A, D like a rope. After flexion, although the muscle is close and linked to the midbone in the concavity between the joints B and C, a line drawn through the middle of the muscle, i.e. its axis, remains away from the angles B and C for two reasons, one being the thickness of the muscle itself, the other the position of points B and C in the depths of the bones of the joints. Two lines drawn similarly from the extremities in the concavity ABCD and FHIG have the same ends F and G. The enveloped line FHIG thus is shorter than the enveloping line ABCD. But the normal length of the muscle FEG was equal to the line ABCD. Consequently, when the joints are flexed, the muscle is longer than the line FHIG and it thus must be loose or at least less tight than it was in extension AED.

Proposition XI

When the joints are flexed, the muscles in the concavity exert a smaller force or even no force at all.

The vital action of the motive power of the muscles is exclusively a contraction. This contraction is small, attaining two finger breadths, seldom three or four as appears in vivisection. On the contrary, loosening of the muscles in extreme flexion of the joints is considerable, sometimes greater than three or four finger breadths. When the muscles are relaxed they obviously retain the same length as the bones into which they are inserted in the concavity of the joints. Their greatest voluntary contraction is not more important than their excess of length resulting from the flexion of the joints. Then the bone and a suspended weight can only be pulled minimally. Consequently, voluntary contraction of the muscle does not evoke any force on the suspended bone. It is as if there were no pull since a loose rope is unable to carry a weight attached to it. Contraction thus will produce no pull, exactly as if there was no motive force.

What common sense suggests is confirmed by an obvious experiment. Anatomy shows that the flexor digitorum muscles have their origin in the distal epiphysis of the humerus. They insert into the palmar aspect of the distal phalanges of the fingers and are connected with all the elbow, wrist, metacarpophalangeal, proximal and distal interphalangeal joints by ligaments and membranes. If all these joints are flexed completely, then the muscles do not shorten sufficiently and become loose and redundant. As shown by experience, the tips of the fingers then exert very little pressure against the palm of the hand despite strong attempts at such compression. The same is observed with many other muscles as we shall see in due time. Therefore, there can be no doubt about the truth of this proposition.

Chapter VI

Mechanical lemmas useful to explain the power or the moment of the muscles.

Because of its looseness, the motive force of the muscle is weakened. This force is also changed for another reason which can either increase or decrease its moment. In order to proceed cautiously and scientifically in this matter, some lemmas must be presented.

Proposition XII

No force, whatever its magnitude, if passing across the point of application of a lever i.e. its fulcrum, can balance a load, however small, acting at the end of the lever.
Tab. II, Fig. 5.

The point of application or fulcrum of the lever AB is C. A load R, even very small, is suspended from the extremity B of the lever. A huge force D pulls the lever at point C. The line of action EC of force D passes across the fulcrum. I claim that the moment of load R is larger than that of the pulling force D. The load D cannot overcome and not even balance the resistance R. The ratio of the moments of forces R and D results from the ratio of the absolute magnitudes of these forces and the ratio of the distances of their lines of action from the fulcrum. The distance of R from the fulcrum is the lever BC, that of the rope EC is 0. Therefore, the ratio of the moment of force R to that of force D is equal to R/0. Consequently, the moment of the resistance R is always larger than that of force D and the resistance R, however small it may be, will never be counterbalanced by force D however large this may be. Q.E.D.

Proposition XIII

If two opposite forces are applied at an extremity of a lever, one acting at right angles and the other pulling obliquely, and if they have the same moments, the ratio of the magnitude of the force pulling obliquely to that acting at right angles is equal to the ratio of the length of the lever to the distance of the oblique line of action from the fulcrum.
Tab. II, Fig. 6.

Two opposite forces C and E are applied at the end B of the lever AB the fulcrum of which is A. Their moments are equal. Therefore, one balances the other. Force C pulls directly. Its line of action CB is at right angles to the lever BA. Force E pulls obliquely. Its line of action HB forms an acute or an obtuse angle ABH with the lever. The distance of this line HB from the fulcrum A is AH. Thus AH is at right angles to HB. I claim that the ratio of the magnitudes of the forces E/C is equal to the ratio of BA/AH. The line AH is prolonged by a distance GA equal to BA. A force I equal to C is applied at G. The line of action GD of I is at right angles to GA. Since the equal forces C and I act with equal lever arms AB and AG and at right angles to these lever arms, their moments are equal. But the moments of E and C are also equal. Therefore, the moments of forces I and E are equal. On the angulated scales with equal arms GAB the fulcrum of which is A, two forces E and I with equal moments are thus applied. I pulls at right angles. E is oblique, acting in a direction HB parallel to GD (force B being removed or neutralized). Consequently, as we showed in the Book on the Force of Percussion[1], the ratio of the magnitudes of the forces E/I and the ratio of the magnitudes of the forces E/C are equal to the ratio AG/AH = AB/AH. Q.E.D.

---

[1] Prop. 39

Proposition XIV

If two opposite forces are applied obliquely at the same end of scales or of a lever and have equal moments, their magnitudes are in inverse proportion to the distances of their lines of action from the fulcrum.
Tab. II, Fig. 7.

The fulcrum or centre of rotation of the lever AB is A. At its extremity B two forces E and D are applied pulling obliquely the lever in the directions HB and GB. The moments of these forces are equal. Pulled by opposite forces the lever, of course, remains immobile. The forces balance each other. Perpendiculars AH and AG are drawn from the centre of rotation A to the lines of action of the forces. I claim that the ratio of the magnitudes of E and D is equal to the ratio of GA/AH. The lever BA is prolonged to F in such a way that AF is equal to AB. At the extremities B and F, two forces C and K are suspended pulling at right angles to the scales. Their moments are equal to the moments of any of the forces E or D. Two forces E and C of equal moments are applied at the extremity B of the lever AB (forces D and K being momentarily removed or neutralized). One pulls at right angles, the other obliquely to AB. The ratio of the magnitudes E/C or E/K thus is equal to AB/AH[2]. On the other hand, two forces K and D of equal moments are applied to the scales BAF the arms of which are equal. K pulls at right angles, D pulls obliquely to the scales. Therefore, the ratio of the magnitudes K/D is equal to the ratio of the lever arms GA/AB[3]. Hence, after exchanging the terms of the equations[4], the ratio of the magnitudes E/D is equal to the ratio of the lever arms GA/AH. Q.E.D.

Proposition XV

If two opposite forces are applied at two points of the same lever and have equal moments, their magnitudes are inversely proportional to the distances of their lines of action from the fulcrum.
Tab. II, Fig. 8.

The fulcrum of the lever AB is B. Two opposite forces H and E are applied at points A and C in such a way that the lever remains immobile. In other words, the moments of the forces are equal. Their lines of action are AH and CF. Two perpendiculars BI and BF are drawn from the fulcrum to these lines of action. I claim that the magnitudes E and H are inversely proportional to the lever arms BI and BF. Two opposite forces D and N are applied at points C and A. Their lines of action AN and CG are at right angles to the lever BCA. The moment of force D is equal to the moment of force N, to the moment of force H and to that of force E. The ratio of the magnitudes D/N is equal to AB/BC.

[2] According to Prop. 13
[3] According to Prop. 13
[4] Prop. 20 lib. 3. Eucl. rest. Auct.

At the end C of the lever BC the fulcrum of which is B, two forces of equal moments are applied, one D at right angles, the other E obliquely. Therefore, the ratio of the magnitudes of the forces E/D[5] is equal to BC/BF. Hence, exchanging the terms of the equations, the ratio of the magnitudes E/N is equal to BA/BF. Similarly, the ratio of the magnitude of force N pulling at right angles to the magnitude of force H pulling obliquely is equal to IB/AB. Consequently, exchanging the terms of the equations[6], the ratio of the magnitudes of the forces E/N is equal to the ratio of the lever arms IB/BF. Q.E.D.

Proposition XVI

If two opposite forces are applied at the extremities of an angulated balance the fulcrum of which is at the apex of the angulation and if these two forces have equal moments, their magnitudes are inversely proportional to the lengths of the lever arms. Reciprocally, if two forces are inversely proportional to the lengths of their lever arms, their moments are equal.
Tab. II, Figs. 9 and 10.

The fulcrum of an angulated lever or balance ABC is B. Two opposite forces H and E the lines of action of which are IA and CD are applied at points A and C and ensure equilibrium. The forces thus have equal moments. Their lever arms are BI and BF. I claim that the magnitudes of these forces H and E are inversely proportional to BF/BI. CB is prolonged to the right so that BN = BI. A force O equal to H is applied at N at right angles BNL to BN. The two equal forces H and O pulling with equal lever arms BI and BN have equal moments. According to the hypothesis, the moments of forces E and H are also equal. Therefore, two forces O and E having equal moments are applied at the extremities of the balance NC the fulcrum of which is B (force H being removed or neutralized) and their lever arms are BF and BN. Consequently, the ratio of the magnitude of force E to that of force O or of force H which is equal to O is inverse to the ratio of BN or BI to BF.

On the other hand, with the same data, the ratio of the magnitudes E/H (Fig. 10, Tab. II) is equal to the ratio of the lever arms BI/BF. I claim that the moments of the forces E and H are equal. The same construction is built as in Fig. 9. In the imaginary scales NBC the ratio of the magnitude of E to that of O or of H which is equal to O, is equal to the ratio of NB or its equal IB to BF which are the lever arms. Thus, the moments of E and O are equal. But the moment of H is equal to that of O since they are two equal forces acting at right angles with equal lever arms. Consequently, the moments of forces E and H are equal. Q.E.D.

[5] Prop. 13

[6] Prop. 20 lib. 3. Eucl. rest. Auct.

Chapter VII

The tendons of the muscles must not be attached to the extremities of the articulated bones but well on tuberosities in the vicinity of the joint with their direction oblique to the longitudinal axis of the bone.

After these premisses, I notice that Nature in all her works operates with admirable wisdom and cunning and never without reason. Nobody could have arranged the tendons of the muscles better than they are in animals. Since the tendons are never inserted in the extremities of the articulated bones but sometimes on lateral tubercles and more frequently in the vicinity of the joint, it was worth-while to examine why Nature was forced to use such an arrangement of the muscles and what considerable advantage huge epiphyses developed at the extremities of the bones provide.

Proposition XVII

If the line of action of a muscle passes across the centre of rotation of a joint, the muscle will be unable either to flex the limb against resistance or even to maintain the limb in its position. Even if there is no resistance the muscle will be unable to flex the joint.
Tab. II, Fig. 11.

The bone AB presents with a round-shaped epiphysis AE. The centre of rotation or fulcrum is in the middle of the mobile epiphysis. This epiphysis is contained in the articular cavity EFA of another bone GF. The head D of the muscle DE is inserted in the extremity G of the bone and its tendon E is attached to the end E of the epiphysis AE. The tendon passes exactly in front of the centre C of the epiphysis. DEC is a straight line. AB is horizontal. I claim that, whatever its power, the muscle cannot flex the joint or even resist the weight of the bone AB because the action of the muscle is nothing else than its own contraction. The muscle acts as a force pulling the bone AB in the direction CED from C towards D by way of a rope DE. The fulcrum of the horizontal lever AB is C. The limb AB is pushed downwards by its own weight at its centre of gravity H in a vertical direction HI. Force D is applied at the fulcrum of the lever and pulls the limb AB in the direction CED. Consequently[1], the force exerted by the muscle D, whatever its magnitude, can never overcome the resistance of the bone AB pulled downwards at H and thus cannot flex the joint. The muscle cannot even prevent the bone AB from falling down. It cannot maintain the bone straight. If the bones GF, AB are vertical, the bone AB maintained by the ligaments or supported by the cavity of the bone FA will not move either as long as its centre of gravity is just above C because, on the one

[1] Prop. 12

hand, the bone AB does not flex spontaneously if the lever arm of its centre of gravity is 0 and, on the other hand, the muscle applied at the centre of rotation exerts no moment against the resistance of the bone which it does not support. Consequently, the bone AB does not flex.

Proposition XVIII

If the tendon of a muscle is inserted in the lateral surface of the epiphysis of a bone and its direction does not pass across the centre of rotation of the joint, the limb can flex. The ratio of the force of the muscle to the resistance which it must overcome is greater than the ratio of the distance of the line of action of the resistance to that of the muscular tendon from the centre of rotation of the joint.
Tab. II, Fig. 12.

The tendon E of a muscle is inserted at the side of the epiphysis K in such a way that the prolongation of the muscle DEK does not pass across the centre C of the epiphysis. A perpendicular CK is drawn from this centre C to the prolongation of the muscle DEK. This perpendicular is the distance of the line of action of the muscle from the centre C or the lever arm of the muscle. CH is the distance of the line of action of the resistance, i.e. the lever arm of the weight of the limb AB, from the same centre. I claim that the limb AB can be flexed by the force of the pulling muscle. The ratio of the muscular force to the resistance of the limb is greater than the ratio of the lever arms HC/CK. The solid bone CAB can flex about the centre C, moved by the forces applied at E and H. ECH is an angulated lever the fulcrum of which is C. Rupture of the equilibrium and flexing the lever by the muscular force pulling in the direction KED require that the ratio of the muscular force to the resistance of the limb be greater than the ratio of their distances from the centre CH/CK. Such was the proposition.

It must be noted that the limb AB can flex upwards and rotate until the line of action of the muscle DE passes across the centre C, i.e. as long as there persists an angle DEC and the two lines DE and CE do not form one straight line. This results from the two preceding propositions.

Proposition XIX

If the end of a tendon is inserted in a tuberosity of a bone beyond the epiphysis, in the vicinity of the latter, the joint can move more than half a circle.
Tab. II, Fig. 13.

The data are the same. The two bones BA and GF are flexed to a maximum. The tendon of the muscle DEK is inserted in the vicinity of the epiphysis or a little further towards B such as in O. The tendon OKE turns round the epiphysis as does a rope round a pulley. The straight tendinous rope DE touches the round-shaped epiphysis at one point E. The radius CE is the distance between the line of action of the muscle and the centre of the joint. If the muscle keeps pulling, the tendon moves more and more over the joint, the contact proceeding from E to K and from K to O. All the time the tendon remains in contact somewhere over the whole surface EOK and the distance of the line of action of the muscle from the centre C remains the radius of the epiphysis. Only after the tendon is in contact with the joint at point O is this distance decreased: when two straight lines DE and CO form an obtuse angle, what amount does this angular inclination attain? The tendon EKO can surround the surface of the epiphysis more than over a half circle. Therefore, the joint AB can flex more than 180°. As long as the distance of the line of action of the muscle from the centre remains unchanged or is not completely nullified, flexion of the joint remains possible.

If the tendon is inserted further from the epiphysis O, towards B, for example at M, it also surrounds the epiphysis. It will be able to flex the limb to beyond 180° like a rope round a pulley. While the contact continuously changes, the distance of the line of action of the tendon from the centre of rotation remains the same, equal to the radius of the epiphysis. However, this arrangement is different from the previous one because the distance of the tendon from the centre of rotation is increased when the two straight lines DE and CM form a right angle. Then the tendon is separated from the epiphysis. Afterwards, if the joint keeps flexing, the distance of the tendon from the centre of rotation decreases again until it becomes 0.

The wonderful advantage of using a bony epiphysis is thus obvious. Besides providing the bones with strength and protection against fracture and dislocation, as mentioned above, it also ensures contact with the tendon over the periphery of the epiphysis during all the movement of the limb. Henceforth, the tendon is always distant from the centre of rotation by the radius of the epiphysis. This appears with the deltoid muscle. Its distal tendon is inserted in the vicinity of the proximal epiphysis of the humerus. Therefore, allowing for elevation and circumduction of the arm beyond 180°, the line of action of the muscle is always at the same distance from the centre of rotation of the humeral head. The same arrangement exists for the glutei muscles which move the femur and in other animal joints. Consequently, because of the lever, in this instance, the pulling power of the muscle is always of the same magnitude.

Proposition XX

A tendon must not be inserted in the very extremity of a bone but close to the joint.
Tab. II, Fig. 14.

Two bones AB and GF are articulated at AF so that AB can move about the centre C of the joint, as do the bones of the arm. The muscle DE is inserted in the extremity G of the humerus GF. Its distal end is inserted in the extremity E of the ulna AB which must flex about the centre of the joint C, describing the arc BH. I claim that Nature could not and should not attach the distal tendon E to the very extremity B of the bone AB. If the insertion E is supposed to be close to the extremity B, i.e. close to the carpus, then the tendon and the muscle DE either are or can be separated from the joint and the bones DAB, or they are retained by a hook or some ligament. In the first instance, the bone AB cannot flex towards FG up to the position AH without being pulled by the muscle DE which must contract. Then the length of DE shortens in DM to an eighth of DE. In the arm the shortening would be more than one and a half feet. This would be not only inconvenient but even impossible. This would be inconvenient since the thickness of the arm would increase up to a diameter CM equal to CE. As a result the arm would be as thick as the abdomen of the animal. This monstruous thickness would impede any other motions of the arm and even of the animal. Moreover, by its structure, a muscle can seldom contract by more than two or three finger breadths. Such insertion which would require an enormous contraction of the muscle, more than one and a half feet, would be completely impossible. Its absurdity appears obvious if we suppose the bone AB to be the left humerus which must rotate about the centre C of the scapula. This bone, to move towards the chest, ought to be pulled by a muscle ED attached to the right side D of the chest. Another elevator muscle would have to be attached to the top of the head going down to the inferior aspect of the abdomen. These muscles with the muscles of the right upper arm would require as much space as a big barrel. The same would occur with the muscles of the legs. Such a bulky construction would be ridiculous and unsuited for movement and efficient contractions. It must thus be totally rejected.

On the other hand, the tendon DE could be retained close to the joint AF by a hook or a ligament R acting as a pulley. Then a three finger breadths contraction of the muscle DE could flex the ulna AB over the arc BH. However, this arrangement is not essentially different from that in which a muscle half as long would be inserted in the vicinity of the joint C, towards A. Therefore, the additional length RE of the tendon would be not only useless but also inconvenient. This would be against the practice of Nature which hates superfluity and seeks simplicity everywhere. Consequently, the tendon of a muscle had to be inserted in the vicinity of the extremity of the mobile bone, near the joint. Q.E.D.

Comment

However, there are ligaments to maintain the flexor digitorum muscles close to the joints of the hands and feet but this results from another necessity of Nature. These muscles are certainly as thick and as long as are the ulna and the tibia. They could not be located in the small spaces provided by the palm of the hand or the sole of the foot or between the finger joints. To allow easy touching and all the other actions of the hands, these had to be flat and thin rather than bulky and swollen.

Chapter VIII

On the motive power and on the moments of the flexor muscles of the elbow.

Mechanically, the magnitudes of forces which balance each other are proportional to the velocities or to the distances which would be travelled by the forces during the same time if they moved.

A muscle can be attached to the bone in two ways, either directly or through a pulley or as in a balance with two equal arms. In the latter instance, the displacement of the resistance corresponds to the shortening of the muscle. The construction can also be like a machine with unequal arms acting as levers. Then the contraction of the muscle is not equal to the displacement of the resistance during the same time. In this instance the force of the muscle is certainly not equal to the resistance although their moments are equal. In the first instance the magnitudes of the forces and their moments ought to be equal. Actually, at closer analysis the muscle seems to be a machine (which we will show in due time). Therefore, the magnitude of the force of contraction of the muscles is not equal to the weight of the resistance and, for an equal displacement, the weight which is raised does not indicate the magnitude of the force exerted by the muscle which contracts.

However, to avoid confusion, we shall temporarily disregard this inequality. To start more easily, we shall call apparent magnitude of the force of the contracting muscle the force equal to the weight elevated by this muscle in an even displacement. The magnitude of the muscular force when it works using a contrivance which makes the velocities unequal with equal moments of the forces, will be called the modified force of the muscle.

Moreover, the true or apparent magnitude of the force, as well as the modified force, can be exerted, or rather applied, totally in such a way that no part remains inactive. It is then called the total force of the muscle. When the muscle does not exert all its possible force and does not carry all the weight which it could carry, then the force of the muscle will be designated as partial.

The contrivances through which the muscles act are levers either simple or complicated or multiple and connected in different ways. But we shall start from the easiest and simplest operations. We shall give some clear examples of important muscles, neglecting the less important which by their nature are similar, to avoid swelling this book to an exaggerated volume.

Proposition XXI

The magnitude of the force of any animal muscle necessarily must be greater than the weight which the muscle raises and never smaller.

Wise Nature made the animal of different organs so that it could move from one place to another and carry out different operations required to stay alive. This cannot be done by giving the animal a circular shape like that of a truffle.

The animal had to be composed of articulated parts such as the arms and the legs to enable it to walk and to touch objects. Such limbs could not move at the joints without muscles which pull them by force of their contraction. We will show that the motive force cannot be less than the weight and the resistance of the limbs but must necessarily be greater. A limb will be considered, for example an arm. The arm must of course move in any direction about the scapula in order to be able to pull, to carry and to push resistances such as its own weight and external objects which must be grasped. These operations require adequate shape, adequate forces and instruments. The arm must obviously be long like a lever articulated about a centre which is the firm fulcrum of the shoulder. In a lever one must also consider the points of application of the motive force and of the resistance. The motive force contracts the muscles which must be attached in the vicinity of the fulcrum of the lever as mentioned above[1] and the resistance acts at the extremity of the lever. The ratio of the muscular force to the resistance is always equal to the ratio of the longer distance of the resistance to the shorter distance of the muscular force from the fulcrum. Therefore, the motive force is always greater than the resistance.

Proposition XXII

First analysis of the apparent magnitude of the force which can be exerted by two muscles, the biceps and the brachialis anticus, flexing the forearm horizontally in supine position. This force is more than twenty times the weight which is carried and is greater than 560 pounds.
Tab. II, Fig. 15.

The upper arm EA, the forearm and the hand AB form almost a horizontal straight line. The hand is supinated. A rope GB is tied at the extremity B of the fingers which are extended as is the hand. A weight R hangs from this rope at G. This weight will be progressively increased until the motive force of the muscles DC can no longer carry it. A weight heavier than R would not be supported. The muscular force is just enough to carry the weight. We surmise that the moment of the muscular force and that of the weight R are equal since none of these forces gives way to the other. Experiments show that in a strong young man the weight R is not more than 26 pounds to which must be added the weight of the forearm and hand which is about 4 pounds. The weight of the forearm and hand does not act at the extremity B of the lever but at mid-length H where the centre of gravity is. Therefore, we add in B another weight of 2 pounds which is in the same ratio to the weight of the forearm and hand as the ratio of the distances OH/OB. We have then a straight and weightless lever. Twenty-eight pounds suspended from B thus represent the summation of the weight of 26 pounds and that corresponding to the forearm and hand. The line of action CD of the pulling muscle forms a very acute angle with the line CO, since the tendon is exactly tangential to the articular head A. A perpendicular

[1] Prop. 20

OI is drawn from the centre O to the line of action CI of the tendon. Then, as demonstrated above[2], the ratio of the force of the pulling muscle DC to the resistance R together with the forearm and hand is equal to the ratio of the distances OB/IO. After careful investigation the length OB of the forearm and hand was found to be more than twenty times the radius of the epiphysis IO. Consequently, the power of the muscle DC is more than twenty times the resistance of the weight R together with the forearm and hand. This resistance was 28 pounds. Therefore, the magnitude of the apparent absolute force exerted by the muscle which pulls and tends to flex the forearm is more than 560 pounds.

Proposition XXIII

Force exerted by the same muscles when the humerus is vertical and the ulna horizontal.
Tab. II, Fig. 16.

The humerus EA and the ulna AB now form a right angle. The ulna remains horizontal and the humerus is vertical. In this position the length of the lever OB remains the same. At its extremity B the greatest weight which can be sustained by the same muscles DC is 33 pounds, as shown by experiments. But the angle ICO formed by the tendon and the radius OC is less acute than in the previous position when the bones were both horizontal. When the humerus EA is flexed in relation to the ulna AB, the tendon DC flexes with the humerus. However, the angle ICO does not become a right angle because the tendon is tightly connected to the joint at I by membranous ligaments and by the skin. These envelopes play the role of a pulley in the angle formed by the joint A. But the tendon IC is not so firmly fixed at I that it cannot be somewhat raised. Therefore, the straight line OI perpendicular to the direction of the tendon CI becomes significantly longer than in the previous instance. We can feel that on our own arms. Consequently, the ratio of OB/IO is smaller than what was found in the first position but the ratio of these distances OB/IO is equal to the inverse ratio of the muscular force DC to its counterpart, the resistance exerted by the weight R and the weight of the forearm. Thus the muscular force is less than twenty times the resistance. From the preceding research, the maximum force of the biceps and brachialis anticus muscles is assumed to be 560 pounds. If the maximum weight R is 33 pounds which gives a resistance of 35 pounds with the weight of the forearm, the distance OI would be the sixteenth of the distance BO whereas it was the twentieth in the previous experiment. As a result of a significant increase of the distance IO, a maximum weight of 35 pounds can be raised by the same muscles.

However, it must be noted that, when the joint EAB is flexed, the muscles DIC are not as tight as previously but must be somewhat loosened. But the

[2] Prop. 16

motive force of contraction of each of the two muscles is not decreased because both muscles DC are not attached to the tip of the humerus. The biceps is inserted in the extremity L of the scapula HLE. The brachialis anticus is inserted in mid-humerus. Since the scapula HEL remains fixed in the same position, the humerus EA rotates about the centre E of the joint. The scapula forms an angle LEO less acute when the humerus moves further downwards while the proximal origin of the biceps muscle D is more raised and further increases its distance from the humeral head E because the distance LDI subtending the angle LEO increases. This tightens the biceps muscle the more the further the humerus is moved downwards. Thus, although the brachialis anticus muscle is loosened as a result of the closing of the angle EOB, the tightening of the tendon of the biceps due to the elevation of its extremity D above the humeral head is much more important.

Proposition XXIV

The apparent magnitude of the force exerted by the biceps muscle can be evaluated approximately at 300 pounds, that of the brachialis anticus at 260 pounds.
Tab. III, Fig. 1.

The humerus EO is moved backwards in such a way that the angle HEO is as acute as possible. At the same time the ulna is also flexed to remain parallel to the upper border LH of the scapula. Then the alternate angles HDI and CID are equal. Therefore, loosening of the biceps muscle DIC inside the acute angle COE is exactly compensated by its tension over the apex of the angle HDO. Consequently, the normal tension of the biceps muscle is unchanged and the muscle keeps the length it had when all the arm was horizontal. The muscle thus does not loosen and maintains the same force of contraction as in the horizontal position. But the brachialis anticus muscle FI does not enjoy the same advantage. Its origin is at mid-humerus F and it inserts in the vicinity I of the proximal epiphysis of the ulna. Since EOC is an acute angle, the brachialis anticus muscle FI is maximally loosened and thus exerts only little or almost no force. In these conditions it is possible to determine the force of the biceps muscle alone, as long as the distance OI of the tendon from the centre of the elbow joint does not change, by suspending a weight R from the fingers. In this position R and the weight of the forearm make 25 pounds. The distance IO is about a twelfth of the length BO of the forearm and hand. Therefore, the magnitude of the force of the biceps muscle is twelve times the suspended weight R and the weight of the forearm. This magnitude is 300 pounds if the brachialis anticus muscle is not involved as a result of its maximum loosening. In the first experiment the magnitude of the force exerted by both the biceps and brachialis anticus muscles was 560 pounds. Subtracting the magnitude of the force found for the biceps muscle alone, 300 pounds, gives 260 pounds. This is the magnitude of the force exerted by the brachialis anticus muscle. This was the object of the research.

Proposition XXV

Force exerted by the same muscles when the humerus is vertical and the ulna points downwards.
Tab. III, Fig. 2.

The humerus EA and the ulna AB are vertical. The greatest weight R which can be suspended from B would be enormous if the ligaments were strong enough to withstand this weight.

If, afterwards, the ulna AB is flexed for a short time towards the horizontal, forming an obtuse angle EAB with the vertical humerus and an acute angle BAK with the horizontal line OK, then the maximum weight R must increase considerably because, a perpendicular BK being drawn from B to the horizontal AK, the weight R pulls obliquely the lever AB as if it hung at point K of the lever OK. The new lever OK is shorter than OB. The force of the muscle raising the lever pulls from point I with a lever arm IO. The ratio of the magnitude of the force exerted by the muscle, which remains constant, to the resistance of the weight R is equal to the ratio of KO/IO. If KO is only twice as long as OI, the weight R supported in this position is necessarily equal to half the magnitude of the motive force. R is then 250 pounds. If the distance OK is shorter than OI, the weight R is even greater than the magnitude of the force exerted by the muscles.

Consequently, during flexion of the elbow with elevation of the ulna AB, the efficacy of the force of the muscles decreases as a result of the lengthening of the lever OK. Therefore, the weight R must also be diminished.

Proposition XXVI

Force exerted by the muscles with the arm prone and horizontal.
Tab. III, Fig. 3.

The force of the flexor muscles of the forearm can be exerted in another way. The forearm AB is supine and horizontal. It must be flexed downwards towards G by the muscles DC which are underneath the bone. The rope BLG turns round a drum or a pulley ML rotating about a fixed axle M. When the hand B comes down the weight R is raised. The forearm can be compared to a lever AB with its fulcrum at O. The weight R pulls the extremity B of the lever upwards towards L. The force exerted by the muscles DC pulls the lever AB downwards from I towards D. If what was said above is verified here also, there is, however, a difference. Previously, the extremity B was pulled downwards not only by the weight R but also by the weight of the forearm and hand. Here, on the contrary, the weight AB of the forearm and hand no longer opposes but rather combines with the force of the pulling muscles. Therefore, in this position, the muscles pull the forearm downwards as does the forearm by its own weight. The sum of these two forces balances the moment of the weight R. The weight of the forearm and hand, whereas in the first instance it would be added to the resistance R, now is added to the force of the muscles. The maximum force of the muscles DC was shown to be equal to 560 pounds[3]. Thus, if the lever AB was weightless, since the distance OB is assumed to be twenty times the distance OI, the weight R should be 28 pounds. Two pounds are added which balance the weight of the forearm and hand AB and make the lever weightless. Thus, in this position, 30 pounds is the maximum which can be supported by the force exerted by the muscles DC.

The same conclusion is arrived at by another reasoning. A weight of 2 pounds corresponding to the forearm pulling downwards at B is balanced by a tension force of 40 pounds at I, the ratio 1/20 being inverse of that of the lever arms. The force exerted by the muscles DCI was 560 pounds. Consequently, if the force applied at I and replacing the weight of the forearm and hand is added, the sum is 600 pounds.

Chapter IX

On the flexor muscles of the lower leg and their action about the knee.
Tab. III, Fig. 4.

The tibia flexes about the round-shaped epiphysis of the femur as does the ulna about the cylindrical epiphysis of the humerus. There exists, however, a difference. In the forearm the flexor muscles are inserted almost only in a median point in the vicinity of the proximal extremity of the ulna. The muscles flexing the lower leg are inserted at two points on the opposite sides of the proximal epiphyses of the tibia and fibula like the reins of horses. Three muscles, the semitendinosus, semimembranosus and gracilis, are inserted on the medial side of the tibia and one, the biceps, laterally in the proximal end of the fibula. But this discrepancy does not disturb the nature and properties of the lever. The main discrepancy with the elbow is completely different and consists in the difference of structure of the knee joint CB and elbow joint. In the elbow the proximal extremity of the ulna tightly surrounds the distal epiphysis of the humerus about which it rotates. The knee is different. The femur AB ends distally in a large basis BH which is not really cylindrical. It is somewhat flattened resulting in an elliptical or spiral curvature. This epiphysis is not tightly contained in the tibial epiphysis CH which is as large as the femoral extremity but flatter. The femoral condyles are not deeply buried in corresponding cavities of the tibia. They are separated from the tibia by thick and hard semilunar cartilages which act as a cushion and provide cavities with some depth. The semilunar cartilages are attached externally to the articular capsule which inserts in the epiphysis of the tibia. Internally they decrease in thickness and end in thin and circular blades. The semilunar cartilages move over the smooth articular

[3] Prop. 22

surface of the tibia which is kept wet by the synovial fluid thus enabling their sliding back and forth. They are pulled forwards when the knee is extended and backwards in flexion. Therefore, in extension the axis of rotation F of the knee is displaced forwards towards C. In flexion the axis F does not remain at the same place in the femoral epiphysis BH but moves backwards towards H. The distance FH between the axis of rotation and the line of action GI of the flexor muscles decreases more and more while the tibia flexes and altogether moves backwards. The tendon GIH is tangent to the posterior aspect H of the distal epiphysis of the femur. During flexion point H moves inside the joint towards C. The tendon GIH is no longer pushed backwards as it was in extension and is not displaced away from the articular cavity of the tibia and from the axis of rotation F. Therefore, the distance FH decreases. The tibia CD being extended on the femur AB in a horizontal plane, a weight R is suspended from the heel. This weight is as heavy as can be supported by the flexor muscles LG of the lower leg and attains 65 pounds. If the tibia CD is horizontal and the femur AB is vertical, the maximum weight R attached to the heel D, which can be raised by the same muscles is not more than 33 pounds.

Proposition XXVII

First analysis of the motive power exerted by the four muscles flexing the lower leg, which is thirty times greater than the suspended weight and is more than 949 pounds.
Tab. III, Fig. 4.

The trunk, thigh and lower leg are horizontal, the subject lying face down. The centre of rotation of the knee is F. FH is the distance of the line of action of the tendon GH from the centre of rotation F. The distance FH of the line of action of the flexor tendons from the centre or axis of flexion F is about the thirteenth of the distance between the centre of rotation F and the tuberosity of the os calcis D from which the weight R is suspended. The force of the muscles balances the weight R and the weight, approximately 10 pounds, of the lower leg which hangs freely. Acting at point D the lower leg would exert a force of 5 pounds. Therefore, according to the laws of the levers, the ratio of the muscular force EI to the weight R together with the weight of the lower leg is equal to the ratio of 13/1. The weight R and that corresponding to the lower leg make 73 pounds. Consequently, the force exerted by the flexor muscles is more than 949 pounds.

Proposition XXVIII

Explanation of the fact that less weight is raised by the same muscles when the trunk and thigh are upright and the lower leg horizontal.
Tab. III, Fig. 5.

When the forearm is extended on the arm and both are kept horizontal, the maximum weight which can be suspended from the extremity of the hand is not more than 26 pounds[1]. If the forearm is kept horizontal and the upper arm and trunk are vertical, a heavier weight can be supported from the extremity of the hand, i.e. 33 pounds. The opposite is observed for the flexor muscles of the lower leg. When the femur and tibia are horizontal the maximum weight suspended from the heel is 67 pounds. If the femur is vertical with the tibia flexed at 90° and horizontal the maximum weight suspended from the heel is not more than 31 pounds. This difference in similar operations seems very surprising. Its cause must be sought for. When the femur and tibia are at right angles to each other, the maximum weight suspended from the heel D is not more than 31 pounds although the same muscles can exert a force of 949 pounds as appears from the previous proposition. According to the laws of equilibrium, the ratio of the length of the tibia CD to the distance FH of the line of action of the muscles from the centre F should be equal to the ratio of 949 pounds to 36 pounds, i.e. a ratio of more than 26/1. Hence, the distance FH would be the 1/26 of the distance FD. There is actually no such diminution of the distance FH. Therefore, the diminution of the moment of the muscular force must have another and unknown cause. Actually, the flexor muscles EG exert less force than previously because they are somewhat loosened since the distance EG is shortened by flexion of the knee joint. Therefore, these muscles probably exert less force, 500 pounds, which would be in about the same ratio to a suspended weight of 36 pounds as 14/1. The length of the lever FD remains almost the same. Thus, the distance FH of the line of action of the tendons from the centre F must be significantly reduced. This is what happens since, during flexion, the axis of rotation of the tibia is displaced from the anterior to the posterior aspect of the distal femoral epiphysis and can be brought closer to the tendons GH. This is the explanation if the force of the muscles is not decreased much more by their loosening which is unlikely.

Proposition XXIX

Explanation of the fact that, when the thigh is vertical and the lower leg and the trunk are horizontal, more weight can be suspended from the heel than when the trunk is upright but less than when the thigh and lower leg are both horizontal.

Tab. III, Fig. 6.

The trunk LMEN is horizontal and parallel to the lower leg CD. As shown by experiments, the maximum weight R suspended from the heel D is 46 pounds. Addition of a weight corresponding to the lower leg makes 51 pounds. In the previous position it was 36 and in the first 73 pounds. The cause of the increment of the motive force must be sought for.

Four muscles flex the tibia, the semitendinosus, semimembranosus, gracilis and biceps. The first three originate from the ischial bone at L. The biceps muscle originates partly from the ischial bone L and partly from the femur

[1] Prop. 22

itself at E. Flexion of the knee joint with formation of the angle AED by the femur and tibia loosens the four muscles. Flexion of the trunk stretches the muscle LEG over the apex of the angle MLB. The origins of the muscles must be displaced towards L which compensates for their loosening IG due to the flexion of the knee. The muscles thus remain as tense as in the first position with the tibia extended on the femur and horizontal. Therefore, they are not loosened and can exert the same force. However, this is not true for the short head of the biceps EG which is inserted at mid-femur F. It can be assumed that the short head of the biceps provides about the seventh of the force exerted by the four muscles. Therefore, the total maximum force which was equal to 949 pounds will be reduced to 813 pounds as a consequence of the loosening of the short head of the biceps. Thus, the ratio of the muscular force to the resistance of the suspended weight R with the weight of the lower leg, which amounts to 51 pounds, is equal to the ratio of FD/FH, i.e. 16/1, because the distance FH of the line of action of the tendon from the centre F is somewhat shorter than what was found in the previous calculation.

Remark

It must be noted that the method of analysis which we have used throughout this chapter is not totally accurate because it does not comprise the whole apparent force of the muscles. However, it had to be presented at first because of its facility. It will be easier to determine the whole apparent force of the muscles later.

Chapter X

On the double increase of the forces of the muscles flexing the forearm and the lower leg.

As clearly enunciated by divine Plato, Geometry and Arithmetics are two wings by which we ascend to heaven since they enable us to understand the arcanes of Astronomy. We can also claim that Geometry and Mechanics are the ladders by which we climb to the wonderful knowledge of the movements of animals. Who could deny this fact which has been demonstrated in the two previous chapters? However, since we were not more used to mechanics, we missed the truth by far, not because we tolerated error and what has been said hitherto must be rejected but because, although some of it is true, we did not follow more exact science. For better understanding some mechanical lemmas must be presented which, as far as I know, have not been considered hitherto.

Proposition XXX

If two forces pull along the same straight line the two ends of a rigid or contractile rope and their moments are equal to the moment of the resistance of the rope, the force with which the rope resists tension is equal to the sum of the pulling forces which are themselves equal to each other.
Tab. III, Fig. 7.

The extremities A and B of a hard or soft and contractile rope AB are pulled in opposite directions by two forces or two weights R and S in such a way that there is equilibrium and the forces are equal to the moment with which the rope resists the tension, i.e. one force does not prevail over the other. I claim that the forces R and S are equal and the sum of both forces is equal to the force with which the rope resists tension. S and R are weights and pull the rope AB through two pulleys F and G. It is understood that the rope is weightless. The two forces or the weights R and S which pull the rope in opposite directions through the pulleys F and G are in equilibrium. Therefore, according to mechanics, the forces or the weights R and S are equal. XZ is the force with which the rope AB evenly thick and strong resists the tensions exerted by the two equal forces R and S (no matter whether the resistance is due to the glue or ties by which the elements of the rope are connected or whether it is due to the force with which these elements contract, retract and are brought closer to each other). Since the force of resistance and contraction is exerted with the same effort by all the elements of the rope AB which is of even thickness and strength throughout, a half X of the power of the rope resists the force R by contracting a half AC of the rope over the distance AD and the other half Z of the power acts against the force S which is equal to R, by contracting the other half BC of the rope over the distance EB equal to AD. The displacement of the half rope AC which is shortened by the distance AD is opposed by the force X because this force pulls the rope, and by the resistance R which is raised the same distance during the same time. Therefore, the force X contracts the half rope AC and the weight R is elevated with equal velocities. Since both forces balance each other, their moments are equal. The magnitude of force X thus is equal to the resistance R. Similarly, the magnitude of force Z is equal to the resistance S. Consequently, two forces XZ, or rather the force with which the rope AB resists the tension is equal to the sum of the two equal resistances R and S. Q.E.D.

Comment

When a rigid rod (Fig. 8, Tab. III) is compressed longitudinally by two forces, the force with which the rod resists compression is equal to the sum of the two forces which compress it. This can be demonstrated in the same way.

Proposition XXXI

If an extremity of the same weightless rope is now attached to a strong nail and the other extremity pulled by a weight or some other force does not move, the force with which the rope resists tension is twice the force which pulls.

Tab. III, Fig. 9.

A weightless rope AB is attached to a strong nail at B and pulled by a weight or a force R at A. The force with which the rope resists tension is XZ. Forces XZ and R are in equilibrium. It must be demonstrated that the magnitude of the force XZ is twice that of the resistance R. After the nail has been removed, it is replaced by the force S exerted either by the hand or by a weight acting through the pulley BE and pulling the extremity B of the rope in the direction AB from A to B. The force S resists the tension exerted by the force R so that the rope remains immobile. According to the previous proposition, it is obvious that the weight or the power of the hand S is equal to the resistance R and the force with which the rope resists opposite tensions must be equal to the sum of the two forces R and S or to twice the force R. The nail at B exerts the same force as S. Thus, the nail resists the tension of the weight R with an equal moment since the rope pulled by the force R is prevented from moving down by the nail. The rope is maintained in the same position as if it was retained by the force S. Therefore, the force exerted by the nail is equal to force S or R. Consequently, the force XZ with which the rope AB resists the opposite tensions of R and the nail is equal to the sum of the two forces or to twice the force R. Q.E.D.

Proposition XXXII

If a weightless rigid rod is compressed from above by a weight or by any force and rests inferiorly on the hard ground, the force with which the rod resists compression is twice the compressive force.

Tab. III, Fig. 10.

A weightless rigid rod stands perpendicular on a hard surface LB at B and is compressed by a weight or any force R at A. The line of action of force R is AB, from A to B. We demonstrate in the same way that the force with which the rod resists compression is twice the compressive force R. The hard ground is replaced by the hand S or the extremity B of a balance EB with equal arms. The rod AB supporting the weight R is placed at B and is balanced by the counterweight X. The hard ground LB resists the compression of force R and prevents its coming down with the same force as the force exerted by the subjacent hand or the weight X. The resistance of the hard ground LB is thus equal to the compressive force R and the rod is compressed by the two of them. Consequently, the force with which the rod AB resists compression is twice the weight or the force R. Q.E.D.

Proposition XXXIII

Another demonstration of the same claim.
Tab. III, Fig. 11.

The data are the same but the forces are opposite, i.e. the rope AB is contracted by a force XZ so that its end A rises up to H where it remains. The rope AB is divided in two equal parts at C as is the segment AH at D. The force is divided in two equal parts X and Z. CF equals DH. It appears[1] that the half force X shortens one half of the rope CA and the other half force Z shortens the other half of the rope CB. The resistance R is raised by the force X alone (not considering Z) shortening the half rope CA by the distance AD. Thus, force X is moved by the distance AD, in the same direction, by shortening the half rope and raising the resistance R. Therefore, the velocities of force X and resistance R are equal and their moments are equal at the end of the contraction of the half rope when these forces are in equilibrium. Consequently, the magnitude of force X is equal to the resistance R. Moreover, while force X acts, the other half force Z is not inactive but exerts its own action contracting the other half CB of the rope exactly as much as the first half AC was contracted by force X. The extremity B fixed by the strong nail S cannot move towards the midpoint C of the rope. Consequently, contraction forces point C to move towards the nail B, travelling a distance CF equal to AD. Actually the midpoint C of the rope cannot ascend up to F except if the weight R hanging from the segment CD shortened by force X is raised from point D to H by a distance equal to the elevation CF of the midpoint C. On the other hand, the action of the force Z is not helped by force X which is exhausted in the contraction of the rope AC and in pulling the weight R from A to D. Afterwards, force X no longer acts except to maintain the shortening of the rope CD as a knot in this rope would do. Consequently, by another and distinct action, the force Z raises the same weight R by a movement of equal velocity and in the same direction. At the end of the second contraction their moments are equal. The magnitude of force Z is equal to the weight R. But initially the magnitude of force X was equal to the same force R. Therefore, the sum of the magnitudes of forces X and Z, i.e. the magnitude of the force exerted by the whole rope AB, is twice the magnitude of the resistance R. Q.E.D.

Proposition XXXIV

If an extremity of a contractile rope is fixed by a nail and the rope is pulled by some weight until the forces are in equilibrium, the velocity of the pulling weight is twice the velocity of contraction of the rope.
Tab. III, Figs. 12 and 13.

The data are the same. It is assumed firstly that the rope BA is shortened not by its own contraction but by a transverse tension exerted by a force XZ round

[1] Prop. 30

a second nail E deviating the rope thus folded BGE from EB to G (Fig. 12, Tab. III). This tension raises the weight R from A to H. The rope is supposed not to be contractile. The length AB of the rope is equal to the length HEGB of the pulled rope. Subtracting the common segment EH, the segment AH is equal to the folded segment EGB but the latter is twice the length BG. This is the displacement from B to G caused by the force XZ. During the same time the weight has travelled the distance HA. This distance is twice that carried out by the force XZ from B to G. Therefore, the velocity of force R is twice that of force XZ. The distance AB of the weight from the nail is shortened in the same way by an actual contraction of the rope in Fig. 13, Tab. III as it is by the transverse tension of the rope in Fig. 12, Tab. III in which part of the rope BGE is folded and pulled sideways and remains there. In Fig. 13, Tab. III, the segment of rope LK completely disappears as a result of contraction. This is similar to what we observe with a spring or with the strings of a zither. Such a shortening results from opposite displacements. All the elements of the half rope AC contract and come nearer to point C at mid-rope. All the elements of the half rope BC also contract and move nearer to the same midpoint C. The displacement due to the shortening of the half rope AC occurs like in a spring moving from A to C over a distance KC (Fig. 13, Tab. III). At the same time the opposite movement of the half rope BC is hindered by the nail and C moves towards B over a distance LC equal to KC. But the forces X and Z by shortening the half ropes are displaced in opposite directions by exactly as much as these half ropes shorten. The forces X and Z both thus have the same velocity and the resultant of these two forces shortens the whole rope with the same velocity as one force X alone shortened the half rope AC. Similarly, two horses travelling the same distance during the same time will also travel the same distance when teamed under the same yoke. Thus, the team formed by the two horses will have the same velocity as one of the horses alone. During the time in which the rope shortens, the weight R is raised exactly as much as the rope AB is shortened by a distance KL and the force XZ is moved shortening the rope with the same velocity with which force X carries out the contraction LC, which is half the distance KL. Consequently, the velocity of the weight R is twice that with which the force XZ shortens the rope AB. Such was the proposition.

Comment

And here admiration stops, overwhelmed by the requirements of mechanics: the force shortening a rope attached to a nail must be twice the weight pulling the rope. Whenever the moments of the forces are equal the forces are necessarily inversely proportional to their velocities. Since the velocity with which the force contracts the rope attached to a nail is half the velocity with which the weight is raised, the force shortening the rope equals twice the weight which pulls the rope.

The following general rule can be deduced: whenever a rope or any other machine has an extremity fixed, the force with which the rope or the machine resists tension, i.e. the force which is exerted to support the resistance, is the double of the force which is sustained. The force which resists compression in the base of any column or of any other body or of any fixed obstacle is always the double of the compressive force.

After these premises, the apparent forces of the flexor muscles of the forearm and lower leg can be determined again.

Proposition XXXV

Second analysis of the forces exerted by two flexor muscles of the forearm, the biceps and brachialis anticus, which are twice as great as what we determined in chapter VIII. The force exerted by the biceps muscle alone is more than 600 pounds and that exerted by the brachialis anticus muscle is more than 520 pounds.
Tab. II, Fig. 15.

As for proposition 22, the humerus EA is fixed and articulates with the ulna AB at A. The two bones EA and AB are supine and horizontal and the elbow is extended. The extremity B of the hand supports a weight R of 26 pounds. Considering the weight of the forearm, the total weight hanging from the extremity B of the horizontal lever AB is 28 pounds. The length of the lever AB is twenty times the distance IO of the tendon IC from the fulcrum. We deduced that the force with which the muscle DC pulls the forearm and tends to raise it with the suspended weight R is greater than 560 pounds[2]. Actually the muscle DC is attached to the fixed extremity D of the immobile bone EA as it would be to a nail and the other extremity of the tendon C is pulled downwards by the weight R and that of the forearm. Therefore[3], the force of contraction of the muscle DC is twice the resistance. But, because of the lever, the same muscular force is twenty times the weight suspended from B. Consequently, the magnitude of the force exerted by this same muscle is equal to 1120 pounds, i.e. the double of which was found in proposition 22.

The force exerted by the biceps muscle alone was found[4] to be equal to 300 pounds and that of the brachialis anticus to 260 pounds.

Consequently[5], the apparent force exerted by the biceps muscle alone is equal to 600 pounds and that exerted by the brachialis anticus is equal to 520 pounds.

[2] Prop. 22
[3] Prop. 31
[4] Prop. 24
[5] From Prop. 31

Proposition XXXVI

Second analysis of the forces exerted by four flexor muscles of the lower leg, which are twice what was determined in chapter IX and are thus equal to 1898 pounds.
Tab. III, Fig. 4.

The femur AB and the tibia CD are aligned, prone and horizontal like in proposition 26. A weight of 73 pounds is suspended from the heel. This weight is supported by four flexor muscles of the lower leg LG and EG. The ratio of the length of the lever FD to the distance FH of the tendons IG from the centre of rotation or fulcrum F is equal to 13/1. We thus deduced that[6] the force exerted by the four flexor muscles of the lower leg was equal to 949 pounds.

The proximal extremities L and E of the muscles LG and EG are attached to the ischial bone and femur. The distal tendons G are pulled downwards by the weight R and that of the lower leg, by way of the lever FD. Therefore[7], the force of contraction of the muscles EG is twice the resistance R. But, according to the laws of the levers, the ratio of the force of these muscles to the resistance R is equal to 13/1. Consequently, the apparent force of the flexor muscles of the lower leg is greater than 1898 pounds.

Comment

Since all the muscles of the animal, with few exceptions, are attached at both their extremities and one of these extremities is usually fixed, in contracting these muscles Nature is forced to exert a force equal to twice the resistance which She must overcome. Doubling the forces is of little importance, considering the other enormous multiplication which will be demonstrated later.

Chapter XI

On the moment and apparent magnitude of the motive power exerted by the extensor muscles of the lower leg.

Order would require that we study the forces of the other muscles which flex the arm about the shoulder, the thigh about the pelvis, the fingers and toes about the metacarpals and metatarsals. But these muscles are formed differently and constitute a machine by far different from the one which we de-

---
[6] Prop. 27
[7] From Prop. 31

scribed. Therefore, their study must be postponed until the lemmas which are useful to their description have been mentioned. After recalling some lemmas, we shall discuss only the other muscles which have the same simple structure with which we dealt above and which operate by way of a double lever.

Proposition XXXVII

Two levers are articulated about a point of one of their extremities and attached by a rope inside or outside the bow they form. The two forces flexing or extending the joint are equal to each other. The ratio of the force exerted by the connecting rope to the sum of these forces is equal to the ratio of the double of the common distance of the flexing or extending forces to the sum of the distances of the rope from the common centre of rotation. Tab. IV, Figs. 1 and 2.

Two levers AB and EB are mobile about a common point B and form a bow. A rope DCF is attached at D and F and passes either externally or internally over the pulley C. Tension of the rope thus either opens or closes the angle ABE. Two forces R and S are applied at A and E. These forces act in the direction AE and tend to close or to open the arms of the bow. I claim firstly that the forces R and S are equal. A perpendicular BG is drawn from the centre B to the line of action AE. The two levers connected by the rope constitute an angulated balance ABE mobile about the fulcrum B. Two forces R and S pushing obliquely the free extremities AE of the scales in a common direction AGE balance each other since one does not give way to the other. Therefore[1], the ratio of the magnitudes of R and S is equal to the ratio of BG/BG since BG is the distance of the line of action of both forces R and S from the centre B.

Then (Fig. 2, Tab. IV) two perpendiculars BH and BI are drawn from the same centre B to the lines of action CD and CF of the rope. The two levers AB and EB are in contact at B and can flex about the same centre of rotation B, pushed together or pulled away from each other by the forces R and S. The extremities A and E of the levers would be displaced towards BG or away from BG if they were not prevented from doing so by the resistance of the rope DCF. Since the tension of the rope DCF opposes the action of the two forces R and S, one part CD of the rope must resist the action of force R and the other part CF must resist the action of force S. Point B is immobile: it is maintained in the same position by the opposite forces. Therefore, AB and EB are two levers mobile about the same fulcrum B. In both the force balances the resistance. The force R tends to flex the extremity A of the lever AB in the direction AG, the distance of which is BG from the centre B. The resistance of the rope DC or force of contraction pulls the lever from D in the direction DH the distance of which is BH from the centre B. One force does not give way to the

[1] From Prop. 16

other since the lever remains in the same position. Consequently[2], the magnitudes of the forces DC and R are in inverse proportion to the distances BG and BH. The same reasoning shows that the ratio of the magnitudes of the forces CF/S is equal to the ratio of the distances BG/BI. Therefore, the ratio of the magnitude of the force exerted by the whole rope DCF to the sum of the two forces R and S, or twice S, is equal to the ratio of twice the distance BG to the sum of the distances BH and BI. Q.E.D.

Proposition XXXVIII

If the same weightless bow is fixed by a nail or stands upright on the ground and is compressed by one weight alone or pulled by one force, the ratio of the force of the rope to the force acting on the bow is equal to the ratio of the distance of the acting force to the quarter of the sum of the two distances of the rope from the centre.
Tab. IV, Fig. 3.

The same bow ABE as in the previous proposition is considered again. It is assumed to be weightless. Its extremity E is fixed by a nail S or the bow stands upright on a firm horizontal plane SL. It is pushed by a weight or another force R. For the rest the arrangement is the same as in the previous proposition. I claim that the ratio of the magnitude of the force exerted by the rope DCF to the pushing force R is equal to the ratio of the distance GB of the line of action AGE from the centre B to the quarter of the sum of the two distances BH and HI from the same centre. The weight R can stay in the high position above the upright bow ABE only if the straight line AGE drawn from the centre of gravity of R to the support E (which is the line of gravity of R) is perpendicular to the horizontal plane SL. The resistance of the nail S or of the unyielding ground LS which acts exactly in the same way pulls or pushes the extremity E of the lever BE from or towards A with the same energy as if the force of the nail or of the ground SL were replaced at E by the hand or another force which would act with the same energy as R. The force R and the resistance of the nail or of the ground are equal. Their sum opposes and balances the force of the rope DCF. Therefore, the ratio of the magnitude of the force exerted by the rope to the sum of the two forces compressing or stretching the bow and equal to each other (R and the resistance of the nail S or of the ground) is equal to the ratio of twice GB to twice the sum of the distances BH and BI or the ratio of GB to the sum of the distances BH and BI. Finally the ratio of the force exerted by the rope DCF to the weight R alone is equal to the ratio of GB to the quarter of the sum of the distances BH and BI. Q.E.D.

[2] From Prop. 16

Corollary

Consequently, the moment of the force exerted by the rope DCF is equal to twice the compressive or tensile force R although they seem to balance each other. It thus appears that the force of the rope, in these circumstances, could act with the same levers against twice the force R. Half the moment of the rope has to be exerted against the force of the resistance of the ground or of the nail S.

Proposition XXXIX

With the same arrangement but a weighing bow, research of the ratio of the force exerted by the rope opening or closing the bow to the force acting on the extremity of the latter.
Tab. IV, Fig. 4.

The angle ABE has some weight. The rope DCF and the pulley C have some weight also. The ratio of the force exerted by the rope to the weight which compresses or stretches the bow must be found. In this hypothesis, the same bow ABE is subjected not only to the weight R but also to its own weight. Both compress or stretch the bow. Similarly, the nail S or the ground is forced to support not only the load R but also the weight of the bow and rope. Therefore, the common centre of gravity must be found. M is the centre of gravity of all the machine ABCE. The centre of gravity of the load R is at point R. A straight line RM is drawn and divided at N in inverse proportion to the weights. The ratio of the weight R to the weight of the contrivance ABCE is equal to MN/NR. Point N is the common centre of gravity of R and the machine ABCE. A straight line NB is drawn and a new and weightless bow NBE appears. The centre of gravity of the whole is N and the line of action of its weight is NGE which is vertical. The line of action of the rope DC remains unchanged as does its lever arm BH. It is demonstrated in the same way as above that the ratio of the magnitude of the force exerted by the rope DCF to the weight R together with that of the weighing machine ABCE is equal to the ratio of BG to the quarter of the sum of BH and BI. That was the object of the research.

Proposition XL

Analysis of the forces exerted by the extensor muscles of the lower leg, which is six times the weight supported and is about 2280 pounds.
Tab. IV, Fig. 5.

Two rods ABE form a bow corresponding to that formed by the thigh AB and the lower leg BE which articulate at the knee B. The foot SL is supported by the ground XZ. The opposite foot is raised from the ground. The subject who weighs 180 pounds thus stands on one foot with the knee flexed moderately and carries on his neck a weight of 200 pounds as stevedores do when they bend down to pass through the door of a house. Undoubtedly the mass formed by the body VABE and the carried load R is usually more than 380 pounds. It acts at the common centre of gravity N, thus exerting a force. This force must be supported by the bow ABE. Its line of action NAE is vertical and passes through the distal extremity E of the lower leg above the foot LS. Obviously the single flexible bow by its nature would collapse about its centre B under such a load and the two rods AB and BE would fall and become parallel to the horizontal plane XZ. This collapse is efficiently opposed by the force of the three extensor muscles of the lower leg DCF which tend to open the bow ABE. Actually, the bow ABE rests on the ground XZ at L and is compressed at A by a load heavier than 380 pounds. The line of action of this load is NAE. The distance GB of this line of action from the centre of rotation B is about three times the distance BH of the muscle DC or the tendon CF from the centre of rotation B[3]. The ratio of the force exerted by the rope or the muscle DC to the weight compressing the bow is equal to the ratio of the distance GB to half the distance BH. Therefore, the force of the muscles DCF is six times that of the compressive weight. Consequently, the force exerted by the three extensor muscles DCF of the lower leg is a little less than 2280 pounds. This was the object of the research.

Comment

In this analysis we neglected the rectus femoris muscle since, in this experiment, its action is almost nil as is the action of the tensor fasciae latae. Because they are attached to the iliac spine they are loosened by the forwards bending of the trunk. Therefore, they are unable to take part significantly in the powerful extension of the knee.

Proposition XLI

Analysis of the force exerted by the soleus muscle which plantar-flexes the foot. This force is three times the compressive weight and is thus greater than 1140 pounds.
Tab. IV, Fig. 6.

Experience shows that a stevedore with a load R carried on his neck can raise a foot. Not only can he stand on the sole of the opposite foot with the leg bent, as said above, but in walking he can even be supported on the ball of his foot, the heel being raised. Then another bow ABE appears formed by the tibia AB articulated with the foot EBC about the centre of rotation B. The plantar sole CS rises towards the heel C. Only the ball of the foot E touches the ground at

[3] From Prop. 38

LS. Then the soleus muscle DHC supports not only the load R carried on the neck but also the weight of the whole subject, including his foot CBE, which was not the case in the previous proposition. The line of action NAE of the total weight is vertical. Its distance GB from the centre of rotation B is longer than the distance BH of the tendon CH from the same centre B and smaller than twice this distance. We assume GB to be one and a half times BH. The ratio of the muscular force DC to the weight compressing the bow is equal to the ratio of GH to half the distance BH[4]. Therefore, the force exerted by the muscle DC is three times the weight compressing the bow. The latter was more than 380 pounds. Consequently, the force exerted by the soleus muscle is more than 1140 pounds. This was the object of the research.

Comment

We neglected the gastrocnemius muscle which also plantar-flexes the sole because it is not usually involved in this experiment. It inserts in the distal extremity of the femur. Being loosened by flexion of the knee it can no longer pull the os calcis and, therefore, does not help the plantar flexion of the foot. It must be remarked that the quadrupeds cannot stand on one foot. They stand at least on either the hind feet or the forefeet when they fall or jump. When they are erect, they must stand on their plantar soles lying flat on the ground, as do bears, apes, dogs and others. Horses cannot stand erect at all. They can at best stand for a short while on their heels. Then the forces exerted by the soleus muscles which plantar-flex both feet can be easily calculated, using the method described above. The distance of the line of action of the total weight of the horse with its load from the centre of the ankle[5] seems to be six times the distance of the tendon of the soleus muscle from the same centre. The ratio of the force exerted by the two soleus muscles to the compressive weight is equal to 12/1. The horse weighs more than 250 pounds and it carries more than 450 pounds. The force exerted by the two soleus muscles thus is 12 times 700 pounds, which makes 8400 pounds. Birds can carry a load equal to their own weight. The force exerted by their soleus muscles is easier to find since they always stand with the heels raised and when walking they are forced to stand on one foot. The swan usually does not weigh more than eleven pounds. The distance of the perpendicular through its centre of gravity from the centre of its ankle is about ten times the distance of the line of action of the tendon of the soleus muscle from the same centre. Therefore, the force exerted by the soleus muscle (which is tripartite in the swan) is sixteen times the weight of the swan and its load, twenty-two pounds. The muscular force thus is equal to 352 pounds.

This is not a precise measure. But in jumping the force exerted by the soleus muscles is much more as we shall show.

---

[4] Prop. 38

[5] From Prop. 40

Chapter XII

On the greater increase of the force which, to carry the same load with the same bones, is required from the muscles which are involved in the operation.

The fact that the force exerted by the muscles is greater than the supported load thus has been demonstrated. This rightfully appears as surprising. Even more surprising is its increase which we shall now describe after presenting some lemmas of mechanics.

Proposition XLII

A weightless lever is divided and its segments are linked by a flexible tie. A weight is hung from one of its extremities and two forces pull the two segments of the lever obliquely to the same side. These forces are unable to straighten the lever and to balance the weight.
Tab. IV, Fig. 7.

The lever AC is divided and articulated at B. Its segments AB and BC can flex either at the joint B or at the fixed fulcrum A. They are weightless. A weight R hangs from C. Two forces F and K applied at points G and H of the rods pull these upwards along two parallel directions GE and HO oblique to the same side. I claim that the lever ABC can neither be straight nor remain horizontal and the forces F and K cannot balance the weight R. The lever AB mobile about the fulcrum A is pulled upwards by force F from G in the direction GE. This lever AB is not pushed downwards by any load since it is supposed to be weightless. It is not pulled downwards by the weight R as long as it extends horizontally in ABC, because of the flexion of the rod BC about the joint B. Therefore, the joint B must move upwards, pulled along the arc BM with radius AB. Point B cannot be stopped anywhere but must rise along the arc BM. The two extremities of the rod BC can move in opposite directions. C is pulled downwards by the weight R and B upwards by the force F raising the joint B along the arc BM. Whether the midpoint H is maintained in its initial position or is pulled upwards along the direction HO, the rod BC would have to follow, C moving downwards to c and B upwards to b, until the rod BC has become vertical. Thus, the two rods AB and BC articulated at B, flex and, as a result, none of the forces F or K can balance the weight R with the lever horizontal. Q.E.D.

Proposition XLIII

The data are the same. Two forces can be applied to the rods to maintain the lever horizontal and balance the suspended weight.
Tab. IV, Fig. 8.

The construction of the lever ABC is the same. I claim that the assertion of the title is possible. A rope is attached at the extremity H of the rod CB. It is led through a hook or a pulley O belonging to the other rod AB and pulled by a force K. Tension of this rope can maintain the rods AB and BC aligned as a knot or a tie would do. Then the moment of the force K which pulls the rope from H to O must be equal to the moment of the weight R. If the action of the force K was smaller than the moment of the weight R the rod BC would move downwards along the arc Cn about the centre B. If it was greater than the moment of the weight R the rod would move upwards along the arc CN. Moreover, the two rods AB and BC being linked by force K constitute one straight and rigid lever. But this force K cannot prevent the whole lever AC from moving downwards pulled by the weight R about the centre A, because the entire force K is used to maintain the two rods aligned. This alignment can be maintained whatever the inclination of the lever AC, for example in the position Abc. Therefore, to maintain horizontal the lever AC thus aligned, another force F is required which pulls at G. The moment of this force F must be equal to the moment of the weight R. Consequently, both the moment of K and that of F must be equal to the moment of the weight R. These two forces K and F maintain the rods AB and BC aligned. This was the object of the research.

Corollary

Consequently, the sum of the moments of the two equal forces F and K is equal to twice the moment of the weight R.

Proposition XLIV

A weighing lever is formed by several articulated rods. A weight is suspended from its extremity. The lever is aligned and supported by as many ropes attached to the fixed extremity as there are rods. Then the ratio of the sum of the forces exerted by the supporting ropes to the suspended weight taken as many times as there are rods, with the half weight of the first rod taken once, that of the second rod taken twice, that of the third rod taken three times, and so on, is equal to the ratio of the sum of the lengths of all the rods, the first taken once, the second twice, the third three times, and so on, to the sum of the half-distances of the lines of action of the ropes from their fulcra. But all the successive terms must be proportional to each other.
Tab. IV, Fig. 9.

The lever AD is made of several rods AB, BC, CD of even thickness and weight, articulated at A, B and C. A weight R hangs vertically from the furthest extremity D. The rods are aligned and supported by the forces exerted by the ropes EM, HGF and LKI which are attached at L, H and E. The distances CD, BD and AD are in the same proportions as half the distances CP, BN, AO of the ropes from the fulcra C, B and A. I claim that the ratio of the sum of all the forces exerted by the ropes EM, HGF and LKI to three times the weight R with half weights of the rods, one for the rod AD, two for the rod BC and three for the rod CD, is equal to the ratio of the sum of the lengths of once the first rod AB, twice the second rod BC and three times the third rod CD, to the sum of half the distances AO, BN, CP of the ropes from the fulcra A, B, C. We assume that the two rods AB and BC are maintained aligned and horizontal by their ropes. Then C is the fulcrum of the lever CD pulled with equal moments downwards by the weight R and upwards by the rope LKI attached at I and deviated by the hook K. The ratio of the force exerted by the rope LKI to the weight R with that of the lever CD or rather with half the weight of the rod CD considered as acting at D is equal to the ratio of the length of the lever CD to half the distance CP[1]. The force exerted by the rope LKI only maintains the rods BC and CD aligned and horizontal. The rod AB is maintained horizontal so that the lever BD can move about the fulcrum B. Again, the lever BD is pulled downwards about the centre B by the weight R and the weight of all the lever BD or by half this weight if acting at D. The lever BD is also pulled upwards by the rope HGF, deviated by the pulley G and attached at H. The moment of the supported weight and that of the force exerted by the rope are equal. Therefore, the ratio of the magnitude of the force exerted by the rope HGF to the weight R with half the weight of the lever BD considered as suspended at D is equal to the ratio of the length of the lever BD to half the distance BN of the rope from the centre B. Here also the only action of the rope HGF consists of linking AB and BD and of maintaining these rods horizontal. Therefore, we remain with the lever AD mobile about the centre of rotation A. This lever is pulled downwards by the weight R and by its own weight or by half its weight if this is considered as acting at D. The lever is also pulled upwards at M by the force exerted by the force EM attached at E. The moments of the supported weight and of the force exerted by the rope are equal. Thus, the ratio of the magnitude of the force exerted by the rope EM to the weight R with the weight of half the lever AD considered as acting at D is equal to the ratio of the length of the lever AD to half the distance AO. The distances CD, BD, AD are in the same proportion as the half distances CP, BN, AO. The different terms thus collected are also proportional to each other. Therefore, the ratio of the sum of the magnitudes of the forces exerted by the ropes LKI, HGF and EM to the sum of three times the weight R and the half weights of CD, BD and AD, one for AB, two for BC and three for CD, is equal to the ratio of the sum of the lengths of the levers CD, BD and AD, or once AD plus twice BC plus three

[1] From Comment Prop. 34

times CD to the sum of half the distances AO, BN and CP of the lines of action of the ropes from the centres. If there are more than three rods, the same proposition is true.

To compensate for the inconvenience of this arduous demonstration, before presenting the other lemmas referring to this matter, the reward of the work will be to collect savoury fruits.

Proposition XLV

The human arm supinated down to the tip of the fingers
is extended horizontally and at the extremity of the last four fingers
a weight is suspended, as heavy as can be supported in this position.
The apparent force which Nature exerts in contracting all the muscles
of the arm which combine to support the weight
is more than 209 times the suspended weight.
Tab. IV, Fig. 10.

The lever AG made of six rods articulated at six joints represents a human arm extended horizontally and supinated. Its direction is not given. But the elbow B and the fingers DEF are a little flexed so that they can move a little downwards and the extremities of the last four fingers of unequal lengths are brought to the same level. From the extremities G of the last four fingers a weight R is hung as heavy as can be supported by a strong young man. This is less than 9.5 pounds. Since the straight lever AG is made of six articulated rods, the forces of all the ropes or muscles IP, PK, KL, QM, KM, KN and KO are required to support the lever AG and the suspended weight R, as was demonstrated. We shall investigate the magnitude of the forces developed by all the muscles. Their total is equal to 209 times the weight R. Let us consider the segment AF of the lever. This segment is maintained aligned by its ropes. F is the fulcrum of the lever FG pulled downwards by the weight R and upwards by part of the force of the flexor digitorum profundus muscle KO the tendon OH of which is connected to the joint EF by a ligament acting as a pulley. The distance of the line of action of the tendon HO from the centre F is equal to half the thickness of the joint. The thickness of this joint is about a quarter of the length of the bone FG. The length of the lever, i.e. the distal phalanx FG is thus four times the distance FH. Consequently, the ratio of the involved part of the magnitude of the force of the muscle KO to the load R with half the weight of the four distal phalanges FG (the weight being considered as acting at G)[2] is equal to the ratio of the distance FG to half the distance FH which is the quarter of the thickness of the distal joint. Therefore, the part of the muscular force KO which opposes the load R is equal to eight times the weight R, the small weight of the distal phalanges being neglected. The load R was 9.5 pounds. This part

[2] From Comment Prop. 34

of the muscular force KO is thus equal to 76 pounds. The lever EG is also maintained aligned by the muscle KO. Its fulcrum is E. The load R acts on it with half the weight of the eight distal and middle phalanges which can be neglected. The length EG is more than eight times the distance EN of the line of action of the tendon from the centre of the joint. Therefore, the force exerted by the muscle KN is sixteen times the weight R and is equal to 152 pounds.

The lever DG formed by the whole fingers is maintained extended by the muscles KO and KN and can move about the centre of rotation D. This lever is pulled downwards by the same load R with half the weight of the twelve phalanges which can also be neglected. The lever is pulled upwards by the lumbricales muscles QM helped by the muscle KO. The distance DG is more than ten times the distance DM of the line of action of the tendon from the centre D. This distance DM is equal to half the thickness of the metacarpo-phalangeal joint. Therefore, the ratio of the force exerted by the muscles QM and KO to the weight R with that of the fingers is more than 20/1. Consequently, the force exerted by these muscles is more than 190 pounds.

The lever CG formed by the hand is maintained extended by the muscles KO, KN and QM. It is pulled downwards by the weight R and a half pound which is half the weight of the hand. The lever is also pulled upwards by the muscle KL about the centre C. The length of the hand is more than ten times half the thickness of the wrist. Therefore, the ratio of the force exerted by the muscle KL to the weight R with half the weight of the hand is greater than 24/1. Consequently, the force exerted by the muscle KL is more than 240 pounds.

The lever BG made of the forearm and hand is maintained aligned by the muscles KO, KN, QM and KL. This lever is pulled downwards by the weight R with half the weight of the forearm and hand, 2 pounds. It is pulled upwards by the muscles PK about the centre B. The ratio of the length of the forearm and hand to half the thickness of the elbow is 24/1. Therefore, the ratio of the force exerted by the muscles PK to the weight of 11.5 pounds (the weight R and half the weight of the forearm and hand) is more than 48/1. Consequently, the force exerted by the muscles PK is greater than 552 pounds.

Finally, the lever AG made of the upper arm, forearm and hand is maintained aligned horizontally by the muscles PK, KL KN, QM and KO. This lever is pulled downwards by the weight R and half the weight of the arm and hand, i.e. 3.5 pounds. It is pulled upwards by the deltoid muscle IP about the centre of the humeral head A. The length of the arm AG is more than thirty times the radius of the humeral head, which is the distance of the line of action of the deltoid muscle from the centre of rotation of the humeral head. Therefore, the force exerted by the muscle IP is more than sixty times 13 pounds which corresponds to the load R with half the weight of the arm. Consequently, the force exerted by the deltoid muscle in the operation is equal to 780 pounds.

When added, the forces exerted by all the muscles involved in sustaining the weight R attain 1990 pounds. Nature thus is forced to exert a force 209 times greater than the weight R. Otherwise it would not be possible to support this weight at arm's length. Q.E.D.

Comment

It is worth-while noting that in the operation thus described, not all the muscles involved in the support of the weight R at arm's length in the supine position must exert all their power. It is sufficient if one of them exerts its maximum force to balance the weight R. Then the force of all the other muscles may be greater than the moment of the resistance. However, they are forced to use part of their force, the remainder being inactive. All must oppose a moment equal to that of the resistance, as demonstrated above. The maximum force exerted by two flexor muscles of the forearm was shown in the previous reasoning to be more than 1120 pounds. But here it has been found[3] to be much less, i.e. 552 pounds. The reason why in this instance the same muscles do not exert all their power is that the flexor digitorum profundus is weak and cannot sustain a weight heavier than 9.5 pounds. To support such a light weight only part of the force of the flexor muscles of the forearm equal to 552 pounds is sufficient.

This can be subject to experiment. The arm is extended horizontally and pronated, the thorax being inclined to the side. The maximum weight which can be supported is then hung from the tips of the fingers. This weight is small, much less than 6 pounds, although the force of the same deltoid muscle is able to support a weight much heavier, as appeared in the preceding experiment.

Proposition XLVI

If the extremities of the two parallel arms of a Z contrivance are subjected to opposite forces acting along the same line, the opposite forces are inversely proportional to the lengths of the arms of the Z.

Tab. V, Fig. 1.

The contrivance ABCD is made of three rods connected by two alternate angles B and C. AB and CD are parallel. Two forces R and S are applied at A and D. They act along the same line AED in opposite directions and tend to close or to open the angles B and C. I claim that the ratio of force R to force S is equal to the ratio of DC/AB. In the contrivance ABCD the middle rod BC is not subjected to the forces R and S. The other two levers are pushed in opposite directions about points B and C. Therefore, the middle rod BC behaves like the fulcrum of a balance the arms of which are AB and DC. The two forces R and S act on the two levers AB and DC about the fulcra B and C along the same line AED but the levers remain parallel. Therefore, the attempt of force R at moving the lever AB is measured by the straight line AE and the attempt of force S at moving the lever DC is measured by the straight line DE. Consequently, the ratio of the velocity of force R to that of force S is equal to AE/DE or to AB/DC since the triangles ABE and DCE are similar. The moments of R and S are equal since they balance each other and one does not yield to the other. Consequently, the forces being in inverse proportion to their velocities, the ratio of the magnitude of force R to that of force S is equal to DC/AB. Q.E.D.

---

[3] Prop. 35

Proposition XLVII

If several weightless and articulated rods, connected by ropes passing round the joints form an alternate contrivance, the forces acting on the contrivance along the same line of action are in inverse proportion to the distances of their line of action from the centres of rotation.
Tab. V, Figs. 2 and 3.

Firstly, three rods AB, BD and DE (Fig. 2, Tab. V) are articulated at B, D and connected by ropes GFH, IKL which span the joints B and D. The rods form an alternate contrivance which is supposed to be weightless. Two forces R and S act on the contrivance, along the same line ACE but in opposite directions in an attempt to close the angles if the ropes pass about the convexity of the joints or to open the angles if the ropes pass in the concavity of the joints. Perpendiculars BM and DP are drawn from the centres of the joints to the common line of action of the two forces. I claim that the magnitudes of forces R and S are in inverse proportion to DP/BM. Force R acts on the lever AB at A in the direction AM as if it was acting at the extremity M of the lever BM. Similarly, force S behaves as if it was acting at the extremity P of the lever PD. Thus two forces R and S of equal moments, since one balances the other, act on the alternate contrivance BM, DP along the same line and BM and DP are parallel since both are at right angles to the same straight line ACE. Consequently, according to the preceding proposition, the ratio of the forces R/S is equal to PD/BM.

Secondly, the contrivance ABVDE (Fig. 3, Tab. V) is made of four rods by adding the intermediate angle V spanned by the rope NOQ. The contrivance is supposed to be divided at X where a force Z is applied. This force Z acts on the contrivance ABVX opposing force R with an equal moment. As results from the first part of this proposition, the ratio of the forces R/Z is equal to the ratio of the distances TV/BM. If the contrivance is divided at C the rope NOQ opposes two forces which compress the bow CVX from C and X. These forces would be equal. Therefore, the force C which would compress the bow CVX ought to be equal to Z. The whole contrivance CVDE is then considered. Force Z acting at C and force S have the same moment since S as well as Z is balanced by A. The ratio of these forces Z/S is equal to the ratio of the distances DP/TV. But the ratio of the forces R/Z was equal to the ratio of the distances TV/BM. Consequently, exchanging the equalities, the ratio of the magnitude of force R to that of force S is equal to the ratio of the distances DP/BM. If the contrivance was made of any number of rods, it would be demonstrated by the same process that the forces acting on the end rods are in inverse proportion to the distances of their common line of action from the joints. Q.E.D.

Proposition XLVIII

The data are the same. The moments of the forces exerted by all the ropes which attempt at closing the angles of the contrivance are equal to twice the sum of the moments of the external forces which act on the contrivance. There are as many external forces as there are ropes and the forces opposing each other are considered.
Tab. V, Figs. 3 and 4.

The data are the same. The contrivance is made of more than three rods and is imagined to be divided at X where force Z is applied which opposes force R, their moments being equal. Both forces Z and R act on the three-rod contrivance ABVX against the resistance of the ropes GFH and NOQ. Obviously, force Z acts directly against the force exerted by the rope NOQ as force R acts against the resistance of the rope GFH. Since the rod BCV is supposed to be rigid, its inferior segment VC resists the tension or compression exerted by the bow ABC as if this bow CBA was hanging from a nail C or as if this bow was standing on a solid plane passing through C. Therefore, the bow ABC is not different from the bow which is compressed by the force R and is attached to a nail C or stands on a solid plane at C. Thus[4], the moment of the force exerted by the rope GFH is equal to twice the moment of force R. Then two opposite bows ABC and XVC resist each other or act on each other with equal moments at C since one does not yield to the other as a result of the continuity and rigidity of the rod BCV. Consequently, since, as said previously, the bow XVC is not different from a bow subjected to a force Z and attached to a nail C or supported by the ground at C[5], the moment of the force exerted by the rope NOQ is equal to twice the moment of force Z. Finally, since the bow EDX is not different from a bow subjected to force S and attached to a fixed nail X or standing on the ground at X, the moment of the force exerted by the rope IKL is equal to twice the moment of force S[6]. Thus the moment of the forces exerted by the rope GFH equals twice the moment of force R, the moment of the force exerted by the rope NOQ equals twice the moment of force S and the moment of the force exerted by the rope IKL equals twice the moment of force Z. We demonstrate the same if there are more than three rods.

Corollary

Consequently, if the contrivance made of alternately folded rods attached to a nail or standing on the ground is subjected to one force alone, all the ropes which span the angles exert the same force as if they were subjected to as many forces as there are ropes. This ensures balance.

---

[4] From Coroll. Prop. 38
[5] From Coroll. Prop. 38
[6] From Coroll. Prop. 38

If the contrivance ABVDE attached to a nail or standing on the ground at S is subjected to one force R alone all the ropes GFH, NOQ and IKL are balanced and exert the same force as if they were opposing six forces of equal moment. These six forces would be R and another force equal and opposite to R, two forces X, Z and two forces equal to the resistance of the nail or of the ground S.

Proposition XLIX

The data are the same. If the magnitude of one force acting on the folded contrivance and the distances of the line of action of this force and those of the ropes from all the centres of rotation are known, the magnitudes of the forces exerted by all the ropes can be found.
Tab. V, Figs. 3 and 4.

The data and the diagrams are the same. The magnitude of force R exerted on the weightless contrivance against the resistance S of the nail or of the ground is given as well as the distances BM, TV, PD of the line of action of the force R from the centres B, V, D and the distances of the lines of action of all the ropes GFH, NOQ, IKL from the centres. Bα is a quarter of the two distances of the rope GFH from the centre B, Vβ a quarter of the two distances of the rope NOQ from the centre V and Dθ a quarter of the two distances of the rope IKL from the centre D. Since all the distances of the ropes from the corresponding centres of rotation are given, their quarters Bα, Vβ, Dθ are also known. (1) The ratio of force R to the force exerted by the rope GFH is equal to Bα/BM. The three quantities R, Bα and BM are given. Therefore, the fourth member of the proportion, force GFH, is known. (2) The ratio of the forces R/Z is equal to TV/BM. The three quantities R, TV and BM are given. Therefore, force Z is known. (3) The ratio of force Z to the force exerted by the rope NOQ is equal to Vβ/TV. (4) The ratio of the forces Z/S is equal to DP/TV. The first three of these quantities being known, the fourth S is known. (5) Finally, the ratio of force S to the force exerted by the rope IKL is equal to Dθ/DP. Since S, Dθ and DP are given, force IKL is known. Adding the three magnitudes of the forces exerted by the ropes thus found gives in one figure the sum of the forces exerted by all the ropes which balance force R. Q.E.D.

If the contrivance has some weight, however, the element ABC is subjected to force R and to its own weight considered as acting at the common centre of gravity A. The sum of both balances the opposite force C. Similarly, the double bow ABVX is subjected to R and to its own weight considered as acting at A. The triple bow ABVDE is subjected to R and to its own weight considered as acting at A. The ratio of R + the contrivance ABVX to Z thus is equal to the ratio of TV/BM. The ratio of R + the contrivance ABVDE to force S is equal to the ratio of DP/BM. Consequently, the forces exerted by the ropes can be found using the method thus described.

Proposition L

A weightless double alternate contrivance is subjected to one force alone. A rope passes in the concavity of an angle and over the apex of the other angle and another rope passes over the apex of the first angle, thus at the opposite side of the first rope. The moments of the forces exerted by the two ropes are equal to the sum of twice the moment of the force acting on the arms of the angle surrounded by two ropes, and four times the moment of the force acting on the arms of the angle spanned by one rope. Tab. V, Figs. 5 and 6.

The weightless contrivance ABDE folded at B and D is compressed by a force R against the resistance S of a nail or of the ground. The rope IKL attached to the arm AB at I and to the non-contiguous arm ED at L passes in the concavity of angle B and over the convexity of angle D. Another rope GFH passes over the apex of angle B, thus on the side opposite to the first rope. I claim that the sum of the moments of the two ropes GFH and IKL is equal to the sum of twice the moment of force R acting on the bow or angle ABC and four times the moment of force S exerted on the bow EDC. We assume the junction B to be rigid. The rods AB and BC remain inclined to each other by the angle IBC. Then it does not matter whether the rope KI is attached to the arm BA or to the arm BC. Whatever its attachment proximally, the rope IKL spans the angle D of the bow BDE[7]. The moment of the force exerted by the rope IKL is equal to twice the moment of force S acting on the bow CDE. But the angle IBC is not rigid and can move. Therefore, the rope IKL when acting and thus pulling its extremity I towards KL cannot open the angle EDC but would rather flex the arm BA towards BC. In order to enable the angle EDC to remain open it is necessary not to close the angle IBC. This is not possible without adding a new rope GFH opposite, exerting a force to open the angle IBC and thus maintaining tension in the rope IKL. Then, at equilibrium, it is obvious that the rope GFH achieves two tasks: firstly it maintains the rope IKL tense; secondly, it opposes force R and its opposite force at C[8]. Therefore, the moment of the force exerted by the rope GFH is equal not only to twice the moment of force R but also to the moment with which the force exerted by the rope IKL resists the tension evoked by the rope GFH and which is necessary for IKL to remain tense[9]. The moment of the force exerted by the rope IKL is equal to twice the moment of the resistance S of the nail or of the ground. Consequently, the sum of the moments of the forces exerted by the two ropes GFH and IKL is equal to the sum of twice the moment of force R and four times the moment of S. Q.E.D.

---

[7] Prop. 48 and from Coroll. Prop. 38

[8] Coroll. Prop. 38

[9] Coroll. Prop. 38

Proposition LI

A contrivance made of several alternate bows is subjected to one force and ropes span two adjacent angles, one in the concavity and the other over their apex, and the last rope spans the one angle close to the point of application of the external force. The sum of the moments of the forces exerted by all the ropes is equal to the sum of twice the moment of the force acting on the arms of the first angle thus spanned on both sides, four times the moment of the force acting on the arms of the second angle and six times the moment of the force acting on the third angle and so on.

Tab. V, Figs. 7 and 8.

The contrivance is made of three or more alternate bows ABVDE and is compressed by a force R against a nail or against the ground S. Two ropes LKI and NOQ each span two angles, one in a concavity, the other over an apex. Moreover, the rope GFH spans the angle B on the side opposite to the rope NOQ. Force Z is imagined acting on the intermediate bow CVX. I claim that the sum of the moments of all the forces exerted by the ropes IKL, NOQ and HFG is equal to the sum of twice the moment of R, four times the moment of Z and six times the moment of S, and so on in the same order if there are more elements in the contrivance. The contrivance is imagined as divided at C and the rope NO as attached at Y in the middle of the rod VC. The three-rod contrivance CVDE is spanned by the rope LKI attached to two non-adjacent rods and the rope NOY spans the angle XVC only, on the opposite side. Therefore[10], the moment of the rope LKI is equal to twice the moment of the resistance S of the nail or of the ground and the moment of the rope NOY is equal to the sum of twice the moment of the force Z and twice the moment of the force S. If the rope NOY is detached from Y and attached at Q, then the rope GFH acts against two resistances, i.e. against the double of force R and against the tension of the rope QO. The force GFH is balanced by all these forces. Thus[11], the moment of the force exerted by the rope GFH is equal to the sum of twice the moment of force R and the moment of the rope QON, i.e. the sum of twice the moment of force Z and twice the moment of the resistance S. Consequently, the sum of the moments of the three ropes GFH, QON and IKL is equal to the sum of twice the moment of force R, four times the moment of force Z and six times the moment of force S. If there were more rods and angles, the same process would go on. Q.E.D.

---

[10] From Prop. 50
[11] From Prop. 50
[11] From Prop. 49

Proposition LII

The data are the same. If the force acting on the contrivance, the distances of the line of action of the force and the distances of all the lines of action of the ropes from the centres are known, the apparent magnitudes of the forces exerted by the ropes can be found.
Tab. V, Figs. 7 and 8.

The data are the same. The force R is given as well as the distances of its line of action from the centres B, M, T, V, D, P and the quarters of the distances of the lines of action of the ropes from the centres, Bα, Vβ, Dθ. The distances BQ of the rope QO from the centre B and VI of the rope KI from the centre V are also given. From these data, the forces exerted by the ropes must be found. The contrivance is supposed to be weightless. The ratio of the forces R/S is equal to DP/BM. The ratio of the forces S/Z is equal to TV/DP. The ratio of the forces S/Z is also equal to Dθ/PD. The ratio of force S just found to the force exerted by the rope LKI is equal to Dθ/PD. The ratio of force Z also found to the force exerted by the first part of the rope NOQ is equal to Vβ/TV. The rope NOQ not only opposes twice the force Z but also pulls and maintains the rope IKL tight. Therefore, the rope NOQ is balanced by the double not only of the force Z but also of the resistance of the rope IKL. The force exerted by the rope NOQ can be divided into two parts. One of these parts is balanced by twice the force Z and has been found. The remainder of the force exerted by the rope NOQ is balanced by the resistance of the rope IKL and has to be found. In the angulated balance IVD the forces are applied in the vicinity of the centre V: the force exerted by the rope NOQ part of which is applied at Y and that exerted by the rope IKL applied at I. Therefore, the ratio of the force exerted by the rope IKL which has been found to the second part of the force exerted by the rope NOQ is equal to βV/VI. The total force exerted by the rope NOQ is equal to the sum of its two parts which have been found. The ratio of force R which is known to the first part of the force exerted by the rope GFH is equal to Bα/BM. The other part of the force GFH is balanced by the resistance of the rope QON. The ratio of all the force NOQ to this second part of the force exerted by the rope GFH is equal to Bα/BQ. The force exerted by the single rope LKI, the two parts of the force exerted by the rope NOQ and the two parts of the force exerted by the rope GFH are added. Their sum represents the forces exerted by all the ropes opposing the force R which is given. Q.E.D.

If the contrivance has some weight, the ratio of the sum of the weight R and the weight of the whole contrivance ABVDE imagined as acting at A to the force of resistance S of the nail or of the ground is equal to DP/BM. Similarly, the ratio of the sum of the weight R and that of the element ABVX considered as acting at A to the resistance Z is equal to TV/BM. From the three known forces R, S and Z, using the method described above, one can find the force exerted by the rope LKI, the two parts of the force exerted by the rope NOQ and the two parts of the force exerted by the rope GFH.

Consequently, as we suggested in proposition 45, to support some weight, one muscle alone which appears directly to oppose this weight is not enough. Several other muscles combine in carrying out the operation. Thus, after showing why wise Nature needs so much force to sustain light weights, it is worthwhile to find out up to where its effort increases.

Proposition LIII

When a stevedore carrying a load on his neck and holding it with his arms stands on the ball of one foot with flexed hip, knee and ankle, and heel raised, the force exerted by Nature in the extensor muscles of the leg combining to support the load may be more than forty-four times the carried load.
Tab. VI, Fig. 1.

The contrivance ABVDE is made of four rods articulated at B, V and D which can flex in alternate directions. Such is the human body carrying a load R of 120 pounds on its neck and holding it with the arms. The head and the vertebral spine are bent forwards down to the coccyx AB. The hip B, knee V and ankle D joints are flexed and the heel K is raised. The total mass RABVDE is supported by the ball ES of one foot at S. This position occurs in walking when stepping over the doorstep of a house. Then the entire human machine is maintained erect by the forces exerted by the glutei muscles GFH, the vasti muscles YON and the soleus muscle LK. All these muscles combine to support the load R. If one of them failed to exert its force, the load could certainly not be carried. If the muscles GFH did not contract and exert their force, the rope GFH would be loose. Then the rod AB and the load R would tip downwards towards the knee V. If the muscles YON were inactive, the whole mass RABV would fall despite the bow ABC being maintained tight. Finally, despite the contraction of the muscles GFH and YON which would maintain the bows ABC and CVX tight and open, and support the machine XVBAR in the erect position, if only the soleus muscle LK failed, the whole mass would tip not only because of the collapse of the angle VDE but also because the segment DV would align with the ground and the perpendicular line of action ACX of the weight of the whole mass may fall beyond the knee V. Then the whole machine would tip forwards. We shall analyse all the muscles GFH, YON and LK and compare them with the load R which they support. Firstly, the position of the line of action AE of the total weight and its distance from every centre must be found. To this end, a line is drawn from the extremity E of the bent body on the ground, which supports the whole machine in the erect position, to the common centre of gravity A. This line is vertical, passes across the points C and X thus intersecting the femur and the tibia. The distances MB, TV, PD of this line from the centres of the joints are then measured. It appears that the distance MB is more than four times the distance of the glutei muscles from the centre of the head of the femur, the distance TV more than three times the distance of the tendon of the vasti muscles from the centre of the knee and the distance DP more than one and a half times the distance of the Achilles tendon from the centre of the ankle. The weight of the man is more than 150 pounds and his centre of gravity is in the vicinity of the hip B. The weight AB from the head to the hips is 75 pounds. Adding the part BC down to mid-thigh and the opposite leg which is raised, the segment ABC can be evaluated at 122 pounds and the segment ABVX down to mid-lower leg at 142 pounds. The ratio of the sum of the weight R and the weight of the element ABVX to force Z, i.e. the ratio of 242 to 322.66 pounds which is the weight or force Z[12], is equal to the ratio of the distances TV/BM = 3/4. The ratio of the sum of the weight R and that of the whole contrivance ABVDE to force S, i.e. the ratio of 270 to 720 pounds which is the weight or force S, is equal to the ratio of the distances DP/BM = 3/8. The bow ABC is subjected to the weight R and its own weight. That makes 230 pounds. The ratio of half the distance of the rope GF from the centre B to the distance MB is 1/8. Therefore, the force exerted by the glutei muscles FGH to balance the weight R with the weight of the element ABVX (242 pounds) is 1936 pounds in this instance[13]. The ratio of half the distance of the line of action of the rope YO to the distance TV is 1/6. Therefore, the force exerted by the muscles NOY to balance force Z of 306.66 pounds is 1840 pounds. The ratio of half the distance of the line of action of the rope KL to the distance DP from the centre of the ankle is 1/3. Therefore, the force S of 520 pounds is balanced by a force of 1560 pounds exerted by the muscle KL. The sum of the forces exerted by the muscles GFH, NOY and KL is 5336 pounds whereas the supported weight R is 120 pounds. Consequently, the force exerted by all the muscles involved in supporting the weight R is more than forty-four times the supported weight R. Such was the proposition.

Proposition LIV

The data are the same. The forces exerted by the rectus femoris and gastrocnemius muscles which are used in carrying the load, are found. Adding these forces to the previous ones gives a force higher than fifty times the supported load.

Tab. VI, Fig. 1.

The data and the diagram are the same. But, besides the vasti muscles NOY, the rectus femoris muscle OQ is also considered. Its origin Q is in the inferior anterior spine of the ilium and its insertion is at N below the knee. Besides the soleus muscle LK, the two parts of the gastrocnemius IK are also considered. They originate from the femoral condyles and are inserted in the os calcis. Although these muscles appear to be useless, on thinking it over, they may be as active as if the subject had all his joints extended. Closing the angle ABV loosens the muscular rope QO exactly as much as it tightens and pulls the other part of the rope ON over the apex of the angle BVX. The rope QON thus keeps

[12] From Prop. 47

[13] From Prop. 38

the same tension as if the contrivance ABVX was aligned. In extension the muscle QON can exert its force advantageously. When the hip and knee are flexed it can also contract and thus help extension of the lower leg. Similarly, the muscular rope IK will be exactly as tight when the knee CVX is flexed and the ankle XDE is dorsiflexed as if the three rods CV, VD and DE were aligned. Therefore, the muscle IK can contract and help plantar flexion of the foot. The forces exerted by the muscles QO and IK thus must be found. We shall show that the load R is smaller than a fiftieth of the forces exerted by all the muscles by which this load is supported. Being equal to twice the moment of S, the force exerted by the muscles LK is 1560 pounds. A part of this force must be attributed to the soleus muscle LK and the remainder to the gastrocnemius IK. It is assumed that the forces exerted by the muscles are proportional to the number of their fibres or to their cross section. This seems sensible. As we shall see, the cross section of the gastrocnemius is more than half that of the soleus muscle. Therefore, the force exerted by the gastrocnemius IK is more than a third of the total muscular force, i.e. 520 pounds. On the other hand, the force exerted by the muscles NOY and OQ is equal to three times the double of the sum of Z, i.e. 1840 pounds, and the force exerted by the rope KI, i.e. 520 pounds. The total force exerted by the muscles NOY and OQ thus is 2360 pounds and the cross section of the rectus femoris muscle OQ is a quarter of that of the two vasti muscles NOY. Therefore, the force exerted by the rectus femoris muscle OQ is a fifth of the total muscular force and is equal to 572 pounds. Finally, the glutei muscles GFH balance twice the weight of the load R with the bow ABC and also the force exerted by the rope OQ, the latter being 572 pounds and the former 1548 pounds. Therefore, the total force exerted by the glutei muscles in this instance is 2120 pounds. The sum of the forces exerted by the muscles LK, IK, OQ, GFH involved in supporting the load R thus is 6040 pounds. The load R of 120 pounds is only the fiftieth of this force.

We shall show later that the forces exerted by the extensor muscles of the thigh, lower leg and foot, when their joints and the back are flexed, are not sufficient to support this load carried on the neck. Much greater forces are required. They are provided by other muscles involved in carrying out the operation. To demonstrate this, it is necessary to describe a new machine which wise Nature uses to carry out this work.

Proposition LV

A bow made of several articulated rods with ropes spanning the joints stands on the ground with its lower extremity and supports a load at its upper extremity. The ratio of the force exerted by each rope to the sum of the weight of the load and that of the corresponding part of the bow is equal to the ratio of the distance of the line of action of the load to half the distance of the line of action of this rope from the centre of the joint which this rope spans.

Tab. VI, Fig. 2.

The bow ABCDFE is made of the rods AB, BC, CD, DE, EF connected by the ropes ZY, IX, LK, HG which span the joints B, C, D, F over the convexity of the bow. This is fixed to the ground S by its lowest rod FE and compressed by a load R at A. The common centre of gravity M of the load R and of the bow ABCDF is found. From this centre of gravity a perpendicular MS is drawn which is the line of action of the common weight. The distance of this line from the centre of the joint F is MF. A perpendicular NS is also drawn from the common centre of gravity N of the load R and the element ABCD of the bow. Its distance from the centre of the joint D is ND. Another perpendicular OS is drawn from the common centre of gravity O of the load R and the element ABC of the bow. Its distance from the centre of the joint C is OC, and so on. I claim that the ratio of the force exerted by any rope LK to the sum of the weight of the load R and that of the above element of the bow DCBA is equal to the ratio of the distance ND to half the distance DK of the line of action of the rope KL from the centre of the joint D, and so on for the other elements. The rod FE is assumed to be fixed to the ground. The bow ABCDF is maintained by the contraction of the ropes HG, KL, etc. as if it was uninterrupted and rigid. Therefore, the bow AFE can be considered to be made of two rigid elements AF and EF one of which EF is fixed to the ground and the other AF articulating about the centre F, is maintained erect by the force exerted by the rope GH and is compressed by the load R and its own weight. The line of action of the common weight is MS. Thus[14], the ratio of the force exerted by the rope GH to the sum of the weight of the load R and that of the element FA is equal to the ratio of the distance MF to half the distance FH of the line of action of the rope from the centre of the joint F.

Similarly, the rod DF is maintained in its position by the force exerted by the rope GH. It is as if the rod DF was fixed to the wall at F. The bow DC, BA is maintained by the ropes XI, YZ and thus can be considered as one rigid and uninterrupted rod DA. Therefore, the bow is made again of two rods one of which DF is supposed to be fixed to the wall and the other DA is compressed by the load R and its own weight. This compression is balanced by the force exerted by the rope LK. Thus[15], the ratio of the force exerted by the rope LK to the sum of the weight R and that of the element DA is equal to the ratio of the distance ND to half the distance DK. The same process is repeated for the other ropes and the proposition is demonstrated.

---

[14] From Prop. 38

[15] From Prop. 38

Proposition LVI

If the bow formed by the vertebral spine bent forwards above the sacrum is compressed by a load carried on the neck, the sum of the forces exerted by the ropes which extend the spine is equal to the sum of as many times the weight of the supported load as there are vertebrae and the weight of the portion of the human body comprised between the horizontal planes across the considered joint and across the next one distally, taken twice for the distal portion, four times for the next one, six times for the third portion, and so on by increments of two.

Tab. VI, Fig. 2.

The same bow ABCDFE is constituted by the vertebrae FD, DC, CB, BA, etc. of the spine bent forwards and supported by the sacrum EF. The spine is compressed by a load R carried on the neck. The vertebrae are connected, as were the rods, by the posterior muscles HG, KL, XI, YZ, and also by cartilages. When the spine is bent the cartilages have the ability to contract as springs do. They thus help the action of the muscles to straighten the spine or to maintain it erect when compresssed by the load R. Therefore, we shall conceive as ropes straightening the back the forces of the muscles together with those of the cartilages. Horizontal planes FM, DN, CO, BP are drawn across the intervertebral joints. They divide the body into segments FDNM, DCON, etc., which will be called cylindrical segments appended to the vertebrae. Each of these, such as CDNO, is assumed to be rigidly connected to its vertebra so as to form with this vertebra one solid body as if it was a wooden cylinder. We shall show that the muscular forces HG, KL, XI, YZ are equal to the sum of twice the weight R multiplied by the number of vertebrae considered (thus three times, four times, etc.), and twice the weight of the cylindrical segment FDNM, four times the weight of the cylindrical segment DCON, six times the weight of the cylindrical segment CBOP, and so on, by increments of two, up to thirty-four times the weight of the segment adjacent to the neck.

The bow FCA is maintained by the tension of the ropes KL, XI, YZ. This bow thus is equivalent to a rigid rod FA articulated at F and the rod EF is fixed to the plane SE. Therefore, the two-part bow AFE is compressed by the load R and the weight of the rod AF corresponding to the segment ABFM of the human body. The line of action MS of the total weight passes across the common centre of gravity. Its distance from the centre F is MF. The force exerted by the rope GH is equal to the sum of twice the weight R and twice that of the body segment ABFM. On the other hand, the rod DF is fixed by the rope GH and this gives a new two-part bow ACD. Therefore, the force exerted by the rope LK is equal for a second time to the sum of twice the weight R and twice the weight of the segment ABDN. Similarly, the force exerted by the rope XI is equal for a third time to the sum of twice the weight R and twice the weight of the segment ABCO, and so on up to the seventeenth vertebra which is the first thoracic. Twice ABFM + twice ABDN + twice ABCO + twice ABP are equal to twice the distal cylindrical segment DFMN + four times the next segment CDNO + six times the third segment BCON + eight times the fourth cylindrical segment ABP. Q.E.D.

Proposition LVII

Evaluation as precise as possible of the weights of the segments of the trunk appended to the vertebrae.
Tab. VI, Fig. 2.

Using the same diagram, the weight of each cylindrical segment appended to a vertebra, such as CDNO, must be sought for. All the elements of this segment are connected to form with the vertebra CD one solid body as if it was a solid piece of wood. If all the viscera in the abdominal cavity with the fluids which they contain were solid and firmly bound to the lumbar and thoracic vertebrae, then the cylindrical segment CDNO would constitute a solid lever DN, articulated about the fulcrum D and balanced by two equal and opposite forces. Its own weight acting at its centre of gravity N would pull the lever downwards towards S and the force exerted by the rope LK by its contraction would pull the lever upwards. But the viscera are soft, slippery and mobile for their most part. They are enclosed in a kind of bag formed by the peritoneum, the abdominal muscles and skin. When the man is erect, they are supported by the bottom of this bag, the pelvis made of the iliac, ischial, pubic and sacral bones which are bound to each other. Therefore, an abdominal cylindrical segment delineated by two parallel planes, one along the upper aspect of a vertebra and the other along its lower aspect, cannot be considered as a solid lever. It is thus necessary to remove from the cylindrical segment CDNO the part which is supported by the pelvis EFS. The other part which is attached to the vertebra CD will be attributed to the cylindrical segment and the lever DN.

In the thorax the cylindrical segments appended to the vertebrae can be considered as solid without risk of error because the ribs are firmly bound to their vertebra and to the sternum, and the viscera contained in the chest, maintained by many ligaments, make this area evenly compact and solid as if it was one homogeneous mass separated from the abdomen by the strong diaphragm muscle. At regular intervals of time, during inspiration of air, it is compressed by the abdomen but not so that the weight of the thorax is decreased and that of the abdomen increased. Consequently, the actual weight of the cylindrical segments appended to the five lumbar vertebrae remains to be examined and particularly the part of this weight which must be subtracted from the segments for the sake of accuracy. However, it is not possible to give more than an approximate evaluation based on the most reliable calculation, as we are used to. The average weight of the trunk and head of a non-obese person is 75 pounds. Fifteen pounds are subtracted for the head and neck. The remaining 60 pounds is the weight of the trunk itself. The lumbar vertebrae are wider and longer than the thoracic vertebrae and all decrease regularly from the bottom upwards. The ratio of the five lumbar to the twelve thoracic vertebrae is 5/9. The ratio of the sum of the lumbar cylindrical segments to that of the thoracic segments is the same. The weight of 60 pounds thus is divided in the ratio of 5/9. The weight of the five lumbar segments is less than 22 pounds and each of the segments weighs certainly less than 5 pounds. The largest part of the abdominal viscera is supported by the pelvis and only a small part is attached to the vertebrae and to the ribs. We assume that 9/10 of the abdomen is supported by the pelvis and only 1/10 is attached to the lumbar vertebrae and the ribs. Therefore, the minimum weight which can be attributed to the five lumbar cylindrical segments is five pounds and to each of them one half pound. A weight of three pounds approximately can be attributed to each of the cylindrical segments appended to the thoracic vertebrae. This was the object of the research.

Proposition LVIII

Analysis of the expedients in the structure of the spine.

Obviously, the multiplicity of bones and joints was created by the wisdom of the Divine Architect to enable the animal to move in different ways. To enable the different joints to perform as conveniently and easily as possible, the articular surfaces of the bones were covered by light and slippery cartilage. Convexities and concavities of the joints were actually provided with distinct layers of cartilage in such a way that one bone can move and rotate on the other. This is a rule of action instituted by wise Nature. It is, therefore, surprising to see how much this rule is disturbed in the vertebral joints. Here the bony extremities are not round-shaped and convex or concave, and smooth, as would be required by rotatory movements. They are flat and rough. Moreover, they are not covered by smooth and slippery cartilage clearly separated from the cartilage of the adjacent bone. Two adjacent vertebrae are strongly connected by one common cartilaginous ligament which is interposed. It is thus worth enquiring why and to what end Nature has designed this new structure. Firstly, the substructure of the animal body, stable and strong like the hull of a ship, had to be bony. In man it has to support the body like a column. Therefore, the spine is roughly cylindrical but it is wider inferiorly than superiorly.

Secondly, the body of the animal must not be rigid but flexible. Therefore, its hull, the spine, had to be divided into different segments articulated with each other. But to provide stability and to prevent the risk of dislocation, the bases of the vertebrae had to be wide and flat and their joints had to be strongly held together.

Thirdly, the bundle of medullary nervous fibres had to be led through the spine, to diffuse the animal faculty from the brain to the whole body. This medullary bundle could not suffer contusion, stretching or angulation. Therefore, Nature had to provide that the back bends without sensible angulation, i.e. that the medulla undergoes only slight bending with a very long radius of curvature. This was achieved excellently by dividing the length of the vertebral spine into many small vertebrae. Two adjacent vertebrae can form only a very obtuse angle. Therefore, all the vertebrae describe a very slight curvature such as can be conveniently shaped by a polygon with many sides.

Actually, to achieve flexion of the vertebrae with such a long radius of curvature and simultaneoulsy provide stability and prevent the risk of dislocation, the vertebrae must not be connected by distinct and separate cartilages. It was enough that the bases of two adjacent vertebrae be linked firmly by one common soft cartilage. This cartilage by its softness would be used as a cushion preventing attrition of the bone, by its strength it would impede dislocation, by its elasticity it would permit small movement of the vertebrae in all directions.

Comment

After these premisses, it must be noted that if the intervertebral disk is compressed obliquely by the superior vertebra, part of this cartilage is compressed considerably and part of it is loose and stretched. Like springs and machines this cartilage tends to retract when stretched more than is permitted by its structure. In so doing, it pulls down the part of the superior bone which is too separated from the inferior bone. The part of the cartilage which is too compressed tends to expand. In so doing it drives back that part of the superior bone which is closer to the inferior bone. This must occur because the substance of such cartilage is of very strong and resistant consistency, although somewhat soft. Therefore, it acts as a spring as shown by experience.

Moreover, the tensile fibres which constitute the intervertebral disk are stronger than the fibres of the muscles of the back even when these are contracted. Consequently, when the intervertebal disks and the muscles of the back combine with all their strength to carry a weight, most of the weight is supported by the disks and the muscles exert only a smaller force. After these premisses, we shall demonstrate this lemma.

Proposition LIX

The balance AB (Fig. 3, Tab. VI) carries the two weights R and S and is supported by the three rods CD, GH, EF which act as springs. One of them CD, too much compressed, tends to expand and to raise the arm GA with which it is linked. The other EF, too much stretched, tends to contract and pulls the arm BG with which it is linked. The balance AB, being in equilibrium, remains immobile. I claim that the ratio of the weight R to the sum of the counterweight S and the forces of the two springs is equal to the ratio of the distances BG/GA. The ratio of the force exerted by the rod EF while it tends to contract, to the weight X, is equal to BG/GE. Similarly, the ratio of the force exerted by the rod CD while it tends to expand, to the weight Z, is equal to BG/GC. The moments of the springs CD and EF are equal to the moments of the weights Z and X hanging from B. The balance AB is supported by the rods or springs CD, GH and EF. EF being too stretched pulls the arm EG, and CD being too compressed pushes away the arm CG. Therefore, the middle spring GH, in an average state of extension, plays the role of a fulcrum. Point G is the centre of the balance. Finally, the ratio of the weights SXZ/V is equal to AG/GB. Therefore, if the weight R and the forces of the two springs CD, EF are removed and the weight V is suspended from A, the balance AB is in equilibrium about its centre G. The moments of the springs exerting their forces at C and E are equal to the moments of the weights X and Z hanging from B. Consequently, the ratio of the sum of the weight S and the forces of the springs considered as acting at B to the weight V hanging from A is equal to AG/GB. The balance thus is in equilibrium if the weight V is removed and the weight R is put again at A. Consequently, the weights R and V are equal. Therefore, the ratio of the sum of the weight S and the forces exerted by the springs CD, EF considered as acting at B to the weight R is equal to AG/GB. Q.E.D.

Proposition LX

Although the force exerted by the intervertebral disk during flexion of the spine is assumed to be greater than the muscular force compressing the vertebrae, the moments of these two forces may be equal.

The resistance to tension of the intervertebral disks is greater than that of muscles of the same cross section. The disks can support heavier weight than the muscles. On the other hand, the normal force of contraction of the muscles is not greater than the resistance to tension of their muscular fibres. Otherwise the muscles would be disrupted when contracting. Consequently, the resistance to tension of the disks is greater than the normal force of contraction of the muscles. Therefore, a heavier weight can be suspended from a disk than can be pulled by a muscle of the same cross section. The muscles which control the spine have fibres looser and less dense than the disks and their cross section is not much larger than the intervertebral disks. It can thus be assumed that the force exerted by the disks when opposing flexion and separation of the vertebrae is not smaller than three times the force of the lumbar muscles.

On the other hand, the lumbar muscles are inserted in the extremities of the spinal and transverse processes. That is where they exert their force. On the contrary, the force of the disks is exerted over all the articular surfaces of the vertebral bodies, in an area located between the centre of the base of the vertebra and its periphery. It appears that the distance of the line of action of the muscles from the centre of the base of the vertebra is more than three times the distance of the line of action of the force exerted by the disks from the same centre. Since the magnitudes of the forces are in inverse proportion to the distances of their lines of action from the common fulcrum, the moments of the force exerted by the muscles may be equal to the moments of the force exerted by the disks. Q.E.D.

Proposition LXI

If the spine of a stevedore is bent and supports a load of 120 pounds carried on the neck, the force exerted by Nature in the interverterbal disks and in the extensor muscles of the spine is equal to 25 585 pounds. The force exerted by the muscles alone is not less than 6 404 pounds. Tab. VI, Figs. 1 and 2.

The bow ABCDE used in proposition 55 represents the bent spine of a stevedore carrying a load R of 120 pounds on the base of his neck, close to the thorax. The whole back is bent and compressed by the load. The back is supported by the very strong intervertebral disks and by the longissimus dorsi, sacro-spinalis, sacro-lumbar, semispinatus, splenius and complexus muscles. These muscles are inserted in the transverse and spinal processes of the vertebrae as well as in the sacral, iliac and occipital bones. The forces exerted by the intervertebral disks and by the muscles will be analysed. Any cylindrical segment appended to a vertebra in the human body constitutes a balance such as HFM. Its fulcrum F is a point in the middle of the base of the vertebra. This balance carries two forces the moments of which are equal. The load R of 120 pounds and the weight of the body ABFM tend to flex the arm FM of the balance. The other arm HF of the balance is pulled by the muscles HG and by the intervertebral disk from its centre to its periphery opposite the fulcrum. The distance MF is about seven times half the diameter of the fifth lumbar vertebra since the vertical across the centre of gravity of the body bent forwards and carrying the load R falls outside the pubic bone. The distance HF is less than three times half the diameter of the vertebra. A third of the force exerted by the disk applied at H is equal to the moment of the muscles. The sum of these two moments is equal to the moment of the weight acting at M. Half the human body weighs 75 pounds on average. The weight of the viscera supported by the pelvis, 18 pounds, is subtracted. The 57 pounds which remain are added to the 120 pounds of the load. The total weight which compresses the arm FM of the balance at M thus is 177 pounds. The ratio of this weight to the muscular forces HG combined with a third of the forces exerted by the intervertebral disk is equal to the ratio of half HF to FM[16], or 3/14. Therefore, the sum of the muscular forces GH which control the fifth lumbar vertebra and a third of the resistance of the intervertebral disk is equal to 826 pounds. The muscular forces are equal to 413 pounds[17] and the forces exerted by the disk are equal to 1 239 pounds.

For the calculation of the cylindrical segments appended to the other lumbar and thoracic vertebrae, it must be noted that the ratios of the arms of the balance are little altered. The fulcra are always points in the middle of the bases of the vertebrae. They are displaced anteriorly much more than the centres of gravity R, etc. as a result of the thoracic bending. The longer arm DN, CO, etc. of the balance is progressively shortened. But the shorter arm DK, CX, etc. is

[16] From Comment Prop. 34
[17] From Prop. 6

Table 1

Vertebrae		Force exerted by the muscles	Force exerted by the disks
Lumbar	5	413	1239
	4	411	1232
	3	408	1225
	2	404	1212
	1	403	1211
	12	401	1204
	11	394	1183
	10	387	1162
	9	380	1141
	8	373	1120
	7	366	1099
Thoracic	6	359	1078
	5	352	1057
	4	345	1036
	3	338	1015
	2	331	994
	1	324	973
Pounds		6404	19181
			6404
			25585

also shortened because the processes of the vertebrae become the smaller the closer they are to the neck. Moreover, the weight of a number of cylindrical segments must be subtracted from the total weight depending on the position of the vertebra considered[18]. Calculation was carried out taking these remarks into account. The results are given in Table 1. The force exerted by the extensor muscles spanning the 17 thoracic and lumbar vertebrae is equal to 6404 pounds and that exerted by all the intervertebral disks between the same vertebrae is equal to 19181 pounds. Thus the total of the forces exerted by Nature in these muscles and intervertebral disks is no less than 25585 pounds. This is the effort of Nature to carry a load of 120 pounds and half the body weight, 75 pounds.

[18] Prop. 57

Proposition LXII

If a stevedore carrying a load of 120 pounds on his neck, with spine, hip, knee and ankle flexed and heel raised, stands on the ball of one foot, the force exerted by Nature in the extensor muscles of the back, thigh, lower leg and foot to balance the weight is equal to 13 766 pounds. Tab. VI, Figs. 1 and 2.

The position is the same as for propositions 53, 54 and 61 and the diagrams are also the same. To carry a load R of 120 pounds on the neck, the forces of the soleus, gastrocnemius, rectus femoris, vasti, glutei and spine extensor muscles are involved[19]. The extensor muscles of the leg exert forces equal to 7 362 pounds and[20] the extensor muscles of the spine no less than 6 404 pounds. This makes a force of 13 766 pounds. Q.E.D.

Comment

We shall not discuss this analysis at greater length. It must only be pointed out that, to carry a weight of 120 pounds on the neck, the forces exerted by the muscles thus mentioned are not sufficient. Nature is compelled to enlist auxiliary forces. When the neck is bent forwards the splenius and complexus with the scalenius and transversalis muscles have to support the weight of the head multiplied by seven, as a result of the number of cervical vertebrae and of the lengthening of the lever from which the weight of the head is supported due to its bending forwards. Besides, the human frame thus bent and folded cannot remain supported by the ball of the right foot if the left leg is not raised from the ground. To achieve this position all the flexor muscles of the raised thigh, lower leg and foot must be used. Nature is compelled to use the forces of these muscles although secondarily, to enable the weight carried on the neck to be supported in this position. I do not mention the forces which breathing in the meantime requires from the thoracic intercostal muscles, from the diaphragm and from others, and which we perceive. Finally, the forces and efforts which Nature exerts in the muscles appear to increase more and more, almost ad infinitum. Now, before going further, we must describe another structure of the muscles which constitutes a mechanical organ different from what has been described so far. An understanding of it will result from the next lemmas.

[19] From Prop. 53 and 54
[20] From Prop. 61

Chapter XIII

Lemmas necessary to analyse the motive power of the muscles the fibres of which are not parallel and pull obliquely.

Proposition LXIII

If two opposite forces pulling a rope in different directions oblique to each other, about one or several fixed points, have equal moments, the magnitudes of the forces are equal.
Tab. VI, Figs. 4, 5 and 6.

The forces R and T pull the rope ACB in any directions AC and CB over the fixed point C or over several fixed points C and H (Fig. 5, Tab. VI) in such a way that the rope can move smoothly and without hindrance. R exerts a force only in the direction CA and T only in the direction CB. The moment of the force R is equal to that of the resistance T so that neither of these forces yields to the other. I claim that the magnitude of R is equal to the magnitude of T. Point C is fixed. The rope ACB turns round the fixed point C as round a pulley mobile about its centre C. Two forces R and T thus acting in the directions AC and CB only would move with equal velocities in these directions round the fixed point. The nearer the resistance T is brought towards C, the further the force R is brought away from C, exactly by the same distance. The moments of these forces are assumed to be equal. Consequently, the magnitudes of R and T are equal. The same deduction can be drawn when the rope passes round several fixed points C and H (Fig. 5, Tab. VI). Q.E.D.

Comment

At first glance, the truth of this proposition can be questioned. The two forces R and T (Fig. 7, Tab. VI) are weights and pull the rope ACB round a fixed point or a nail or a pulley C located at the top D of the right-angled triangle DEG upright on a horizontal plane EG. The right angle of the triangle is at G. Then the weight R supported by the slope DE pulls the rope in the direction AC parallel to DE and the weight T in the direction CB at right angles to the horizontal base EG. The forces R and T can balance each other although R is greater than T, according to the ratio of the hypotenuse ED to the perpendicular side DG. This theorem which is generally accepted seems to contradict our proposition in which we claimed that the two forces R and T must be equal.

Actually, considering this matter with attention, it appears that the two cases are different. A perpendicular AF (Fig. 8, Tab. VI) is drawn to the slope DE and to the direction of the rope CA which is parallel to the slope DE. A parallel AK to the horizontal plane EG is also drawn as well as a line AB bisecting the angle FAK. This line AB is prolonged until it intersects the vertical rope CB at B. The straight line BFH parallel to the horizontal AK intersects AF at F.

Finally, AH is drawn, perpendicular to BFH. The alternate angles FBA and BAK are equal and BAK is equal to BAF. Therefore, the sides AF, BF of the triangle AFB are equal. Then, the load R exerting its force vertically is supported by the slope DE as if it was carried by the arm AF of a balance about the fulcrum F. In both instances, the weight R is forced to move along the inclined direction CA tangential to a circle of radius FA. On the contrary, the weight T hangs freely and can move vertically whether it hangs from the rope CB or whether it is attached to the horizontal arm FB of the balance. Finally, the weights R and T carry out opposite movements whether they are tied to the rope ACB which turns about the nail C or whether they are carried by the angulated balance AFB the fulcrum of which is F. In both hypotheses, the weights act similarly. In the angulated balance BFA with arms of equal length the ratio of the magnitude of R to its moment (or the moment of T which is equal for reasons of equilibrium) is equal to the ratio of the arm FA of the balance (or FB) to the distance FH of the line of action AH from the fulcrum F. The weight T is equal to its own moment because it acts at right angles to the arm FB of the balance. The ratio of R/T thus is equal to BF/FH or FA/FH. The two triangles FHA and DBH or DGE, also a right-angled triangle, are similar to the right-angled triangle FLH. Therefore, the sides of the angles F and D are proportional, i.e. the ratio of AF or BF to FH is equal to the ratio of ED/DG. Consequently, the ratio of the weights R/T is equal to ED/DG.

In the case of the preceding proposition 63, although pulling the rope AC obliquely (Fig. 9, Tab. VI) the force R of the hand does not compress vertically the slope HA. Therefore, the forces R and T behave as if they were pulling the equal radii IC, HC of the same pulley or the equal arms of an angulated balance ICH in directions perpendicular to these arms. The ratio of the moment of force R to the equal moment of the resistance T thus is equal to the ratio of the product of the magnitude of R and its velocity depending on the arm CH, to the product of the magnitude of T and its velocity depending on the arm CI. The arms CH and CI are equal. Consequently, the magnitudes of forces R and T are in the same ratio as are the moments of R and T and the lever arms CH and CI.

Proposition LXIV

If the moment of a force pulling a rope obliquely is equal to the moment of the resistance pulling the other end of the rope vertically and if the intersection of the forces is mobile vertically, the ratio of the magnitude of the force pulling obliquely to the resistance is equal to the ratio of the length of the oblique direction to its height.
Tab. VI, Fig. 10.

Force R exerted on the rope ACE which is oblique to the line DCE, pulls the resistance T in such a way that the intersection C of the two directions AC and CE is not fixed but mobile along the perpendicular DCE to the horizontal plane GCI. This can occur in different ways, either by movement of point C in a smooth groove cut in the column DE, or by fastening point C to the extremity of the horizontal lever GC mobile about the fulcrum G or to the extremity of the lever IC mobile about the fulcrum I, or by holding point C with the hand or something else H so as to prevent the intersection point C to follow another direction than DCE. The moment of R is equal to the moment of T. A perpendicular AD is drawn to DCE. I claim that the ratio of the magnitude of the force R to the resistance T is equal to AC/CD. GF is drawn, at right angles to CA. The angle C of the rope is not fixed but is retained by either the lever or the groove in the column or a force H and can move only along the line DE depending on whether the rope is pulled with the weight T or whether the force R lets it loose. In any case, T hangs from C as if it was suspended at the extremity of the lever CG mobile about a fixed fulcrum G. Point C is thus forced to move along DE which is tangential to the circle of radius GC. Forces R and T of equal moments thus pull the extremity C of the lever GC about the centre G. T pulls at right angles to the lever in the direction CE whereas R pulls obliquely along CF. Therefore, the ratio of the force R to the resistance T[1] is equal to the ratio of the length CG of the lever to the distance GF. But the triangles ADC and CGF are similar: the alternate angles GCF and CAD are equal since the lines AD and GC are parallel, and F and D are right angles. Consequently, GC/GF or the ratio of the magnitude of force R to the resistance T is equal to AC/CD. Q.E.D.

Corollary

If the directions of the forces in equilibrium are aligned, their magnitudes are equal. In Fig. 11, Tab. VI, the line of action CA of force R and its height CD in relation to the plane CG drawn through any point C on the common line of action and at right angles to this line of action ED, coincide. Therefore, the height DC is equal to the line of action CA.

Proposition LXV

The data are the same. No finite force can raise or retain a weight, however small it may be, in horizontal position.
Tab. VI, Fig. 12.

The force R can be very great and the weight T very small but mobile along ED. I claim that force R pulling obliquely the rope AC can never raise the weight T up to D in the horizontal plane DA. The ratio of the force R to T is equal to T/S. And $(R\text{-}S)/S = (AD)^2/(DH)^2$. It must be shown firstly that the weight T can be raised up to H and not higher. In the triangle ADH, with its right angle at D, the square of AH is equal to the sum of the squares of HD and

[1] From Prop. 13

DA. Therefore, the ratio of the sum of the squares of AD and DH, i.e. the square of AH, to the square of HD is equal to R/S. Their square roots are in the same ratio. Thus, AH/DH = R/T. Consequently, when the weight T is at H, the forces R and T balance each other[2]. If the weight T was pulled higher, such as to O, then the ratio of AO/OD would be greater than AH/HD (as can be easily demonstrated) or the ratio of AO/OD would be greater than that of the forces R/T. Therefore, the moment of R would be smaller than the moment of T. Consequently, force R could not balance and even less raise T up to O. The weight T cannot be raised or retained at the horizontal level DA because at D, T can move only along DE tangential to a circle of radius AD and then the line of action AD would pass across the fulcrum A of the lever DA. Therefore[3], force R would be unable to support the small weight T. Q.E.D.

Proposition LXVI

If two forces applied at the extremities of a balance are in equilibrium, the moment of one of the forces opposes the sum of the moment of a part of the fulcrum and that of the opposite force.
Tab. VI, Fig. 13.

The balance AB with the weights R and S, the centre of gravity of which is C, is supported by a fulcrum T or carried by the hand which prevents the fall of the balance AB with the weights. I claim that the moment of force R opposes not only the moment of a part of the fulcrum or the hand T, but opposes also the moment of force S. One action does not impede the other. The centre of gravity C of the balance tends to move downwards along the same line of action CT and with the same velocity as it is pulled upwards by the force T of the hand or supported by the fulcrum. One force does not prevail on the other since the balance remains immobile. Therefore, the force T exerted by the fulcrum or the hand is equal to the sum of the weights R and S. Thus, the moment of the weight R is equal to the moment of a part of the force T exerted by the hand. The balance AB is in equilibrium about its centre of gravity C and the two weights are immobile not as a result of inertia but because of the actions of the forces R and S. The weight R acts on the arm CA of the balance with as much force as the weight S tends to move the arm CB downwards. Thus the moment of the weight R alone is exerted against the counterweight S and equally against a part of the force T exerted by the hand. One action does not prevent the other.

If several or even a countless number of balances are used in an order such that the centre of one hangs from the extremity of an arm of another, it can be shown that the moment of one weight R is exerted not only against the counterweight S but also against parts of the countless fulcra CD, etc.

---

[2] From Prop. 64
[3] From Prop. 12

Proposition LXVII

If the same weight is suspended from the adjacent extremities of two balances and is balanced by two counterweights at the opposite extremities of the balances, any of the counterweights is equal to the moment of part of the weight.
Tab. VI, Fig. 14.

Two balances AC, DC are adjacent at C. Their fulcra are B and E. The same weight V hangs from their adjacent extremities C. Two counterweights R and S are suspended from the opposite extremities A and D. The sum of the moments of the counterweights is equal to the moment of the common weight V so that both balances AC and CD are horizontal and in equilibrium. I claim that the moment of the counterweight R is not equal to the moment of the whole weight V but to that of a part of V; S is balanced by the remaining part of V.

R/X = CB/BA and S/Z = CD/ED. The weight V is removed and replaced by X at C. The balance AC supporting the weights R and X about its centre B remains horizontal and in equilibrium[4]. But then the other balance DC cannot remain in equilibrium because the weight X balanced by the weight R cannot pull the arm CE and thus cannot prevent the weight S from falling with the arm ED. To ensure equilibrium of the balance CD, another weight Z must be suspended from C, beside the weight X. Then the two balances are in equilibrium and the sum of the moments of the weights R and S is equal to the combined moment of X and Z. But, by hypothesis, the sum of the moments of R and S was equal to the moment of the weight V. Consequently, the sum of the moments of X and Z is equal to the moment of V and the two weights X and Z hang from the extremities C of the arms BC and EC. The sum of the weights X and Z thus is equal to the weight V. Therefore, the moment of R is indeed equal to the moment of that part of V which is equal to X and the moment of S is equal to the moment of the other part of V which is equal to Z. Q.E.D.

Comment

It results from this proposition that whether R and S (Fig. 15, Tab. VI) are equal or whether they are not, they can be balanced by one weight V. If BC is equal to EC and R is equal to S, the weight V is divided in two parts X and Z. X balances R and Z balances S[5]. Then the ratio of X to R or to S which is equal to R, is equal to AB/BC or AB/CE since CE = BC. Thus, S/Z = CE/ED. Consequently, X/Z = AB/ED, (R + S)/V = BCE/(AB + ED), R/V = BC/(AB + ED).

If the parts X and Z of V (Fig. 1, Tab. VII) balanced by R and S are equal, the ratio of X or Z to R is equal to that of AB to BC or CE which is equal to

[4] Prop. 3 de aequepon. in Archim. Auctoris
[5] Archim. ibidem

BC and S/Z = CE/ED. Thus, exchanging the equalities, S/R = AB/ED, (R + S)/V = (AB + ED)/BCE, R/V = ED/2 CE.

If BC/CE = R/S, then X/Z = AB/ED, (R + S)/V = BCE/(AD + ED) and finally R/V = BC/(AB + ED). It must be noted that, comparing R and V, their moments are not equal in the scales AC in which R balances the weight X. In the other scales the arm BA must be lengthened by a length equal to ED. The weight R must be suspended from the extremity of this longer arm and V from C.

It is very important to remark that in these balances AC, DC (Fig. 3, Tab. VII) adjacent at C, the three weights can vary in a thousand ways and still ensure equilibrium. The weights R and X can be increased or I and K can be decreased. As long as they keep the same proportions, i.e. R/X = BC/BA, they ensure equilibrium. Also in the other balance CD, the weights H and L can be added to or substracted from S and Z. As long as they are in the same ratio as S/Z, equilibrium persists.

Similarly, if the weights remain unchanged, the balances can be modified and remain in equilibrium. For example (Fig. 4, Tab. VII), the equal weights R and S are balanced by the weights XV, VZ in the balances AC, DC. Equilibrium persists if the weight R hangs from f. Then R/Xt = CB/Bf. Also gE/EC = tZ/S and s must be suspended from g.

Proposition LXVIII

If the sum of the moments of two forces pulling obliquely two of three connected ropes is equal to the moment of the common resistance pulling the third rope and the junction of the ropes is mobile along the line of action of the resistance, the moment of any of the forces pulling obliquely is equal to the moment of a part of the common resistance.

Tab. VII, Figs. 5 and 6.

Three ropes AC, BC, EC are connected at C and the weight T pulls the rope CE downwards along the vertical line of action CE perpendicular to the horizontal plane DCL. The weight T is pulled upwards and supported by the equal moments of forces R and S pulling the ropes AC and BC obliquely. The junction C of the ropes is mobile and tends to move along CE. This will happen if a weight T is suspended from C. I claim that the moment of force R is equal, not to the whole moment of T, but only to a part of it and S is balanced by the remainder of T. From two points G and I on the horizontal DCL, at equal distances from the junction C, GF and IK are drawn perpendicular to AC and CB. They delineate DG equal to GF and IL equal to IK. Force R is removed and replaced by an equal weight M at D. Force S also is replaced by an equal weight N at L. The magnitudes of R and M are equal and pull at right angles the equal arms GF and GD. The moments of the forces R and M thus are equal. For the same reason, the moments of the forces S and N are equal[6]. The

[6] From Prop. 15 and 16

sum of the moments of the weights M and N is equal to the sum of the moments of the forces R and S. By hypothesis, the moment of the resistance T is equal to the sum of the moments of the forces R and S. Consequently, the sum of the moments of the weights M and N is also equal to the moment of the resistance T. The two horizontal balances DC and LC thus are in equilibrium[7]. Therefore, the weight M or the force R is balanced by one part X of the resistance T and the weight N or the force S is balanced by the other part Z[8]. Q.E.D.

Comment

The inclination of the ropes being unchanged, the pulling forces and the suspended weight can vary in a thousand ways and still balance each other, as long as R/X = GC/GF = AC/CH (Fig. 5, Tab. VII). Any weight S is balanced by Z if S/Z = IC/IK = BC/CH. Therefore, countless weights, equal or not, can be balanced by two other weights X and Z hanging from C with the ropes similarly inclined.

On the other hand, the same weights (as long as the middle weight ZX is smaller than the sum of the weights at the extremities) can be balanced in many ways with changing inclination of the ropes. The angles ACD, BCD formed by the ropes (Fig. 6, Tab. VII) are assumed to be equal and, in this position, the forces R and S are balanced by the weight XZ. The angle ADC is closed so that aC/CD = R/X. The rope B is moved until bC/CD = S/Z. With this new inclination of the ropes the weights R and S must be balanced by the weight XZ and the scales are in equilibrium.

It is worth noting that, if the levers CG and CI (Fig. 5, Tab. VII) are removed, force R pulls the junction C and the rope BCE retains the same angulation whereas force S pulls the junction C and the rope ACE also retains the same angulation. These opposite tensions retain the junction C along the same line CE perpendicular to the horizontal plane DCL as if point C was attached to the extremities of the two levers CG and CI. Tension of the ropes AC and BC against the fulcra G and I of the levers does not hinder force R to be exerted against the resistance X so that R and X balance each other, and S and Z also balance each other. Thus, the opposite tensions exerted on the junction C by the forces R and S do not prevent these forces to be balanced by equal moments of the resistances X and Z.

---

[7] From Comment Prop. 67

[8] From Prop. 67

Proposition LXIX

The data are the same. The ratio of the sum of the two forces to the resistance is equal to the ratio of the sum of the oblique lengths of the ropes (which are proportional to the forces acting at their end) to the sum of their heights.
Tab. VII, Fig. 7.

The data are the same as for the preceding proposition. The ratio of the lengths AC/CM is equal to that of the forces R/S. Two perpendiculars MO and ABD are drawn to the line of action DCE of the resistance T. I claim that the ratio of the sum of the forces R + S to the resistance T is equal to the ratio of the sum of the two oblique lengths AC + CM to the sum of the heights DC + CO. Two forces R and S pulling obliquely are balanced by the resistance T and the point of junction C is mobile along DCE. Therefore[9], the moment of force R is balanced not by the moment of the whole resistance T but by the moment of a part X of this resistance T. On the other hand, the moment of S is balanced by that of the remaining part Z of the resistance T. The ratio of the magnitude of R to the resistance X, which balances R and is mobile along DCE, thus is equal to the ratio of the length AC to the height CD[10]. For the same reason, the ratio of the magnitude of force S to the resistance Z which balances S is equal to BC/CD or MC/CO since BD and MO are parallel. But AC/CM = R/S. Consequently, the ratio of the sum of the two forces R and S to the sum of the two forces X and Z or the resistance T is equal to (AC + CM)/(DC + OC). Q.E.D.

Corollary

Obviously, the ratio of the part X to Z is equal to DC/CO and the ratio of force R to the resistance T is equal to AC/(DC + CO).

Comment

From the previous propositions it appears that two forces R and S, whether equal or not, can be balanced by a resistance if they pull oblique ropes which form acute angles ACD and BCD, equal or not, with the line of action of the resistance. For any inclination of the rope ACE pulled by force R, a weight X can be found which balances force R. A weight Z can also be found to balance force S, whatever the ratio of R to S and whatever the angles ACD and BCD. The two forces R and S thus can be balanced by the combination of the two weights X and Z. It appears that the three ropes AC, BC and EC can be re-

[9] From Prop. 68
[10] From Prop. 64

tained in one plane and in two planes inclined in relation to each other as long as, in any case, the junction is supposed mobile along DCE. Consequently[11], the ratio of force R to force X which balances R is equal to AC/CD. Similarly, the ratio of force S to force Z is equal to MC/CO, as previously. Q.E.D.

Digression

Since Stevin, Hérigone and other erudite authors think to have demonstrated this proposition in a by far very different way, I am compelled to mention the reasons why I think that their method is not completely reliable and legitimate. This proposition is of Hérigone but is presented differently and more clearly.

Tab. VII, Fig. 8.

An oscillating weight T is suspended from two oblique ropes AC, BC pulled by forces R and S. From any point D of a diameter DCK of the pendulum circle described by the weight T, a parallel DM is drawn to BC and another DN to AC. Hérigone says that the ratio of the magnitudes of the forces R/T is equal to MC/CD; that S/T = NC/CD; that, consequently, R/S = MC/CN and (R + S)/T = (MC + CN)/CD.

To make his demonstration more obvious, Hérigone supposes the weight T to be circular with a centre C and the diameter of the pendulum to be DK. Then, he says, if the weight T hangs in equilibrium from the two oblique ropes AC and BC pulled by forces R and S, it is as if the weight T was supported by two planes or two lines OIG and VIH inclined to the horizon and tangential to the circle at O and V, where the lines of action ACV and BCO of the ropes touch the circle. Then, he says, the force which compresses the plane OIG is equal in magnitude to force S and the force which compresses the plane VIH is equal to force R. If the force S and the plane VIH are removed, the weight T is supported at O by the inclined plane OIG and is retained in this position and prevented from rolling down by the force R pulling the rope CA. CL is drawn parallel to the inclined plane GO, GH horizontally and IP perpendicular to GH. DL is supposed to be parallel to BO and CL to be parallel to the plane GO. The angle DLC thus is equal to the right angle BOG and the angle L is right and equal to the right angle P. Moreover, DC and IP are parallel since both are vertical and LC is also parallel to OIG. Consequently, the angles DCL and GIP are equal. Since the triangle DCL is similar to the right-angled triangle GIP, DC/CL = GI/IP. The weight T is supported by the inclined plane OG. Thus, the ratio of the magnitude of T to its moment in the plane[12] is equal to the ratio of GI to the height IP or DC/CL. But if force R pulled the weight T in the direction CL parallel to the plane GO, force R would be equal to the moment of the weight T considered in this plane OG. Since the weight keeps

[11] From Prop. 64

[12] Comment Prop. 63

pulling the rope in the direction CA, Hérigone and Stevin infer that the ratio of the magnitudes of the forces R/T is equal to MC/CD. I do not know whether they have demonstrated this assertion. Their demonstration, however, can be replaced by the following reasoning. A perpendicular OQ is drawn to CA. The two angles LCM and MCB form a right angle. The two angles MCB, or QCO, and COQ also form a right angle. If the common angle MCO is subtracted (or the equal angles MCB, QCO), two angles LCM and COQ remain which are equal. L and Q are right angles. Therefore, the triangles LCM and QOC are similar and LC/CM = QO/OC. The weight T hangs from the extremity C of the lever CO the fulcrum of which is O. C is mobile along LC, parallel to the inclined plane OG, and is pulled obliquely along CA by the force R. R balances, not the whole force T, but the component of T exerted in the inclined plane. The moment of R is equal to the moment of T measured from CL. Therefore, the ratio of the moment of T to the magnitude of R is equal to the ratio of GQ, the distance from the line of action CA, to CO, the length of the lever, or to LC/CM. The ratio of the weight T to its moment considered in the inclined plane and at G is equal to DC/CL. Therefore, exchanging the equalities, the ratio of the magnitude of the weight T to the force R is equal to DC/CM. By the same process, it is shown that the ratio of the weight T to the force S is equal to DC/NC. Thus R/S = MC/NC and the ratio of the weight T to the sum of the two forces R + S is equal to DC/(MC + CN).

Modern geometry demonstrates the same proposition in a different and remarkable way. A parallelogram DMCN (Fig. 9, Tab. VII) is drawn about the diagonal DC. Perpendiculars AE and BF are drawn from A and B to the ropes BC, AC. The two forces R, S and the weight T are in equilibrium. The line or segment CA, as long as it is firmly held at A, does not fall as if, the force R being removed, this segment CA was attached by a nail at A. The extremity C of the segment AC is pulled downwards by the weight T in the direction DC and upwards by the force S in the direction BCE. The forces S and T are in equilibrium. Their moments are equal. The ratio of the magnitude of T to its moment or to the equal moment of S is equal to CA/DA[13]. The ratio of the moment of S to its magnitude is equal to EA/AC. By exchanging the equalities, the ratio of the magnitudes of the forces T/S is equal to EA/DA; the ratio of the sine of angle ACE or DNC to the sine of angle ACD or CDN is equal to DC/CN. By the same reasoning, it appears that the ratio of the magnitudes R/T is equal to MC/CD. Therefore, the ratio of the forces R/S is equal to MC/CN. The ratio of the sum of the forces R + S to the weight T is equal to (MC + NC)/DC.

It results from these two demonstrations that, when two forces support a weight by pulling two ropes in oblique directions and they are in equilibrium, the ratio of any of the two forces to the suspended weight must be equal to the ratio of the corresponding side of the parallelogram formed by the ropes about the diagonal of the parallelogram to the same diagonal of the parallelogram.

---

[13] Prop. 13

Conversely, when this parallelogram is supposed to be built and the ratio of any of the forces to the suspended weight is equal to the ratio of the corresponding side of the parallelogram to the diagonal of the parallelogram, then the two forces and the suspended weight must be in equilibrium.

I think that the first of these propositions is erroneous in its totality without determination of several moments. The second is absolutely correct as I shall demonstrate using my own method (Tab. VII, Fig. 10).

A parallelogram DGCH is drawn (Fig. 10, Tab. VII) about the diagonal DC on the line of action of the weight T. Two perpendiculars GL and ADB are drawn to DC. The ratio of the force R to the weight T is assumed to be equal to GC/CD and the ratio of the force S to the weight T to be equal to CH/DC = GD/DC. I claim that the two forces R and S support the weight T in this position by way of the oblique ropes AC, CB and the sum of their moments is equal to the moment of T. The weight T is removed and replaced at E by a weight X which balances the force R. Another weight Z which balances the force S is added at E. It appears that R/X = AC/CD = GC/CL since AD and GL are parallel[14]. Similarly, S/Z = BC/CD = CH/DL = GD/DL since the triangles BDC and GLD are similar. Therefore, the ratio of the force R to the sum of the two resistances X and Z is equal to GC/(CL + LD) or GC/CD. But by hypothesis R/T = GC/CD. Thus, R/(X + Z) = R/T and T = X + Z. But the forces R and S are balanced by the weights X and Z. Consequently, in the same position, the forces R and S balance the weight T, equal to the sum of X and Z and attached to the same rope CE. Q.E.D.

However, although this particular proposition is true the same cannot be said of its reciprocal. This appears from the demonstration of proposition 68 the validity of which nobody can question[15]. Actually, it was shown that two forces R and S can balance a weight by supporting it obliquely, whatever the ratio of R/S and whether this ratio is greater or smaller than GC/CH or whether or not the ratio of the sum of the magnitudes of the forces R and S to the weight T is different from (GC + CH)/CD.

My opinion not only is based on the reliability of the demonstration but it can also be confirmed by experiments (Tab. VII, Fig. 11).

A rope pulled at its extremities by two equal weights R and S is passed round two smooth and slippery nails or round two pulleys A and B attached to the horizontal AD (Fig. 11, Tab. VII). Another rope EC tied to the first at C pulls it downwards. The angle ACD formed by the first rope and the prolongation CD of the second rope EC perpendicular to AB is smaller than the angle BCD. A parallelogram DGCH is drawn about the diagonal CD. The alternate angles GDC and DCH are equal. Each of them is greater than the angle GCD. Therefore, in the parallelogram GH, the side GC is longer than the side GD or CH. From the end of the segment CF which is equal to GC and from the end of GC, FO and GL are drawn parallel to ADB. The weight T is found since R/T = GC/(LC + OC). This weight T hangs from the extremity E of the rope

[14] From Prop. 64

[15] Prop. 68 and Comment

Table 2

Angles ACD		BCD	
Degrees	Minutes	Degrees	Minutes
14	. 4	88	. 17
34	. 55	79	. 38
44	. 16	73	. 30
45	. 44	72	. 25
54	. 54	64	. 51
60	. 0	60	. 0

EC. The experiment shows that the three weights are in equilibrium. This could not appear from the demonstration of Hérigone. Indeed the weight R would have to be greater than S in the ratio GC/CH. Moreover, the weight T would have to be smaller than it is actually since its measure depends on DC and not on LC and OC. The assertion of Hérigone can be refuted experimentally in six hundred other ways. If the weight T is equal to R or to S, equilibrium is obtained with unequal angles as appears in Table 2. The conclusion is that the method of Hérigone is erroneous.

It is worth-while to point out the cause and the origin of this error and to show that it is false and impossible to assume that any of the two extremities A or B of the rope can be used as the fulcrum of the lever and that one of the forces R or S is equal to the moment of the whole resistance T. We shall demonstrate that this is erroneous in the next discussion (Tab. VII, Fig. 12).

A is the fixed point of the rope of a pendulum or of an iron rod AC (Fig. 12, Tab. VII). The weight T attached at C can move along the circumference of the circle with the radius AC. The weight T behaves as if it was supported by the inclined plane NIC, tangential to the circle at C. A perpendicular IL is drawn to the horizontal line LC. Then[16] the ratio of the weight T to its moment in such an inclined plane is equal to IC/IL, and the ratio of T to the force with which T is supported by and compresses the plane IC is equal to IC/LC[17]. But the force with which the weight T is supported by the plane IC is equal to the force R which, by supporting T in the same position, replaces the supporting plane. Therefore, the ratio of the weight T to the force R is equal to IC/LC. The angles LCD and ICK are equal since they are right angles. Subtracting the common angle ICD, the angles ICL and DCK are equal. If DK is drawn perpendicular to AC, the angles L and K are right angles. The triangles ILC and DKC thus are similar. Therefore, IC/IL = DC/DK and IC/LC = DC/CK. DC measures the magnitude of the weight T and DK its moment, whereas CK indicates the force R.

[16] From Comment Prop. 63

[17] Prop. 42 de vi percussionis

Force S acts, not against the whole weight T, but against the moment DK which T exerts in the inclined plane IC, and pulls in the oblique direction CB. Thus, as we showed in recalling the proposition of Hérigone, the ratio of the magnitude of force S to the resistance T considered in the inclined plane IC or the ratio of S to the moment DK of T is equal to OC/CN, CN being parallel and equal to DK. Therefore, CO measures the magnitude of S, CK the magnitude of R and CD the magnitude of T.

Then (Fig. 13, Tab. VII) DM is drawn parallel to BC, and DP perpendicular to BC. It appears firstly that, using this method, with point B supposed to be fixed, force S is measured by PC and not by OC which is longer, as occurred in the first case. Force R is measured by MC and not by KC which is shorter, as occurred previously. Secondly, in Hérigone's demonstration, force S was measured by OC and force R by MC. With both methods, DC remains the measure of the weight T.

The result is also obtained by assuming that the extremities A and B of the ropes AC or BC are fixed, one after the other, and that the ropes are levers or rods mobile about successively a nail A and a nail B.

We assume that the same weight T (Fig. 14, Tab. VII) hangs from two ropes AC and BC which are attached to nails at A and B. In these circumstances, it is as if the weight T was supported by two inclined planes CK and CG tangential respectively to the circles with the radii AC and BC. Then the load T which tends to move along the two inclined lines CK and CG, is compelled to move or to exert a force along the line CO bisecting the angle GCK. It must thus be assumed that the load T is supported by the inclined plane CO on which it exerts its weight and compression. Consequently, the ratio of the magnitude of the weight T to its moment in the inclined plane CO is equal to CO/CP. The ratio of the magnitude of the weight T to the force with which this weight compresses the plane CO is equal to CO/OP or to DC/DX if a perpendicular DX is drawn to OCX[18]. The force exerted on the plane CO by the compression of the weight T is equal to the sum of the forces R and S which, supporting the weight in this position, replace the inclined plane CO. Therefore, the ratio of the magnitude of the weight T to the sum of the two forces R and S is equal to CO/OP or to DC/DX. This is not only obviously false but also against the opinion of the authors thus mentioned who think that the ratio of the weight T to the sum of the two forces R and S is equal to DC/(MC + CN). MC + CN is much longer than DX. This can be easily shown.

If this process was legitimate and since all use the same hypothesis that points A and B are fixed, either successively or at the same time, and that the ropes are in the same position and support the same weight T as levers would do, they should arrive at the same conclusions concerning the ratio of the forces R and S to each other and to the weight T. Since this does not occur, some fault must exist in the processes. This fault does not originate from fallacious argument and nothing has been assumed against the principles of me-

[18] From Prop. 42 de vi percussionis

chanics. Therefore, the assumption itself must be impossible and erroneous, that the extremities A and B of the ropes can be used successively or at the same time as the fixed fulcra of the levers and that either the force R alone or the force S alone can be equal to the moment of all the resistance T.

Certainly, when the weight T is supported and balanced by the oblique tension of forces R and S, the two ropes are pulled by opposite forces. Therefore, although the forces are in equilibrium and even the declivity cannot compel them to move, this equilibrium is not inert but results from the opposite tensions imparting potential movement. This equilibrium thus cannot be conceived as an absolute immobility as I showed elsewhere. As a result of this potential movement, point C, the junction of the ropes, also tends to move along the line of action CE of the weight T (Fig. 10, Tab. VII). Because of this tendency of the junction C to move along CE, force R alone or force S alone cannot balance the totality of the resistance T, as Hérigone supposed, but only part of it[19]. The edifice built on such erroneous basis thus is unstable. After this lengthy digression I revert to our subject.

Proposition LXX

The same weight is supported with equal moments by more than two forces pulling simultaneously obliquely in the same or in different planes, and the junction of the ropes is mobile along the line of action of the resistance. The ratio of the sum of the forces to the resistance is equal to the ratio of the sum of the lengths of the ropes, proportional to the corresponding forces, to the sum of their heights.

Tab. VIII, Fig. 1.

The weight T is supported with equal moments by the forces R, S and Q which pull obliquely the ropes AC, BC and FC in the same plane or in different planes. The junction C of the ropes tends to move in the direction DCT. $R/S = AC/CE$, $R/Q = AC/CG$. From A, E and G perpendiculars AD, EI and GH are drawn to the line of action DCT of the weight T. I claim that the ratio of the sum of the forces R, S and Q to the resistance T is equal to $(AC + EC + GC)/(CD + CI + CH)$. Since all the forces R, S and Q support the weight T with equal moments and the junction C of the ropes is mobile in the direction DCT, any of the forces is balanced by a part of T, R by X, S by V and Q by Z[20]. Therefore[21], $R/X = AC/CD$; $S/V = EC/CI$; $Q/Z = GC/CH$. The numerators are proportional, R, S and Q to AC, EC and GC. Consequently, $(R + S + Q)/(X + V + Z) = (AC + EC + GC)/(DC + IC + HC)$ or, since $X + V + Z = T$, $(R + S + Q)/T = (AC + EC + GC)/(DC + IC + CH)$. Q.E.D.

[19] Prop. 68

[20] Prop. 68

[21] Prop. 66 and 69

Corollary

If all the forces and their inclinations are equal, then the ratio of their sum to the resistance is equal to the ratio of the length of one rope to its height. Since the inclinations of all the ropes are equal, all the angles ACD, BCD and FCD are equal. Therefore, all the ropes are on the surface of a cone the axis of which is CD, the direction of the rope of the resistance T. Moreover, since all the forces R, S and Q are supposed to be equal, the lengths of the ropes which are proportional to the forces, are equal. Their heights are then also equal to CD because the angles are equal. Consequently, the ratio of the sum of the lengths of the ropes to the sum of their heights, or the ratio of the sum of the forces R + S + Q to the resistance T, is equal to the ratio of the length of one rope AC to its height CD.

Proposition LXXI

The same weight is supported with equal moments by four forces or more pulling ropes on both sides of a vertical rope to which they are attached, from which the weight hangs and which is mobile vertically, and the forces on each side are equal to each other and pull parallel ropes. The ratio of the sum of the forces to the resistance is equal to the ratio of the sum of the lengths of two oblique ropes proportional to the forces, one from each side, to the sum of their two heights.

Tab. VIII, Fig. 2.

The weight T is supported by several forces R, V, Z, S, X and Y. The moment of the forces is equal to that of the weight. Forces R, V, Z act on one side and are equal. They pull the ropes AC, HE, KF parallel to each other and attached to the rope DCF from which the resistance hangs. The other forces S, X, Y are also equal to each other. They pull the parallel ropes BC, IE, etc., on the other side and are attached to the vertical rope at the same points C, E and F as their counterparts. The rope CEF is mobile in the direction DCFT. The lengths of the oblique ropes are proportional to the magnitudes of the forces:
AC/CB = R/S. Perpendiculars AD and BG are drawn to DCF. I claim that the ratio of the sum of all the forces R + V + Z + S + X + Y to the resistance T is equal to the ratio of the sum of the lengths of the ropes AC + BC to the sum of their heights DC + GC. All the forces R, S, V, X, Z and Y together support the weight T and there is equilibrum. The junctions C, E and F on the rope CFT, are mobile in the same direction. Therefore[22], each of the forces is balanced by a part of T, for example R by L, V by M, Z by N, B by O, X by P, Y by Q and so on. Consequently[23], R/L = AC/CD. The segment HE is equal to AC. Hg is drawn perpendicular to DCF. The triangles ACD and HEg thus are similar because AC and HE are parallel as are AD and Hg. Therefore,

[22] From Prop. 68
[23] Prop. 66 and 69

HE/Eg = AC/CD. DC and Eg are equal and V/M = HE/Eg. Also, V/M = AC/DC. For the same reason Z/N = AC/CD since Z is equal to A and KF is parallel to AC. Therefore, the ratio of the sum of the forces R + V + Z to the sum of the resistances L + M + N is equal to AC/CD. On the other hand, the forces S, X and Y are assumed to be equal and their lines of action BC, IE, aF are parallel. Thus, S/O = X/P = Y/Q and the ratio of the sum of the forces S + X + Y to the sum of the resistances O + P + Q is equal to the ratio of one of the forces S to one of the resistances O and S/O = BC/CG. Thus, (S + X + Y)/(O + P + Q) = BC/CG and the numerators are proportional. Consequently, the ratio of the sum of all the forces R, V, Z, S, X, Y to the sum of all the resistances L, M, N, O, P, Q or their total T is equal to the ratio of the sum of the lengths of two ropes AC + CB, one from each side, to the sum of their heights DC + GC.
(R + V + Z + S + X + Y)/T = (AC + CB)/(DC + BC).
The ratio of the sum of the forces R, V, Z of one side to the resistance T is equal to AC/(DC + GC). Q.E.D.

Corollary

Consequently, if all the forces on both sides are equal and their lines of action are equally inclined, the ratio of their sum to the common resistance is equal to the ratio of the length of one rope to its height.

If the different forces R, V, Z of one side are equal not only to each other but also to the forces S, X, Y of the other side and if the inclination of their lines of action is also equal so that ACD is equal to BCD and so on, the ratio of any of the forces R to the part of the resistance which this force balances is obviously equal to AC/CD. Therefore, the ratio of the sum of the forces to the sum of the resistances is equal to the ratio of one force to one resistance. Hence, the next proposition.

Proposition LXXII

A homogeneous weighing rod is supported horizontally with equal moments by several forces pulling ropes in the same plane, and half of the ropes are equally inclined to one side and pulled by equal forces and the other half are equally inclined to the opposite side and pulled by equal forces. The ratio of the sum of the forces to the resistance is equal to the ratio of the sum of the lengths, proportional to the forces, of two ropes differently inclined to the sum of their heights.

Tab. VIII, Fig. 3.

The column TV is homogeneous, of even width throughout and mobile at right angles to its length TV which is horizontal. The different points c, c, c of the length are the centres of equal cylinders into which the whole column must be thought to be divided. Each of these points is supported by two ropes Ac and

Bc. All the ropes are in the same vertical plane. All the forces R, r, r are equal and all the ropes Ac, ac, ac which they pull are parallel and equally inclined to F. Similarly, all the forces S, s, s are equal and they pull in the same plane as their counterparts, the ropes Bc, bc, bc which are equal and equally inclined to H. The ratio of one of the forces R to one of the forces S is equal to the ratio of the length of one of the parallel ropes AC to the length of one of the parallel ropes BC. Perpendiculars AF and BH are drawn from A and B to the length of the column FcH. I claim that the ratio of the sum of all the forces R, r, r + S, s, s to the common resistance of the column TV is equal to the ratio of the sum of the lengths of two ropes, one from each side, Ac + BC, to the sum of the heights of two ropes, one from each side, AF + BH. Any set of two forces R and S pulls two oblique ropes AC, Bc which are attached to the same point in the middle of a small cylinder c and supports this small cylinder, the moments being equal[24]. Point c is mobile vertically, perpendicularly to FcH. Therefore[25], the ratio of the sum of the two forces R + S to the weight c of the small cylinder which they support is equal to the ratio of the sum of the lengths Ac + BC to the sum of the heights AF + BH. The same is true for the other forces which balance the other small cylinders. Consequently, the ratio of the sum of all the pairs of forces to the sum of the weights of all the small cylinders which is the weight of the column TV is equal to the ratio of the sum of a set of two forces to one part of the weight, i.e. (AC + BC)/(AF + BH). Q.E.D.

Corollary

If all the forces and the inclinations of the ropes on both sides are equal, the ratio of the sum of the forces to the weight of the column is equal to the ratio of the length of one rope to its height.

If the different forces R, r, r pulling to the left are equal to each other and also to the different forces S, s, s pulling to the right, and the different angles ACF are equal to each other and to the different angles BCH, then the ratio of any force R or S to half the weight of the small cylinder C which this force balances is equal to the ratio of one length Ac to its height AF or of a length BC equal to Ac to a height BH equal to AF. Consequently, the ratio of the sum of the forces R, r, r and S, s, s to the weight of the column TV is equal to the ratio of one length AC to one height AF.

[24] From Prop. 69
[25] Prop. 69

Proposition LXXIII

A weight is supported with equal moments by several equal forces pulling ropes radiating to the periphery of a quadrant and the junction of the ropes is mobile along the line of action of the weight. The ratio of the sum of the forces to the weight is equal to the ratio of the sum of the lengths of the equal ropes to the sum of their heights.
Tab. VIII. Fig. 4.

The weight T is supported by several equal forces R, V, X, Z, S, etc. pulling the ropes AG, BH, FI, ZK, ML, etc., from the periphery GKL of a quadrant to another quadrant ADM concentric to the first. The junction of the ropes is at C at the end of the line CE. Perpendiculars AN, VO, XP, etc. are drawn to the line of action ECD of the weight T. NC, OC, PC, etc. are the heights of the ropes of equal lengths. I claim that the ratio of the sum of all the equal forces R, V, X, etc. to the weight T which they balance is equal to the ratio of the sum of the lengths of the ropes AC, VC, XC, etc. to the sum of their heights NC, OC, PC, etc. This is demonstrated easily as in proposition 70.

Proposition LXXIV

A weight is supported with equal moments by several equal forces pulling ropes radiating to the periphery of a quadrantal segment of a sphere and the junction of the ropes is mobile along the line of action of the weight. The ratio of the sum of all the forces to the weight is equal to the ratio of the sum of the equal lengths of the ropes to the sum of their heights.

This is demonstrated in the same way as the preceding proposition.

Chapter XIV

On the muscles pulling obliquely, their varying structure and their action.

So far we have considered flexion of limbs pulled directly by muscles made of parallel fibres. Flexion carried out by radial muscles must also be considered. Their function is precisely perceived only by very few people.

Proposition LXXV

If the tendon of a radial muscle cannot be maintained in its position, the contraction of the fibres will not move the resistance along the same direction.
Tab. VIII. Fig. 5.

A weight or resistance R is supported by a radial muscle ACEG. The fleshy or tendinous extremity BDFG of the muscle either is broad or, if the tendon GI is cylindrical, it is neither retained by ligaments as in a sheath nor moved round a pulley, but is free to move to and fro. I claim that if all the fibres AB, CD, EF contract together or the fibres AB and EF together or the fibres CD alone, the others remaining at rest, the load R will be displaced straight along the same direction IGDC. If the lateral fibres AB alone contract, the others remaining inactive, the load R will be displaced obliquely in a direction parallel to the fibres AB. If the fibres EF alone contract, the load R is raised obliquely parallel to the fibres EF. Firstly, if the fibres CD which are in the alignment of the tendon GI contract alone and the others remain loose, the weight R can be raised only in the direction ICG in which the tension is carried out, unimpeded by the loose fibres AB and EF.

Secondly, if the oblique fibres AB and EF contract, the fibres CD remaining at rest, and if the tensions are equal, i.e. if the shortening of the fibres AB and EF is equal and their inclination symmetrical, the point of junction G cannot be pulled more towards A than towards E. The junction thus moves along IGC which bisects the angle AGE.

Thirdly, the same occurs if the fibres CD contract together with the lateral fibres because the action of the fibres CD does not disturb but rather strengthens the action of the fibres AB and EF.

Fourthly, if the fibres AB alone contract and the fibres CD and EF remain loose, the junction G and the weight R must be pulled upwards obliquely from G to A in the direction GBA because the tendon GI is neither retained nor prevented from being moved anywhere. Similarly, contraction of the fibres EF alone raises the junction G and the weight R obliquely from G to E in the direction GFE. Q.E.D.

Proposition LXXVI

If the cylindrical tendon of a radial muscle is contained in a groove or a sheath or if it is retained in a given position by a pulley or ligaments, then the resistance is always displaced along the same direction, whether all the fibres or only some of the lateral fibres contract.

Tab. VIII, Fig. 6.

A resistance R is supported with equal moments by a radial muscle ACEG pulling the cylindrical tendon DGI. This tendon is retained in the same position DGI by the ligaments G, O. I claim that if the fibres AB alone contract and the muscles CD and EF remain inactive, or if the fibres CD alone, or the fibres EF alone or all the fibres together contract, the resistance R will move always in the same direction DGI. The tendon DGI is retained in the same position by the hooks or the ligaments G and O. When pulled by the contraction either of all the muscles or of one AB alone, the tendon thus cannot leave the groove or the sheath GO. Therefore, whether all the fibres AB, CD, EF contract or whether only the fibres AB contract whereas the others remain loose, the resistance R always moves in the same direction GI.

This is what happens with the quadriceps muscle. This muscle is broad and its fibres radiate obliquely on both sides. Whether they all contract or whether only some of them contract and the others remain inactive, this always results in one straight motion which is an extension of the lower leg. This is true also for the temporalis muscle the fibres of which radiate towards the periphery of a segment of sphere and always pull the lower jaw in the same direction even if only some fibres are supposed to contract. The reason for this behaviour is that the tendons are firmly maintained in the knee or beneath the zygomaticus arcus as if it were round a pulley. The same occurs for all the other muscles the tendons of which pass round pulleys or through rings or ligaments as if it were through sheaths. This applies to the flexor and extensor muscles of the lower leg and fingers or toes mainly in birds and testaceans. In these animals the muscles are inserted inside cavities in the bones and their fibres either radial or multipennate originate from one straight tendon.

It thus can be inferred that a heavier weight can be carried when all the fibres of such radial muscles contract than when only some of the fibres exert a force and the others remain inactive, providing that all of these fibres exert a force of the same determined magnitude. On the other hand, it may be possible that the same fibres exert more or less force by will or by necessity. For example, the flexor muscles of the forearm can support a weight of 20 pounds or a weight of only one ounce, in the same position, and both weights are supported by the same muscles. But this matter will be discussed later.

Proposition LXXVII

Analysis of the structure, action of and forces exerted
by the multipennate muscles.
Tab. VIII, Fig. 7.

Multipennate muscles are most obvious at autopsy in crabs and crayfish. Their shape ABFDCH is represented in Fig. 7, Tab. VIII. Their perimeter BAHCD is hard, either bony or cartilaginous and in other animals tendinous or membranous. Muscular fibres of two types originate from the periphery, left HABFG and right HCDFG. They insert in a central structure FEG prolonged towards I and H, and form angles BFG and DFG, mostly equal to each other. Their sides HABFG and HCDFG are made of parallel fibres. All the fibres are columnar, thick, adjacent to each other, of even consistency. A weight, or the resistance of a bone of a limb which is pulled and supported by the force of the muscle, is attached to the extremity E of the mobile tendon GFE. Although the upper extremities AHCI of the multipennate muscles appear as if they were constituted by radial fibres, at closer examination this is not really true. The central tendon EFG to which the collateral fibres are attached is not prolonged up to the very extremity H of the muscle but ends at I. The remaining part IH is not tendinous but fleshy. Two types of fibres are also attached to it, almost parallel to the other rhomboidal fibres as in the feathers of birds. The whole triangular or four-sided figure AICA is made of fibres and pulls the extremity I of the tendon IFE upwards.

The action of the whole multipennate muscle is very different from the action of the muscles described so far. The latter constitute one bundle of parallel fibres and when they contract, they pull the resistance in the direction of these fibres. The multipennate muscles contract in opposite directions, from FGI towards BA and towards DC. When loose, their fibres form acute angles AGC and BFD. When contracted they form less acute angles[1] and, by their transverse movement, they pull the mobile tendon IGF towards IH together with the suspended resistance R. We shall explain in due time why Nature gives up her simplicity and facility of behaviour in these multipennate muscles.

We shall also show the method for analysing the forces exerted by these muscles. The moment of the forces resulting from the contraction of the fibres of a multipennate muscle is equal to the moment of the resistance when there is equilibrium, i.e. when one force does not prevail on the other. As long as the muscular fibres contract, opening the acute angles BFD, ACG, and the resistance is raised, the moment of the muscle is greater than that of the resistance. In the loose inclination AGC of the fibres, the ratio of the force of contraction of the muscle to the resistance R is greater than the ratio of the length AG or CG of the fibres to their height GK. At the end of the movement the fibres are contracted at the most and are in the position AIC. Then the weight or the resistance R is supported by as many forces as there are fibres pulling the resistance obliquely from both sides and the tendon FGI is mobile in the direction EFI. The ratio of all the forces exerted by the fibres (i.e. the force of contraction of the multipennate muscle) to the resistance R thus[2] is equal to the ratio of the length AI of the contracted fibre to its height KI, the forces exerted by the different fibres being equal and their inclination the same. But the angle AIK formed by the contracted fibres and the mobile tendon EFI is always smaller than 30°. If the length of the fibre AI is ten, its height is about nine. Therefore, if the apparent magnitude of the force exerted by a multipennate muscle when it contracts is ten, its moment or the resistance R is nine.

Now to analyse the structure and action of the radial muscles, some lemmas must be presented.

Proposition LXXVIII

A weight is pulled by two oblique ropes each of which is bifurcate; the four branches of the bifurcation are pulled by oblique forces, and the moments of the weight and of the forces are equal. The ratio of the sum of the forces to the weight is equal to the ratio of the sum of the lengths of the four divisions of the ropes, proportional to the forces, to the sum of their heights, multiplied by the ratio of the sum of the lengths of the two ropes pulling directly, proportional to the moments with which they pull, to the sum of their heights.

Tab. VIII, Fig. 8.

[1] From Prop. 71

[2] From Coroll. Prop. 72

The weight T is attached to two oblique ropes BE and HE. The rope EB is in turn divided into two other ropes BA and BC. The rope EH also is divided into two other oblique ropes HG and HF. Four forces A, C, G and F pull the weight T in oblique directions and balance this weight. The ratio of the lengths of the ropes AB/BC is equal to the ratio of the forces A/C. Similarly, BC/HG = C/G and HG/HF = G/F. From points A, C, G and F, perpendiculars AI, CO, GL, FN are drawn to the directions of the ropes EBOI, EHNL. Then BE/EH = (IB + BO)/(LH + HN). Perpendiculars BD and HK are drawn to the direction PED. I claim that the ratio of the sum of all the forces A, C, G, F to the weight T is equal to the ratio of the sum of the lengths of the four ropes AB, BC, GH, HF to the sum of their heights IB, BO, LH, HN, multiplied by the ratio of the sum of the lengths BE, EH to the sum of their heights DE, EK.

(A + C + G + F)/T = (AB + BC + GH + HF)/(IB + BO + LH + HN) × (BE + EH)/(DE + EK)

The weight T is balanced by all the forces A, C, G, F which pull the ropes BE, EH. The moments of the weight and of the ropes are equal. Therefore, the two forces A, C balance a part X of T and the two forces G, F balance the remainder Z of T. The resistance X is removed and replaced by a resistance M pulling the rope BE in the direction IBE and balanced by the forces A, C. The junction B is mobile along IBE. Then AB/BC = A/C. The ratio of the sum of the two forces A + C to the resistance M thus[3] is equal to ABC/IBO. But the resistance X, which was replaced by M, is also balanced by the same forces A + C and the moment of M is equal to the moment of X. Both pull the angulated rope BEP with the point E mobile in the direction DEP. Therefore[4], M/X = bE/dE = BE/ED. The ratio of the sum of the forces A + C to the resistance X is equal to the product of the ratio of the sum of the forces A + C to M, and the ratio of M/X.

(A + C)/X = (A + C)/M × M/X.

Thus, (A + C)/X = ABC/IBO × BE/ED. For the same reason, the ratio of the sum of the forces G + F to Z is equal to the product of the ratio of GHF/LHN, and the ratio of HE to its height EK.

(G + F)/Z = GHF/LHN × HE/EK.

Then AB/BC = A/C, BC/GH = C/G, GH/HF = G/F and the ratio of the moment of the forces A, C or M to the moment of the forces G, F or Q is equal to BE/EH. Consequently, the ratio of the sum of the four forces A + C + G + F to the resistances X + Z = T is equal to the product of the ratio of the sum of the lengths of the four ropes AB + CB + GH + FH to the sum of their heights IB + OB + LH + NH, and the ratio of the sum of the lengths of the two ropes BE + EH to the sum of their heights DE + KE. Q.E.D.

If the weight T is supported from point E by more than two ropes, the conclusion is the same.

---

[3] From Prop. 69

[4] Prop. 64

Proposition LXXIX

The radial muscles cannot be constituted by fibres extending all from the end of the tendon as if they were radiating from a centre.
Tab. VIII, Fig. 9.

The radial muscle ADGH is inserted over the surface of a bone or cartilage ADG. A resistance T is suspended from the extremity H of its tendon IH. I claim that it is impossible that all the fibres of the muscle originate from the centre H and reach the periphery ADC as do the radii of a circle or a sphere. If it was possible, the fibres AH, BA, CH, etc. would converge from all the points of the periphery ADG towards H like to the centre of a circle or of a sphere. But the muscular fibres are not long immaterial lines. They are material and columnar of even thickness throughout, adjacent to each other, and they must swell evenly throughout their whole length, whereas the space about the centre H is narrow if not indivisible. It would thus be necessary for the fibrous columns to interpenetrate each other near the centre H and this is impossible.

We can assume that fibres do not reach to the centre H although they are orientated towards this point but insert in a large tendinous surface H; that in more remote areas ABG they separate widely from each other and the interstices are broader and broader as required by the even thickness of the columnar bodies. But this is false and is against the evidence. We actually see that the columnar fibres of the muscles are almost contiguous and form bundles evenly full and tight. It is thus impossible that the fibres AH, BH, CH, etc. converge towards H as do the radii of a circle or of a sphere.

Proposition LXXX

The radial muscles necessarily must be made up of several adjacent multipennate muscles, either flat or thick.
Tab. VIII, Fig. 10.

The radial muscle AEGLP has a broad origin, either circular or spherical AEGL, and ends up in the small extremity P of a tendon PQ to which a resistance T is attached. The fibres which pull the resistance hanging from the cylindrical tendon must originate from all the periphery of the origin of the muscle AEGL. It is impossible that the fibres originating from the broad origin AEGL unite and converge to the centre P as do the radii of a circle or a sphere[5]. On the other hand, the columnar fibres must be able to dilate and swell easily and be adjacent to each other taking advantage of the unequal empty spaces which separate them. Consequently, for the fibres to be arranged so as always to be parallel and contiguous, there is no other way than the muscle being multipennate. A radial muscle thus must be made of several multipennate muscles with the following structure. Several tendinous radii PF, PK, PO radiate from the

[5] From Prop. 79

centre P of the circle AEGL but they do not reach the periphery AEGL. Thin fleshy fibres originate from both sides of each radius, such as CFAB and CFED. They constitute one multipennate muscle. Its tiny fibres actually must be attached to the two tendons AB and ED the ends of which are firmly tied at A and E. Then the columnar fibres can contract and swell evenly keeping their mutual contact and parallelism while pulling the tendon FCP in the direction CP. Additional tendons GH and LM attach at G and L. Two types of fibres HE, HG and NG, NL are connected by the intermediate areas ED, GH. This arrangement results in two more multipennate muscles which pull point P in the directions PI, PN. Consequently, all pull the resistance T in the same direction QP. Q.E.D.

Comment

This structure of the radial muscles which mechanical reasoning suggested to me could not be confirmed experimentally except, imperfeclty, in sea and river crayfish. Therefore, I am very happy to see that the famous anatomists Steno and Lorrerius observed exactly the same structure in the human deltoid muscle and described it most accurately.

Proposition LXXXI

Determination of the forces exerted by radial muscles, the resistance and the inclinations of the tendons and fibres being given.

In the same diagram, the resistance T is known. It is supported by the forces resulting from the contraction of all the fibres AEGLNIB. The moment of these forces is equal to the moment of the resistance. The angles formed by the fibres BCF and DCF and all the others which are usually equal to each other, in most instances do not seem to be more than 8°. The angles formed by the multipennate tendons and the line of action of the resistance CPI, NPI, etc. are also given. The average between the largest and the smallest does not seem to be more than 45° and not less than 26°. The force resulting from the contraction of all the muscle AELP must be found. Since the angles of the fibres BCF, DCF, etc. are not more than 8° [6], the ratio of the length of any of the fibres to its height is equal to the ratio of the trigonometrical radius to the cosine of the angle BCF or 100/99. If the angles formed by the multipennate tendons and the line of action of the resistance, such as CPI, are averaged at 45°, the ratio of the length of the tendon to its height is 100/71. If this angle is 30°, the height of the tendon is 87. The ratio of the sum of the forces exerted by all the fibres of the multipennate muscles ACE, EIG, GNL, equal to each other, to the resistance T is equal to the product of the ratio of the length of one of the fibres BC to its height (100/99)[7], and the ratio of the sum of the lengths of the ten-

[6] Prop. 64
[7] Prop. 78

dons CP, etc., (proportional to the moments with which they are pulled) to the sum of their heights (99/70 or 99/86). Therefore, the ratio of the force exerted by all the fibres of the muscle AEGLP to the resistance T is equal to 100/70 if the angles of the tendons are 45°, or equal to 100/87 if the angles of the tendons are 30°, or equal to 100/89 if the angles of the tendons are 26°. That was the object of the research.

Chapter XV

Analysis of the forces exerted by the radial muscles described so far.

In chapters X and XII we considered the apparent forces exerted by several muscles, the flexor and extensor muscles of the carpus and of the second and third phalanges, the deltoid and the glutei, assuming that these muscles were made of bundles of parallel fibres and pulled their tendon and the resistance thus attached in a direction parallel to the length of these fibres. This is not exactly true since these muscles are made of radial fibres. Therefore, the forces which they exert must be determined more accurately. It will apppear that their contraction requires a greater effort.

Proposition LXXXII

Determination of the forces exerted by flexor digitorum sublimis and profundus, flexor carpi and deltoid muscles in the case of proposition 45. Tab. IV, Fig. 10.

In the diagram illustrating proposition 45, the muscle KO flexes the third phalanx FG of the fingers about the centre F. Together with the lumbricales muscles, it also flexes the lever DG formed by the three phalanges. The part of the force of the muscle KO which flexes the third phalanx of the fingers FG was shown to be equal to 76 pounds. This, however, would be true only if the fibres of the muscle KO were parallel to the direction of its straight tendon HO. But the fibres are radial and constitute a multipennate muscle KO. The force of the muscle thus must be increased because of the inclination of the fibres. This inclination is not more than 14° as far as can be judged. Therefore, the ratio of the force of that part of the muscle to the resistance R[1] is equal to the ratio of the length of the fibres to their height, that is 100/97 or approximately 78/76. Consequently, the force exerted by that part of the muscle KO which flexes the third phalanx FG is approximately 78 pounds.

[1] Prop. 64

A part of the muscle KO, together with the lumbricales muscles MQ, flexes the lever DG formed by the three phalanges. It was assumed that, if the muscles were made of fibres parallel to the direction BM or KO, the force which they exert ought to be 190 pounds. Let us assume that the force exerted by the lumbricales muscles, because of their smallness, is smaller than that exerted by that part of the muscle KO which flexes the vector DG. The force exerted by that part of the muscle KO is evaluated at 95 pounds. Because of the inclination of the fibres it must be increased to about 98 pounds. The force of 78 pounds exerted by the other portion of the muscle KO which pulls the lever FG must be added. Consequently, the force exerted by the whole muscle KO is not less than 176 pounds.

The muscle KN flexes the second phalanx of the fingers. If it was made of fibres parallel to the direction of its tendon KN, its force would be 152 pounds, as was shown. But the muscle is multipennate and the inclination of its fibres does not seem to be more than 14°. Therefore, the force exerted by these oblique fibres is equal to approximately 157 pounds.

If the muscle KL, flexor carpi, was made of fibres parallel to the direction of its tendon KL, the force exerted by this muscle against the resistance acting on the hand CG would be 240 pounds, as was shown. But this muscle is multipennate and the inclination of its fibres does not appear to be more than 14°. Therefore, the force exerted by these oblique fibres is more than 247 pounds.

The flexor muscles AK of the forearm exerted a force greater than 552 pounds. As a result of the little inclination of their fibres this force is only little increased. It becomes at least 558 pounds. In the case of proposition 45, the deltoid muscle IP moves the whole arm in extension AG. If the muscle was made of fibres parallel to the direction of its tendon IP, the force exerted against the resistance compressing the horizontal lever AG would be 780 pounds. But the deltoid muscle is radial and made of several multipennate components. The inclination of the fibres in each multipennate component does not seem to be more than 8° and the average inclination of the tendons is not more than 30°. Therefore[2], the force exerted by the radial fibres of the deltoid muscle is equal to about 907 pounds.

The increase of the forces exerted by these muscles as a result of the inclination of their fibres is equal to 150 pounds. But the force exerted by all the muscles maintaining the arm horizontal and supine with a weight R was found previously to be equal to 1990 pounds. Consequently, the apparent force exerted by these muscles is now equal to about 2140 pounds. Q.E.D.

[2] From Prop. 81

Proposition LXXXIII

Determination of the force exerted by the glutei muscles
in the case of proposition 53.
Tab. VI, Fig. 1.

The bow ABC is composed of the spine AB bent forwards from the neck down to the coccyx and the femur BC flexed on the pelvis and supported at mid-femur C. This bow is compressed by a load R and is maintained in this position by the force exerted by the glutei muscles GFH[3], as said above. This force must now be determined since the glutei muscles have a radial structure and we assumed in proposition 53 that their fibres were straight and parallel to the tendon FH. In this condition the force which they exerted to sustain the load R and the body ABC was equal to 1840 pounds. But the glutei muscles are radial, made of several multipennate components with long fibres. The inclination of these fibres in every multipennate component appears to be more than 8°. The average inclination of the tendons seems to be 45°. Consequently[4], the force exerted by the radial fibres of the glutei muscles is more than 2621 pounds.

Proposition LXXXIV

The force exerted by the deltoid muscle is about twice greater
than what we determined in proposition 82.
Tab. VIII, Fig. 11.

In the comment to proposition 45 we indicated that the muscles of the arm involved in the raising of a weight hanging from the tip of the fingers do not exert the whole of their force but must adapt to the weakest, the flexor digitorum profundus. In this operation, the partial force exerted by the deltoid muscle was shown to be equal to 907 pounds[5]. We wish to find the greatest force which the radial deltoid muscle can exert. Therefore, we must consider an experiment in which the deltoid muscle alone acts. The humerus AB in horizontal position is directly subjected to the heaviest weight possible R and to the weight of the arm, and is retained by the force of the deltoid muscle CD. The maximum load is more than 55 pounds. The ratio of the length of the lever AB to the distance of the line of action of the muscle CD from the centre A of the joint or the radius of the humeral head is 14/1. Therefore, the muscular force acting on the lever is equal to 770 pounds. But, since the muscle CD is attached at C, its force must be doubled[6] and thus is equal to 1540 pounds. Finally, because of the obliquity of its fibres[7], the total force exerted by the deltoid muscle CD is 1750 pounds. This is a little less than twice the magnitude found in proposition 82. Q.E.D.

[3] Prop. 53
[4] From Prop. 81
[5] Prop. 82
[6] From Prop. 34 and its Comment, and Prop. 35
[7] From Prop. 81

Proposition LXXXV

The total force exerted by the glutei muscles is greater than twice what we determined in proposition 83.
Tab. VIII, Fig. 12.

We showed in proposition 83 that the force exerted by the glutei muscles in the case of proposition 53 was equal to 2621 pounds. It appears that this is not the maximum force of these muscles. In another experiment the glutei muscles act directly. The leg, thigh AB and lower leg BC, are prone and horizontal. The heaviest weight possible R is suspended from the heel C. This weight is more than 65 pounds. The ratio of the length of the horizontal lever AC to the distance of the line of action of the muscles from the centre A of the hip joint or the radius of the femoral head is 31/1. Therefore, the force exerted by the glutei muscles is equal to 2015 pounds. Because of the firm insertion of the muscles in the iliac bone at F the actual force is twice as much and is equal to 4030 pounds. Finally, since the muscular fibres are very oblique, the total force exerted by the glutei muscles FD is 6000 pounds. This is more than twice the 2621 pounds which were found in proposition 83.

Proposition LXXXVI

Determination of the force exerted by the multipennate flexor pollicis longus muscle.
Tab. IX, Fig. 1.

The forearm AB hangs vertically and a weight R is suspended from the extremity E of the distal phalanx* of the thumb DE held horizontally. The greatest weight which can be suspended in this position is not more than 20 pounds. The ratio of half the diameter of the interphalangeal joint to the length of the distal phalanx is 1/3. The distance of the tendon of the flexor pollicis longus muscle FC from the centre of the interphalangeal joint D of the thumb is equal to half the diameter of this joint. Therefore, the force exerted by the muscle FC is equal to 60 pounds. Since the muscle is attached to the ulna at F, this force must be doubled[8]. That gives 120 pounds. Finally, the muscle FC is multipennate. Its fibres are attached to a central tendon with which they form angles of not less than 14° during maximum contraction. They insert in the side of the ulna and in fixed ligaments. Therefore[9], the total force exerted by the muscle FC is equal to 124 pounds approximately.

* Borelli seems to attribute three phalanges to the thumb! ("termino E tertii articuli pollicis DE")
[8] From Prop. 34 and its Comment, and Prop. 35
[9] From Prop. 81

Proposition LXXXVII

Description of the structure and contrivance by which the muscles move the lower jaw and method for finding their motive force.
Tab. IX, Fig. 2.

Among the mechanical instruments used by wise Nature in the animal, the structure of the tool which is called pliers seems to me to be deserving our interest. Its most elegant expression is found in the jaws of animals. It is true, however, that the pliers used in the lower jaw are different from ordinary pliers. Although both are levers, the lower jaw does not overcome a great resistance using a small force as ordinary pliers do but, on the contrary, the lower jaw requires a motive force greater than the resistance. The jaws are round-shaped representing in a way half ellipses the inferior one of which revolves about two cylindrical small heads which are mobile in cavities of the temporal bone to which they are attached. Their two halves can be considered as levers linked together at the chin. In the diagram, the rounded bone ACa is the mandible, made of two levers AC and aC, connected at C. Both can flex and rotate together about the two small heads A and a. Two pairs of muscles are attached to them one on the right side, the other on the left. One of the muscles is the temporalis muscle F which originates from the whole cavity of the temporal bone and ends in a tendon attached to the coronoid process D. The other muscle is the masseter EB which originates from the arcus zygomaticus and the zygoma, and ends inferiorly in the angle of the mandible B. The opposite part of the pliers is the upper jaw which is stable and receives the blows delivered by the lower jaw acting as a hammer. Teeth are implanted in the lower jaw, incisors which cut the hard food and molars which are used to grind. The resistance which the force exerted by the muscles must overcome thus is the toughness of the food which is ground in the mouth and prepared for deglutition and easier digestion and fermentation in the stomach.

A comparison is sought between the motive power of the muscles moving the jaw and the resistance of the bodies which are cut and ground by the teeth. Therefore, the force exerted by the contrivance in this operation must firstly be considered. Secondly, the structure, position and arrangement of the muscles which move the mandible must be looked at. Thirdly, the force of the resistance which has to be overcome must be determined. As far as the first point is concerned, it is obvious that the said pliers are reduced to a simple lever. The two small heads A, a of the lower jaw are the fulcrum. The resistance is applied either at the anterior teeth C or at the molars G. The motive force is applied at D and at B where the two muscles, temporalis and masseter, are inserted. In cutting, carried out by the incisors, another contrivance, very different from pliers, is used. This contrivance is a wedge. Indeed the teeth are like as many wedges. Mechanically, the ratio of the force pushing the wedge to the resistance is equal to the ratio of the basis of the wedge to its side. Actually, if simple compression or grinding is considered, as carried out by the molars, like a press, there is no wedge action. Then a simple lever is used. Concerning the second point, the muscles do not appear to be made of bundles of fibres parallel to each other and to the tendons. They are radial. This is most obvious for the temporalis muscle. In this muscle the fibres radiate from a cylindrical tendon and subdivide into several solid multipennate components extending throughout the whole cavity of the temporal bone which they fill. The inclination of the muscular fibres does not seem to be more than 8°. The average inclination of the tendons in the temporalis muscle does not seem to exceed 25°. As far as the third point is concerned, the magnitude of the motive force of the muscles can be known only from its effect, i.e. from the resistance which they overcome. It is thus necessary to measure this resistance experimentally. This is easily done. The greatest weight R which can be supported by the human muscles is attached to the last molar teeth G. This weight appears to be enormous. I remember that some mountebank raised from the ground a weight of 160 pounds and reliable men told me that they saw a weight much heavier than 300 pounds being raised by people who had been sitting and stood up. The force which the muscles can develop in closing the jaws, to overcome and raise such huge weights, is used to divide and grind hard food. We thus surmise that the resistance which is overcome by the temporalis and masseter muscles is more than 160 pounds.

Proposition LXXXVIII

The force exerted by the temporalis and masseter muscles
is equal to approximately 534 pounds.
Tab. IX, Fig. 2.

In the same diagram, the weight R suspended from the last molar teeth is equal to about 200 pounds and is supported by the simultaneous action of the masseter and temporalis muscles. The volume of the masseter muscle appears bigger than that of the temporalis muscle and the latter is more radial than the former. By averaging, we can surmise that half the weight R is supported by the masseter muscles and the other half by the temporalis muscles. The masseter muscles are somewhat radial but the inclination of their fibres does not exceed 8°. Therefore, the ratio of the magnitude of the force exerted by the masseter muscle to its moment, as a result of the obliquity of the fibres, is about 100/99. In the temporalis muscle, however, besides the obliquity of the fibres, there is an obliquity of the multipennate tendons. The average inclination of these tendons does not exceed 25°. Therefore, the ratio of the magnitude of the force of the temporalis muscle to its moment is 100/90. The ratio of the sum of the magnitudes of the forces exerted by the temporalis and masseter muscles to their moment thus is 200/189. Then, the lever formed by the lower jaw is considered. Perpendiculars aM and AM are drawn from the fulcra A, a to the lines of action EB and Fd of the muscles. They seem to be equal to four fifths of the distance aG of the last molar teeth from which the weight R is suspended, from the centre of the joint. Therefore, the ratio of the magnitude of the force exerted by the temporalis and masseter muscles to the resistance R is

100/75 = 4/3. But since these muscles are firmly inserted in the bones[10], their magnitude must be doubled[11]. The ratio of the magnitude of the force exerted by the muscles to the resistance thus is 8/3. The load R weighed about 200 pounds. Consequently, the force exerted by the muscles is equal to approximately 534 pounds.

Comment

It seems surprising that small muscles which together do not weigh more than one pound in man can exert such a motive force. The force exerted by the same muscles in dogs, wolves, bears and lions is even more surprising. In these animals, the lever is much longer, elongated like a cutwater. This lever is the distance between the most anterior teeth and the centre of rotation of the joint of the mandible. This distance is somewhat four or five times the distance of the line of action of the muscles closing the lower jaw, from the centre of the joint. Therefore, the moment of the force exerted by these muscles is eight or ten times the resistance. Taking into account the obliquity of their fibres, the magnitude of the force exerted by the muscles is about nine times the resistance. Although this looks huge, big dogs pull and raise a calf or a boar, thus a weight of more than 150 pounds. Consequently, the magnitude of the force exerted by the muscles closing the lower jaw is more than 1350 pounds. How huge this force is in bears and lions can be observed from the fact that these animals pull a cow with their teeth. This enables one to surmise the hugeness of the force exerted by such small muscles.

However, the muscles closing the lower jaw do not exert their whole power when the weights mentioned above are seized and raised with the teeth. Several other muscles are involved in this operation, the muscles which raise the head and those which straighten and maintain the cervical and thoracic spine. As said above, if one of these two groups of muscles acts more weakly and is able to support less weight than the temporalis and masseter muscles would if they could exert their whole power, then it is impossible to measure this force, using the weight supported by the molar and incisor teeth.

The truth of our deduction is confirmed by the breaking and grinding of the hardest stones of fruit, bones and other bodies by the teeth of dogs, bears and lions or by the pincers of crayfish or scorpions. These matters sometimes cannot be broken and crushed even when subjected to a compression of 1000 pounds, using metal pliers. It is thus surmised that the total force exerted by these muscles is much more than the above evaluation.

[10] From Prop. 22

[11] From Comment Prop. 34

Proposition LXXXIX

Structure and contrivance by which the intercostal muscles operate and method to find the force which they exert.

The ribs appear as bones similar to the lower jaw. Both are round-shaped and form half ellipses. Both are levers mobile about two small heads at their extremities. These two machines, however, are different from each other.

Firstly, the mandible pivots easily and smoothly about its two small heads articulated in the cavities of the temporal bones, whereas the ribs can carry out some small movement with difficulty about their fulcra and their joints are not at all stable. The sternum to which the anterior small head of the rib is attached is not fixed and immobile as are the vertebrae with which the posterior head of the rib is linked.

Secondly, the two sides of the lower jaw do not come closer together or go away from each other and, therefore, constitute a bicephalous lever. The ribs do not keep the same curvature. But their sides expand a little like the arms of a spring and, therefore, not only do they work as bicephalous levers but also they present the force and advantage of a spring.

Thirdly, in the mandible, the tendons of the muscles are inserted in a well-determined place, i.e. in the coronoid process and its base. Muscles are implanted all over the circular length of the ribs. Therefore, the motive force is not applied at one point of the lever where one tendon is inserted but is rather applied over the crest of the rib through all the surface of insertion of the intercostal muscle made of countless fibres. Hence, the centre of application of the force made of as many components as there are fibres must be found.

Fourthly, the fibres of the muscles closing the lower jaw are radial, whereas the intercostal fibres are arranged in two ways. They are parallel to each other and part of them intersect the other part almost at right angles.

Fifthly, the resistance which is overcome by the muscles of the lower jaw was found by suspending a weight from the teeth. It is hardly possible to do the same for the ribs. It is not difficult, however, to apply in another way a weight which the force of the intercostal muscles balances. A column can be laid over the ribs of a man in recumbent position. The weight of this column is at the limit of the tolerable but still can be raised for a little while by an energetic inspiration. This weight is not less than 50 pounds. But it does not represent an accurate measure of the resistance since the intercostal muscles must also overcome the resistance of the ribs and their connection with the vertebrae. Both oppose movements as does a tight spring. This resistance is considerable, as we shall see, and may be equal to the weight of the supported column. Therefore, the total resistance is more than 100 pounds.

Proposition XC

Approximate evaluation of the force exerted by the intercostal muscles.

If the effect of inspiration which is the introduction of air into the chest, resulted from an impulse or insufflation, it would no doubt require little force since air presents almost no resistance and is indifferent to movement. Inspiration would be carried out with the same little force if the thorax did not oppose distension and was indifferent to its expansion or constriction as is a waterskin. But matters are different. The thorax is like bellows. Once compressed it is maintained by as many strong machines as there are ribs and it can expand only when pulled by the machines which the ribs constitute. The force exerted by the machine to stretch the spring is equal to the resistance of the spring. Thus, the moment of the force exerted by the intercostal muscles must be equal to the force with which all the ribs oppose stretching. It is not possible to measure the degree of this resistance with certainty. However, relying on a double assumption, we shall demonstrate that this force is considerable. One results from the experiment carried out recently in England. A weight of 100 pounds was set upon a bladder of pig half filled with air. The neck of the bladder was ligated to a tube with a valve. Blowing air from his thorax a man could swell the thin bladder and raise the weight considerably. It is true, however, that such violent expiratory effort involves also the abdominal muscles and the diaphragm. Therefore, the force exerted by the intercostal muscles cannot be deduced accurately from this experiment. I find the following argument less ambiguous. The number of fibres in all the intercostal muscles appears to be about the double of that in the two temporalis muscles and in the two masseter muscles. Inclination of the intercostal fibres at 45° increases the force by a coefficient of 1.5. The force exerted by the temporalis and masseter muscles is equal to a resistance of more than 534 pounds. Consequently, the force exerted by the intercostal muscles is more than 1068 pounds.

Chapter XVI

Lemmas of mechanics necessary for a more accurate analysis of the motive force exerted by the muscles.

Proposition XCI

Two forces A and B (Fig. 3, Tab. IX) pull the extremities of a stick CD attached to the rods NO, OP. The forces pull in the directions MQ and MF, at right angles to each other, and are in equilibrium. Their moments are equal. I claim that the ratio of the forces A/B is equal to CM/MD, i.e. the ratio of the corresponding segments of the two main rods intercepted by the stick CD. Perpendiculars GC, GD are drawn from C and D to the main rods. They intersect at G. From G a perpendicular GH is drawn to CD at H. From H perpendiculars HK and HL are drawn to GC, DG. A force I is applied at H. Its moment is equal to the moment of A or of B. Force B is removed. Since the moments of A and I are equal and the intersection C is mobile along CF[1], the ratio of the forces A/I is equal to the ratio of the height HK to the length CH or equal to CM/CD as a result of the similarity of the triangles.

The ratio of the forces I/B is equal to DH/HL and, as a result of the similarity of the triangles, is also equal to DC/DM. Consequently, A/B = CM/MD.

Proposition XCII

The straight line DEH (Fig. 4, Tab. IX) is divided into two equal parts at E. A rope ACB is attached to the rod DH. Two forces X and Z acting at D and H pull the rope in opposite directions along the line AEB and a third force RS pulls the midpoint C of the rope down to C in the direction ECF perpendicular to AB. The ropes move in the grooves DE, HE in such a way that AC, CB are always equal to each other and so are DE and EH. The forces are balanced at C. I claim that the ratio of the force RS to the sum of the forces X + Z is equal to CE/AE.

The force RS opposes the two forces X and Z. Therefore, a half R of this force opposes X and the other half S opposes Z. Their moments are equal. The forces R, X are in equilibrium and pull the extremities A, C of the straight line AC in perpendicular directions. Therefore, the ratio of the forces R/X is equal to CE/EA. Similarly, the ratio of the forces S/Z is equal to CE/EB or to CE/FB or to CE/EA since EA = EB. Consequently, RS/(X + Z) = CE/EA.

Corollary

Two forces XZ, however great but finite, acting against the smallest resistance RS cannot straighten the rope ACB.

This results from the fact that the ratio of the greatest forces X + Z to the smallest resistance RS is equal to AE/EC which cannot be 0. Therefore, the rope ACB must be angulated.

[1] Prop. 64

Proposition XCIII

The data are the same. Research of the greatest angulation which resistance RS is able to achieve.
Tab. IX, Fig. 4.

In the right-angled triangle KIL, the ratio of KI/IL is equal to the ratio of the sum of the forces X and Z to the resistance RS; the ratio of the half rope DE to AE or EB (since EB = AE) is equal to LK/KI. The angle EAC is equal to K. AC meets EF at C where it joins CB. I claim that the rope ACB in this position is in equilibrium.

If the rope is in equilibrium in any other position MNO, the ratio of the sum of the forces X + Z to the resistances R + S is equal to ME/EN. Since the two angles K and EAC are equal and I and E both are right angles, the triangles KIL and AEC are similar. Therefore, KL/IK = CA/AE but DE/AE = LK/KI. Thus, CA = DE. Finally, AE/EC = KI/IL or (X + Z)/(R + S) = ME/EN. The two triangles AEC and MEN thus are similar and the hypotenuse AC is equal to MN since it is equal to half the rope DE. Consequently, EC = EN and the junction N is above C which is against the hypothesis. There is thus no equilibrium except at C.

Proposition XCIV

Two ropes ACB and ADB (Fig. 5, Tab. IX) are equal and join at A and B. They are pulled in opposite directions by the forces X and Z and pulled transversely by the forces G and F until they are separated by a distance CD. There is equilibrium in this position. I claim that the ratio of the sum of the forces G + F to the sum of the forces X + Z is equal to DC/AB.

Force Z pulls the two ropes BC, BD downwards and force X pulls AC, AD upwards. The forces X, Z can be thought of as divided into two in such a way that M and O pull the rope ADB, whereas P and Q pull the rope ACB. The rope ADB is pulled by the forces O, M against the transverse resistance G acting at right angles to AB. There is equilibrium. Therefore, G/(O + M) = DE/EA. Similarly, F/(P + Q) = CE/EA or DE/EA since CE = DE in the rhomb. Consequently,
(G + F)/(O + M + P + Q) = DE/EA = DC/AB, since DC = 2DE and AB = 2EA. Q.E.D.

Proposition XCV

In the same position (Fig. 6, Tab. IX) the ropes are attached at A to a nail X. I claim that the ratio of the sum of the transverse forces F, G to the resistance Z is equal to CD/AE.

The nail A replaces the force X in supporting and pulling the ropes. Therefore, the forces F + G are balanced by the resistance Z and the resistance of the nail. The ratio of the sum of the forces F + G to the resistance X + Z is equal to CD/AB. Therefore, the ratio of the sum of the forces F + G to half the resistance Z is equal to the ratio of CD to AE which is half AB.

Hence, the ropes, even if pulled by a force Z however huge, for example a weight of 10000 pounds, can be deviated by transverse forces however small, for example less than one pound, as long as the moments of the forces are equal. In our example DC would be a 1/10000 of AB.

Proposition XCVI

Two equal and weightless ropes ADBFC (Fig. 7, Tab. IX) and AEBGC, tied at equal intervals at A, B, C, hang from a nail X and are pulled vertically downwards by a force Z. At mid-intervals DEFG they are pulled transversely by equal and opposite forces HIKL in the directions DRE, FSG, at right angles to AC. They thus form equal rhombs ADBE and BFCG in similar positions. The forces are in equilibrium. I claim that the ratio of the sum of the forces which pull horizontally H + I + K + L to the resistance Z is equal to the ratio of the sum of the transverse widenings DE + FG to AR which is half a longitudinal diagonal AB.

The forces H, I deviate the ropes ADB and AEB against two resistances. One is the resistance of the nail X and the other is the force which pulls the junction B downwards. These two resistances are equal. Therefore, the ratio of the sum of the two forces H + I to the force which pulls the junction B downwards is equal to DE/AR. Two forces K, L deviate the ropes BFC and BCG against two resistances. One is the weight Z and the other is the force which pulls the junction B upwards towards A and replaces the nail X. The ratio of the sum of the two forces K + L to the resistance Z is equal to FG/BS or DE/AR since the rhombs are equal. All the forces H, I, K, L thus act against four resistances: the nail X, the weight Z, the force which pulls the junction B downwards and the equal force which pulls it upwards. The two opposed forces at B cancel each other because the force which at B resists the tensions exerted by the forces H, I and prevents the displacement of point B towards R is the tension downwards exerted by the weight Z which acts as if it was hung from B. No other resistance at B acts against the forces K, L which attempt to pull the junction B downwards against the resistance of the nail. Therefore, the intermediate forces pulling the junction B cancel each other. One pulls in one direction as much as the other pulls in the other direction. Consequently, they act to no avail. It is as if they were not acting against the forces H, I, K, L. Therefore, all the forces H + I + K + L are balanced by one force Z, although they actually act against four equal resistances X, Z and two opposite ones at B. The ratio of their sum H + I + K + L to the four resistances X + Z + B + B is equal to (DE + FG)/(AB + BC). Therefore, the ratio of the sum of the four forces H + I + K + L to one resistance Z is equal to (DE + FG)/AR. AR is the quarter of the distance AC.

Proposition XCVII

Same proposition demonstrated in another way.
Tab. IX, Fig. 7.

All the forces H, I, K, L widening the rhombs formed by the ropes are equal. The widenings DE, FG of any of the linked rhombs are also equal. Therefore, the ratio of the sum of the forces H + I + K + L widening all the rhombs to the sum of all the forces K + L widening one rhomb is equal to the ratio of the sum of all the widenings DE + FG to one widening FG. The connected ropes BFC, BGC are maintained at B because the widening of the superior ropes prevents the lowering of the junction B as firmly as if the inferior ropes were fixed by a nail at B. The forces K, L which widen the inferior ropes support the resistance Z by exerting an equal moment since there is equilibrium. Therefore, the ratio of the widening forces K, L to the resistance Z is equal to the ratio of the widening FG to BS which is half BC. Consequently, the ratio of the sum of all the forces H + I + K + L widening all the rhombs to the resistance Z is equal to the ratio of the sum of all the widenings DE + FG to BS. This is the ratio of as many widenings as there are equal and similar rhombs, to AR which is half the vertical diagonal AB of any of these rhombs. Q.E.D.

Thus, although the forces are multiplied to widen more rhombs, they cannot support a resistance greater than Z. The greater the moment of Z, the higher its ascension, as was shown[2].

Proposition XCVIII

The data are the same. Many forces widening several rhombs raise a resistance acting vertically over a distance equal to the upwards displacement resulting in one rhomb from one pair of forces, multiplied by the number of rhombs.
Tab. IX, Fig. 7.

All the rhombs ABDE, BFCG are equal, similar, and arranged similarly. Their corresponding sides and diagonals are thus equal. The difference between the length of the rope ADB and the diagonal AB is equal to the difference between the length of the rope BFC which is equal to ADB, and the diagonal BC. Therefore, the raising of the junction B towards A is equal to the raising of the junction C towards B. After contraction, the interval between Z and A consists of as many shortened diagonals as there are rhombs. Consequently, the raising of the weight Z from C to A is multiplied by the number of rhombs.

[2] Prop. 33

Proposition XCIX

The data are the same. Widenings of the ropes
and their contractions shown with numbers.
Tab. IX, Fig. 7.

The angle DAE which measures the widening of the ropes is given. DAR is half the angle DAE and the triangle DAR is a right-angled triangle, the right angle being R. If DA is the trigonometrical radius, AR is the cosine of angle DAR. RO is the cosine's complement of the angle DAR or the difference between the trigonomical radius and AR. The symmetrical widening DE or FG of the ropes is measured by the diagonal DE which also corresponds to the opening of the angle DAE. It must be shown that the raising of the resistance Z is the double of RO (cosine's complement of angle DAR) multiplied by the number of rhombs.

The cosine's complement OR is the difference between the rope AD and AR. Before the widening of the rhomb the segment of rope AD was in position AO. The extremity D has been displaced up to the level DRE. Therefore, RO is the measure of the shortening of the segment of rope AD. All the sides of the rhombs DF, BF, FC are equal to each other and to AD and are equally inclined to the line ABC. Therefore, OR measures the shortening of the interval and is the same for all the sides of the rhombs. The double of OR is the shortening of the sides ADB of the rhomb and thus of the rhomb. It is the same for the other rhombs. Consequently, the double of OR multiplied by the number of rhombs gives the shortening of the whole rope ADBFC.

This example was calculated for one rhomb ADBE widened by the forces H, I and pulled vertically by the resistance Z. If the rhombs are multiplied as in a chain, the widenings of the ropes remain equal to each other as does their measure. Then the raising of the resistance is equal to the raising of one rhomb multiplied by the number of rhombs.

If there are fifteen adjacent rhombs and the angle DAR is 0.49., then the raising of the resistance Z is fifteen times the raising of any of the rhombs, i.e. 15/142.

If the widening forces H, I are multiplied by fifteen, the resistance Z retains its weight and power. If the forces H, I are one, multiplied by 15 they are 15 and the weight Z is 35.

Using this principle gives Table 3.

Table 3

Angle DAR which is half the opening of the angle DAE	Trigonometrical radius AD	Double DE of the sine DR of the angle DAR	Cosine AR of the angle DAR	Cosine's complement RO of the angle DAR	Widening DE or force HI or rope ADB = 1	Resistance Z	Raising of resistance Z
Degrees							
0.1.10.	100000.00	67.88.	99999.99		1.	1473.	$\frac{1}{100000.00}$
0. 1.	100000.000	58.18.	99999.996	4.	1.	1719.	$\frac{4}{100000.00}$
0. 10.	100000.00	581.78.	99999.58	42.	1.	172.	$\frac{42}{100000.00}$
0. 30.	100000.0	1745 3.	99996.2	39.	1.	57.	$\frac{390}{100000.0}$
1. 00.	100000.	3490.	99985.	15.	1.	29.	$\frac{1500}{100000.0}$
5. 0.	100000.	17431.	99619.	381.00	1.	6. 3.	$\frac{382.00}{100000.0}$
10. 0.	100000.	34730.	98481.	1519.00	1.	3.	$\frac{1519.00}{100000.0}$
20. 0.	100000.	68404.	93969.	6031.	1.	$1.\frac{1}{3}$	$\frac{6031}{100000.}$
15. 0.	100000.	51764.	96592.	3408.	1.	2.	$\frac{3408}{100000.}$
25. 0.	100000.	84524.	90631.	9369.	1.	1.	$\frac{9369}{100000.}$
26. 34.	100000.	89448.	89441.	10559.	1.	1.	$\frac{10559}{100000.}$
30. 0.	100000.	100000.	86602.	13398.	1.	$\frac{86602.}{100000.}$	$\frac{13398}{100000.}$
35. 0.	100000.	114715.	81915.	18085.	1.	$\frac{81915.}{100000.}$	$\frac{18085}{100000.}$
40. 0.	1.00000.	128558.	76604.	23396.	1.	$\frac{76604.}{128558.}$	$\frac{23396}{100000.}$
45. 0.	100000.	141421.	70711.	29289.	1.	$\frac{70711.}{141421.}$	$\frac{29289}{100000.}$
50. 0.	100000.	153209.	64279.	35721.	1.	$\frac{64274.}{153209.}$	$\frac{35721}{100000.}$
55. 0.	100000.	163850.	57358.	42642.	1.	$\frac{57358.}{163850.}$	$\frac{42642}{100000.}$

Proposition C

If two chains of unequal lengths, made of ropes of equal strength and similarly tied together, are widened by apparently equal forces so that all the rhombs are the same, the two weights which ensure equilibrium are equal.
Tab. IX, Fig. 8.

AB is the longer chain and CD the shorter. Both are made of ropes of equal strength and are tied similarly. They hang from nails and are pulled by the weights R and S. Force X widens all the rhombs of chain AB and force Z all those of chain CD in such a way that all the rhombs are similar and the forces X, Z balance the resistances R, S. I claim that the weights R and S are equal. The same resistance[3] which can hang from one rhomb AE widened by the corresponding part of force X is also suspended from the whole series of rhombs AB which are equal and similar to the one pulled by the whole force X. Therefore, the same weight R which is supported by the force X widening the chain AB is also supported with equal moment by a force widening one rhomb AE alone.

Similarly, the weight S is also supported by the force Z widening the whole chain CD as it is by a force widening only one rhomb CF. But all the small machines, i.e. the rhombs AE and CF, are equal and similar and they are widened by equal forces. Therefore, they support equal weights. Consequently, the weights R and S hanging from the unequal chains AB, CD are equal.

Proposition CI

The data are the same. The ratio of the lengths of the chains AB/CD is equal to the ratio of the forces X/Z.
Tab. IX, Fig. 8.

All the small machines, the rhombs, are equal, similar and similarly arranged in both chains. Their heights are thus equal. In the unequal chains AB, CD the ratio of the lengths AB/CD is equal to the ratio of the numbers of rhombs in the chains AB and CD. The ratio of the lengths is also equal to the ratio of the sums of the widenings or of the transverse diagonals of the rhombs constituting AB and CD or to the ratio of the forces X/Z.

Consequently, equal weights will be raised proportionally to the lengths of the unequal chains and to the forces pulling the unequal chains. There are as many individual vertical shortenings (pairs of cosine's complements of half angles) as there are rhombs. But the lengths of the chains are proportional to the numbers of rhombs since the chains result from the addition of the lengths of the rhombs. Similarly, the number of widening forces is proportional to the number of rhombs and thus to the length of the chain. Therefore, equal weights are raised proportionally to the lengths of the chains and to the forces which contract the chains.

[3] From Prop. 97

Proposition CII

The data are the same. The chains AB, CD (Fig. 9, Tab. IX) are equal and the magnitudes of the forces LX, KZ are apparently unequal but they raise the weights R, S to the same heights BG, DH. The forces are proportional to the weights.

The rhombs AE, CF are small machines and their contracting forces are LM, KN. The chains AB, CD, equal and similar, are made of an equal number of rhombs. After equal contractions the shortened chains AG, CH remain equal and hence their individual rhombs AE, CF also remain equal, similar and similarly arranged. The same is true for all the other rhombs. Therefore, their widths and half their heights are equal. But[4] the ratio of the force LX to the resistance R is equal to the ratio of the sum of the widenings of all the rhombs of the chain AG to half the height of one rhomb AE, or to the ratio of the sum of the widenings of all the rhombs CH to half the height of one rhomb CF (since they are equal, similar and similarly arranged in both chains), or to the ratio of the force KZ to the resistance S. Therefore, the ratio of the force LX to the weight R is equal to the ratio of the force KZ to the weight S because of the relation between the sum of the widenings of all the rhombs of the chains AG or CH to half the height of one rhomb AE or CF. Changing equalities gives LX/KZ = R/S.

Proposition CIII

Chains made of ropes similarly tied such as AB, CD, EF, etc. (Fig. 10, Tab. IX) are equal, similar, parallel and adjacent. The bundle which they constitute is contracted by forces X, Z, T of apparently equal magnitude so that their rhombs are equal, similar and arranged in alternate order like a network, and they are balanced by the weight RV which they support. R is the weight supported by one chain AB contracted by the force X. The moments of R and X are equal. I claim that the ratio of the forces XT/X and that of the weights RV/R are equal to the ratio of the sum of the chains of the bundle ABGH to one chain AB.

The weight RV is raised by several chains contracted by individual forces seemingly of equal magnitude so that no chain remains inactive. Therefore, any chain with its own particular force raises its part of the total weight RV. All the chains thus act together and support the weight RV as if this weight was subdivided into as many parts as there are chains. The chains are assumed to be equal and similar and they are contracted by equal forces since they appear to be of equal strength. Therefore, the different forces can support equal weights

[4] Prop. 96

with equal moments and there are as many forces as there are chains. Consequently, the ratio of all the chains of the bundle ABGH to one chain AB is equal to the ratio of the total of the forces XT to one force X, and to the ratio of the total weight RV to one weight R.

Proposition CIV

The data are the same (Fig. 10, Tab. IX). The ratio of the force XT of the whole bundle ABGA to the weight RV suspended in equilibrium is equal to the ratio of the sum of the widenings of all the rhombs of one chain AB to half the height of one rhomb.

The individual force X contracting the chain AB supports the part R of the weight. The moments of X and R are equal. Therefore[5], the ratio of the force X to the weight R is equal to the ratio of the sum of the widenings of the rhombs of the whole chain AB to half the height of one of the rhombs. But thc total of the chains contained in the bundle ABGH is also a multiple of one chain AB, the total force XT is the same multiple of the partial force X and all the weight RV is the same multiple of the partial weight R. The different parts when multiplied by the same factor remain in the same proportions. Consequently, the ratio of the forces XT to the weight RV is equal to the ratio of the sum of the widenings of all the rhombs of one chain to half the height of one rhomb of this chain.

Proposition CV

Two bundles AC and EG (Fig. 11, Tab. IX) are made of chains of equal width but of unequal heights. The weights R and S are supported with equal moments by the forces XZ and TV which are apparently equal. The forces and the raisings of the weights are proportional to the lengths of the bundles AB and EF.

The component X of the force XZ contracts the chain AB alone and supports the component I of the total weight R in equilibrium. Similarly, the component T of the force TV contracts the chain EF and supports the component O of the total weight S in equilibrium.

The ratio of the forces XZ/X and that of the weights R/I are equal to the ratio of the number of equal chains in the bundle AC to one chain AB. Similarly, the ratio of the forces TV/T and that of the weights S/O are equal to the ratio of the number of chains in the bundle EG to one chain EF. The numbers of chains are equal in both bundles. Therefore, the ratio of all the chains in the bundle EG to one chain EF is equal to the ratio of all the chains in the bundle

[5] Prop. 96

AC to one chain AB. Therefore, XZ/Z = TV/T. Similarly, R/I = S/O. The weights I and O are equal. Thus, the ratio of the weights R/S, that of the forces XZ/TV, that of the shortening of the bundle AC to the shortening of the bundle EG are equal to the ratio of the length AB of the bundle AC to the length EF of the bundle EG.

It appears thus that the same weight R which is supported by the entire bundle VC whatever its length is also supported by a smaller bundle BC formed by the union of all the lowest rhombs of the bundle ABCD.

Proposition CVI

The two bundles AC and EG (Fig. 12, Tab. IX) are of the same height but the former AD is wider than the latter EH. I claim that the ratio of the forces XZ/TV and that of the weights R/S are equal to the ratio of the widths of the bundles AD/EH. Raising of the weights is equal.

The chains AB and EF are equal and similar. The magnitudes of the small forces X and T are apparently equal. The weights I and O thus are also equal since they are supported by equal forces exerting equal moments. The ratio of the force XZ to its small component X or T, since X = T, as well as the ratio of the weight R to its small component I or O, since I = O, is equal to the ratio of all the equal chains of the bundle AC to one chain AB or EF. Similarly, the ratio of the forces T/TV and the ratio of the small weight O to the weight S are equal to the ratio of one chain EF to all the chains of the bundle EG. Consequently, changing and arranging the equalities, the ratio of the forces XZ/TV and that of the weights R/S are equal to the ratio of all the chains of the bundle AC to all the chains of the bundle EG, or to the ratio of the width AB of the bundle AC to the width EH of the bundle EG since, in both bundles, the chains are equal, similar, similarly contracted and contiguous.

Moreover, since all the chains are equal, similar and widened by equal forces, and must shorten equally, the bundles AC, EG contract equally.

Proposition CVII

The data are the same. The heights and widths of the bundles are unequal. I claim that the ratio of the weights R/S (Fig. 1, Tab. X) is equal to the ratio of the width AD of the bundle AC to the width EH of the bundle EG; that the ratio of the raising of the weight R to that of the weight S is equal to the ratio of the heights AB/EF and to the ratio of the forces XZ/TV.

If a section is carried out separating KB, so that AK = EF, then the ratio of the forces XZ/MZ is equal to the ratio of the heights AB/KB. The force MZ is that force which contracts all the chains of the bundle KBCL.

The weight R is supported as well by the force XZ contracting the bundle AC as by the force MZ which contracts the bundle KC. The bundles KC, EG are of equal heights. Therefore[6], the ratio of the forces MZ/TV and that of the weights R/S are equal to the ratio of the widths BC/FG.

The bundles KC, EG are equally shortened and raise the suspended weights R, S to the same heights[7]. The weight R is raised by the bundles AC and KC of even width to altitudes proportional to the lengths AB/KB or AB/EF.

Finally, the ratio of the forces XZ/MZ is equal to that of the heights AB/KB. The ratio of the forces MZ/TV is equal to that of the widths BC/FG. The ratio of XZ/TV results from the ratio of XZ/MZ and the ratio of MZ/TV. Therefore, the ratio of the forces XZ/TV results from the ratio of the heights AB/EF and the ratio of the widths BC/FG.

Proposition CVIII

One extremity of a chain AC (Fig. 2, Tab. X) in oblique position is attached to a nail A and the other extremity C is pulled by a weight R acting in the direction CH inclined to CA. The point of junction C is mobile along BCH. The moment of the weight is equal to that of the force XZ contracting the chain AC. I claim that the ratio of the magnitude of the force XZ to the suspended weight R is equal to the ratio of the sum of all the widenings of the rhombs of the chain AC to the height CH of half the length CE of one rhomb.

The weight R is removed and replaced by a force S which opposes with the same moment the direct tension of the chain SCA produced by the same force XZ which widens all the rhombs of the chain. The moments of R and S are equal to each other and to the moment of the force XZ. Moreover[8], the ratio of the force XZ to the resistance S pulled directly or the force VZ pulled by the lowest rhomb (and equal to S) is equal to the ratio of the sum of all the widenings of the rhombs of the chain AC to half the height CE of one rhomb. After removal of force XV, the force VZ remains or the force S which balances the weight R by pulling the height CE as a rope. The junction C of the lines of action is mobile along CH. Therefore[9], the ratio of the force VZ, or S, to the weight R is equal to CE/CH. Consequently, the ratio of the force XZ to the weight R is equal to the ratio of the sum of the widenings of all the rhombs of the chain AC to the height CH.

---

[6] Prop. 106
[7] Prop. 105
[8] Prop. 97 and 104
[9] Prop. 64

Proposition CIX

In the right-angled triangle AHB (Fig. 3, Tab. X), from the acute angle A a line AC is drawn inside the triangle and intersects the upright BH. I claim that the ratio of the segment HC of the upright to the hypotenuse CA thus drawn is smaller than the ratio of the difference BD between the hypotenuses to the remaining segment BC of the upright.

A circle CDEF with radius AC is drawn about A. BCH and BA are prolonged to the opposite periphery FE of the circle. BG is taken such that BG/BC = BC/BD. The two lines BF, BE intersect the circle and each other. The rectangle EBD is equal to the rectangle FBC. Therefore, FB/EB = BD/BC = BC/BG = FC/EG. But GE is smaller than ED. Consequently, CF/ED = CH/AD = CH/CA < CF/EG = BD/BC.

Proposition CX

The chain AB (Fig. 4, Tab. X) is contracted by a force XZ. At first the chain raises a weight S directly. Then pulling obliquely it raises a weight R mobile along BCH. I claim that the weight S is heavier than the weight R. S is raised less than R so that the ratio of the weights S/R is smaller than the ratio of the raising of R to that of S.

The ratio of the force XZ to the weight S pulled directly[10] is equal to the ratio of the sum of all the widenings of the chain AC to half the height CE of one of its rhombs and inversely. The ratio of the force XZ to the weight R is equal[11] to the ratio of the sum of all the widenings of the rhombs of the chain AC to GC the vertical projection of half the height of a rhomb CE. Therefore, S/R = CE/CG and GC is smaller than CE. Thus S is greater than R. Secondly, the weight S pulled directly is raised towards A by a distance equal to the shortening of the chain AC. This shortening is the difference DC between AB and AC. The raising of S thus is DC. The raising of R is the displacement BC in the direction or through the groove BCH. The difference DC between the two sides AB, BC is smaller than the base BC of the triangle ABC. Therefore, the heavier weight S is raised less than the lighter weight R. Thirdly, BI is drawn parallel to CA. BI is shorter than BA and, consequently, IB/BH < AB/BH; S/R < EC/CG = IB/BH because of the similarity of the triangles HBI and GCE. Therefore, S/R < IB/BH; IB/BH < BC/CD[12]. Consequently, the ratio of S/R is smaller than the ratio of BC raising of R to CD raising of S.

---

[10] Prop. 96
[11] Prop. 108
[12] From Prop. 109

Proposition CXI

The data are the same. The angulation of the chain is given as well as its contraction. The proportions of the raisings of unequal weights can be expressed with numbers.
Tab. X, Fig. 4.

The angle ACH is 30°. The shortening CD is a fifth of the length of the chain CA. Then if AD or AB is equal to 100, CD is equal to 25 and HB is equal to 86.6, since it is the cosine of 30°. AH is 50. In the right-angled triangle AHC, AH is 50, the hypotenuse CA is 125 and CH is 114.6. Therefore, the difference BC of HC and HB is 28. Then BH/BA = 25/26.5 approximately. Consequently[13], the ratio of the weights S/R is equal to AB/BH = 26.5/25. The ratio of the raising of S to that of R is equal to CD/BC = 25/28.

Proposition CXII

Why Nature uses radial multipennate muscles in animals.
Tab. X, Figs. 5 and 6.

The same weight certainly requires less force to be pulled and carried when it is moved directly than when it is moved obliquely by way of ropes inclined to the direction of the movement of the weight. The inclined fibres of the muscles are ropes which by their contraction act in their own directions. Thus a resistance is pulled by inclined fibres at the price of a greater force than if it was pulled directly in the direction of the fibres. Since Nature, always economical, pursues conciseness and simplicity in her operations, it is surprising that oblique fibres were adopted in muscles. Careful attention actually shows that Nature was forced by material necessity to choose a long and more difficult way. To understand the necessity in this matter, it must be noted that some muscles have to exert considerable force such as the glutei, the vasti and others. To exert such huge motive force there is no other way than multiplying the number of fibres or of chains made of small rhomboidal machines. Any fibre can pull a certain resistance and, therefore, a large number of fibres can support a heavy weight. But, if so many fibres were adjacent, parallel and at right angles to their base, they would constitute a right-angled parallelipiped such as ABCD (Fig. 5, Tab. X). The width BC of such a bundle would be exaggeratedly bulging and would require much space in which to be housed and to move. Such wide spaces are not available about the bones and cannot be since several other muscles must also be housed, which are aimed at other movements. Moreover, the wide bases AD, BC of this muscle shaped like a right-angled parallelipiped should be connected by wide tendons and thick bony tubercles which would be very inconvenient. No other muscle could be attached to these tubercles and,

[13] Prop. 64

therefore, the other movements of the limb would be impeded. Actually, Nature provided for all these necessities most wisely by setting in narrow spaces muscles with a huge abundance of fibres, such as FH, IKL (Fig. 6, Tab. X). These muscles are not too large. Their tendons are thin N, M. Their fibres do not insert in a wide and flat base but over all the circumference of the tendon NOI. Similarly, several men, one behind the other, use to pull with their hands a long rope to which a heavy weight is attached. They are thus positioned in a long row and all of them can exert their force.

Almost all the muscles are shaped this way so as to be conveniently placed and adapted to the restricted spaces on the sides of the bones in which a long and thin muscle can take place without inconvenience to others. This rule of Nature is so necessary that in the fingers the motor muscles are not inserted in the phalanges but at a distance, in the proximal epiphysis of the ulna. The bundles of fibres extend from the elbow to the wrist. They pull the small bones of the fingers by way of tendons acting as belts.

The obliquity of the fibres must not disturb you. In most instances this obliquity is minimal, being not more than five or six degrees. Therefore, the magnitude of the force exerted by the fibres must be only little increased beyond the moment, as was shown. This spending of forces cannot be avoided and is necessary. Moreover, it is largely compensated by a considerable gain in convenience.

Chapter XVII

More accurate analysis of the motive power of the muscles described above.

So far we have discussed the excess of motive force by which Nature carries and moves resistances and weights using muscles. But this is not all. Nature exerts even far much more force for other mechanical reasons as we shall show.

Proposition CXIII

The tendons and nerves which, after being stretched, contract, must be made of many small machines linked longitudinally like a chain of contractile springs.

Guided by Nature, we must proceed from the known to the unknown. We search by which contrivances and which mechanism tendinous fibres and contractile wires contract either spontaneously after violent stretching or by an external cause. We thus must consider similar operations perceptible to the senses. These are the operations which we observe in curved contrivances, such as springs and rings. In these, indeed, the extreme parts, when too much stretched, come back closer to each other. Therefore, the tendinous fibres must contract and shorten because of a structure similar to that of these contrivances. We see that contraction of the fibre as well as that of a metallic wire does not occur in one of its parts only but in all of them. This appears from the following experiment. A steel wire is divided into equal segments which are painted in white and black alternately like the divisions of an astrolabe. All the segments of such a wire contract equally after violent stretching. The same occurs in flesh fibres in which the small particles cannot come closer together without the force of a machine. We are thus compelled to admit that two adjacent particles in such a thread are moved one towards the other by the force of a machine similar to a spring. One machine alone is not sufficient to carry out the uniform contraction. We observe that the two arms of a spring do not actually shorten but, keeping their initial lengths, they bend one towards the other. As a result, their extremities are brought closer together. It is different for the contraction of a thread, however, in which the components of any pair of particles distributed all over the length of the thread come closer together. Consequently, the fibre of flesh or the steel wire is made of as many very small machines similar to springs as there are links between pairs of particles constituting the length of the thread. Therefore, we can compare the fibre of flesh and the steel wire with a chain formed by a long row of connected links or springs. In such a structure one can verify the remarkable property of the fibres which can be bent and folded everywhere. Such deformations would not be possible without the connections between the links of the chain.

Proposition CXIV

Any muscular fibre is similar to a chain made of many rhombs which can contract as springs do.

The muscles are bundles of very thin and strong tendinous threads which contract as do the strings of a zither or steel wires when stretched. The fibres can contract with much more force by order of the will. Such contraction cannot be understood without a machine like a spring and the contraction of all the elements of a fibre cannot be conceived without a continuous series of small machines like a chain, as was said. Therefore, any muscular fibre is similar to a chain of small machines. These small machines are similar to rings or springs. They can be neither hard nor rigid since the flesh fibres are soft and flexible over their whole length. Therefore, they are rather similar to rhombs made of threads tied together which are widened by a motive faculty and thus contract. Consequently, one must conceive the muscular fibres as a series of small machines of porous or rhomboidal shape like a chain made of rhombs of filaments, as discussed in the preceding chapter.

Proposition CXV

The small machines or rhomboidal pores made of flesh fibres must be so small that their length does not exceed 1/20 of a finger breadth.

The muscular fibres are thinner than female hair. The pores or rhomboidal cavities of the small machines which exist in these fibres must be smaller than the width of the flesh threads. Therefore, the row of rhombs continuous over the whole length of the fibre is at least as long as the product of the width of one flesh thread and the number of rhombs in this row. But fifty contiguous flesh fibres are smaller than a finger breadth. Consequently, the small machines or the rhomboidal pores are so small that a series of fifty of them does not exceed one finger breadth. We tentatively assume that twenty small machines in a row are equal to one finger breadth.

Proposition CXVI

The texture of the muscle is similar to a fascicular network made of adjacent chains.

One can feel that the flesh fibres when inactive are soft and elongated like loose zither strings. The internal sides of their pores must be adjacent and in close contact. Their heights are elongated. When they contract, they harden and swell, and their height decreases. But fibrous filaments cannot be stretched. Therefore, contraction and swelling of a fibrous column can be conceived only if the transverse diameters of the pores increase and their longitudinal diameters decrease. During such swelling, the fibres must look like a chain made of rhombs. The bundles of the muscle made of adjacent parallel columns are clustered together. Consequently, the small machines or the swollen segments of one fibre must be adjacent to other small machines similarly contracted and they must constitute the same texture as a fascicular network made of small rhomboidal machines.

Proposition CXVII

The ratio of the motive force which contracts one small machine of a muscular fibre to a suspended weight is equal to the ratio of the widening of this rhomboidal machine to half its height.
Tab. IX, Fig. 6.

The small machines of which the fleshy and tendinous filaments of the muscles are made are like the rhomb ADBC formed by the ropes ADB and ACB, linked at A and B, attached to a nail X and supporting the weight Z. There are two forces G and F which pull the ropes and widen the rhomb. The ratio of the sum of these forces to the weight Z is equal to the ratio of the widening DC of the ropes to AE half the diagonal AB. Any motive force contracting the one small machine formed by the muscular fibre ADB certainly shortens the distance AB and raises the weight Z. This, however, cannot be conceived without widening of the ropes which are angulated and folded like a spring. Therefore, similarly, the ratio of the motive force contracting a small machine, like a rhomb hanging from a fixed point X, to the weight Z is equal to the ratio of the widening DC of the rhomb to AE half the diagonal AB.

Corollary

A motive force, however small, is able to raise a suspended weight, however heavy. For example a weight of 10000 pounds can be somewhat raised by a motive force of 1/100 pound since the widening CD can be 1/1000000 of AE. Then the forces are in equilibrium.

Proposition CXVIII

The ratio of the displacement of a force contracting one small fibrous machine to the raising of the resistance is equal to the ratio of the sine of half the angular widening of the fibres of the rhomb to twice the cosine's complement of half the angular widening.
Tab. IX, Fig. 6.

Two forces G and F contract the rhomboidal machine ACDB, pulling the mid-ropes from E towards C and D. The displacement of each of these forces C or D is measured by the lines EC, ED. EC = ED is the sine of half the angle CAD. The displacement of the resistance Z is measured by the decrease of the distance AB or by the difference of the total length of the ropes ACB or ADB and the longitudinal diagonal AB. This difference is equal to twice the cosine's complement of angle CAE. Therefore[1], the ratio of the displacement of the force GF to the raising of the resistance is equal to the ratio of the sine CE of the angle CAE to twice the cosine's complement of the same angle CAE.

But it appears in Table 3 (proposition 99) that up to an angle of 51° widening of the rhomb exerts a force smaller than the resistance. Beyond 51°, the force always exceeds the resistance. The displacement of the forces or the widening of the rhombs is always more than the raising of the resistance.

[1] Prop. 99

Proposition CXIX

The ratio of the motive force which contracts the series of small machines constituting one fleshy filament to the resistance of a suspended weight is equal to the ratio of the sum of the widenings of all the small machines to half the height of one rhomb.

It was shown[2] that, in a chain made of rhombs of ropes, the ratio of the sum of all the forces widening the rhombs to the suspended resistance is equal to the ratio of the sum of the transverse diagonals of all the rhombs to half a longitudinal diagonal of one rhomb. Such a chain is similar to the muscular fibre constituted by small machines. Its operation is the same. Therefore, similarly, the ratio of the sum of all the forces which contract the individual small machines to the suspended resistance is equal to the ratio of the sum of the widenings of all the small machines or rhombs to half the longitudinal diagonal of one rhomb or contracted small machine.

Corollary

Consequently, the same weight which is supported by countless forces contracting all the small machines of the fleshy fibre is also supported by one force contracting one small machine.

The same is true in a bundle made of many fleshy fibres and in a muscle. The weight which is supported by countless forces contracting all the layers of the small machines constituting the bundle is also supported by the forces which contract one layer of small machines. If there are a thousand times as many small machines in a whole muscle as there are in one layer, the motive force which contracted one layer of small machines, after having been multiplied by a thousand, will not support a weight a thousand times heavier.

Proposition CXX

The motive force contracting a series of small machines contained in the fibres of a muscular bundle raises the same weight at a level as many times higher than the motive force which contracts one layer of small machines, as the latter force is contained in the former and as there are layers of small machines.

It was shown that in a chain made of small machines a weight is raised to a height which is equal to the raising achieved by one small machine alone multiplied by the number of small machines in the chain and the forces elevating the small machines are in the same ratio. Similarly, with muscular bundles made of fibres, or with chains made of small machines, the suspended weight must be raised by a motive force which is the product of the force contracting one fibre and the number of small machines in any singular fibre. Q.E.D.

[2] Prop. 96

Proposition CXXI

If two muscles of the same animal are equally thick, i.e. made of an equal number of fibres, but their lengths are unequal, and they support equal weights, the motive forces and the raisings of the weights are proportional to the lengths of the muscles.

This results[3] from the fact that, when two bundles of equal thickness and unequal lengths, made of chains of small machines, as described above, support equal weights, the motive forces and the amounts of raising are proportional to the lengths of the bundles. The bundles of the muscle are equally thick and unequally long, made of muscular fibres made in turn of small machines. The proposition thus is demonstrated.

Proposition CXXII

If the lengths of the muscles are equal but their thicknesses are unequal, the motive forces and the suspended weights are proportional to the thickness of the muscles but the weights are raised to the same height.

Consequently[4], in two bundles of equal length but unequal thicknesses, made of contractile chains, the forces and the suspended weights are proportional to the thickness of the bundles and the weights are raised to the same height. Thus, two muscles of equal length and unequal thicknesses raise the weights to the same height and the weights are proportional to the forces and to the thickness of the muscles.

Proposition CXXIII

If the lengths and thicknesses of muscles are unequal, the suspended weights are proportional to the thicknesses of the muscles. Their degrees of raising are proportional to the lengths of the muscles. The ratio of the forces depends on the ratio of the thicknesses and the ratio of the lengths.

If the lengths and thicknesses of two bundles of contractile chains are unequal[5], the suspended weights are proportional to the thicknesses, the raisings to the lengths. The ratio of the forces depends on the ratio of the lengths and the ratio of the thicknesses. The same is true for the muscles.

---

[3] From Prop. 105
[4] From Prop. 106
[5] From Prop. 107

Corollary

Hence the necessity appears which forces Nature to use excesses of motive force in long muscles.

The weights must be raised considerably. The pores or small machines in the muscular fibres are small and narrow. They do not exceed the breadth of one fibre. A suspended weight could not be raised more than half the height of one rhomboidal small machine. Therefore, the small machines had to be multiplied and set in long rows. Then all the small raisings result in large displacements of the bones, which are necessary for the movements of the animal.

After these premisses, the forces exerted by the muscle can be analysed more accurately.

Proposition CXXIV

More accurate analysis of the forces exerted by the flexor digitorum profundus and sublimis, and deltoid muscles.

We showed[6] that the flexor digitorum profundus muscle exerts a force equal to 176 pounds as a consequence of the lengths of the levers and of the inclination of the multipennate fibres. The oblique fibres of this muscle are longer than two finger breadths. Over a finger breadth of fibre there are more than twenty small machines which constitute a fibrous chain. Thus, in individual fibres of this muscle there are more than forty small machines. The same resistance is supported by the more distal layer of small machines of the muscle as by the whole muscle. Supporting the resistance of 9.5 pounds required a force of 176 pounds. This force is exerted by the distal layer of small machines, but a force forty times greater is exerted by all the fibrous chains of the muscle.

Consequently, the force which is exerted by the flexor digitorum profundus is more than 7040 pounds.

Similarly, the force exerted by the flexor digitorum sublimis was shown to be equal to 157 pounds. As a result of the length of the fibres or chains made of small machines, this force must be multiplied by forty. Consequently, the force exerted by the muscle is 6280 pounds.

Finally, the motive force exerted by the deltoid muscle in the described position was shown to be equal to 907 pounds and the total force exerted by the muscle was 1540 pounds. But[7], as a result of the length of the fibres and the number of small machines, this force exerted by the deltoid muscle must be multiplied by more than 40. That gives 61600 pounds.

[6] Prop. 82
[7] Prop. 84

Proposition CXXV

More accurate determination of the force exerted by the glutei muscles in the case of proposition 53.

We showed that the force exerted by the glutei muscles was equal to 1840 pounds as a result of the levers. Because of the inclination of the radial fibres[8], the force exerted by these muscles was equal to 2621 pounds[9] and the total force of the same muscles was 6257 pounds[10]. Finally, these radial fibres are long. They seem to be longer than four finger breadths. They are assumed to be more than three finger breadths. Therefore, the fibrous chains are made of more than sixty small machines. Consequently, the force mentioned above must be multiplied by more than sixty. That gives a force greater than 375420 pounds.

Proposition CXXVI

More accurate determination of the motive force exerted by the flexor pollicis longus muscle.

We showed[11] that the force exerted by the flexor pollicis longus muscle is approximately 124 pounds. The fibres of this multipennate muscle seem to be longer than one and a half finger breadths. Therefore, the force must be multiplied by more than thirty. The more accurate force exerted by this muscle is thus about 3720 pounds.

Proposition CXXVII

More accurate evaluation of the motive force exerted by the temporalis and masseter muscles.

It was shown[12] that the motive force exerted by the temporalis and masseter muscles was approximately equal to 534 pounds. The length of the fibres of these radial and multipennate muscles is not more than one and a half finger breadths. Therefore, the force must be multiplied by at least thirty. The force exerted by these muscles thus is not less than 16020 pounds.

---

[8] Prop. 53
[9] Prop. 83
[10] Prop. 85
[11] Prop. 86
[12] Prop. 88

Proposition CXXVIII

More accurate analysis of the motive force exerted
by the intercostal muscles.

In proposition 90 it was shown that the motive force exerted by the intercostal muscles was approximately 1068 pounds. The length of the fibres seems to be one and a half finger breadths. Therefore, the force must be multiplied by about thirty.

Consequently, the magnitude of the force exerted by the intercostal muscles is not less than 32040 pounds.

The magnitudes of the forces exerted by other muscles in man or in animals can be found by the same method. But this is enough for our subject.

Chapter XVIII

On the stance of the animals.

So far we discussed at length the enormous motive force which Nature exerts in different muscles when performing the simplest movements to raise the limbs themselves and other heavy weights which they carry. The operations of the limbs and of all the animal, i.e. the different movements of the limbs and their results must be considered now in order to understand the mechanism of these operations. Firstly, we shall discuss the stance, i.e. the action of standing. This discussion not only is useful and pleasant to know for itself, but must be dealt with because walking occurs in standing and cannot be understood without understanding the stance.

Proposition CXXIX

The normal position of the joints is not extension but slight flexion.

We know from personal experience that, when lying in bed, we rest more pleasantly when the joints are slightly flexed. This seems convenient for the following reason. Extreme positions of the joints, either extension or flexion, require violent stretching, compression of the tendons, membranes and muscles which by themselves are unpleasant and painful. Therefore, a middle position between full extension and full flexion is normal and comfortable. But it must be shown that slight flexion of the joints is normal. The limbs are made to extend and to flex. In full extension, the limbs are aligned. In full flexion, they are bent as much as possible. Mid-position lies between these extremes. Therefore, intermediate flexion, between alignment of the limbs and complete bending, is the normal position of the joints.

Proposition CXXX

The flexor muscles of a limb are shorter than the extensor muscles, but both contract equally.
Tab. X, Fig. 7.

From what was said in the preceding proposition, it is easy to detect the error of those who say that the flexor muscles of a limb are longer than their antagonistic muscles, the extensors. They call length of a muscle not only the muscular body but also the tendons which are attached to bones. They pretend that flexion of a limb is much larger than its extension. They also say that the flexor muscles are longer and contract more than the length of the limb. It will be shown that this concept is erroneous.

AB is the humerus, FO the ulna and DC the radius. They articulate at the elbow B. ED is a flexor muscle of the forearm inserted in the medial aspect D of the radius DC and in the proximal extremity E of the humerus. FG is an extensor muscle inserted in the olecranon F of the ulna FO and in the proximal extremity G of the humerus. The forearm is extended to H so that the upper arm AB and the forearm FC are aligned. The limb thus is fully extended. The forearm cannot be rotated beyond H towards I without dislocation of the elbow. Therefore, the extremity H of the straight line ABH is the end of extension and the beginning of flexion. Inversely, L is the end of flexion and the beginning of extension. The forearm FO rotates through the arc HCL, a little less than 180°. The same arc is described from H to L when the limb flexes and from L to H when the limb extends. Consequently, the limb flexes exactly as much as it extends.

Secondly, in full extension the border D of the articular cavity of the radius is at its greatest distance from E but does not go over the convexity of the humeral epiphysis B. Therefore, the maximum length of the flexor muscle ED is equal to the length of the humerus AB. In full flexion, when the coronoid process F of the ulna FO is close to the humerus AB, the olecranon F must pass over the epiphysis B towards D. Therefore, the maximum length of the extensor muscle FG is equal to the length of the humerus GB and the half circle BFD circumscribing the joint. Consequently, the extensor muscle GFD, rather than being shorter, is longer than the flexor muscle DE by the semicircular distance BFD.

Thirdly, during the rotation of the ulna by about a half circle HCL, the borders D and F of its articular cavity remain at equal distance from the centre of the humeral epiphysis and are forced to move along equal half circles during extension as well as during flexion. Then the muscles ED and FG shorten exactly as much as their extremities D and F come closer to their origins during the rotation. Therefore, the muscles ED and GF shorten equally. Consequently, it is erroneous to claim that the flexor muscle DE contracts more than the extensor muscle GF.

Proposition CXXXI

Maintaining a limb in extension does not result from the tone of the antagonistic muscles.
Tab. X, Fig. 7.

The joint ABC is in normal midflexion. The angle ABC is a little more than 90°. Then neither the flexor muscle ED nor the extensor muscle GF act since they are in their normal degree of extension. If the limb is extended and the ulna is moved to FH, it is obvious that the muscle ED is elongated more than normal by stretching of its small machines and inversely the muscle GF is loosened since the distance between F and G has shortened. Therefore, considering the material force of the small machines, it is clear that the muscular chain DE thus stretched spontaneously exerts some effort tending to come back to its initial contraction. The extensor muscle GF is loose and does not exert any force of contraction. But this faculty of the material of the small machines seems so insignificant that it is impeded by any minimal weight or other resistance. Therefore, the animal contracting power of the muscles must be involved. In such instance, I claim that the extensor muscle GF alone must contract, become tight and swell by order of the will without any involvement of the flexor muscle ED. Whatever the contraction of the extensor muscle GF, the limb FC never moves towards I. Otherwise the bones would dislocate. Movement of the ulna towards I is prevented by the bony processes and contraction of the flexor muscle DE is not necessary. The position of the limb thus can be maintained without tonic tension of the antagonistic muscles but by the pull of the extensor muscle only.

Proposition CXXXII

A hard weighing body rests on the ground if the line of support, i.e. the straight line joining the centre of gravity of the body and its support on the ground, is vertical. Otherwise the body tips towards the side to which the line of support is inclined.
Tab. X, Fig. 8.

A weighing hard body ABC of any shape is considered. This body rests on a horizontal plane RS at V. A straight line GV is drawn from its centre of gravity G to the support V. I claim that, if GV is vertical, the body ABC remains standing in the same position but, if the angle GVR is acute, the body necessarily must tip towards R.

The centre of gravity is the point such that if a weighing body is suspended at this point it remains immobile. The point of suspension must be on the line along which force and movement of gravity are exerted. This line is vertical. Therefore, the body ABC suspended at its centre of gravity G and exerting its weight in the direction of the rope DG at right angles to RS, remains immobile in this position. The weighing body ABC is supported by the ground V along the vertical direction GV as if it was suspended by a rope GD. Consequently, the weighing body ABC supported at V remains immobile. But, if the angle GVR is acute, ABC must tip towards R. GX is drawn at right angles to RS and the point X is between V and R. The body AB is supported by the oblique lever GC about the centre of rotation or the fulcrum V and its weight is exerted along GX which is vertical. The body AB when supported by the inclined lever GC is in the same position as if it was on the plane GH perpendicular to GC. This plane GH is inclined to the horizontal plane RS. Therefore, the body AB must roll downwards along this plane GH. Actually it must fall with the lever GC, towards X.

This is true when a weighing body is in contact with the ground at a single point V. If the support is larger, such as VT, then the body can remain immobile in different positions as long as the vertical drawn from the centre of gravity G falls inside the support VT. The body will tip as soon as the perpendicular CX falls outside the support VT.

Proposition CXXXIII

Different ways of preventing the fall of a weighing body standing on the ground when the line of support is not vertical.
Tab. X, Figs. 9, 10, 11.

The line of support GV of the weighing body ABC (Fig. 9, Tab. X) is drawn from the centre of gravity G to the contact with the ground V and forms with the horizontal RS an acute angle GVR. Can anything be done to prevent the body from falling?

Firstly, another body D is added at C in the obtuse angle GVS so that the common centre of gravity O of the two bodies ABC and D is on the straight line GD and divides this line in inverse proportion to the weights of the two bodies. A straight line OV is drawn from O. If this line OV is perpendicular to RS, the bodies ABC and D are undoubtedly at rest in their actual position. OV is the line of gravity of the two bodies since the two bodies tend to fall along this vertical line.

Secondly, fall can be prevented without adding a new weight by prolonging the part BC (Fig. 10, Tab. X), like a human arm extended laterally, in such a way that the centre of gravity of the part BC is displaced to D, further away from point G than it was initially. As a consequence, the common centre of gravity of the elongated figure ABC is O, on the vertical OV or line of gravity, like in the first instance.

Thirdly, without altering the shape of the body ABC (Fig. 11, Tab. X), fall can be prevented by addition of a strut XZ which forms a triangle GVX with the line of support GV so that the vertical GT drawn from the centre of gravity G intersects the basis XV at any point T between the points X and V.

Proposition CXXXIV

The centre of gravity of the human body, straight, lies between the buttocks and the pelvis.
Tab. X, Fig. 12.

A naked man lay straight on a flat table AB the middle of which I set on the edge of a prismatic support. The centre of gravity of the table was on the vertical through this edge.The table was in equilibrium when point C between the buttocks and the pelvis of the subject lying on the table was just above the edge of the support. Therefore, the centre of gravity of this man lying straight was in this location C.

Proposition CXXXV

Positions and actions during which men can stand erect.

The human machine is made of several bony columns linked by round-shaped, smooth and slippery joints. One bony column cannot be set and maintained vertically upon another without involving joints, tendons and muscles. Because of the round shape of the bases of the bony columns the support occurs at points rather than across flat areas, contrary to columns made of squared stones which stand by stable contact of their flat bases.

However, the bones of the limbs of the animal obviously are used as columns and, moreover, as levers. When erect, they carry the weight of the whole animal by their strength, as do columns, without needing muscles. But when rods are inclined and compressed by their own weight and the weight of additional loads they cannot remain in such position without an animal force preventing their tipping and falling by pulling the muscular ropes to which they are attached.

Men can stand erect, their whole body remaining vertical. Then the bony columns constituted by the tibiae, femora and spine stand above each other. But the line of support must fall vertically from the centre of gravity of the whole body to between the plantar soles or through one sole. Otherwise, the body could not stand and would fall to the side towards which the line joining its centre of gravity and the support is inclined. In the erect position the bones behave like columns: their strength ensures support and there should be neither fatigue nor effort. But this erect position is unstable as a result of the slipperiness of the joints. This explains why standing men continuously oscillate. They need their muscles to correct displacements and prevent falling.

Proposition CXXXVI

Man is not maintained erect by the tone of the antagonistic muscles of all the limbs.

We must find with which muscles men control their oscillations when standing erect.

First of all we must reject the error of those who think that the erect position of man is maintained by the simultaneous action of all the extensor and flexor muscles. Oscillation and fall can occur forwards, backwards or to the sides. Falling forwards results from anterior flexion of the lower legs about the feet and anterior flexion of the spine towards the knees. The femur, however, cannot flex anteriorly about the knee without dislocation. Anterior flexion of the leg can be prevented only by the extensor muscles of the leg, the soleus and the glutei. The flexor muscles of the leg are not involved. They would rather enhance the fall. Therefore, they remain inactive. Consequently, flexion and fall forwards are impeded by the extensor muscles alone and not by the simultaneous tone of both extensor and flexor muscles acting together.

Fall backwards results from too much plantar flexion of the ankle, flexion of the femur about the knee and backwards flexion of the spine. Flexion of the knee can be controlled by the extensor muscles of the knee alone and not by the flexor muscles. Exaggerated plantar flexion of the ankle and extension of the hip backwards are prevented by the dorsiflexor muscles of the ankle and the flexor muscles of the hip.

Similarly, sidewards oscillation can be checked by the muscles on the opposite side. Consequently, standing erect does not involve the tonic pull of all the antagonistic muscles but only extensor and some flexor muscles whereas some flexor muscles of the knees remain inactive. Muscular tone may be present at ankle and hip level but not at knee level.

Proposition CXXXVII

Men cannot stand only on one heel or the ball of a foot. They stand with difficulty on one plantar sole alone but easily on the two feet.

When the foot is dorsiflexed, the contact with the ground is the heel. Since the heel is round-shaped it must contact the ground almost at one point only. The line of support joining the centre of gravity of the human body and the support on the ground passes across this point. For the man to stand erect this line must be vertical. It seems impossible to maintain this line vertical firmly when standing on one foot because the human machine with its solid and fluid components can never remain absolutely immobile. Breathing, the flow of humours and a thousand external factors shake it permanently. Man thus is in perpetual oscillation when standing on the ball of one foot or on a stone or on a wooden wedge.

Standing on the whole plantar sole of one foot is not impossible but fairly difficult. Using the force of his muscles, man can turn, bend or stand up and thus maintain vertical his line of support which oscillates over the large contact with the ground constituted by the plantar sole and can intersect any point of this support. Then the man stands erect.

However, erect position is the easiest on both feet. Then the line of gravity can move inside the large quadrangular area constituted by the two plantar soles and the surface of the ground between them. Even if oscillating, the man thus can stand erect without falling.

Proposition CXXXVIII

Determination of the magnitude of the forces exerted
by each of the feet when man stands erect.
Tab. X, Fig. 13.

The centre of gravity of the human body R is A. The body R is supported by the two oblique columns of the legs BA and CA. The line of gravity is ADH. A segment AG is taken on AC such that the ratio BA/AG is equal to the ratio of the force exerted by the strut BA to that exerted by the strut AC. GI is drawn parallel to the horizontal BC. The lines BA, CA are prolonged and intersect FHE parallel to CB. I claim that the ratio of the weight R to the force exerted by the strut of the leg AB is equal to (DA + AI)/AB; the ratio of the force exerted by the strut AB to the force exerted by the strut AC is equal to AB/AG. The weight R is carried by the struts BA and CA with the same force as if it was suspended by the ropes AE and AF inclined as are BA and CA. The ratio of the forces exerted by the ropes EA/FA or the ratio of the forces exerted by the struts BA/CA thus is equal to BA/AG. Therefore[1], the force exerted by the strut BA is measured by the length of the line BA and the force exerted by the strut AC is measured by the length of the segment AG. The weight R of the whole body is measured by the sum of the lines AD + AI. Consequently, if the weight of the body is known, the magnitude of the force exerted by each leg is known.

Proposition CXXXIX

Oscillation of a man standing on his two feet flat on the ground
is controlled with very little effort.
Tab. X, Fig. 14.

The human body stands on a horizontal plane ST. The thigh and lower leg BD are inclined so that the perpendicular BV drawn from the centre of gravity B arrives at point V beneath the plantar sole. The man stands and is maintained

[1] Prop. 69

in this slightly inclined position by the extensor muscles of the leg. It must be shown that small effort is required to make the direction BD of the leg perpendicular to the ground ST. The bow BDV is compressed by two opposite forces, the weight R and the ground reaction force, which have the same line of action BV. Their moments are equal since they balance each other. The ground reaction force prevents downfall of the lever DV. Things occur as if an angulated balance BDC with equal arms was supported by the fulcrum D. At the end of the horizontal arm C a counterweight X would balance the weight R. R acts at the end of the arm BD in the direction BV perpendicular to the horizontal arm CD. The weight X acts at right angles in compression on the arm DC. The ratio of the magnitude of R to its moment or to the counterweight X is equal to the ratio of CD or DB (= CD) to DV[2].

In the beginning of the oscillation the distance DV is small as compared to BD or DC. Then the moment of the whole human machine R is a small part of the magnitude of its weight. Therefore, the extensor muscles of the leg which replace the weight X must exert little force to bring the lever DB towards the vertical.

Proposition CXL

When the line of gravity of the human body is outside the plantar sole of the one supporting foot or outside the quadrangle delineated by the two supporting feet, no muscle can prevent the body from falling. Tab. X, Fig. 15.

The human body R stands on the ground ST with all the plantar sole BC. The angle ABC formed by the leg and the ground is obtuse so that the perpendicular AV falls outside the plantar sole. I claim that no effort of muscles can prevent the body from falling. The body R can be prevented from falling towards V only by inclining the lever AB towards S or, in other words, by closing the angle ABS. The angle B being decreased and made acute by the muscles of the leg, the foot CB must be brought closer to the leg AB. This occurs by dorsiflexing the foot CB to BD. But the weight of the whole body R acting at A cannot yield to the small weight of the foot CB which is not attached to the ground ST but is only in contact with it. In such an instance, the whole machine RABD is supported by the heel B and the total weight tips from A towards V.

Secondly, if the perpendicular line of gravity AV lies in front of the acute angle ABC beyond the tip C of the foot, falling also follows inescapably. Falling cannot be prevented without the plantar flexor muscles of the foot opening the angle B. This brings the support to the tip of the foot C and thus the line of support AC is still inclined to the subjacent horizontal plane. Consequently, the weight R falls towards the perpendicular through V.

[2] De vi percussionis Prop. 39

Proposition CXLI

What can prevent from falling, in the same position.
Tab. XI, Fig. 1.

From what was said so far, it appears that man standing erect is unstable if his legs are not vertical. However, the line of gravity drawn from the centre of gravity to the ground, can move to and fro as long as it does not leave the one supporting foot or the quadrangle delineated by the two supporting feet. The oscillation then is easily controlled by the flexor or the extensor muscles of the feet and legs.

But when the line of gravity is outside the support delineated by the feet, falling will follow except if the falling body is supported by mechanical aids. We shall explain how this is possible.

When the line of support is oblique the body falls. At the beginning the falling is slow and develops little force. Therefore, something can be done quickly to avoid this inconvenience. As soon as the line of support AB (Fig. 1, Tab. XI) is no longer vertical, the body R exerting its weight at A starts a circular movement about the centre B, at the extremity of the radius AB. As long as this radius is close to the vertical, the movement is slow because of the small declivity from the top of the circle. This is pure mechanics. Therefore, as a result of an animal force, the head or the chest R are quickly flexed by the muscles towards DE opposite to the inclination ABV until the common centre of gravity A is moved from A to E so that EB becomes the line of gravity perpendicular to the horizontal line TS. Falling thus is prevented.

Moreover, at the beginning of the fall, the moment exerted by the weight R of the column is very small because the lever AB is little inclined. The ratio of this moment to the magnitude of the weight R is equal to the ratio of the distance VB to the radius or lever AB[3]. Therefore, a small displacement of the head or of the chest is sufficient to restore balance.

This can also be obtained by extending the leg AH or the arm CD towards the side opposite to the falling. This lengthens the lever of the weight of the arm. This weight, despite its smallness, then exerts a larger moment because of the length of the lever CD and the centre of gravity is moved easily from A to E, which prevents falling.

Often falling is prevented by support on and beating of ambient air. Birds beating air with their right wing turn to the left. Similarly, when starting to fall to the right T, man beats the air quickly with extended arm and opened hand CD through the arc DF. Air resistance repels him towards S. If the energy of the beating is sufficient to move the centre of gravity to E so that the line of support becomes vertical EB, falling is prevented. This is obvious with tightrope walkers who keep balance by using a pole which is moved either to the right or to the left as a lever. When they are close to falling, they quickly and violently beat the air on the side of falling whith their pole and thus regain balance in the erect position.

Such skilful movements result from mechanical practice acquired from childhood. Therefore, they are carried out without our knowing although they are not reflexes.

[3] Prop. 13

Proposition CXLII

How people standing on one foot or on two feet with the body bent can remain erect and bend the body further without falling.
Tab. XI, Fig. 2.

The hips C and knees D are flexed and the ankles E are dorsiflexed. The man stands on his feet. He remains erect as long as the vertical drawn from the centre of gravity of the body intersects the plantar sole of the supporting foot or the area delineated by the two supporting feet. The centre of gravity of the head and trunk is A, that of the thighs is G and that of the lower legs is H. The straight line joining the centres of gravity A and G is divided at I[4] in inverse proportion to the weights. The straight line joining I and H also is divided at L in inverse proportion to the weights. L is the centre of gravity of the whole body thus bent. If the vertical LB drawn from L to the ground ST intersects any point of the foot FE or between the two feet, the body thus bent remains erect without falling. This is always verified, in movement as well as at rest, whatever the degree of bending. The more the lever CR formed by the head and trunk moves forwards, the more the buttocks C move backwards so that the line of gravity LB always remains central and perpendicular across the plantar sole, as said above.

It is wonderful that this law of equilibrium of Nature is observed in such a variety of movements, in running, jumping, dancing. Every time it is transgressed, a fall necessarily ensues.

This is why, when standing with the back to a wall, it is not possible to bend the chest far anteriorly without falling. It is necessary that the line of gravity falls perpendicularly between the feet for man to be able to remain erect.

This is also why, when sitting with the trunk and the legs aligned horizontally it is not possible to stand up from the seat. Then the centre of gravity of the trunk and thighs is far behind the feet. Raising is possible only if the head and thorax are moved forwards considerably, or if the feet are brought backwards or are attached firmly by some ties like by a hook. Then the displacement of the centre of gravity or the action of the muscles enables the knee to be extended, the buttocks and chest to be raised and the trunk to bend forwards.

Proposition CXLIII

Why standing alternately on one foot with the other loaded vertically is less fatiguing than standing on both feet simultaneously.
Tab. XI, Fig. 3.

Experience shows that prolonged standing on both feet B and C in symmetrical position induces fatigue which we use to moderate and reduce by putting the

---

[4] Prop. 3. De aequip. in Archimed. Auctoris

weight of the body on one leg AB held vertical whereas the other leg AC is almost not loaded. Then, changing position, the leg which rested carries all the load. This alternate loading is less fatiguing than carrying the body all the time on the two legs in symmetrical position.

Some authors have solved this problem by saying that, when we stand on both feet, all the muscles of both legs, flexors as well as extensors, are involved by their tone; on the contrary, when we stand on one foot, only the muscles of this side are active and the others are at rest. The work carried out by all the muscles is twice that carried out by half of them. Therefore, we can stand on one leg more easily and with less fatigue than on the two legs.

These distinguished authors, however, overlook the falsity of their assumption. It is true that less work is required to carry a weight of ten pounds with the right hand alone than to carry the same with the right hand and another ten pounds with the left hand. Then the two hands carry a weight of 20 pounds, twice what one hand only carries. But it is not true that the same weight of 20 pounds is supported more easily by one hand than if it was divided into two 10-pound loads supported each by one hand. Similarly, the muscles of one leg supporting all the weight of the human body ought to fatigue more than when this weight is carried by the two legs. Each leg then ought to carry only half the weight.

Indeed we observe that it may be very difficult to carry a whole load whereas transporting this load in several parts is easy. The muscular forces are sufficient to overcome a smaller resistance. The muscular fibres then are less stretched and suffer less. But we notice that the main cause of fatigue and degradation of the animal faculty is continuous action of the muscles. We can support heavier loads more easily by an intermittent action interrupted by pauses. We also incur more fatigue when standing than when walking at leisure. It appears that alternating position by standing now on one foot and then on the other is somewhat similar to walking. The principal cause of this phenomenon is given in chapter XIX. Changing of position, and rest are so useful that, when sitting, we find it inconvenient to lean on both feet and amazingly we relax by crossing the legs for a short period and then crossing them the other way. In so doing each leg alternately supports its own weight and that of the opposite leg. However, this change of position is pleasant.

Proposition CXLIV

How the legs of birds differ from those of men.
Tab. XI, Figs. 3, 4 and 6.

Birds also stand and walk on their two legs but differently from men. This discrepancy results from a difference in the structure of the legs. The number of bones is different, the shape of the bones is different, the distribution and structure of the muscles are different, and the segments of the legs are different.

Firstly, in birds RS (Fig. 4, Tab. XI) the femur BC is shorter than the lower leg CD, contrary to what exists in men. The femur is attached proximally to the iliac bone HS which corresponds to the human pelvis, and distally to the tibia at the knee C. In men, the lower leg is made of two bones of equal length, the tibia and fibula. In birds the tibia CD is alone and very long. Instead of fibula the birds have a short, thin and pointed ossicle. In men the tibia is articulated with the foot made of several bones. These bones form the plantar sole which rests on the ground through its whole length. In birds the lower leg is articulated with a third bone DE, long and round-shaped which is maintained above the ground and is called tarso-metatarsus. Finally, in men the foot ends with the toes which are in contact with the ground as well as the plantar sole. In birds the digits EG are longer and radiate in a sphere from a centre. The bird touches the ground only through them and stands on them. This is as if we used to stand and walk on the tips of our toes only, with raised heels.

Secondly, the shape of the legs appears to be very different. In men the first joint above the ground is the knee M (Fig. 3, Tab. XI) which faces forwards and flexes backwards. In birds the first joint D above the ground (Fig. 4, Tab. XI) faces backwards and flexes forwards. To be accurate, the first joint is the ankle and not the knee since the femur and the tibia are articulated at C (Fig. 4) usually with a patella. This joint corresponds to the human knee M (Fig. 3). In birds it is hidden by several muscles and faces forwards. The next joint D in birds (Fig. 4) is that of the tibia and the tarso-metatarsus. The latter corresponds to the raised os calcis L (Fig. 3) when man stands on tip toes.

Thirdly, the distribution and shape of the muscles are also different. In men all the flexor muscles of the toes originate either from the os calcis B (Fig. 3, Tab. XI) or from the lower leg MB. None is prolonged above the knee M. In birds no flexor muscle of the toes originates either in the foot itself EG (Fig. 6) or in the tarso-metatarsus ED or at its proximal extremity D corresponding to the human os calcis. Some flexor muscles of the toes reach the proximal extremity of the tibia C, or span the knee and reach the femur BC or even originate from the ilium, above the hip at HS. Such is the digastric muscle. A long tendon of this muscle passes in a groove round the convexity of the distal epiphysis of the femur, through a tendinous sheath in the knee, through a cartilaginous sheath round the posterior convexity of the tarso-metatarsus and finally through an aponeurosis pulley. It ends at the extremity of the plantar sole and is subdivided to the tips of the digits which it flexes.

Fourthly, there are four apparent joints in birds' legs. The first is at the upper extremity of the femur B (Fig. 4, Tab. XI). The second is the knee C. The third is the raised ankle D. The fourth is at the bases of the digits E. In men this fourth joint also exists but does not appear because of the support over the whole length of the foot, including the os calcis and the metatarso-phalangeal joints.

It must be noted that the joints B, C, D, E of the birds are orientated alternately except for the distal two. The first and most proximal B flexes anteriorly; the second, the knee C, flexes posteriorly; the third, the ankle D, flexes anteriorly; the fourth E, that of the four digits, also flexes anteriorly.

Fifthly, the position of the proximal joint of the leg in birds must be noted. The body RS of the bird is almost horizontal and must be supported at its mid-length B whereas the trunk of man in erect position must be supported at its lower end (Fig. 3, Tab. XI).

Proposition CXLV

In birds the extensor muscles of the legs are much longer in comparison with their antagonistic flexor muscles than in men.

Almost all birds seek their food and pebbles in the fields with their bills. Therefore, they cannot stand and walk separated from the ground by the length of their legs but must be close to the ground. On the other hand, they cannot begin to fly without a huge jump to avoid breaking their extended wings by beating the ground. Such a jump is possible only by lengthening the legs considerably. The legs of birds thus had to be able to be elongated for flying and also to be shortened for satisfying their hunger close to the ground. This required long bones of the legs which could be flexed considerably. Birds feed often and for long periods of time. Then and when flying or when sleeping, their legs must be flexed and shortened. Consequently, the muscles had to be arranged to maintain the legs flexed rather than to extend them. The normal position of the legs in birds thus is much more flexed than in men, as appears in Fig. 4, Tab. XI, where the bones form acute angles. The normal structure of the muscles had to conform to the normal arrangement of the bones. Therefore, the extensor muscles which circumscribe the convexity of the joints[5] had to be longer than the flexor muscles which describe a shorter trajectory in the concavity of the joints. The necessity of such an inequality can be understood from the following fact: among several parallel curves having their extremities at the same level, the ones which are most inside, in the concavity, are shorter than those which are outside, at the convexity.

Proposition CXLVI

Analysis of the birds' stance.

We observe that the legs of birds are straight neither in stance nor in walking. Their joints do not extend completely to constitute straight lines perpendicular to the ground like columns, as occurs in men. The bones of the legs are always folded at acute angles to each other and support the centre of gravity of the bird inclined forwards. Since the bird thus supported does not fall, it must be deduced that the line of gravity drawn vertically from its centre of gravity falls in the area occupied by one foot or delineated by the two feet.

[5] Prop. 130

Moreover, as a result of the flexion of the four joints, none of the joints is maintained in flexion by the tone of the muscles, i.e. by antagonistic muscles acting with equal force. Necessarily, the extensor muscles alone ensure the support and the flexor muscles remain inactive. This can be deduced from what was said.

Proposition CXLVII

Why birds stand more easily on one foot than men.
Tab. XI, Figs. 3 and 4.

The reason why men stand uneasily on one foot results from the difficulty of maintaining their line of gravity AB (Fig. 3, Tab. XI) in the narrow area occupied by one foot, despite oscillation. In birds the exit of the line of gravity AE (Fig. 4, Tab. XI) from the area occupied by the long digits EG of one foot can be prevented much more easily because the oscillations are controlled more easily. Firstly, a long horizontal pole suspended at its midpoint is more easily balanced than a pole supported vertically on the palm of a hand. Similarly, the body of a bird in prone position and supported at mid-length is more easily balanced than the human body in erect position. Secondly, the line of gravity AE is very short in birds and very long in men. Therefore, the oscillations occur at the extremities of radii of inequal lengths. They are thus more easily controlled in birds oscillating at the extremity of a short radius than in men. Thirdly, the body of a bird is light whereas that of man is heavy. The oscillations of the light and rare body of a bird are more easily controlled than those of man. Fourthly, the human toes are short and parallel to the axis of the foot whereas the digits of a bird are long, articulated and extend like the radii of a sphere. Therefore, birds can prevent oscillations, either forwards or backwards, or even sideways, more easily because their separate digits resting on the ground act as struts to support the falling body. For all these reasons, birds can stand on one foot more easily than men.

Proposition CXLVIII

How a rope can be pulled by flexion of a rod.
Tab. XI, Fig. 5.

Two rigid rods AB and BC are linked by a joint B. A rope DEF is attached to the extremity D of the rod AB, passes over a pulley F and is pulled by a suspended weight R. When the two rods are aligned the length of the rope DF obviously is equal to that of the two rods measured by their axes AB, BC since the opposite sides AD, FC of a parallelogram are equal. The rods are flexed about the joint B so that BC is brought into the position BC, and the rods form an angle ABC. Then the length of the rope DEH circumscribing the convexity must be greater than that of the circumscribed rods measured by their axes or the initial length of the rope DEF. The circumscribing rope DEH thus is longer than DEF. But the rope was supposed to be tensed and pulled by the weight R. Therefore, flexion of the rods raises the weight R. The more the angle ABC is closed, the more the weight is elevated. If the two rods are flexed completely and touch each other, elevation of the weight R is equal to half the circumference of the joint B.

Proposition CXLIX

Mechanically, flexion of the legs of birds must result in flexion of the digits. Tab. XI, Fig. 6.

The proximal end of the femur CD is articulated in the acetabulum C in the extremity AB of the iliac bone of the bird. The femur is articulated with the tibia DL at the knee ID and the tibia is articulated with the tarso-metatarsus LF at the ankle joint E which is raised. The digits FG are attached at F.

The principal digastric flexor digitorum muscle originates from the extremity of the iliac bone at HC. The strong and round tendon is attached to the patella I in some birds, but in the swan, the eagle and the falcon, it enters a sheath in a groove dug in the knee I, passes through a loop K, crosses the tibia laterally and arrives posteriorly at the distal extremity of the tibia. The other flexor muscles of the digits originate from the distal extremity D of the femur and the proximal extremity of the tibia. Their tendons join at M with that of the digastric muscle and, united, arrive at E. They pass the ankle in a smooth and slippery cartilaginous sheath to the heel F where they enter another sheath or aponeurosis canal. Finally they subdivide and arrive at the bases G, G, G of the talons through small tendinous canals close to the ground.

I observed that, when the joints I and L are extended and the three bones CD, DL, LF are aligned or nearly aligned, the tendons HIKEF are loose and thus the digits FG remain open and extended. But if the iliac bone AB is compressed and the joints are flexed completely so that the upper aspect IC of the femur CD touches the iliac bone AC, the upper aspect of the tibia DE touches the lower aspect of the femur DC and the upper aspect of the tarso-metatarsus EF touches the lower aspect of the tibia IL (as shown in Tab. XI, Fig. 7), then the joints of the foot FG are in full flexion. The digits are closed with such force that I could not drive a wooden wedge between the digits of a dead eagle nor between those of a dead falcon. Such tightness obviously results from the mechanism described above[6], i.e. after flexion of the joints, the tendon HIKEF is shorter than the circumscribed bones.

Since the tendons are strong and inextensible, their extremities FG pulled through F towards E must necessarily compensate their stretching over the convexity of the flexed joints. Therefore, the digits FG must flex and be tight against each other. Our problem thus can be solved easily.

[6] Prop. 148

Proposition CL

Why the birds rest and sleep standing on and grasping tree branches, without falling.
Tab. XI, Fig. 7.

During sleep, the motive faculty of the muscles is inactive and when the muscles of the legs do not act, the joints cannot be set or maintained at a given degree of flexion. It thus seems impossible that sleeping birds RS do not fall as would do terrestrial animals and that they can sleep, hiding their head O beneath a wing, standing on thin twigs in trees GF, without falling even despite blowing wind. We often notice this firm stance of birds when they are sleeping. The cause of this surprising operation is worth analysing. The bird RS is supported by the resistance of the thin twig GF which it touches with its breast bone. The bird thus has three points of support on the twig: its breast bone and its two legs. However, the principal support of the birds' weight is the breast bone. The two legs are used as struts preventing the bird from falling to the right or to the left. But the most difficult question must still be answered: why does the bird not fall forwards towards R or backwards towards S since the support is not a wide area of flat ground but a single transverse line GF at the end of a twig? The bird can be retained firmly only by its digits grasping the twig GF tightly and strongly like pliers and linking the bird and the twig rigidly. The adherence must be firm enough to resist the oscillations of the bird itself and gusts of wind. The grasping of the digits E must result from some motive force which violently contracts or pulls the flexor muscles of the digits. But at rest, during sleep, and even more so after death, the muscles do not contract. Therefore, tension of the tendons must result from some other cause which is inherent to the structure itself of the muscles.

Actually, the bird RS, when awake, arrives on a twig of the tree GF with extended legs and digits. Then the joints of the legs BCD are flexed until the breast bone is supported by the twig. The shortness of the tendons of the extensor muscles HILE of the legs forces the digits E to flex with considerable strength[7] so as to grasp the twig of the tree FG like pliers. This thus does not result from the motive force exerted by the muscles but simply from the weight of the bird which when lowering its body must fold all the segments of its legs. Therefore, without voluntary movement of the muscles and without effort, the bird strongly grasps the twig by which it is supported, even at rest or during sleep. This grasp prevents oscillation and falling. The bird thus remains at rest without any effort.

[7] Prop. 150

Proposition CLI

Quadrupeds can stand neither on one foot nor even on two feet.
Tab. XI, Fig. 8.

Quadrupeds differ from men. Firstly, their body is prone and they walk in this position whereas men are erect. Secondly, they have thin feet whereas those of man are broader and longer. Quadrupeds also differ from birds not only by the width of the plantar soles but also by the fact that the centre of gravity of birds is above their feet whereas the centre of gravity of the quadrupeds is not at the vertical of any of their feet. Hence, quadrupeds cannot stand on their two forelegs AB only. Their long body cannot be erected sufficiently to bring the centre and the line of gravity GH between the two forefeet AB. If they stand up violently, because of the small dimension of their feet, the line of gravity GH is easily beyond the plantar soles AB on one side or the other and they must fall. For the same reason, quadrupeds cannot remain standing on their hind legs CD alone for any period of time. However, if thus erect, they rest the tarsus on the ground, as men do, then they can stand for some time. We know that bears and dogs sometimes do that.

Their body being prone, they cannot stand on the two left legs AD. Because of the narrowness of the feet, the line of gravity GH soon leaves the line AD joining the two left feet or that joining the two right feet, as a result of oscillation, and the animal falls.

If support occurs on two legs AC diagonally opposite, the line of gravity GH also leaves the line AC joining the supporting feet as a result of oscillation and the animal falls.

If quadrupeds are unable to stand on two legs with their body prone, even less can they stand on one.

Proposition CLII

Quadrupeds with their body prone can stand only on four or three legs.
Tab. XI, Fig. 8.

The centre of gravity G of a quadruped is at its mid-length, in the abdomen between the fore- and hind legs. In prone position, the vertical line of gravity GH necessarily falls into the quadrangle ABCD comprised by the four feet. The body of a quadruped is conveniently supported by its four legs which are perpendicular to its length and act as columns. Oscillation cannot be so large as to move the gravity line GH outside the quadrangle ABCD delineated by the feet. The animal thus stands without falling.

The same occurs when one leg B is raised from the ground. The three other feet ACD then delineate a triangular area in which the line of gravity GH falls. The body of the animal thus stands on three columns.

Proposition CLIII

Force exerted by each leg of the quadrupeds when standing.
Tab. XI, Figs. 8, 9 and 10.

Firstly, a quadruped stands on three legs ADC (Fig. 8, Tab. XI). The vertical line of gravity Gh falls at h. The body R of the animal is supported by three struts ADC as if the mass R was suspended at h, the extremity A of the lever AE was supported by a force A and the extremity E was supported by the two forces D and C acting on the lever DC. Then the weight R (Fig. 10, Tab. XI) is divided into two parts X and Z such that X/Z = Ah/hE (Fig. 8). The weight X (Fig. 10) is divided into two parts M, O such that M/O = DE/EC (Fig. 8). The weight Z (Fig. 10) is carried by force A (Fig. 8), the weight M (Fig. 10) by force D (Fig. 8), and the weight O (Fig. 10) by force C (Fig. 8).

Secondly, the weight R (Fig. 8) is supported by the four legs ABCD and the centre of gravity is at the vertical of H. The lines AB and CD are linked. FE is divided at H in inverse proportion to the forces exerted by the feet. The weight R (Fig. 9, Tab. XI) is divided into parts X and Z such that X/Z = FH/HE (Fig. 8). Z (Fig. 9) is divided into parts M and N such that M/N = AF/FB (Fig. 8). Similarly, X (Fig. 9) is divided into parts O and P such that O/P = DE/EC (Fig. 8). The weight M (Fig. 9) is supported by the leg A (Fig. 8), N (Fig. 9) by the leg B (Fig. 8), O (Fig. 9) by the leg D (Fig. 8) and P (Fig. 9) by the leg C (Fig. 8). However, when the legs are inclined to the horizontal the magnitudes of the forces thus exerted are greater than their moments in a ratio equal to that of the inclined length of the leg to its height. The proportions mentioned above are modified accordingly.

Moreover, most of the force exerted by the legs to carry the body of the animal is due to the strength of the bones of the legs rather than to the action of the muscles. The effort thus is not hard.

The forelegs of quadrupeds are provided with fewer and weaker muscles. This may be due to the fact that these animals use to stand on their forelegs extended and perpendicular to the ground. The forelegs thus can carry the weight of the animal by their own strength, like columns. On the contrary, the hind legs are never extended but always flexed while the animal stands. They thus must be maintained by muscles.

Proposition CLIV

Animals provided with six legs or more require more effort
to stand than quadrupeds.

Animals with multiple legs must stand according to the same mechanical laws as quadrupeds do but this requires more effort. The legs of insects are never straight like columns but are always flexed at acute angles when they stand on the ground. Therefore, they must be supported by the force of muscles. This, however, is hardly inconvenient for them. Because their body is so small their weight is decreased proportionally to the square root of their volume, as Galileo elegantly demonstrated. Therefore, elephants keep their hind legs straight and perpendicular to the ground like columns.

Chapter XIX

On the walking of bipeds.

Aristotle and some modern authors wrote on the walking of animals. However, if I am right, they did not describe properly the mechanism and the cause of such movement, as will appear in the following.

Proposition CLV

When man walks not all his mass is raised from the ground
by the force of muscles but only part of it, less than a quarter.
Tab. XI, Fig. 11.

Jumping differs from walking in that, when jumping, all the body loses contact with the ground, the two legs are raised at the same time and all the machine is pushed upwards and forwards like a missile. During walking the human body is always in contact with the ground, supported alternately by one leg and the other. During this alternate support it seems that each half of the body weight is in turn raised and moved. It is as if a lying column AB was resting on the ground with its extremity B and its extremity A could be elevated and moved over the arc AC about the centre B. In such an instance, the force raising the extremity A of the column is equal to half the weight of the column. The other half is supported by the ground. Then the column being in BC is moved about the centre C. Its extremity B is raised and moved over the arc BD towards CD by a force equal to half the weight of the column, and so on. But wise Nature seeks the greatest economy during walking. Less than a quarter of the weight of the human mass is raised from the ground alternately by the force of muscles.

The human body R (Fig. 13, Tab. X) stands erect on the two legs BC. The line of gravity AD drawn from the centre of gravity A falls between the two feet at D. The triangle ABC is isoceles. Distance DC is obviously less than a quarter of the length of the leg BA or AC. The weight of the man R is supported by a double force, the strength of the bones of the columns or struts AB, AC and the force of the muscles which maintain the columns extended and prevent the bones to give way and flex about the joints. The force exerted by the strength of the columns AC, AB is measured by the perpendicular AD. The force exerted by the muscles of one leg is equal to the moment of half the weight R supported by the pole or lever AC mobile about the centre C. This moment of half the weight R is measured by the line DC. Therefore, the moment of the motive force of the muscles which prevents flexion of the legs AC, AB is less than a quarter of the weight R when the man stands.

During human walking, the body R can proceed towards K only if the isoceles triangle ABC is transformed into a right-angled triangle and then into a triangle with an obtuse angle so that the angle ABC becomes at first a right angle and then an obtuse angle. This can occur only if the length of the side AC is increased and that of the side AB is shortened. This operation is easy by plantar flexing C. Then the ball of the foot touches the ground and the strut AC is lengthened. Simultaneously, the opposite knee is flexed and the ankle B dorsiflexed, shortening the pole AB. As a result the body R moves forwards until the line of gravity AD coincides with AB, i.e. until the line of support AB has become perpendicular to the ground. Then the force which lengthens the leg AC while the opposite leg AB rests on the ground, is less than a quarter of the weight R. When the body is erect in the vertical position RAB, all the weight is carried by the strength of the bony column AB. Then the pole AC is useless and does not support the body R. The foot C thus can be raised easily from the ground without danger of the man falling. Since the whole leg AC is less than a quarter of the body, it must be suspended by the force exerted by the flexor muscles of the leg which is less than a quarter of the weight of the man. Then the same foot is moved forwards simply by falling and loosening of the muscles. The same process repeats itself alternately with each leg. Consequently, less than a quarter of the weight R must be carried alternately by the force of the muscles during human walking.

Proposition CLVI

The way of moving the mass of the human body forwards during walking varies.
Tab. XI, Fig. 12.

Man would not move forwards if he raised the feet from the ground alternately and put them back in the same place where they were initially. To progress horizontally it is necessary to move the mass of the body forwards. It must thus be sought by which organs and which operations this occurs. At first glance, human walking appears similar to the movement of a pair of compasses BAC held erect over a horizontal plane. In standing they form an isoceles triangle ABC. Their line of gravity AD falls exactly between the two legs B and C. After raising the leg C from the ground until the line of gravity AD coincides with the line of support AB and is perpendicular, by a rotation about the axis AB, the leg AC describes a conical surface ACE. Then, with the leg AC taking support at E, the pair of compasses is raised again until the leg AE is perpendicular and the other leg AB rotates and describes the arc BF, and so on. The pair of compasses thus is moved by alternate raisings and rotations. Actually, progressing in such a deformed way is inconvenient. But Nature displaces the human body more easily and elegantly. During stance, the legs resting on the ground certainly form an isoceles triangle ABC. Then Nature carries out several circular movements simultaneously which result in forwards displacement. The anterior leg BA rotates about the centre B in a vertical plane (Fig. 13, Tab. X). At the same time the body R moves forwards towards K. This displacement occurs as follows. The foot LC is plantar flexed by the soleus muscle and the leg forms an obtuse angle ALC with the foot. The tip of the foot touches the ground at C. The length of the leg thus is increased by the length of the foot CL. The isoceles triangle is transformed and becomes at first a right-angled triangle when the leg AB is vertical. During this operation, the body R is supported by the two feet and can be easily inclined somewhat so that the leg AB becomes vertical. Moreover, as a result of the plantar flexion of the foot and the lengthening of the leg AC, the ground is pushed by the ball C of the foot and, by reaction, the body R is moved forwards towards K. Similarly, a boat pushed by a mariner with a pole leaves the waterside. This propulsion is admirably facilitated by inclining the head and the chest slightly forwards towards K. Hence, the centre of gravity of the body and the line of gravity are brought a little beyond the foot BO resulting in a tendency to fall. The weight of the body R thus is displaced forwards spontaneously and then, falling is immediately prevented by raising the foot LC and displacing it quickly to K beyond the line of gravity. Double support then occurs again with more stability. By this mechanism the human body moves forwards.

Proposition CLVII

Man cannot walk exactly along a straight line.

The human body can move forwards only by alternate supports of the feet on the ground. This support occurs by shifting the line of gravity successively to the right and to the left foot. The feet do not move along one straight line but along two parallel lines. This is obvious. The tracks of the feet of a man walking in dust or in mud are parallel but do not constitute one straight line. Moreover, geese and heavy-set men with short legs not only move their feet along two parallel lines but also go forwards with oscillations, standing successively on the right and on the left foot and displacing their centre of gravity accordingly. All other men actually do the same but it is less obvious. Two vertical stakes, one white and one black, are set in a flat surface, separated by some distance. A man stands in the alignment of the two stakes, closer to the black one which hides the white which is further away. Whatever his effort to walk exactly straight, the subject never succeeds and sees the white stake alternately to the right and to the left side of the black one. This demonstrates that man cannot walk along a straight line. His line of gravity describes a serpentine path between two parallel lines and thus the human body cannot proceed along one simple straight line.

Proposition CLVIII

Description of all the movements occurring during human walking.

During walking the human body is always supported by the ground. It is carried by the strength of the legs acting as bony columns at the price of little effort of the muscles and little inconvenience of the sensation due to compression of tendons and stretching of membranes. Moreover, during double support, forwards displacement of the centre of gravity results from lengthening of the posterior leg by plantar flexion and backwards impulse on the ground. The entire body is then above the anterior leg and pushed a little in front of it. It moves forwards horizontally. Then the lengthened posterior leg takes off the ground with flexion of the hip and knee and dorsiflexion of the ankle. This movement is carried out by the muscles of the swing leg, which raise less than a quarter of the body weight. The leg is moved forwards beyond the anterior foot by the impetus resulting from the impulse of push-off and by the forwards flexion of the head and trunk. Then another double support occurs and the leg which has become posterior acts in turn. And so walking goes on.

Proposition CLIX

Walking on a horizontal surface is usually easy, hardly fatiguing and sometimes pleasant.

On a horizontal surface there are neither bumps nor holes, neither acclivities nor declivities. Therefore, during walking, the legs when swinging forwards can be flexed just enough for the foot not to impinge the ground. Hence the flexor muscles make little effort. Moreover, the feet are not forced into holes or on uneven ground. They are thus not twisted violently. The bones of the feet are not subjected to excessive stretching and twisting since they do not walk on acute stones or in holes. Therefore, they are not hurt. Consequently, walking on a flat surface requires little effort and is painless. On the other hand, movements of the limbs, neither too hard nor unpleasant, wonderfully enhance blood circulation, insensible perspiration and metabolism. That is why such walking, if not exaggerated, usually induces a pleasant feeling.

Proposition CLX

Why walking up a slope is fatiguing and unpleasant.

Experience shows that going up stairs is very fatiguing and unpleasant. It is not difficult to find the cause of this inconvenience if the action of the limbs when going up is considered closely. When standing with the two feet on the lowest stair, firstly the right foot, for example, must be raised almost vertically, higher than the height of the next stair. Such elevation is not possible without considerable flexion of the joints of the leg. This loosens the flexor muscles[1]. Their voluntary contraction thus requires an effort. Although the weight of the leg is less than a quarter of that of the whole body, because of the looseness of the muscles, the animal force must overcome a weight much heavier than this quarter and altogether compensate for the unfavourable arrangement of the levers.

Secondly, the centre of gravity of the human body is moved forwards by the impulse of the right foot on the first stair before it leaves the latter, and by anterior inclination of the head and chest. Therefore, when the right foot takes support on the second stair, the line of gravity is across the foot and the second stair. Thirdly, the left leg is raised by shortening and flexion of its segments. Therefore, its resistance which is equal to a quarter of the body weight must be overcome again. Fourthly, the right leg which is flexed must be made perpendicular to the foot which is supported by the second stair. This requires full extension of the hip, knee and ankle (plantar flexion). This operation is laborious because the whole body weight must be elevated vertically and the mechanical difficulties described elsewhere thus must be overcome. Consequently, to climb one stair, the muscular force must raise a quarter of the body weight twice and altogether the whole body weight. This is the more fatiguing and the more unpleasant, the quicker the stairs are ascended.

It must be pointed out that, during walking over a flat horizontal surface, the whole mass of the human body has never to be raised by the force of the muscles since the body is always supported by one leg or the other. When climbing stairs the weight of the body is raised by the force of the muscles and not by the strength of bony columns. This implies hard work and fatigue.

Going up a slope is fatiguing for the same reason. Moreover, the feet are dorsiflexed. In this abnormal position the plantor flexor muscles are stretched more than normally and support is not stable. Therefore, it is hard and fatiguing.

Proposition CLXI

Going down a slope is somewhat more fatiguing
than walking over a horizontal plane.

When going down stairs or a slope, neither the legs nor the whole human body must be raised and pushed forwards. On the contrary, the body goes down by its own weight. Therefore, people generally think that going down is easier. After closer consideration, however, it appears that going down requires some effort. When going down stairs the right leg, for example, must be raised briefly and moved forwards. Then if the centre of gravity of the body was also pushed forwards, the whole body would fall until the anterior foot would meet the next stair with some damage and pain. This inconvenience is avoided by retaining

[1] Prop. 11

the centre of gravity in its initial position. The line of gravity thus is maintained perpendicular to the foot on the upper stair. Then flexion of the leg on the upper stair progressively brings down the mass of the body. The column constituted by the right leg is shortened and the opposite leg is extended and moved forwards until the body is supported by both feet resting on adjacent stairs. Then the line of gravity is moved towards the foot resting on the lower stair and which plays the role of a column. The left leg leaves the upper stair and is moved downwards by the same process.

Lowering the centre of gravity thus is not a fall but the muscles sustain the body and move it progressively downwards. Therefore, without an effort of the extensor muscles which give way little by little, this lowering of the centre of gravity is impossible. Such effort is the more obvious the more slowly we go down the stairs. Walking over a flat ground does not require such support by muscles since the support is provided by the strength of the bony columns alternately. Consequently, going down is more fatiguing than walking over a flat ground.

Going down a slope, the feet are plantar flexed. This abnormal position is also fatiguing.

Proposition CLXII

Why the foot undergoes great concussion when, going up or down stairs in the dark, we think there is one more stair.

When going up stairs, if we think that there remains one stair which we must go up, the centre of gravity of the body is maintained above the supporting foot while the opposite foot is raised to take support on the stair which we think is next. To this end, the centre of gravity, and the line of gravity, are moved forwards beyond the supporting foot. This results in an unforeseen fall. Consequently, the raised foot strikes the ground with great concussion of the whole body and the leg is hurt by the shock.

When arriving downstairs, if we think there is one stair left to go down, the leg supported by the ground starts shortening by flexion of its segments and the opposite leg extends to be able to reach the next imagined stair. Because of these two movements the foot strikes the ground unexpectedly and violently. The pain is similar to that which we usually perceive when striking the ground with our heels only.

Proposition CLXIII

Walking of birds differs somewhat from human walking.

Birds walk by the same mechanical operations as men do. They are different as far as the application of the motive force is concerned. Man is supported by the strength of the bones of his legs like by alternate columns, whereas birds never stand on extended legs. Their legs are always flexed. Therefore, the body of birds is always maintained erect by the muscles. The muscles of the swing leg rest and work less than those of the supporting leg. When the bird stands on its two legs, the extensor muscles of each leg support half the weight of the bird. When a leg is raised its extensor muscles do not work and only the flexor muscles raise the weight of the swing leg. This weight is less than a sixth of the weight of the bird.

Proposition CLXIV

How men walk on ice.
Tab. XII, Fig. 1.

Steel skates are attached to the feet AB, CD. The subject stands erect. His centre of gravity is above the left foot CD. The right foot AB is behind and by a transverse movement pushes on the ice and gives an impulse while scraping to E. As a reaction, the left foot CD moves in the direction of its skate. This displacement is made easy by the smoothness and the slipperiness of ice. Immediately after the impulse is given, the foot AB is raised and the body is balanced on the foot CD which is pushed straight to F by the force of the impetus. Then the right foot AB is moved forwards to G where it supports the centre of gravity of the body. The left foot scrapes back to H and drives the foot G forwards in the direction of its skate before being raised at H. The foot G supporting the body moves to I. Impulses alternate with displacements forwards. This results in a very quick movement on ice, not straight forwards but alternately to the right and to the left following a tortuous trajectory.

The way the skaters stop or change direction seems to be as follows. They stop giving impulses and the impetus ends progressively. Or they bend the leg to the right and the chest to the left for a little while. This forces the skate to scrape the ice sideways and, the centre of gravity being kept backwards, stops the impetus.

The skaters change the direction of their trajectory by pushing with the posterior foot at right angles to the anterior foot and by inclining simultaneously the skate of the supporting foot. They do not slow the displacement of their centre of gravity so as not to impede the movement and not to lose the support of the skate.

Chapter XX

On the walking of quadrupeds.

Particularly on this subject much nonsense has been said not only by ordinary people but even by famous philosophers and anatomists who accepted a false opinion transmitted from generation to generation rather than what they could see with their own eyes.

Proposition CLXV

Quadrupeds do not walk by moving alternately two legs opposed diagonally while the other two are immobile.
Tab. XI, Fig. 8.

According to the prevailing opinion, quadrupeds would walk by displacing forwards two legs alternately while the other two remain standing, like bipeds who swing one leg while the other supports the body. With this prejudice, the ancients observed that the feet of standing quadrupeds delineate a four-sided figure ABCD. They also saw that, when running, their forelegs AB are raised and moved forwards simultaneously (Fig. 7 Tab. XIII), whereas the two hind legs CD stand on the ground EF. Immediately after the forefeet AB are on the ground again, the two hind ones CD are raised and moved close to AB. Running thus results in contraction of the four-sided figure ABCD followed by elongation. This is obvious when horses or dogs run.

During walking neither the two fore- nor the two hind legs are raised and moved forwards together. They do so alternately. When A (Fig. 8 Tab. XI) is moved B stands and, inversely, when B is moved A stands. The same certainly occurs for the hind legs. Actually, it is not easy to distinguish in what order the fore- and hind legs move and whether the two left legs A and D or the two legs A and C move simultaneously. The rapidity of the movement impedes such distinction. The ancients thought they could solve the problem by reasoning. If the two left legs A, D were raised and moved forwards together, the animal would tip to the left. Consequently, the right foreleg B should rather be raised and moved forwards together with the left hind leg D. The diagonally opposed legs would always either be moved or stand simultaneously. From this error of imagination it results that in old and modern brass or marmor equestrial statues two legs diagonally opposite are elevated. The same is seen in paintings.

I wonder that the difficulty and absurdity of such a movement were not perceived. The same authors concede that during movement the animal must be stable and not waver and fall. They thus deny the possibility of moving the two left legs A and D together. The centre of gravity of the huge long mass of the quadruped would have to be just above the straight line BC formed by the support of the two right legs. If it was more towards AD the animal would waver and fall.

If two diagonally opposite legs B, D are raised and moved together, the mass of the animal must also be supported by two legs and its line of gravity cannot fall in a large area but has to be exactly on the line AC. The animal thus will also waver as a result of its unstable position.

Secondly, let us consider the position of the legs after the first movement, when the foot B (Fig. 8 Tab. XI) has been moved to K and D to S. Then the two left feet A and S are close to each other and the right feet KC are very remote from each other. The four feet form a triangular figure with a long basis KC and a short height AB. Therefore, standing would be hardly stable. Subsequent displacement of the feet A and C from this position to I and V would return the animal to a stable position above the quadrangular figure ISVK similar to ABCD. In this way stable and wavering stances would follow each other, imprudently programmed by Nature whereas this inconvenience could be avoided easily.

But we are reasoning whereas experiments are required. If you observe the slow movement of a walking horse you will never see two diagonally opposite feet A and C moved simultaneously. One foot alone is raised from the ground while the other three remain standing. Careful observation of quicker walking of any quadruped shows the same.

Proposition CLXVI

How quadrupeds walk.
Tab. XII, Fig. 2.

A horse can be imagined as a long machine carried by four struts, the legs which stand like columns on the ground where they form a rectangular four-sided figure ABCD. The vertical drawn from the centre of gravity of the horse falls about in the centre E of the four-sided figure. Therefore, the stance of the animal is very stable. Walking starts with one hind foot, for example the left C, which pushes backwards on the ground and thus moves the centre of gravity from E to G. Immediately after that, the foot B is raised and moved forwards to H. This movement is easy since the line of gravity falls at first inside the triangle ABD and afterwards inside the trapezium ABFD. The support thus is either on three or on four columns. Then the other three feet ADF standing firmly circumscribe the centre of gravity G while the left forefoot B is moved forwards to H. The acquired impetus displaces the centre of gravity to I, in the centre of the rhomb AHFD. When the forwards displacement of the two left feet is completed, impulse by and displacement of the right hind foot D begin and then of the right forefoot A. Walking of any quadruped thus always proceeds as described.

Although accurate observation is enough to prove this truth, the philosopher, however, must investigate the utility and necessity of such operation. A perpetual law of Nature consists of acting with the smallest work, the simplest, easiest, most reliable and safest means, avoiding as much as possible inconveniences and redundancies. During walking, animals do not at all move their parts with equal velocity in the same direction as occurs in flying, jumping or crawling. During walking, some parts are displaced which are supported by other parts standing. The animal thus moves forwards while standing. Therefore, stance and displacement which constitute animals' gait must not be wavering but stable. They must be carried out with minimal, although sufficient, work of the muscles. To stand without risk of falling while moving forwards, the animal must be supported by more than two columns, either by three or by four between which the line of gravity falls. We have explained in the present proposition how this occurs in walking.

Secondly, to ensure this support with a minimum of work and inconvenience, it must be carried out by the legs which, as a result of their strength and poor feeling, can easily tolerate weights without notable inconvenience.

Moreover, the animal can be easily moved because its mass is high and must not be raised from the ground. Only one leg has to be raised and moved at a time. Firstly, the ground is pushed behind. If it was pushed by extension of one of the forefeet standing on the ground anteriorly, the centre of gravity of the animal would retrocede. On the contrary, the centre of gravity is pushed forwards by extension of a hind leg like by a pole. As a result, all the mass of the animal is moved forwards on the three flexed erect columns as a chariot moves on its wheels. Secondly, when the hind leg is raised, its joints are flexed by its muscles. Then the forefoot on the same side is moved forwards, as described above. The economy of energy demonstrates the necessity of these operations.

Proposition CLXVII

How quadrupeds attempt to use their two forelegs
like hands to grasp objects.
Tab. XIII, Fig. 11.

Quadrupeds can stand prone only on at least three legs which are used as columns, as said above[1]. But they can raise the fourth leg, either a hind leg or a foreleg by flexing its joints. With this leg they can touch, grasp and pull objects as if it were with a hand. For instance, horses kick or rub their own body. Cats and bears use a foreleg as a hand to grasp or fight with their claws. If their two forelegs are raised at the same time, they cannot easily seize objects except by inclining their chest and with the two hind legs on the ground. Actually, by applying their tarsus bones on the ground, some animals such as dogs, apes and bears, can stand erect like men. Then they can fight and grasp objects with their forelegs but not without inconvenience because the legs and particularly the os calces cannot sustain firmly the weight of the body. The structure of the latter is more appropriate to stand prone on four legs. Hence, dogs standing erect easily fall if they do not lean their back against a wall. But apes and bears

[1] Prop. 156

have feet more similar to human feet. They can stand erect longer with less inconvenience and then can fight with their forelegs.

When dogs stand on their hind feet and horses on their hind hooves, they cannot remain erect and fall easily. However, they can perambulate since their oscillations are corrected by the same movement as that of children who perambulate and run with their legs tied to stilts. But they cannot rest on such struts without falling.

Proposition CLXVIII

Analysis of the gait of six-legged animals.

The long and prone body of insects is supported by six legs and they seem to stand and walk as quadrupeds do. They stand if the vertical drawn from their centre of gravity falls in the space delineated by more than two legs. Then they can raise two or three legs and use them as hands to grasp objects.

Walking is very easy. Firstly, they push on the ground with their hind legs simultaneously or successively to move their centre of gravity forwards. If the forelegs were extended at first, the animal would be moved backwards. Walking thus must start with the hind legs. The other movements must occur as in quadrupeds although their succession may be different since three legs can always remain erect on which the animal stands. If the three legs on one side were moved forwards successively, the movement would start most conveniently with the hind leg, continue with the middle leg and end with the foreleg. In this way three legs would form a parallelogram which is convenient for support and walking of the animal.

However, the gait of six-legged animals seems to be more fatiguing than that of quadrupeds because, in insects, none of the legs is straight like a column. Insects stand and walk always with their joints flexed at acute angles. They thus are not supported by the strength of the bones but by the force of the muscles.

But this inconvenience is easily tolerated by insects because of the small weight of their body in relation to its volume. Their weights decrease proportionally to the square of their size if their shape is the same. This was demonstrated by the famous Galileo.

Proposition CLXIX

How suspended flies and mosquitoes adhere to glass
and walk on it without falling.

However small and rare the body mass of a fly, ant or mosquito and similar animalcules, these animals have some weight exceeding the specific weight of air. Therefore, it is impossible for them to float in air as fishes do in water. They must fall as straw slowly falls through quiet air. Thus, I do not see how these animalcules can adhere to the smooth surface of glass inclined or even horizontal without falling. They are not retained hooked by their claws curved and inserted into the pores of the glass. This is obvious. It seems rather that spongy protuberances at the tip of their legs adhere to the inclined glass and retain the small weight of the animalcule. This does not occur as a result of their fear of vacuum, as commonly believed, but because of their small weight and the elasticity of ambient air, as we said elsewhere[2]. In the same way children applying exactly their palms on a glass cup raise it.

This is confirmed by this most unpleasant corrosive itching provoked by contact of the legs of a fly on our skin in summer time, when rain is imminent and intermediate air is more easily excluded from the surface of our wet skin. Then, the plantar protuberances of the flies adhere tenaciously to our skin and are not easily removed.

On the other hand, this unpleasant corrosive itching results from the simple titillation by the claws, hair and asperities of the legs of the flies touching and rubbing the nervous papillae of our skin, like straw rubbing the skin of the eyelids or of the nostrils produces an unpleasant corrosive feeling.

Flies and mosquitoes are more painful when they pierce the skin with their proboscis and suck blood like leeches. Fleas, lice, and bugs using their many thorny tongs articulated like hands suck blood from the capillary vessels which they have incised.

Chapter XXI

On jumping.

To move from one plane to another, animals not only walk but also jump. Jumping is cleverly dealt with by Nature and must be considered.

Proposition CLXX

Jumping is impossible without flexion of the segments of the legs.

We observe that, as long as men stand erect with their legs extended as if they were columns, whatever their will and effort, they cannot jump. But, after flexion of the joints, energetic contraction of the extensor muscles results in jumping. Some beasts and insects always keep the joints of their legs, at least the hind legs, flexed. They can jump at will. However, if they prepare to jump higher or further, they usually flex their joints even more. Reptiles which have no legs jump only by flexing their spine. This proposition thus is obvious.

[2] De Motion. natur. à gravit. pendent. Chap. 12

Proposition CLXXI

Why a rod or a spring standing on a hard surface, when compressed, bends, resiles and jumps.
Tab. XII, Fig. 3.

A straight and solid rod or a spring FEC standing on the ground RS at C is compressed by the hand or by a weight M until it is bent considerably, acquiring a curved shape ABC. Force M is then removed suddenly. The rod not only returns to its initial straightness but also quickly jumps from the ground. Compression lowers the centre of gravity E of the rod to D. When the force M is removed the force of the spring is released and straightens the two arms BA and BC so that the extremity C is lowered as much as the extremity A must be raised. But the resistance of the ground RS impedes lowering of the extremity C. Consequently, by reaction, the centre of gravity D with the whole spring must be pushed to E by a distance equal to the hindered lowering of C. This straight displacement DE is impossible without the impetus imparted by the force of expansion of the spring. This impetus, once given, is permanent and imperishable as we demonstrated[1]. Therefore, after the rod is straightened, this strong impetus cannot be refrained and raises the spring the centre of gravity of which moves in the direction DE above the extremity E. Consequently, the rod FC leaves the ground like a missile, jumping until the force of gravity which increases progressively is equal to the magnitude of the impetus of the missile. Then there is equilibrium. Ascension is over and is followed by fall as we showed in our book on the force of percussion[2].

Proposition CLXXII

A contrivance made of two rods and standing on the ground jumps when extended by sudden contraction of a rope.
Tab. XII, Fig. 4.

A bow ABC flexed about the joint B stands on the ground at C. Its centre of gravity is D. It is extended by very quick contraction of the rope GQH over its convexity. I claim that the bow will rebound from the ground. As said above[3], contraction of the rope opens the bow pulling its arms upwards and downwards. But, as in Fig. 3, Tab. XII, because of the resistance of the ground RS the centre of gravity D is forced upwards from D to E by reaction. Such a movement is impossible without an impetus being imparted and goes on as long as the impetus is not exhausted. Persistence of the velocity rather increases the continuous movement. When the bow is completely extended the impetus is not exhausted and moves the bow and its centre of gravity from E to F. This cannot occur without a jump.

---

[1] De vi percuss. Chap. 9

[2] Prop. 114

[3] In prop. 171

Proposition CLXXIII

Description of the organs and mechanism of jumping.
Tab. XII, Fig. 5 and Tab. XIII, Fig. 8.

The human body ABE (Fig. 5, Tab. XII) stands erect with straight spine and legs like columns, perpendicular to a horizontal plane RS. The total centre of gravity G is at a distance from the ground equal to the length of the legs. Then, in Fig. 8, Tab. XIII, the legs BCD are flexed at acute angles so that the distance GE of the centre of gravity from the ground is approximately equal to the length of the lower leg, i.e. about half what it was in the erect position. In this position, if the glutei, vasti and soleus muscles contract together, very powerfully and quickly, the three flexed bows ABC, BCD and CDE must be extended with a maximum impetus. Because of the resistance of the ground the centre of gravity G must be pushed upwards to F. This movement is impossible without an impetus being imparted. This acquired impetus by its nature is persistent and cannot remain inactive. It will necessarily remove the human body from its contact with the ground in a jump and push it upwards for some distance until the impetus of gravity progressively increased is equal to the impetus of the jump.

One may think that this theory is contradicted by experience. When the three joints of the legs are flexed and one extends them, standing up in a continuous movement, the centre of gravity must move upwards. However, the impetus which pushes the centre of gravity upwards does not result in a jump.

To meet this objection it must be recalled that, if the movement is slow or interrupted by frequent short delays, although there are very short jumps, these are hidden because subsequent falls due to gravity immediately bring down the body. The eye does not distinguish these minimal jumps. They are also hidden by stretching of the soft parts. Similarly, a small cushion, when pulled upwards, does not leave the ground as long as the movement upwards is not more than the stretching. Consequenlty, for the jumps to become obvious, extension of the joints must be considerable and quick, sufficiently to exert a force of percussion. Then jump must occur.

It thus appears that three springs are at work in man. They do not result from the strength of the material and the tension in the limbs as occurs with wooden or steel rods but from voluntary contraction of the extensor muscles.

Moreover, the structure and arrangement of the three alternate springs are admirable. Firstly, since the centre of gravity is pushed over a three-fold distance, a stronger impetus is imparted to the body of the animal so that it can jump higher. Indeed the imparted velocity increases continuously as long as the joints extend and thus results in a greater impetus, as said above. Secondly, the impulse is exerted along a straight line, either perpendicular or inclined to the horizontal, because the movements of the bones are carried out about all the joints. Therefore, the centres of the joints must move with different velocities in such a way that the different circular movements result in a straight trajectory. Thirdly, the alternate arrangement of the three joints helps in absorbing energy when going down so that the legs are not hurt or dislocated when landing on the ground. By giving way progressively the muscles exhaust the impetus of the fall.

Proposition CLXXIV

If the same body P (Fig. 6, Tab. XII) projected vertically upwards travels different distances C and D, the motive forces A and B which project the body are proportional to the square roots of the distances thus travelled.

The body P is projected upwards over a distance C by the motive force or impetus A. If, initially at rest, the body falls through this same distance C, the impetus acquired at the end of the fall is equal to the impetus A of the projection upwards[4]. Similarly, the distance D described by the same body P projected vertically by a motive force B is equal to the distance through which P would fall from a position of rest to acquire an impetus equal to B. But the distances of natural falls of a same body are proportional to the squares of their velocities[5] or of the impetuses acquired at the end of the movements. Therefore, the ratio of the impetuses or motive forces A/B is equal to the ratio of the square roots of the distances C and D.

Proposition CLXXV

The motive force exerted by man to jump is 2900 times the weight of his body.
Tab. XII, Fig. 4.

In jumping not only the body must be at first flexed and then extended and retained supended but it must be pushed upwards by a force of projection. We thus must find the muscular force necessary for the man simply to stand up and that for his projection upwards. Firstly, the joints of the legs of the human body ABVDE are flexed as required to jump[6]. The weight of the body is supposed to be 150 pounds, its upper part ABC weighing 110 pounds and ABVX 140 pounds. The distance MB is four times the distance of the tendons GQH from the centre B of the joint and the distance TV is three times OV. PD is one and a half times DK. The ratio of the weight 140 pounds of the contrivance ABVX to the force Z of 186.66 pounds is equal to TV/MB = 3/4[7]. The ratio of the weight 150 pounds of the contrivance ABVDE to the force S of 400 pounds is equal to DP/BM = 3/8.

The ratio of the weight ABC of 110 pounds to the force of 880 pounds exerted by the glutei muscles is equal to the ratio of half the distance of the glutei tendons GQH from the centre B to the distance MB, = 1/8[8]. Similarly, the

[4] De vi percuss. Prop. 114
[5] De vi percuss. Prop. 81
[6] Prop. 53
[7] Prop. 47
[8] Prop. 38

ratio of the force Z of 186.66 pounds to the force of 1 120 pounds exerted by the vasti muscles of the two legs NOC is equal to the ratio of half the distance VO to VT, = 1/6. The ratio of the force S of 400 pounds to the force of 1 200 pounds exerted by the soleus muscles is equal to the ratio of half the distance KD to DP, = 1/3. The force exerted by the muscles which extend the three joints of both legs thus is equal to 3 200 pounds. The forces exerted by all the muscles thus mentioned when they raise the human body are equal to more than twenty-one times the body weight. This force must again be multiplied by more than twenty because of the small machines which compose the fibres of the muscles. Therefore[9], the force exerted by the muscles to straighten the human body previously flexed, by taking support on the ground is more than 420 times the weight of the man.

It appears that the motive force exerted by Nature in jumping, to contract the same muscles is about 3 000 times the body weight. Facility of simply standing up as compared with the huge muscular effort and the expenditure in forces necessary to jump sufficiently demonstrates that jumping requires much greater forces than simply standing up. Moreover, we see that man, when standing up, stands on his toes and his mass is supported by the hips, knees and ankles. The body thus is sustained by the resistance of the ground and by his bones. It is different when jumping. The force of projection exerted by the violent contraction of the muscles plays the role of a strut and suspends the mass of the body with its flexed limbs in the air.

To confirm this proposition by a more convincing demonstration, I consider that when man stands up on the balls of his feet he jumps since the movement which pushes the centre of gravity upwards requires an imparted impetus which persists[10]. In complete extension of the joints, this impetus produces its effect, raising the human body from the ground. This is a jump. Because of its smallness this jump is hidden since the body comprises soft parts which, during this short jump, do not lose contact with the ground. In the same way, a cushion projected upwards is stretched and its lowest parts do not leave the ground. Similarly, it must be conceded that, when man stands up, he always jumps at least half a finger breadth approximately, although the feet do not lose contact with the ground because of the stretching of their soft parts.

In a big jump, the human body is sometimes raised more than two feet from the ground, i.e. more than 24 finger breadths. The motive forces which carry out these unequal jumps[11] are proportional to the square roots of the heights of the jumps. Therefore, the motive force exerted actually to jump is six times that required for a hidden jump, i.e. for simply standing up. The latter force was shown to be 420 times the body weight. Therefore, the force exerted in jumping is 2900 times the body weight.

[9] Prop. 125

[10] De vi percuss. Chap. 9

[11] Prop. 174

Proposition CLXXVI

The longer the distal levers of the legs the bigger the jumps.

Violent contraction of all the extensor muscles of the legs is as quick as an explosion of gunpowder. Contraction of the extensor muscles thus extends the bows formed by the segments of the legs at the same time. If the arms of the bow, i.e. the segments, are longer they describe larger circles and thus must move more quickly during the same period of time. Consequently, animals which have longer legs, as compared to the mass of the body, are pushed upwards more quickly when their legs unfold for jumping. Pushed more quickly, they jump higher or further. Animals with longer legs thus make bigger jumps.

Hence grasshoppers, crickets and fleas make long jumps because their hind legs are very long. Their length is equal or even superior to that of the animalcule, unlike those of men and quadrupeds.

Proposition CLXXVII

In the same conditions, smaller and lighter animals make bigger jumps, relative to their body.

Two unequal bodies, if pushed by forces proportional to the bodies, move with equal velocity[12]. For example, if a body is twice as big as another and the force moving the former is twice that moving the latter, both will travel the same distance during the same period of time. Consequently, if the motive forces in the horse and dog were proportional to their body masses, if the weight and mass of a dog is a fiftieth of those of a horse and if the motive forces were evenly and homogeneously distributed in their bodies, the motive force of the dog would be a fiftieth of that of the horse. Therefore, if the other conditions are equal (and, indeed, the limbs and the other organs are in the same proportion), the dog will jump as far as the horse. Actually, both jump eight feet. A jump of eight feet is not more than the length of the horse, whereas it is four times the length of the dog. Consequently, the jump of the dog as compared to its body, is four times more powerful than that of the horse.

However, in jumping, the impetus is not imparted instantly as it is in percussion but by a common movement by which the centre of gravity is propelled at the same time as the limbs are unfolded. The extension of the legs of a dog is less than a quarter of the extension of the legs of a horse. Consequently, the imparted velocity in a dog is much smaller than the imparted impetus in a horse. Hence, a dog should jump less than a horse. The velocity of jumping of the dog thus is increased by a cause other than what appears.

Galileo (On displacement) demonstrated that the weights of animals of smaller and smaller size decrease more quickly than their sizes. The weight is

[12] In Lib. de vi percuss. Prop. 12

proportional to the square of the resistance and strength of the animals. Therefore, the bones of big animals had to be bigger for their strength to be able to withstand the increase in weight. Hence, big and heavy animals are less prompt and agile than smaller animals. It is thus true that lighter animals make bigger jumps, relative to their bodies, than bigger animals.

Proposition CLXXVIII

In jumping at an inclination to the horizon, the trajectory of the jump is parabolic.

When jumping upwards vertically, going up and down occurs along the same perpendicular. The same is true for missiles projected vertically. But if the impulse is inclined to the horizontal, then the trajectory described by the centre of gravity during the jump is a parabolic curve. This curve results from the uniform linear movement due to the projection and the accelerated descent due to the weight of the body. This was demonstrated by Galileo in "On the movement of missiles".

Proposition CLXXIX

Why a jump during running is longer and higher.

When running the animal body acquires some impetus like javelins thrown over a horizontal surface. This impetus is not suppressed by an additional transverse impulse[13] but the movement continues in a new direction. Consequently, the impetus which carries the animal body when running is not suppressed by its deviation due to jumping. This deviation only changes the trajectory of the body. Both impetuses, that of running and that of jumping, reinforce each other. When combined they result in an energy greater than that due to jumping alone. This makes the trajectory of the jump longer. That is why jumping is increased in height and in length when preceded by a run.

Similarly, runners carried by their impetus can climb slopes more easily, as long as their impetus persists.

Proposition CLXXX

When jumping, inclination of the head and chest changes the trajectory of the centre of gravity.

Jumping thus is nothing else than a movement in the direction in which the bows formed by the bones of the legs standing on the ground unfold and push

[13] De vi percuss. Chap. 3. Prop. 5

the centre of gravity of the whole body. The impulse must occur in the direction in which the bows formed by the legs with the supported centre of gravity are inclined. This inclination changes with deviation of the head and chest and their bending forwards or sideways. Therefore, jumping is directed and changed by flexion of the head and chest.

Proposition CLXXXI

How one avoids hurting the feet when falling after a jump.

At the end of a jump the animal body lands under the effect of two impetuses, one due to the projection imparted by the muscles, the other acquired during descent and due to the weight of the body. The body thus is necessarily affected by these velocities and strikes the ground violently. The resistance of the ground could hurt, break or dislocate the feet. To avoid this inconvenience, taught by experience and habit, we use to extend the segments of the legs in such a way that we strike the ground with our toes. Then giving way and progressively flexing the legs we absorb and finally almost exhaust the impetus of the fall.

Moreover, at the end of a jump, we use to slow down the impetus which carries the centre of gravity by bending the head and chest backwards in a reflex action to avoid falling forwards.

Multi-legged animals jump according to the same mechanical rules and operations. There is thus no need to describe them again. They even apply to reptiles although these are deprived of legs. But their spine tortuously bent is equivalent to the articulated legs of quadrupeds. They jump by contracting the muscles which unfold the spine.

Chapter XXII

On flying.

As terrestrial animals walk, birds fly through the air. The movement is carried out by a wonderful mechanism. We shall attempt to explain the theory of flying.

Proposition CLXXXII

Description of the structure of the wings and their elements.
Tab. XII, Fig. 7.

All bipeds and quadrupeds, besides their hind legs, present with two arms articulated at the shoulders. In quadrupeds they are used to walk as are the legs, in men to grasp, in birds to fly. In all these animals the arms are made of the same number of joints and principal bones arranged similarly, namely the scapula, humerus, radius, ulna, and carpal bones.

The scapula is articulated with the extremity of the clavicle which moves about the upper extremity of the sternum. In birds the structure of the scapula is different and more complicated. It is made of two long bones DL, DM forming an acute angle LDM. The upper one LD is linked to the ribs in the back and attached to several paravertebral muscles. The extremity M of the lower bone of the scapula, flat and circular, is attached by a strong tendon to the lateral edge of the sternum. An extremity of the clavicle KN is attached in the angle N of the scapula. In the angle D of the scapula there is a round socket in which the head of the humerus rotates and to which it is attached by a strong tendon.

In the angle formed by the scapula there is a hole I or pulley through which the tendon CIH of the elevator alae muscle passes. It is wonderful to see that foreseeing Nature stabilized the centre of flexion of the wing by two bony struts, the lower extremity of the scapula DM and the clavicle KN, both supported by the edge and the crest of the sternum, so that they can withstand the very strong action of the pectoralis alae depressor muscle BG. The distal extremity of the humerus articulates with two bones, the ulna and the radius, which are longer than the humerus. These bones in turn articulate with the carpal bones which constitute the hand in men and the articulated extremity of the wing in birds. The carpus is a little shorter than the humerus.

The length of the bones of the wings and that of the feathers are not uniform in all birds. I mean that they have not the same proportion as related to the length of the bird. Ostriches have very short and thin wings as compared with their body. Chickens and other birds which fly seldom and little have somewhat longer wings. In doves they are even longer. Eagles, swans, swallows and the other birds which are almost always flying have very long wings. In this latter group the bones of each wing are as long as or longer than the body from the neck to the coccyx. Together with the feathers they are about three times as long as the whole body.

The structure of the bones of birds also appears admirable. They are thin and hard tubes with cavities much larger than those of the long bones in men and quadrupeds. As demonstrated by the famous Galileo in "New mechanical science", this decreases their weight. The tubular structure increases their strength so that the wings of birds are light and, altogether, strong and most appropriate for flying.

The feathers of the wings are also cleverly made by Nature so that they can beat air with powerful percussion very smoothly and strongly like resilient springs. Their substance is cartilaginous but as hard as horn. Their quill is tubular, thus largely hollow. Their curved shaft is filled by very light spongy marrow and is covered by a thin, very hard and convex cortex. Below in the concavity of the feather the cortex is bi-partite, made of two half-cylinders. These are such that after inflexion they both return to their initial shape thanks to their strong resilience, like springs. After expansion, they are again bent and thus beat air violently.

The barbs of the feathers are extraordinarily light, strong and resilient like springs. Their sides are connected by intertwined bushy bristles so that they impede the passage of air like the sails of a boat. To prevent penetration of air even better, smaller feathers are set between the larger ones. All are imbricated like the squames of a fish. All together they simulate the roof of a house made of tiles. Thus interconnected they preclude any fissure.

Proposition CLXXXIII

Movement of birds and of their wings during flying.

To be able to analyse a mechanical contrivance which flies we must exactly observe which parts of the wings move when birds fly, in what order and in what circumstances they move. As long as the birds stand, resting on the ground, they keep their wings folded over their sides. When they are about to fly, firstly they flex their legs and carry out a high jump. Secondly, they unfold their wings which form a straight line perpendicular to the sides of the chest. When open, the wings form a cross with the body of the bird. The wings with their feathers constitute an almost flat blade at some distance from the horizontal ground and are activated by a very energetic vibration. By a movement almost perpendicular to their plane, the wings beat the subjacent air. Subjected to such powerful beating, air, although fluid, resists either by its natural inertia which maintained it immobile or because under this very quick impulse the small machines forming air are condensed and, being elastic, react to compression by resiling, like a hard ground. As a consequence, the bird bounces and carries out another jump in air. Flying thus is nothing else than a movement composed of frequent and repeated jumps through air.

The wing can easily beat the air by a movement almost perpendicular to its flat surface although one of the extremities only of the humerus is attached to the scapula. All the width of the base of the wing remains free since the longest transverse feathers are not attached to the skin of the chest. However, the wing can rotate about its base as well as a fan. The links between the bones, by their protuberances and struts, impede any deviation of the wing. The strong attachment of the feathers prevents their rotation. Tendons acting as do the ropes on the sails of a boat hold the feathers so that they do not open too much.

The cleverness of Nature is not less admirable in the opening and closing of the wings. Actually, the wings close not on the side but by moving their bones upwards where the feathers are implanted. In this way the flat wing can move upwards without being impeded by air, like a sword. Hence they can rise with little force. Then as a result of the transverse tension and the resistance of air, the wings open when they flex, as was suggested and as will appear more clearly later.

Proposition CLXXXIV

Magnitude, arrangement and mode of action of the muscles moving the wings.
Tab. XII, Fig. 7.

The wings are the arms of the birds. Like in men they comprise three segments, humeral, ulnar and carpal. But the wings lack hands subdivided into digits. The number of extensor and flexor muscles also is equal in both animals. But they are very different as far as their magnitude, position and arrangement are concerned.

Firstly, the pectoralis muscles in man flex the humeri. They are small and only slightly fleshy. They are less than the fiftieth or even the seventieth of all the muscles of man. In birds the pectoralis muscles are very big. They are equal to, even exceed and are more important than the sum of all the other muscles of the bird. Hence it can be surmised that their motive force is considerable. This is necessary to vibrate the wings, as we shall demonstrate.

Secondly, the position and attachments of the pectoralis muscles are also noticeable. In men they originate from the flat sternum, cartilages, the middle of the clavicle and the 6th, 7th, and 8th ribs. In birds the sternum is large, hard and heavy, similar to the breastplate of a soldier, with a crest or a partition along its mid-line. The fibres of the huge pectoralis muscles originate from the sides of this crest and from the thorax. These fibres are prolonged by a strong and wide tendon and then by flesh. They insert in the large and round edge of the tuberosity of the proximal epiphysis of the humerus, whereas in men they insert distal to the epiphysis, in the shaft of the humerus. Therefore, in birds the distance of the line of action of the pectoralis muscles from the fulcrum of the lever of the wing is small, equal to half the diameter of the epiphysis which rotates inside a cavity of the scapula. This distance is less than the sixth or the seventh of the length of the humerus, the eighteenth of the length of the whole skeleton of the arm, and about the fortieth of the length of the wing with its feathers.

In men the arm is raised by the deltoid muscle which originates from the lateral aspect of the clavicle and the acromion. The deltoid muscle is antagonistic to the pectoralis muscle. Both pull the humerus in opposite directions, the latter downwards, the former upwards. In birds there is no deltoid muscle. In its stead there is an oblongus teres muscle, small, set in the angle of the breast bone formed by the crest. This muscle, the fibres of which contract in the same direction as those of the pectoralis muscles, has the opposite effect and raises the humerus towards the back and the head whereas the pectoralis muscle lowers and flexes the humerus towards the sternum.

As appears clearly in the diagram, the humerus of the bird is ABD. Its head or proximal epiphysis somewhat flattened BFEC is attached by a strong tendon D in the middle of the cavity of the scapula and in the top C of the edge of the tuberosity of the humerus. Over the length CB the tendon is attached to muscular fibres BG of the pectoralis muscle which pulls down in the direction BG towards G and flexes the humerus BA towards G about the centre of its proximal epiphysis. The tendon HIE of the elevator alae muscle is inserted on the opposite side in the edge E of the tuberosity of the humerus. This tendon passes through the foramen in the angle I of the scapula, as if it were over a pulley. In this way the pull of the elevator alae muscle from I to H results in a rotation in opposite direction, moving the wing away from the chest GB and raising the humerus EA with the whole wing, not laterally but with the edge of the wing backwards like a flag.

Nature was forced to introduce this difference in birds because the big elevator alae muscle could not be attached to the thin and weak bone LD at the top of the scapula and the upper part of the body ought not to be overloaded by so much muscle. The lower part of the chest had rather to be loaded by this adjunction, thus lowering the centre of gravity of the bird to the bottom of the chest. The necessity of such lowering will be discussed.

Besides these two muscles and other smaller ones which help flexion and extension of the humerus, there are flexor and extensor muscles of the forearm and carpus. In birds these muscles are weak and short contrary to what exists in men. In men the hands are bigger and had to carry out more varied and powerful movements which needed more long and strong muscles. Birds had to do without hands. The distal segments of their wings must only be extended and flexed. They thus need fewer muscles.

Proposition CLXXXV

The centre of gravity of the birds must be low.
Tab. XII, Figs. 8 and 9.

The centre of gravity[1] is a point in the middle of the body such that, if suspended from this point, the body would remain in equilibrium, whatever its position. If suspended from any point other than the centre of gravity, the body moves except when it is in a position such that the point of suspension and the centre of gravity are on the same vertical line. The centre of gravity of the body ABC (Fig. 8) is D. The body is suspended from point E. The centre of gravity can rotate over the circumference FGD of a circle. The diameter FD of this circle is vertical. Obviously, if the centre of gravity D is displaced to any point G higher on the circumference, the body ABC, like a pendulum, does not remain immobile and comes down to the lowest position D and there it remains immobile. That is where the line ED is vertical. Inversely, if the pendulum ABC suspended from E is immobile and, after being moved, comes back to this same position such that the side BC is below, the centre of gravity must be a point on the vertical ED, underneath the point of suspension E.

If a body immersed in a fluid remains in the same position its centre of gravity lies in the lower part of the body, closer to the earth. If the ball AB (Fig. 9, Tab. XII) is either in water or in the air and its point B always places

[1] De Aequip. dif. 3 in Arch. Auct.

itself in the lowest position, the centre of gravity of the ball is at some point C different from the centre of the ball and closer to B. This also results from the nature of a pendulum.

Similarly, since birds always fly with their abdomen prone, apparently without any effort, it must be admitted that their centre of gravity is in the lower aspect of their chest and venter.

Although heavier than air, birds are supported by the force of their wings and do not fall. They are suspended from their scapulo-humeral joints which are in the upper aspect of their body. Therefore, their centre of gravity must lie in the lowest aspect of the chest on a vertical from the roots of the wings. This vertical is perpendicular to the long axis of the body.

This assertion is confirmed by the way we use to find the centre of gravity in irregular bodies. If a plucked bird is placed on a horizontal knife's edge in different positions, there is a position for which the bird remains in equilibrium. The centre of gravity is on a straight line drawn perpendicularly to the long axis of the bird and joining the roots of the wings and the middle of the breast bone. Birds sleep in this position resting on small twigs in trees.

The heaviest parts of birds, namely the thickest and biggest bones and the flesh over the chest, are gathered in the lowest part of the body. Indeed the bones of the legs and the sternum are large and thick. On the contrary, the spine and ribs are thin and light. The mass of the muscles of the chest, thighs and lower legs is equal to or even greater than the mass of the viscera and all the other muscles of the bird. To make the lower aspect of the chest much heavier, Nature did not set the levatores alarum muscles above the scapula but hid them in the middle of the pectoralis muscle, as mentioned above. The upper aspect of the thorax is filled with air and thus is much lighter than the lower aspect. The abdominal cavity also is mostly filled by air, above the intestines which are pulled downwards by their own weight and lie over the abdominal muscles. From this arrangement we perceive that Nature took care of placing the centre of gravity of birds beneath the roots of the wings so that their venter is always prone when they fly. In this way, powerful beating of the wings and percussion of the subjacent air enable the bird to fly without falling.

Some observations seem to be opposed to this theory. When birds stand and walk they balance about a point between the scapulo-humeral joints and the hips. A vertical drawn from this point falls between the digits on the ground. The centre of gravity thus is behind the roots of the wings, towards the tail. Moreover, when standing the birds are not exactly horizontal but their body is somewhat inclined.

All this is true but does not oppose the theory. The centre of gravity can be moved from the position thus described and the attitude can vary according to the different requirements of the animal. The neck of the bird is very heavy as a result of the several bony vertebrae and the weight of the head. Therefore, depending on whether the head is retracted or pushed forwards, its moment changes with the length of the lever and the centre of gravity of the bird is moved towards the head or towards the tail. When geese fly, they straighten and extend their neck forwards like the arm of a balance with the head at its extremity. This displaces the centre of gravity towards the head. Moreover, when flying, birds use to fold their legs along the chest. In geese and other birds with a heavy head, the weight of the legs is displaced forwards or backwards to ensure equilibrium. On the other hand, when birds stand on the ground, their wings are folded along the sides and they thus increase the weight of the hind part. When flying, they are open and moved towards the head which increases the weight of the fore-part of the bird. These three causes significantly displace the centre of gravity of the bird forwards until the line which connects this centre and the scapulo-humeral joints is vertical and perpendicular to the long axis of the bird. Then in prone position and beating the air with their wings they can most conveniently fly through the air by successive jumps.

This theory is also confirmed by the solution of the next problem. When sleeping, birds put their head under a wing (Tab. XI, Fig. 7). The first and obvious reason is that voluntary and fatiguing action cannot be carried out during sleep. Then the weight of the neck and head cannot remain suspended and must be supported somewhere. Another reason which relates to our discussion is that retracting the head and resting it towards the tail is like shortening the arm of a balance. This moves the centre of gravity backwards so that the line of gravity falls vertically in the middle of the claws which strongly grasp a branch of a tree, like pliers[2]. In this position the bird is balanced and can easily rest. The following lemmas must facilitate understanding of the mechanism of flying.

Proposition CLXXXVI

The bow EABF (Fig. 10, Tab. XII) is made of three rods EA, AB, BF equally inclined and standing on the ground at E, F. The bow supports the weight D. The equal angles A, B are subtended by the ropes GI, LH. I claim that the ratio of the force exerted by the ropes to the weight D is equal to the ratio of twice the horizontal distance KA between the ends of the extreme rods and AO the distance of the ropes from the joint A.

The weight D acting vertically compresses the ground which resists and there is equilibrium. The weight D thus is equal to the ground reaction force.

The forces exerted by the ropes GI and HL are balanced by the weight D and the ground reaction force. Therefore, half the force exerted by the ropes is balanced by the weight D and the other half by the ground reaction force.

The weight D is supported by two struts AE and BF equally inclined. Therefore, half the weight D is supported by the rope GI and the lever AE. The other half of D is supported by the rope HL and the lever BF.

The lever EA is pushed about the joint A by the ground reaction force E in the direction EK. The distance of EK from the joint is AK. The lever EA is also

[2] Prop. 150

pulled by half the force exerted by the rope GI pulling in the direction GI. The distance of GI from the joint is AO. Therefore, the ratio of half the force exerted by the rope to the ground reaction force at E is AK/AO.

Similarly, the force exerted by the other half of the rope OI is balanced by half the weight D by the same action of the lever AE and the contraction of the rope about the joint A. The directions of the forces are the same. Therefore, the ratio of half the force exerted by the rope GI to half the weight D is equal to AK/AO. Consequently, the ratio of the force exerted by the rope GI to the ground reaction force at E and to half the weight D is equal to 2AK/2AO. The ratio of half the weight D to the ground reaction force is equal to 2AK/AO. Finally, the ratio of the sum of the forces exerted by the ropes GI, HL to the weight D is equal to 4AK/2AO or 2AK/AO.

Proposition CLXXXVII

The data are the same. If the ropes contract with such power that the bow bouncing on the ground jumps obviously, i.e. much higher than after a weak contraction, the ratio of the force contracting the ropes to the supported weight D is equal to the product of the ratio 2AK/AO and the ratio of the square root of the minimal jump to the square root of the visible jump. Tab. XII, Fig. 10.

The bow bouncing on the ground jumps by one finger breadth and then by twenty-five finger breadths. Weak contraction of the ropes forces the weight D upwards but the jump is hardly visible because of its smallness and the extensibility of the bow. Then the ratio of the forces exerted by the ropes to the weight D is equal to 2AK/AO[3]. The ratio of the impetus which results in a jump twenty five times higher to the former impetus is equal to the ratio of the square roots, i.e. 5/1[4]. The ratio of the force exerted by the ropes to the weight D results from the two ratios. Therefore, when the jump is obvious, the ratio of the force exerted by the ropes to the weight is equal to 10 AK/AO.

Proposition CLXXXVIII

The volume of air beaten by a wing when flying is equal to the dihedron described by the length of the wing, acting as a radius, during its movement. Tab. XIII, Fig. 1.

The bird ACBG is flying. Its wing beats the air by slightly flexing. The length AI of the wing rotates about the centre of the humeral head A and describes the sector AEP. However, the wing AE is not a line but a flat surface which beats the air. Its movement thus describes a dihedron. Actually the wing beats only the air which is contained in the dihedron described by its movement. Q.E.D.

[3] Prop. 38
[4] Prop. 174

Proposition CLXXXIX

The dihedron of air beaten by a flying wing exerts a resistance at its centre of gravity.
Tab. XIII, Fig. 1.

The ratio of the arc EP to two-thirds of its chord EP is equal to the ratio of the radius AI to AQ. Point Q is the centre of gravity of the sector AEP of a circle, as was demonstrated by others. Therefore, the centre of gravity of the dihedron described by the movement of the wing lies in the middle of the line intersecting all the parallel flat sectors at their centre of gravity. The centre Q of the dihedron thus divides the length AI of the wing in such a way that AQ is a little less than two-thirds of this length. It must be demonstrated that the dihedron of air AEP exerts its resistance against the beating of the wing, at its centre of gravity Q. The wing AE pushes and beats the air not at one point but over its whole length and width. The beating occurs through circumferences similar to each other which increase proportionally from the root of the wing to its extremity. Their velocities thus are unequal. The dihedron of air is beaten along the same circumferences by which it is described and with the velocities corresponding to these circumferences. But the energy of the beating is equal to the resistance exerted by the air contained in the dihedron. The air thus pushed and condensed resists as if the dihedron were a solid body which would have to be suspended from the wing. Such a solid body exerts resistance at its centre of gravity. The dihedron of air thus opposes also the impulse of the wing at its centre of gravity Q. Therefore, the lever formed by the wing is not the whole length AI but is rather AQ. The extremity Q of this lever is compressed by the elastic force of the dihedron of air. This force, like gravity, opposes the impulse of the wing at its centre of gravity.

Proposition CXC

How air opposes the impulse of the wings.

Air can be displaced and stretched by any motive force, however small, and thus seems to be indifferent to movement. However, it strongly opposes beating by a fan with the more energy the quicker the beating. The cause of this resistance appears to be twofold. Firstly, the beaten air must be displaced, whereas the surrounding air is immobile as required by the nature of fluids. To move only a small volume of air this volume must be mixed into a great mass of air at rest, as if it were inside some vessel. To this end the concave particles of air which were at rest must rub against each other and twist as well as the convex particles which rotate. This mutual rubbing and agitation cannot occur without violence and resistance. Therefore, the impetus of the fan is necessarily opposed and weakened by the flow of air thus displaced.

Secondly, the beaten air opposes the impetus of the fan elastically since the small machines of air thus compressed by the impulse tend to dilate. The fan movement of the wing is opposed not only by these two resistances but also by a reaction. These two resistances increase continuously, the more so, the more the impulse of the fan is continued and accelerated. On the contrary, the impetus of the wing continuously decreases and is slowed by the increasing resistance of the air. Impulse and resistance attain an equilibrium when air is beaten by the wing with a force equal to the resistance opposed by this air.

Proposition CXCI

If the velocity of the flexion of the wings of a flying bird is equal to the velocity with which the beaten air resists, the bird remains at the same height.
Tab. XIII, Fig. 1.

When the bird is flying, the wing AE is flexed towards P with the same velocity as that with which the air subjected to the beating is displaced while resisting. I claim that the bird ACBG flies neither upwards nor downwards. The bird can ascend as long as the bow EABF can carry out a jump by pushing a firm layer of air with its two extreme levers, thus going up. When this layer of air is not firm and retrocedes with the same velocity as that with which it is beaten, then the bird is propelled upwards by its jump and the flexion of the wings exactly as much as it goes down as a result of the lowering of the layer of air which supports the wings. The bird thus remains at the same height.

Proposition CXCII

The data are the same. If the velocity of flexion of the wings is greater than the velocity with which the beaten air retrocedes, the bird flies upwards and its ascension is equal to the difference between the two velocities.
Tab. XIII, Fig. 1.

Using the same diagram, it is assumed that the wing AE is flexed more quickly than the opposing air retrocedes. Thus the arc and its chord EP described by the wing during a given period of time are greater than the distance by which the beaten air moves downwards. But the wing can flex in the air only by describing the same trajectory as the adjacent air. Therefore, the larger movement of the wing must be delayed and shortened by some mechanical expedient to become equal to the smaller movement of the air. This can be achieved by moving the root of the wing A upwards together with the bird. Then, although the extremity E of the wing describes a larger arc EP, its trajectory in space is decreased by the distance by which the root of the wing A moves together with the whole bird. Consequently, the ascension of the bird is equal to the difference between the velocity of flexion of the wing and the escape of the subjacent air.

Proposition CXCIII

The force exerted by the flexor muscles of the wings is more than ten thousand times the weight of the flying bird.

We demonstrated in proposition 75 that the motive force of a jump in man is more than 3000 times his weight. It is possible to show that in birds the motive force of a jump is even greater proportionally to its weight. The weight of the birds is carried on a threefold contrivance and their legs are much longer than in men. These anatomical peculiarities multiply the motive force. This force again is multiplied by more than twenty, as a result of the number of small machines in the muscles of the legs. Finally, the motive force is so increased by the violence and velocity of the jump that it is more than 3000 times the weight of the bird which jumps.

Flying results from repeated jumps carried out by the levers of the wings which are violently flexed by the pectoralis muscles. The motive force exerted by these muscles is proportional to the size of the muscles, as mentioned above. The number and length of the muscular fibres or the mass of the flexor muscles of the wings is denser, stronger and not less voluminous than the mass of all the muscles of the legs together. Therefore, it must be deduced that the force exerted by Nature to flex the wings is greater than that required to jump.

The volume and weight of all the muscles of the legs in men proportionally to the volume and weight of the whole body are smaller than the volume and weight of the pectoralis muscles relative to the volume and weight of the bird. But the motive forces exerted by the muscles to carry the body are proportional to their volume and weight. Therefore, the force exerted by the muscles of the legs in men relative to the body weight is smaller than the force exerted by the wings relative to the weight of the bird.

The jumps made in the air when flying can go on for four hours and more without interruption. We even observe that swallows fly throughout the whole day. Neither men nor quadrupeds nor even plucked birds can jump on the ground with the same frequency for more than half an hour or perhaps an hour. The motive force which acts for a longer time and is sufficient to raise the same weight many times must be greater than the force which can accomplish the same only for a shorter time. Actually, the force exerted by the muscles of the wings is more than four times that exerted by the muscles of the legs. One can be convinced of this fact by the following argument. A man stands at first on solid ground, then on a pillow or on wool or on sand or on a spring. The legs are flexed and the man jumps with the same effort. It is obvious that the jump on solid ground is the highest. The jump is minimal on a yielding ground. For the jumps to be equal, i.e. for the body to jump to the same height, the man standing on a yielding ground must exert a force greater than on solid ground, the more so the less the resilience of the movement as a result of the instability of the ground, since this resilience is the cause of the jump. The layer of air which supports the wings of the birds is fluid and yielding to the blows. Therefore, flying, i.e. carrying out jumps in the air, requires a motive force which is by far more than four times that required to jump on solid ground. But the force required to jump on solid ground is about 3000 times the weight of the jumping animal[5]. Consequently, the force exerted by the pectoralis muscles in flying is not less than 10000 times the weight of the flying bird.

Proposition CXCIV

Causes of the huge motive force of the wings.

Such huge motive force exerted by the pectoralis muscles in birds seems to result above all from the size of these muscles and the stronger organic structure of their fibres. These fibres are indeed thicker and tighter, constituting a dense fleshy mass of the same fibrous consistency as the heart and the fleshy ventricules of birds. The muscles of the legs consist of rare and loose flesh. Hence the pectoralis muscles can contract more powerfully and are able to exert greater force than the muscles of the legs.

Secondly, the action of the wings is enhanced by the small resistance of the body of birds which appear to have a smaller specific weight than that of men and quadrupeds. Relative to their mass, the weight of birds is indeed lighter than that of men and quadrupeds. This results from the fact that the bones of birds are tubular, hollow and thin as are the quills of the feathers; the shoulders, ribs and arms are only slightly fleshy; the chest and abdomen are large cavities full of air; the feathers are very light. The power of the wings thus is increased twofold: by an augmentation of the motive force of the muscles and by a diminution of the weight to be carried.

The resistance is also decreased by the small impetus of descent which results from the weight of the bird flying in the air and tends to bring it down. The impetus of descent is even further diminished by the opening of the wings and tail which slows down the descent. Therefore, the motive force exerted by the wings can more easily repeat jumps in the air since thé resistance due to the impetus of descent is small.

Thirdly, in any jump on the ground, the impetus of the jump is exhausted immediately after the feet strike the ground. Therefore, this impetus must be renewed again and again. In jumps in the air, while birds are flying, the impetus of the jump is not lost in fluid air and helps the subsequent impulses through the air, due to the movement of the wings.

Fourthly, when jumping on the ground, landing hurts the feet which results in fatigue and weakness. This does not happen in jumps in the air. Therefore, the motive force being less weakened can carry out more considerable jumps. All these causes explain the possibility of flying.

The next lemma is necessary for the understanding of the following proposition.

---

[5] Prop. 175

Proposition CXCV

How oblique and transverse impulses can push bodies to move straight ahead.
Tab. XIII, Fig. 3.

The wedge ABC must divide a body into two parts EFG and LMN and separate these parts. The resistant bodies DF and HM are forced to move up over the inclined surfaces CA and CB of the wedge while the latter is driven in the direction of its axis from I towards C. The parts must move transversally in opposite directions over the inclined surfaces CA and CB. If the bodies DF and HM are brought together, they force back the smooth and slippery wedge ABC from C towards I. The wedge is expelled by the compression of the side parts like slippery stones of fruit projected far away by pinching two fingers. This expulsion is carried out with as much force and impetus as the compressing bodies DF and HM squeeze the inclined planes CA and CB. The ratio of the expelling moments to the magnitudes of the compressive forces[6] is equal to the ratio of the heights AI and BI of the triangles to their hypotenuses AC and BC.

Proposition CXCVI

If its wings beat immobile air vertically a flying bird moves horizontally.
Tab. XIII, Fig. 2.

A bird RS is flying and maintains its wings BEA and BCF open. The surfaces of the wings BEA and BCF are beaten by the wind blowing vertically upwards with such force that the bird does not fall. I claim that the bird is pushed horizontally forwards from S towards R. The two bony rods of the wings BC, BE can resist the impulse of the wind by the force of the muscles because of their strength. They thus can remain open. The width of the fan constituted by each wing yields to the impulse of the wind since the flexible feathers can rotate about their insertion on the bony axes BC and BE. Therefore, the extremities A and F of the feathers must be brought closer together. As a result the wings take the shape of a wedge the edge of which points towards AF. Both faces of this wedge are compressed by the ascending wind. Therefore[7], being compressed the wedge is expelled towards its base CBE. This wedge constituted by the wings cannot move forwards without carrying with it the attached body of the bird RS floating in the air and immobile by itself. The body of the bird thus makes place for the air pushed by the wind. Consequently, the bird moves horizontally towards R.

It is assumed that the subjacent air is quiet and is beaten vertically by the fans of the wings. The sails or fans formed by the wings acquire the shape of a

---

[6] From Comment Prop. 63

[7] Prop. 195

wedge with its edge towards the tail. They are subjected to the same force and compression by the air whether they strongly beat the immobile subjacent air or whether, maintained open and immobile, they are beaten by an ascending wind. In both instances, the flexible feathers yield to the impulse and are brought closer together. Consequently, the bird is forced to move forwards towards R, as explained above.

Proposition CXCVII

How birds fly horizontally.

Nature has instituted flying to push birds upwards, maintain them in the air and enable them to move around in space. They could fly upwards only if pushed by frequent jumps or by movements of the wings preventing their fall. Weighing bodies fall vertically. Movements of the flat surfaces of the wings beating the subjacent air thus had to be vertical to maintain birds in the air. Some speak nonsense about their horizontal displacement. They think that such displacement must occur like that of boats which are pushed by the backwards horizontal impulse of oars against immobile water which resists. The reaction of water results in a displacement of the boats forwards. They say that in the same way the wings move horizontally towards the tail and beat the immobile air behind. Resistance of the air results in a displacement forwards. But this is against evidence and common sense. Big birds such as swans, geese and others never move their wings horizontally towards their tail, like oars, when they fly. They always flex their wings downwards describing vertical circles. Moreover, in boats, horizontal movement of the oars can be conveniently carried out. Vertical percussion of water would be useless since the boats must not be prevented from going down, being carried by the water. For birds such horizontal movement would rather thwart flying and would be stupid since it would result in a fall of the bird which must be maintained in the air by continuous vertical vibration of the wings. In her admirable cleverness Nature was forced to adopt a movement which by the same action would maintain the bird in the air and push it horizontally. This occurs by beating the subjacent air vertically but with oblique blows, resulting only from the flexibility of the feathers. Then during beating, the wings acquire the shape of a wedge which is necessary for horizontal displacement of the bird forwards, as said above[8].

---

[8] Prop. 196

Proposition CXCVIII

Flying birds use their tail to inflex their course upwards and downwards but not sideways.
Tab. XIII, Figs. 4 and 5.

The opinion, received from the ancient philosophers, prevails that the tail of birds acts like the rudder in boats. Boats sailing over the horizontal surface of water can be directed only to starboard or to port side by their rudder. Therefore, it is commonly believed that birds use their tail to fly to the right or to the left.

Reasoning and an experiment could easily convince us of the falsity of this opinion. If the rudder of a boat was set as is the tail of birds, i.e. if the surface of the rudder and the rudderpost were attached to the stern not vertically and perpendicular to the surface of the water but horizontally so as to be able to move underwater upwards or downwards in relation to the water surface, the movement of such a rudder obviously could bend the course of the boat neither to starboard nor to port side.

Flying doves, swallows and falcons, when they incurve their horizontal trajectory to the right or to the left, do not open their tail, nor do they flex it upwards or downwards. They keep it straight. This we observe. Finally, doves with a cut tail, like bats which have no tail, turn in the air easily and fly horizontally in a curve. This is enough about a common error.

However, I do not deny that the tail in birds acts as a rudder. But I claim that the tail is used to incurve the trajectory of the bird upwards or downwards. I demonstrate this assertion.

The centre of gravity of the bird AB (Fig. 4) is C. While flying horizontally straight ahead from G towards F, the bird flexes its long tail upwards BH. As long as the bird moves from G towards F in immobile air, its lower face AB can fly freely without any hindrance. It is different for its upper face which is flat and directed upwards by the elevation of the wide and long surface of the tail BH. This strikes the immobile air. The air thus stricken tends to remove the obstacle and would straigthen the tail if it was not maintained firmly in this flexed position by muscular force. The bird thus, being balanced in air, must turn about its centre of gravity C and take the position LK, moving its head upwards from A to L.

Similarly, if the bird flies straight from G to F and bends its tail downwards BI, the tail strikes the subjacent immobile air and rises from B towards N about the centre of gravity C. As a result the head comes down to O.

The result of this mechanical demonstration is confirmed by an experiment (Fig. 5, Tab. XIII).

A long blade of metal AB is immersed in a vessel RST full of water (Fig. 5). This blade is suspended at its centre of gravity C from a thread tied to a cork D in such a way that the blade floating horizontally can be moved. A smaller blade BH similar to the tail of a bird is attached at its posterior extremity. If the smaller blade is flexed upwards and the main blade AB is pulled by a thread CF horizontally towards F, the anterior extremity A is quickly flexed towards L about the centre of gravity C. The blade never rotates horizontally to the right or to the left. The same must occur to birds when they fly.

Proposition CXCIX

How flying birds turn to the right or to the left.

Experience and reasoning show that a small rudder inclined to port side can make a big boat slowly turn in this direction when the boat is sailing straight ahead. But as long as the boat is immobile, if it is not pushed by the wind or by oars, inclination of the rudder does not result in any turning of the boat.

If the rudder is removed and the oars on starboard are used, pushing the water towards the stern, whether the boat is immobile or whether it sails straight ahead, the bows always turn to port side very quickly. The same occurs if the oars on starboard push the water more quickly than those on port side. The reason for this behaviour is obvious and needs no further description. In the same way, when the bird flies in the fluid milieu constituted by air, balanced about its centre of gravity, if only the right wing is flexed downwards and obliquely, pushing the subjacent air towards the tail, the right side must move forwards whereas the left side remains immobile or moves more slowly. As a consequence, the anterior aspect of the bird turns to the left about its centre of gravity. We experience this ourselves when swimming. Moving the right arm with the hand open, towards the buttocks, we turn to the left horizontally. We observe the same in flying doves. Every time they wish to incurve their course to the left, they raise the right wing higher and move it more energetically, beating the air by a movement oblique towards the tail. As a result, the wing and all the right side of the bird rise above the horizontal plane and the left side is lowered below this plane since weaker beating on the left side is not enough to overcome gravity. This turning of the bird flying horizontally is very quick.

Finally, the tail of birds cannot flex sideways as does a rudder. The tail can never be raised vertically but only obliquely. This obliquity has no effect if the feathers of the tail are aligned with the remainder of the body, i.e. if they are in its axis, like a waving flag. But the tail must flex in two ways: deviating from the direction of the axis of the bird and from a straight movement in a horizontal plane. Thus at a maximum it could result in an inflexion of the trajectory between the vertical and the horizontal direction. In flying doves the inclination of the feathers above the horizontal is minimal. The doves thus would move a little sideways but considerably upwards or downwards. But this is against experience. Actually we observe that birds turn very quickly in a horizontal plane.

Birds do not open their tail when they fly and turn to one side. But they open it when they fly up or down and even much more when they brake their impetus to land without shock or injury.

To meet objections the next lemma will be presented.

Proposition CC

If the body of a bird AC (Fig. 6, Tab. XIII) flies horizontally from C towards A on its initial impetus and, during its flight, flexes its long neck and head BA to the left side BI, the trajectory of the whole bird is incurved towards BI.

The centre of gravity of the body alone BC of the bird is D. E is the centre of gravity of the head and long neck AB. Flexing the neck to BI displaces its centre of gravity to F. A line is drawn joining D and F. The ratio of the weights CB/BA = CB/BI is equal to the ratio of the distances EG/GD and also of FH/HD. Before flexion of the neck, the centre of gravity of the whole bird was at G[9]. This centre of gravity was carried straight from D towards E by the imparted impetus. Flexion of the neck deviates the centre of gravity G from its straight trajectory towards H but the impetus from the tail towards the head is retained. These two displacements combine into a transverse intermediate trajectory DI, which is then described by the bird.

Proposition CCI

It does not seem believable that a very quick horizontal deviation of the flight results from lateral flexion of the head and neck of the bird.

As we have seen, a boat can turn to starboard or to port side in two ways: firstly, if the oars on one side push the water towards the stern more energetically than the oars on the other side; secondly, if, when the boat is moving, a vertical rudder attached either at the stern or at the bows is inclined laterally. These two operations are essentially different. The same deviation of the trajectory of the boat results either from a great effort of the oarsmen or from a little force exerted by the helmsman being at the wheel. The deviation does not result from the effort of the helmsman but from the impetus acquired by the boat which the water met by the rudder opposes. Turning of the boat due to the oars on one side is very quick, that due to the rudder is very slow.

These phenomena help to understand a similar operation in the flight of birds. Flexion of the neck can be compared with the use of a rudder.

Firstly, if the neck inclined sideways had the force of the rudder and could bend the trajectory of the bird to the right and to the left, inclination of the neck upwards or downwards should be able to bend the trajectory of the flying bird upwards and downwards. Then the big tail which obviously results in a movement upwards and downwards, acting as a rudder, would have been made by Nature in vain. This is not possible. It must thus be concluded that flexion of the neck does not act with the force of a rudder.

Secondly, eagles, falcons and swallows have a very short neck, and a small and light head. Therefore, their centre of gravity could hardly deviate from the

[9] In Arch. Auctor. de Aequip. Prop. 3

long axis of the bird which should turn laterally very slowly and with great difficulty. But this is not what occurs since these birds turn sharply almost instantly. On the contrary, geese, ducks, swans and other birds of the same type have a very long neck and head, a heavy bill and, when flying horizontally, they turn very slowly. It must thus be concluded that flexion of the neck and head sideways in no way causes any horizontal turn.

Thirdly, if flexion of the neck to one side displaced the centre of gravity of the whole bird significantly from its long axis, the bird would not be able to remain balanced in a prone and horizontal position. The lowered side would have to be violently raised by the wing on this same side. This would result in an effect exactly opposed to the first one and turning as a consequence of a deviation of the centre of gravity would be thwarted. Such useless behaviour would be stupid and unworthy of the cleverness of Nature.

Do not say that the quickest turns of birds are provoked by violent vibration of one wing towards the tail and that slow turns can result from flexion of the neck sideways without additional effort of the wing, as boats turn guided by the rudder without effort of the oarsmen. Indeed, a slow turn by a bird does not need a greater effort of one wing than what both generally exert during usual flight. It is enough that the wing which provokes the turn becomes somewhat deviated towards the tail and pushes air there to carry out a slow turn of the bird sideways without additional effort and with the greatest economy.

Proposition CCII

Why birds can fly for short periods of time horizontally or even obliquely upwards without moving their wings. Tab. XIII, Fig. 9.

Movement of the wings imparts an impetus to the body of the bird as beating of the water by the oars imparts an impetus to a boat. By its nature this impetus is persistent[10] even after the action of the oars has stopped. The boat keeps going on its course until the impetus is neutralized by external obstacles.

Bird and boat propelled by an impetus have the same properties as arrows and other projectiles. As long as a boat moves, if its course is modified by the rudder, its persistent impetus keeps it on the modified trajectory[11]. Similarly, if the acquired impetus pushes a bird straight ahead horizontally ABC and if this course is deviated upwards through BD by the tail acting as a rudder, the persistent impetus must continue the movement upwards along the parabolic curve BEF. However, this ascension soon stops because the weight of the bird always pulls it downwards. As long as the tendency to fall down is smaller than the velocity of the propelling, the bird keeps going up through BE. When the forces are equal at F the bird seems to stream in the air at the same place for a

[10] De vi percuss. Chap. 9

[11] De vi percuss. Chap. 3 and 4

short while, flying with wings open in almost the same horizontal plane. But the bird cannot remain immobile in the same position in air because the flight upwards is never vertical but is oblique over a parabolic curve similar to the trajectory of missiles.

Such an ascension or a horizontal course of the bird does not last long because the impetuses soon become unequal. The impetus due to gravity continuously increases and the impetus of the projection is braked by external obstacles. Hence the necessity arises to repeat the jump in the air by new vibrations of the wings.

The two impetuses, that of propelling and that due to gravity, sometimes cancel or slow each other down as occurs in ascension. Sometimes they enhance each other and result in very quick displacement. Falcons thus fall down like arrows on smaller birds which they strike and tear down with their claws. Some people suspect that birds soar with less effort at very high altitude than closer to the ground. They think that birds weigh less at high altitude because they are less attracted by the magnetism of the earth which, according to them, is the cause of the descent of heavy bodies. Similarly, a piece of iron away from a magnet does not feel its force and is not attracted. Likewise birds far away from the earth would escape the force of gravity. This would be the reason why eagles when flying further away from earth look as if they were almost resting in air, keeping their long wings open and immobile, whereas they are forced to beat the air very frequently and very quickly when they are closer to the earth and start to fly.

But such conjectures seem to be erroneous. Firstly, the hypothesis is difficult to sustain as we showed elsewhere. Secondly, close to the earth where the magnetic force is assumed to be strong and most efficient by these authors, the falcons, after acquiring their impetus, do not beat the air more often but at the same slow tempo and fly as easily as at high altitude. Consequently, they can soar without frequent vibration of the wings at high altitude not because of the absence of telluric magnetic force but because of the acquired impetus.

Thirdly, the region of high altitude is almost never totally quiet but is always agitated by winds which push the clouds. A bird with open wings is easily pushed upwards by the winds or, at least, is enabled to soar down very slowly. In these conditions the impetus due to gravity is delayed by an overextended shape. Similarly, a very thin blade of metal sinks down very slowly in water when not immersed on edge. It must thus be concluded that this is the most significant cause of the facility for the birds to fly at high altitude.

Proposition CCIII

How the impetus acquired by a bird is finally exhausted.

Solid bodies given an impetus cannot be brought into contact with another immobile solid body without percussion. This percussion usually results in explosion or rupture of the parts. It thus had to be provided that, after their flight, birds land without dislocating or breaking their legs. A violent blow which bodies imparted with an impetus are expected to strike had to be prevented. This is not possible without weakening and exhausting the impetus of the bird progressively before landing. Then having lost its impetus the bird can land smoothly.

The means by which the impetus is exhausted at the end of a flight are as follows. The wings and tail are opened so that their concave surfaces are at right angles to the direction of the displacement. In this way the feathers by their size, like the sails of a boat, strike the immobile air and brake the impetus of the bird. To exhaust further its impetus the bird moves its wings forwards quickly and violently, thus inducing a movement opposed to the initial impetus. Such movement exhausts or considerably reduces this impetus. Finally, the remainder of the impetus is braked by the extended legs which do not strike the ground but progressively flex and give way, while the muscles slowly relax.

Proposition CCIV

Men cannot fly artificially by their own forces.

For flying there are three essential points to be considered:

(1) the motive force which must keep the body of the animal in the air;
(2) the adequate instruments which are the wings;
(3) the weight of the body of the animal.

The magnitude of the motive force is judged from the volume and weight of the muscles destined to flex the arms or to move the feathers of the wings. This motive force in birds was shown to be 10000 times greater than the weight of the bird[12]. To give birds such a huge amount of motive force, Nature considerably increased the volume of the pectoralis muscles and cleverly decreased the weight of their body, as mentioned above.

For man to fly by his own power, the motive force exerted by his pectoralis muscles (measured by their volume) should be 10000 times more than the weight of his body together with that of the wings which should be adapted to his arms. But the force of the pectoralis muscles in man is much smaller than what is required for flying. In birds the volume and weight of the flexor muscles of the wings is no less than a sixth of the weight of the whole body. The pectoralis muscles of man thus should weigh more than a sixth of his body. Then the arms with adapted wings could exert a force 10000 times more than the body weight. But this is far from being achieved since the weight of the pectoralis muscles is less than a hundredth of the weight of the human body. Therefore, either the force of the muscles should increase or the weight of the human body should decrease to achieve the same proportion as in birds.

The expedient of Icarus thus appears as a tale since it was impossible. It is indeed impossible either to increase the pectoralis muscles of man or to de-

[12] Prop. 193

crease his weight. Using some machine with levers or another similar contrivance may increase the moment of his power. But the resistance will never be displaced as quickly as the force pushes. Therefore, vibration of the wings provoked by contraction of the muscles will not be able to push the heavy human body upwards with the same velocity as that with which the muscles contract.

The alternative consists of decreasing the weight of the human body. But decreasing the absolute weight of the body is not possible if the machine must remain intact. This weight, however, can be decreased relatively in relation to air. Sheets of lead can float on water by addition of enough cork to make the weight of the complex formed by lead and cork equal to the weight of the volume of the displaced water, according to Archimedes[13]. Nature uses this expedient in fishes the belly of which contains a bladder full of air. This balances the fishes in water where they can remain immobile as if they were masses of water.

Some modern authors thought that the human body can be balanced in air by using the same expedient. Huge vessels with vacuum or very rare fluid inside, would be used, sufficiently big to maintain the human body in the air together with the vessels.

But their hope is vain, as we easily understand. A vessel should be made of some hard metal like brass or bronze. All air should be pumped out from its cavity. But this vessel should be so big that the weight of the volume of air thus displaced would be equal to the weight of the vessel together with the weight of the man who would be attached to it. The vessel thus would occupy more than 22000 cubic feet and the envelope of this sphere should be extremely thin. It would be impossible either to make or to keep such membranous vessel. No pump would be able to create vacuum in it. There would not be enough quicksilver on earth and it could not be handled. Even if vacuum could be achieved in such a huge cavity, the envelope of the brass vessel could not withstand the compression by air which would break the vessel or squeeze it.

Moreover, such a machine with the same specific weight as air could not maintain this equilibrium. Therefore, either it would rise in the air to the limits of the atmosphere, as do clouds, or it would fall. On the other hand, it would be impossible to move such a flying mass because of the resistance of air. Similarly, it is difficult to move feathers or soap bubbles in air. They are rather pushed by some breeze as clouds are moved by the wind.

Thus, it is no wonder that Nature made the swimming of fishes in water so easy and the flying of birds in air so difficult. (Such wondering is usually expressed in all the works on this subject.) We observe indeed that fishes can remain suspended in water easily and without any effort. They can remain in the middle of the water and very easily go up- or downwards. They need the force of their muscles only to move quickly horizontally or obliquely. In contradistinction, birds cannot float in the air. They must remain suspended by exerting continuously their own force, contracting their pectoralis muscles.

[13] De Insid. Flu. Prop. 2 in Arch. Auct.

They repeat frequent jumps through the air and these jumps require considerable force. They do not result from pushing the feet on the ground but from propping up the wings on fluid and inconsistent air.

The way birds fly is not redundant. It is actually the simplest and easiest among different possibilities of flying. The reason why it is different from swimming is that Nature does not attempt impossible enterprises. Actually, water which is heavy can balance fishes but it is impossible for birds made of bones, flesh and fluids which are a thousand times heavier than air, to be balanced by air.

Chapter XXIII

On swimming.

Swimming is a kind of flying. Both are displacements in a fluid milieu. Fishes immersed in this fluid move forwards no less than the birds in air and the terrestrial animals when they walk. The causes and the mechanisms of swimming were not properly considered by the ancients. We shall attempt to clarify that which has been neglected by others.

Proposition CCV

Difference between flying and swimming.

We shall not consider now the difference between air and water as surroundings but rather the operations themselves by which the birds move in air and the fishes in water. From what was said above it appears that flying requires two operations: (1) suspension of the weighing body of the bird in air by successive jumps, by the considerable force of the pectoralis muscles; (2) horizontal displacement of the bird which moves forwards by rowing. Swimming animals also are suspended but by the consistency and density of water itself so that they cannot fall to the bottom. Only to move and proceed through the water do they need the impulse of their own locomotor organs.

Proposition CCVI

Different ways of swimming.

According to Archimedes' theory[1], some bodies float on the surface of water. They are not completely immersed. A part of these bodies emerges above the

[1] De Insid. Flu. Prop. 2 In Arch. Auct.

surface such that the weight of the volume of displaced water is equal to the weight of the whole body, comprising its part immersed and its part emerging. The specific weight of these bodies is less than the specific weight of water. Other bodies when completely immersed in water can remain immobile in any place in the water. The weight of the volume of water thus displaced is equal to the weight of these bodies. They have the same specific weight as water[2]. The bodies which, when immersed, sink to the bottom are heavier than the volume of water thus displaced. Their specific weight is greater than that of water.

All terrestrial animals, as long as they are alive, birds and fishes which breathe and have lungs, are not immersed completely. Part of them spontaneously emerges without any muscular effort. Fishes which breathe and have lungs thus have a specific weight less than that of water. They can move forwards over water like terrestrial animals move on earth. These animals are supported by the density of the water and not by its hardness since it has none. They thus move by using their legs and arms as oars like in boats. This is also the way terrestrial animals, birds, frogs and tortoises swim. Some use one wide oar, moving their tail sideways violently, as one drives a dinghy. Cetaceans move straight over the water using their flat tail like a shovel and pushing water up and down. Dolphins proceed by successive dives and emergences.

All the fishes, with the exception of oysters and a few others, have the same specific weight as water. Therefore, they can remain suspended immobile at any depth. They can move in any direction upwards, downwards and sideways, pushing the water backwards, as we shall see.

But, before going further, the following lemmas will be presented.

Proposition CCVII

Any heterogeneous body immobile or immersed in a fluid places itself in such a way that its centre of gravity comes to lie as low as possible. Tab. XIV, Figs. 1 and 2.

The body PL (Fig. 1) is composite, made part of lead PIH and part of rare and light wood LIH. B is the geometrical centre of the composite body PL and C is its centre of gravity. The body is immersed in water RST. Firstly, its specific weight is equal to that of water. When immersed it does not reach the bottom but remains undisplaced[3]. I claim that it rotates into a position such that the heavier part P comes to lie below.

The volume of water A is equal to the adjacent volume of lead PIH and the volume of water E is equal to the adjacent volume of wood LIH. They constitute two balances AP and EL. In the first the arm from which the lead P hangs turns downwards. In the second the arm from which the lighter wood L hangs turns upwards[4]. The geometrical centre of the composite body PL is neither

[2] De Insid. Flu. Prop. 1 in Arch. Auct.
[3] De Insid. Flu. Prop. 1 in Arch. Auct.
[4] De Insid. Flu. Prop. 8 in Arch. Auct.

raised nor lowered. Consequently, the straight line CB joining the geometrical centre and the centre of gravity of the composite body PL rotates like a pendulum BC about the immobile centre B, describing an arc CD until it reaches BD and is vertical. Then the centre of gravity C is as low as possible, closest to the centre of the earth. P thus occupies the lowest position and L the highest.

Secondly, when the specific weight of the composite body PL is more or less than that of water, in the movement upwards or downwards of the composite body PL the two adjacent balances AP and EL are always constituted. These balances turn about the geometrical centre B as if the body was a boat. In the movement of this boat the two adjacent balances would display their characteristics and the geometrical centre would remain immobile. Consequently, the heavier part P must go down together with the common centre of gravity C to the lowest position, closer to the centre of the earth and the lighter part L must go up, as said above.

Finally, if the composite body PL floats on the water, the same occurs. The sphere EHFG (Fig. 2, Tab. XIV) has its geometrical centre at B[5] and its centre of gravity at C. The part EHF emerges above the surface RS of the water. However the way the composite body rotates about its centre B, the volume of the immersed part FGE is always of the same magnitude since the weight of the volume of water displaced by the immersed part is equal to the weight of the whole composite body[6]. Therefore, the sphere PL remains in the same position as if it were suspended from a nail at its geometrical centre B. But there exists a pendulum BC since the weight of the body PL is concentrated and acts at C. Therefore, the pendulum BC turns about the fixed centre B until it becomes vertical and the centre of gravity lies at D in the lowest possible position.

Corollary

Consequently, the centre of gravity of any heterogeneous body at rest or moving in a fluid lies in the lowest part of this body, closer to the centre of the earth.

Proposition CCVIII

Position of the centre of gravity in animals floating on water.

All terrestrial animals and birds, as long as they are alive, have a smaller specific weight than water. This appears from the fact that they can float. If they are immersed completely in water, they spontaneously go up so that some part of their body emerges. This occurs without any effort by the animal, without movement of the legs which would push the subjacent water.

---

[5] Prop. 19 de Aequip. in Arch. Auct.

[6] De Insid. Flu. Prop. 2 in Arch. Auct.

To find out what part of the animal is heavier and where its centre of gravity lies, one must observe the position of the body of the animal when it floats on water. Quadrupeds and birds, in whatever position they fall or are immersed in water, always come up and emerge prone with the venter below, the back and the head protruding above. Consequently, their centre of gravity is in the middle of the lower abdomen. The upper aspect of the chest, the back and the head are lighter than the other parts. It is in men that this is the least verified because they have a very heavy head. After immersion their head is raised above the water with difficulty, the body being made erect by laborious movement of the legs and arms. In quadrupeds this occurs without any expedient. Their head spontaneously emerges. The cetaceans seem to have their centre of gravity low in their abdomen since they always swim, rest and sleep prone, on water. It is different for the tortoises which sleep supine on water. Their centre of gravity is closer to their back. This results from the broad expanse of the shell which covers their back.

Proposition CCIX

How fishes can rest balanced in water.

According to Archimedes[7], any body immersed in water and not supported by the bottom can remain immobile only if its specific weight is equal to that of water, i.e. if its weight is equal to the weight of the volume of water thus displaced by the body. Fishes can remain immobile anywhere in the depths of water without any effort or impulsion of their tail or of their fins. Consequently, their specific weight is equal to that of water. Hence, fishes are supported by water better and more easily than we are supported by the ground on which we stand. The parts of the fishes below the back and the shoulders are not compressed and fatigued by their own weight as we have shown in our book on the motions resulting from gravity. Fishes do not need legs as do terrestrial animals and birds. They do not become tired and do not perceive the fatigue of standing since their limbs are balanced and thus weightless, and do not compress the subjacent parts. The bodies of fishes can be bigger than those of terrestrial animals, as taught by Galileo, since fishes are not forced to sustain their own weight which exerts no compression, being in equilibrium with water. The way Nature maintains this balance of fishes with water[8] results from the laws of hydrostatics. In the venter of fishes there is a bladder full of air which by its lightness compensates for the excess of weight of the flesh and bones. The volume composed of the solid parts of the fish and the included air thus is made to weigh the same as an equal volume of water. Nature does not use any other means than air contained in the bladder. Oysters, shell-fishes, soles and others which spend their life at the bottom of the sea lack such an air bladder.

[7] De Insid. Flu. Prop. 1
[8] In Arch. Auct. de Insid. Flu. Prop. 1

At the Accademia del Cimento, we observed that a fish, after its air bladder had been disrupted in the vacuum of a tube of Torricelli, could not swim upwards for a month but crawled on the bottom of the pool in which it lived, like a snake.

Moreover, fishes, balanced in water, can move easily sideways, upwards and downwards when they proceed in the water which resists their impulse by its immobility and density. Then the tail leaning on the water, the head and the body are moved forwards by the force of the muscles. Since fishes are in equilibrium everywhere in water as long as all their body is immersed, they can stand and rest in any place either deep or high.

Proposition CCX

How fishes modify their specific weight when necessary.

To maintain the equality of the specific weights of fishes and of the water in which they float it is necessary that the weight and the volume of the fishes remain constant. The density and weight of the water must also remain unchanged. Otherwise the equilibrium would be disturbed which would result in a fall of the fishes to the bottom if their specific weight becomes greater[9] or in their emergence if their specific weight is decreased.

But to maintain the same weight of the fishes and the same density of the water always and everywhere is impossible. Fishes become heavier by eating and lighter by their excrements and perspiration. Water is condensed and made heavier by being mixed with salts, by being disturbed by mud, by ambient cold or lack of sun rays. The same water is made rarer and lighter by being mixed with soft water from streams or rain, by underground heat, ambient warm air and sun rays. These changes occur for different periods of time. At the same time, some parts of the water are enlightened by sun rays and others are in the shadow of clouds or of rocks. The latter parts do not become rarefied like the former. Some areas of the sea are less salty because of the flow of a stream, others not. Moreover, the upper layers of water are lighter than the deeper ones because salt and other earthy particles go down slowly and make the depths of the water muddier.

All these causes disturb the necessary equilibrium between fishes and water. That is why Mother Nature provided the fishes with a mechanism able immediately to adjust their specific weight promptly and easily to a precise equilibrium. Among the elementary bodies none is more expansible or compressible than air. Experience shows that air sent into a pneumatic blunderbuss by the sucker of a pump can be reduced to a fifteenth of the volume which it normally occupies. Air can also be made extremely rare in the vacuum of a Boyle pump or in a Torricelli tube[10].

---

[9] From Arch. Auct. de Insid. Flu. Prop. 1, 2, 4

[10] De Motion. à gravit. pend. Chap. 5. Prop. 121

The following experiment illustrates this operation. The syringe AB is in the same fluid of constant density. The piston CD is tight and adheres smoothly to the internal surface of the syringe. This contains a portion of air DB (Fig. 3, Tab. XIV). This air occupies half the syringe. The nozzle M is obturated. The piston cd is pulled until its base d is brought close to the orifice a and it is fixed in this position by a cotter. The air db inside is rarefied and occupies twice as much volume as it did initially DB.

Then the same volume of air BD is compressed by pushing the piston κδ until it reaches almost the bottom βμ of the syringe and fixing it in this position with the cotter. The volume of air or the space occupied by the same small machines is different in these three positions. CAB is bigger than καβ and smaller than cab. The weight of these three different volumes of air is equal since these volumes are constituted by the same syringe, the same piston and the same portion of air. Consequently, their specific weights are very different. The volume of water EF occupies the space CABM, the volume of water εH the space καβμ and the volume of water eG the space cabm. It is assumed that the three volumes of water ef, EF and φH are equal. The volume EF is bigger than εH and smaller than eG. If the weight of the syringe CABM and that of the volume of water EF are equal, both must have the same specific weight. Therefore, the syringe CABM, when immersed in water, is in equilibrium since its weight is equal to that of the volume of the displaced water[11]. The long syringe cabm is lighter than the volume of the displaced fluid eG since the weight of the water eG is bigger than the weight EF or that of the syringe CB or cb. Therefore[12], the long syringe does not remain immobile in the water but goes up until a part cf emerges above the surface RS. On the contrary, the short syringe κβ is heavier than the displaced volume of fluid εH. Therefore, κβ goes down and rests on the bottom of the vessel.

Imagine that the fish is a machine similar to the syringe CABM (Fig. 4, Tab. XIV and Fig. 10, Tab. XIII). Such a fish contains in its venter a bladder full of air DB. In a middle position this bladder is compressed so as to make the volume of the fish CBM equal to the volume of the water displaced by the fish. Then the fish remains immobile anywhere in the water. If it wants to become lighter to go up, it must loosen its abdominal muscles, thus enabling the elastic air DB to expand in a larger volume db. On the contrary, if it wants to increase its specific weight to go down to the bottom, it must contract the abdominal muscles and tighten the bladder DB like a press so that the air occupies less space δβμ.

Finally, the fluid RTS is heterogeneous with the part VX heavier than VS and lighter than XZ. The fish CABM is in the middle part VX the specific weight of which is equal to that of the fish. The volume EF of fluid is equal to that of the fish CABM and each weighs ten ounces. Then the fish CBM is displaced to the upper and lighter part SV of the fluid. The volume ef of fluid equal to that of the fish CBM is lighter than EF. Some fluid fG must be added so that the total volume eG weighs ten ounces. To remain immobile in this

[11] In Arch. Auct. de Insid. Flu. Prop. 2
[12] In Arch. Auct. de Insid. Flu. Prop. 1

upper part of the fluid the fish CABM weighing ten ounces must increase its volume to make it equal to eG. It can do that easily by expanding the air of its bladder db. The fish CABM is then displaced to the lower level of heavier fluid XZ. The volume $\varepsilon\varphi$ of fluid equal to that of the fish CBM is heavier than EF. A part φH must be removed from this volume so that the remainder εH weighs ten ounces. To remain immobile in the lower layer the fish CABM weighing ten ounces must decrease its volume so that $\kappa\beta\mu$ becomes equal to εH.

Consequently, air in the bladder $\delta\beta$ is compressed and concentrated. Fishes thus rest immobile in equilibrium in the water, go up or down, using this instinctive mechanism. Q.E.D.

Proposition CCXI

Organs and operations by which the equilibrium of fishes in water can be restored after having been altered.

It remains to be investigated how and with which organs the air bladder is constricted or expanded in fishes, whether this air is there at birth, how much air there is, and whether it is renewed, increased and decreased.

The many and strong muscles surrounding the venter of fishes can constrict the air bladder which it contains, thus condensing the air inside and decreasing its volume. But I do not see how this air in the bladder can expand violently to occupy more space. Indeed fishes have no hard ribs but thin and flexible fishbones which easily yield to the compression exerted by the ambient fluid. Therefore, the abdominal cavity cannot expand by the contraction of intercostal muscles as does our chest. The air in the bladder thus must expand and rarefy by some other means which may not be different from what we observe in a goatskin inflated by air. If such a goatskin is strongly bound up the air inside undoubtedly concentrates. When binding up is released the contained air spontaneously returns to its initial volume, as a result of its elasticity. Similarly, the air bladder in fishes may be constricted always more than normal by membranes and muscles acting as usual muscles or as the anal and vesical sphincters which remain constricted and relax only when we wish.

Expansion of the air bladder in fishes, however, seems to be small and would not be sufficient to adapt the equilibrium in areas in which water is soft and light. Then I think that the fishes use their fins, go up to the surface to swallow more air and thus reduce their specific weight. When becoming superfluous in deeper and heavier water, this air is rejected through the mouth. The fishes retain only the adequate quantity of air to remain balanced and be able to stay and rest in the depths without laborious compression.

Air in the bladder of fishes can be increased by absorbing more air and decreased by rejecting the surplus through the mouth, as required by equilibrium. This is suggested by the obvious although thin and narrow duct which connects the air bladder and the fundus of the stomach. This duct cannot have been created to no avail. The bladder is emptied of its air through this duct when the fish is frothing at the mouth in the vacuum of a Torricelli tube.

Proposition CCXII

Fishes do not proceed in water by the impulse of their side fins.

The ancients having observed that boats move pushed by the force of oars and birds fly using their wings, were easily convinced that fishes use their side fins like oars to proceed in water.

This common opinion is obviously erroneous and I wonder that it finds supporters so far. The side fins of fishes indeed are cartilaginous, very flexible, small and narrow as compared to the volume of the fish. The shortness of the levers, the narrowness of the blades and their flexibility make the side fins unable to beat and push water with the velocity and power necessary to move the body of the fish forwards. This is confirmed by the following experiment. If two similar small and flexible oars are adapted to the sides of a boat, these oars having the same proportions in relation to the boat as the side fins in relation to the fish, the boat will move slowly and with difficulty but never like fishes which are as quick as lightning.

Moreover, when fishes move in a pool, their side fins do not beat the water as do oars but remain immobile along their sides. They open only when the course of the fish must be inflexed or stopped. The side fins thus are not used to row but rather to stop by running against the immobile water. Similarly, boats brake their impetus and stop by keeping the blades of their oars in water.

The next experiment is even more convincing. I resected the side fins of living fishes up to their roots using pincers and I put back the fishes thus amputated into the pool. I saw that even without their side fins, the fishes quickly move upwards, downwards and sideways through the water. Consequently, fishes are not moved forwards through water by their side fins but by some other means.

Proposition CCXIII

The two fins underneath the venter of fishes are not used to move but to stand.

The bodies of fishes are inhomogeneous as are those of birds. They are made of heavier elements such as bones and flesh and of very light air, as mentioned above[13]. These different elements are not distributed in birds as they are in fishes. In birds, the heavier part, bones and flesh, occupies the lower aspect of the chest and the lighter part, air, is above. Hence, the centre of gravity of the birds lies in the chest below their geometrical centre. Therefore, when flying, they are prone spontaneously. On the contrary, in fishes, the heaviest element of the bony spine and most of the muscular flesh are in the back, thus above, and the air bladder is in the lower aspect of the venter. Consequently, the cen-

[13] Prop. 112

tre of gravity of fishes is in their back, above their geometrical centre. Therefore, when floating in water they naturally rotate and become supine. Such a position is very inconvenient when swimming and fishes are forced to keep themselves prone artificially. To this end they use the two fins which are on edge underneath their venter. These struts, like the legs in men, by pressing on the water, control the oscillations. To confirm this assertion, I resected all the ventral fins of a live fish using pincers and I put the fish back into the pool where it displayed a peculiar behaviour. It oscillated sideways and could not remain straight, as if it was drunk. This demonstrates our proposition.

Proposition CCXIV

Fishes use their tail to swim.
Tab. XIV, Fig. 5.

Dinghies with one paddle at the stern move straight ahead quickly over the water without side oars. Similarly, fishes are not pushed by their side fins. We observe that they move very quickly in water whenever they paddle with their tail. On the contrary, when their tail remains immobile, they remain immobile at the same place. Consequently, vibration and contortion of the tail is the actual cause of their movement as paddling with one oar at the stern is the cause of the displacement of a dinghy.

This is how it works in both instances. The paddle beats the water behind the dinghy obliquely. This moves the dinghy forwards, although along a tortuous course deviating from a straight path. Deviation actually is immediately corrected either by an opposite movement of the paddle or by an oar maintained obliquely and acting as a rudder. As a consequence, the momentary deviations do not appear and only the straight movement is obvious.

We shall consider the shape and movement of the fish ABC. While the head A and venter B of the fish are straight, its posterior half BFC flexes and vibrates sideways, beating violently the water thanks to the flexibility of the vertebral spine comparable to a strong spring which can be twisted easily to the right and to the left and bounce back. The extremity of the fish is provided with a wide tail DE. This flexible tail is made of cartilaginous rods covered by some thin membrane like the feet of a goose and which can also be tightened or expanded. On the back edge and underneath the venter there are also similar cartilaginous fins which, like the tail, can be flexed, expanded and contracted.

The order of the movement is as follows. The fish ABC being straight starts flexing its tail BFC to the right G. The mobile part BC while rotating about B does not remain straight like the radius of a circle but is incurved twice about B. The side F is moved forwards towards the right G and about F, whereas the extremity D of the tail moves towards the left. This first displacement is not rowing. It is an advancement of the tail similar to that of the legs of a swimming frog. The webbed dorsal and ventral tail fins DE are contracted like the feet of a frog to avoid impeding advancement of the tail BC by striking the water. Displaced in G, the tail is inflexed considerably towards the head A. All the fins are expanded and the segment BG beats the water sideways like an oar, very quickly with all its length, describing an arc which is not part of a circle but rather of an ellipse GD and, in so doing, pushes on the water behind, thus moving the fish forwards from B towards A. The tail has been moved forwards for a second time, from C to H, and beats the water again from H to C. This opposite movement corrects the previous deviation from a straight path and altogether doubles the straight impulse of the fish from B towards A. This is the way fishes move forwards through water.

The tail is also used easily as a rudder to deviate the course to the right or to the left. The two ventral fins on the edge of the venter are used to move towards the depths. When expanded and orientated towards the tail they play the role of a horizontal rudder and act as does the flexed tail of a bird. As a consequence, the head of the moving fish must be inclined to the bottom. On the contrary, the side fins when raised are used to move upwards, like the tail of a bird which when elevated deviates the trajectory of the bird upwards.

Perhaps the tail fins of the fishes, when inclined obliquely upwards or downwards, can act in the same way. This is suggested by the fact that fishes, starting from a position of rest, move upwards or downwards immediately. This quick deviation cannot result from the small force exerted by the side fins acting as a rudder. Similarly, the rudder cannot change the position of an immobile boat.

Proposition CCXV

More motive force is required from the muscles for fishes to swim than for birds to fly.

Using the same method by which we have so far investigated the forces exerted by Nature to carry out the movements of the animals, we can surmise that fishes need more force to swim than birds to fly. This results from an examination of the muscles by which these displacements are carried out. The forces ascribed by Nature to these displacements can be deduced from the quantity of muscles and the abundance of their fibres. Muscular tissue is much more abundant in fishes than in birds and fishes have fewer bones than birds. Therefore, the motive forces exerted by fishes to carry out all their movements are more than those required from birds. Moreover, almost all the muscles in fishes are used to vibrate the spine and the tail. Legs must not be moved since they are missing. Some force is used to constrict the venter and to move the jaws, very little to flex the thin fins. Consequently, almost all the motive force of the fishes is used for swimming which results from vibration of the spine and tail. On the contrary, in birds, the pectoralis muscles which move the wings represent half the flesh of the bird. Thus, Nature uses half the motive force of a bird for flying. We showed that this force is more than 10000 times the weight of the bird. Consequently, the motive force required from a fish to swim is almost twice as much, proportionally, as that required from a bird to fly, and is a little less than 20000 times the weight of the fish.

Proposition CCXVI

Why Nature was forced to exert such great motive force for fishes to swim.

The main difficulty in flying results from the necessity to keep the body of the bird in the air by successive and frequent jumps. Fishes must not be carried in water since the water itself, acting as a stevedore, supports their weight. The fishes thus are in equilibrium as oblivious of any movement sideways, upwards or downwards, as if they were weightless. Their body thus is balanced and can be moved by any motive force, however small, as we showed in our book on the force of percussion[14]. Fishes in water can be pushed and moved by any very small force. They thus do not need such a muscular apparatus. But it is wrong to ascribe to Nature such clumsiness as to provide fishes with so many muscles and so much force for swimming, to no purpose and without necessity. Therefore, it must be recognized that Nature needs so much force for some reason which escapes our notice. Closer consideration shows that, to move bodies in equilibrium, some force is sufficient however small[15]. But this movement is very slow. If it must be quicker, more force is required. A boat in stagnant water can be pulled by a hair but very slowly. If the boat must sail very quickly, the force of a hundred oars is not enough. The difficulty originates in the inertia and weight of the water which must be expelled to leave room for the boat and then be moved back to fill the space left behind by the boat. This displacement of water is not possible without some motive force being imparted to the huge mass of water which must be agitated continuously. This force must be the greater, the more powerful and the quicker the impetus to be imparted must be.

To surmise the magnitude of the motive force of fishes from the analogy with the displacement of a boat, I observe that most of the boat emerges and moves in air and only its small lowest part is surrounded by water. If it was completely immersed, the boat would require a three or four times greater impulse, to move with the same velocity, than previously when it was pushed over the surface and air offered little or no resistance to the displacement of its emerged part. The resistance of the volume of water thus displaced would be three or four times greater in immersion.

Moreover, air opposes no resistance to the movement of the oars whereas water does as a result of its consistency and weight. Therefore, rowers move the oars forwards on edge in the water thus reducing their cutting surface, as frogs and geese decrease the surface of their feet by tightening their fingers and webs when they move them forwards. They expand them only when they push the water backwards with a great impetus. The fact that Nature seeks economy sufficiently shows that it is very difficult to move the feet and the oars forwards in water. Therefore, although fishes do not exert any effort to remain balanced in water and can move slowly in it very easily, because of the inertia of water they need a huge motive force to move very quickly as they are compelled to. This force is about twice that of flying birds.

[14] De vi percuss. Prop. 16
[15] De vi percuss. Prop. 16

Proposition CCXVII

Why birds and quadrupeds swim instinctively.

Since birds and quadrupeds, when alive, have a smaller specific weight than water, some part of their body must emerge when they are in water[16]. The emerging part must be the lightest and the heavier part in which the centre of gravity of the animal lies must be below, like a pendulum[17], and thus immersed. Since the animals cannot live without breathing continuously, it is necessary that the lightest part thus emerging be the head with the mouth which can inhale air. No doubt, if the legs emerged and the head was immersed, the animal would drown.

Nature has taken care that, in birds and beasts, the head, bill and mouth are very light compared with the viscera and limbs. As a result, in water, the venter is lowered like a pendulum and the head and mouth emerge above the surface. Thus breathing, these animals can survive.

Moreover, the long hull of their body spontaneously places itself horizontally because the chest full of air after breathing in occupies the dorsal area. Beasts thus can retain the same position when swimming as when they walk on the earth. When they walk on the earth, they stand alternately on their legs by a habit acquired from birth. Similarly, in water they use their legs alternately as usual and move without difficulty. They swim.

Proposition CCXVIII

Why men cannot swim instinctively and of what consists the ability to swim.

There is no doubt that men, as long as they are alive, have a smaller specific weight than water since part of their body always emerges above the surface of water. This occurs naturally, without effort of the limbs and muscles, as divers experience when they are projected from the bottom to the surface like pieces of wood, by water itself. On the contrary, they cannot dive to great depths without an impetus acquired in the air when falling or without muscular force when they swim to the bottom or without increasing the weight of their body by addition of rocks or lead. They cannot remain immobile on the bottom without taking hold of some rocks. Bathers extending their arms and legs horizontally in water spontaneously go up to the surface without exerting any force, until part of their body emerges.

Although, in that respect we are like the quadrupeds, we differ considerably from them in the structure and configuration of our body and in the position of its centre of gravity and of the centres of gravity of the body parts. In men the head is heavier as compared with the remainder of the body. It is filled with a

[16] In Arch. Auct. de Insid. Flu. Prop. 2

[17] Prop. 210

big heavy brain and made of bones and flesh so that it contains no empty cavities or cavities with air. When the head is immersed and air is expelled, the small cavities of the nostrils and ears fill immediately with water. On the contrary, the head in beasts is lighter as compared with the body and with water. They have a smaller brain. Their face is cancellous and contains long and multiple ducts full of air.

Consequently, the mouth of man does not remain above the surface to breathe. Firstly, the weight of the head immerses the mouth, contrary to what occurs with beasts. Secondly, the part of the animal which must emerge is small as compared to the whole body. The mouth must be in this emerging part to be able to inspire air and, to be able to breathe more conveniently without risk of drowning, the mouth must be some distance above the surface of the water. This is what happens for beasts but not for men. In beasts, the nostrils are above at the extremity of the head, on the long pointed mouth. As a consequence, the mouth necessarily must emerge above the surface like a stick. The nostrils thus being well away from the water, the beasts can breathe easily and without any hindrance. On the contrary, whatever their position in water, men either are unable to breathe or breathe with difficulty. If they are prone, a part of the back, neck and head emerges but the mouth and nose are immersed and they cannot breathe.

If standing erect at right angles to the surface their head does not emerge completely but only the vertex as a result of the small difference between the specific weights of man and water. Then the mouth and nose are below the water surface and cannot breathe. If a man lies supine in water, horizontally or otherwise, his head which is heavier is lower than the chest inflated by air. This is according to the laws of hydrostatics. The mouth and nose being immersed or washed by the waves do not inhale air but water and the man drowns.

Because of the unfitness of their anatomy men cannot take spontaneously in water the position required to keep their mouth and nose emerged. Therefore, they cannot live or move in water.

The ability to swim which is acquired by training essentially consists of raising the head so that the mouth and nose always emerge. This is achieved by a laborious movement leaning on and beating the water alternately with the arms and legs as if they were oars. Similarly, the tight-rope walker remains erect and balanced by beating the air with a pole.

With effort and fluttering of arms and legs we not only maintain the head erect but, extending the neck backwards as much as possible and thus imitating the attitude of quadrupeds, we keep our mouth and nose above the surface of the water. To make this easier, while we swim, we use to maintain the head completely emerged by successive jumps and the body little inclined to the surface so that breathing is not impeded by the waves and we move forwards by fluttering our arms and legs. This refutes the common error of those who assert that beasts swim because they do not perceive the danger to their life. How do they know how the beasts feel about fear? Certainly not from a description by the beasts. But external signs betray their emotions. We claim that beasts are very afraid of suffocating since they shriek, resist and try to escape when we attempt to throw them in water.

Moreover, simple and innocent children audaciously throw themselves in water. Not only do they not swim like beasts but they drown. The beasts thus can swim instinctively not from a lack of fear but because of an appropriate anatomy. Men by nature are unskillful at swimming and can only do so by moving their arms and legs artificially. They lack the ability to raise their head above the surface to breathe.

It remains to be investigated why man unable to swim when falling in water dives and then emerges three or four times before he drowns and remains in the depths. After some days the cadaver emerges and floats like a cork.

There are three causes for man to dive initially. The first is the fall. Man, like wood and all bodies with a specific weight less than that of water, does not remain immobile in water until equilibrium is attained and a well-determined part of his body is immersed. If he is higher he falls down and, while falling, acquires a new impetus which immerses him further than normal. Similarly, a piece of wood falling in water is immersed completely. The second cause of the immersion of man is the disordered and unsuited agitation of his arms and legs which if not carried out in the required order may enhance drowning. The third cause is the increase of the specific weight during immersion as far as air is breathed out and replaced by as much water inhaled in the lungs.

Emergence and repeated immersion occur for the same reason that a piece of wood immersed more than normal repeats ascensions and descents like a pendulum. These movements are enhanced by the causes described above.

After expulsion of all the air, the chest is filled by water. Then the specific weight of the man is higher than that of water. It is thus no wonder that he remains at the bottom like a rock. Finally, after some days, the humours and ingested food in the vessels and intestines ferment and putrefy. They resolve into gases which swell the vessels. Hence the specific weight of the body becomes again less than that of water. Consequently, the corpse must emerge, following the laws of hydrostatics.

Proposition CCXIX

Men immersed for a long time, like frogs and cetaceans, cannot survive without breathing.

Frogs and cetaceans which have lungs spend most of their life in water. They breathe in and out and retain their vital movement as do terrestrial animals. But they easily dive and remain for remarkable periods of time without breathing in again. Meanwhile their heart keeps beating and their blood keeps coursing through the lungs. It must thus be admitted that the exchanges of the blood flowing through the narrowest pulmonary vessels take place with the same air retained in the lungs by the modifications of pressure in the thorax. This appears in the pulmonary vesicles of frogs, which when compressed concentrate the air which they contain. This air in turn compresses the vessels by its elasticity and this results in exchanges with the blood. Indian divers who retrieve pearls from the bottom of the sea would survive in the same way if what is said about them is true, that they can remain a whole hour under water. I know that such a violent effort to keep the inhaled air is painful and harmful. When we dive for some seconds we experience pain. Cetaceans and frogs cannot survive if they remain under water always or for a very long time. Men, being not used to such a regimen, cannot survive under water even for some hours. Although training during their whole life, cetaceans cannot acquire the ability to remain and live permanently under water without interruption despite their natural structure. It thus appears even more impossible for men, however they train, to live under water for a long time.

Proposition CCXX

By using mechanical artefacts, men can breathe when immersed in water. Tab. XIV, Figs. 6 and 7.

So far two machines have been imagined to enable divers to breathe under water for a long time. The first is a cylindrical vessel ABC (Fig. 6, Tab. XIV) hollow and full of air, closed at its upper extremity AE and open at its lower extremity BC. This diving-bell is pulled downwards under water RS by the weight of a big stone P and hung by a rope EF. Experience shows that the air ABC contained in the bell does not escape through the aperture CB as long as the sides AC and EB of the cylinder are vertical. This air becomes the more condensed the deeper the diving-bell is immersed. When divers can no longer tolerate the absence of air, they introduce their head into the bell and recover by breathing. Such a machine is difficult to make and to move. Moreover, it does not seem to be totally able to achieve its goal since in the depths of the sea the air contained in the diving-bell ABC becomes condensed and compressed by the weight of the incumbent water and may become unsuited for breathing. Besides, vapours rising continuously from the water together with those exhaled by the diver may vitiate the air and result in suffocation.

Another machine was devised and consists of a narrow tube AB (Fig. 7, Tab. XIV) made of goatskin and maintained open by a wire spiral. The head of the diver is introduced into the lower orifice BC of the tube, the edge of which is adapted round his neck and chest so as to prevent water to flow in through the junction or through the seams. The upper extremity A of this tube emerges above the surface of the water RS. The immersed man can easily breathe in and out through the tube. This machine, however, is cumbersome since it cannot be displaced in water without external help.

Proposition CCXXI

A man with his head enclosed in a closed container can breathe and survive for several hours if the container is often opened to renew the air.

Man must suffocate after a while if his head is kept in a closed container. This is commonly thought to be similar to the flame which is extinguished by being enclosed in a tight vessel. Some people think that the flame of life thus enclosed must be extinguished rapidly. But this comparison appears to be irrelevant as we showed elsewhere. Maintaining a fire and a flame consists of a very quick expression of igneous exhalations from ambient air. Firstly, this expression cannot occur without circulation of air which requires large quarters. Secondly, the air surrounding the flame must not be vitiated by sooty smoke or moist which both obturate the pores of the embers and of the wick and impede the transpiration of the igneous exhalations as happens when water or even oil is thrown on a fire. In a tiny dwelling air not only is impeded to move but, from moist and sooty fumes, acquires the consistency which prevents the exit of igneous particles. It is no wonder then that the fire is suffocated and extinguished. Actually there is no flame in the heart and lungs of animals. But breathing and the movement of air maintain life in another way and for a very different reason, as we shall see in due time. Experience proves that there is no flame with the properties thus described, in the heart. The lungs of frogs are closed vesicles full of air and their metaphoric vital flame is not extinguished, any more than the fact that the ventricles of the heart are enclosed entails death and suffocation. The dense air made of vapours of water which remains in the vesicles of the lung for a long time and is renewed only after a long interval, must be considerably moistened since the internal walls of the alveoli of Malpighi are always wet. This moisture is changed into vapour by the permanent warmth of the chest as appears when breathing out on a mirror on which countless droplets form like dew. Air in the pulmonary cavities is always damp but does not provoke suffocation. The air in a small container enclosing the head of a man is less moistened by inspiration than the air which is in the lungs. How would this air because of the smallness of the container induce suffocation as a result of the same degree of humidity as it had previously in the chest? Moreover, against coughing and pulmonary conditions, vapours of medicinal waters and dense and abundant fumes are inhaled without causing suffocation. Consequently, when the head of a man is enclosed in a container, suffocation does not result from vitiation of the ambient air by an abundance of vapour.

Moreover, in the winter, children sleep beneath their blankets and do not suffocate.

I do not deny, however, that air can finally provoke unconsciousness and suffocation if too damp and too hot or vitiated by soot. Therefore, if the air contained in a container enclosing the head attained a high degree of warmth, dampness and vitiation, it would easily lead to unconsciousness and death.

There are chemists who use some ointment or nitre which maintains the vital flame by its properties. Hence some people think that animals enclosed in tiny dwellings soon suffocate because the air much inhaled and exhaled has been deprived of this ointment or nitre and thus is made unable to vivify the flame of the heart.

If this was true, frogs, cetaceans and Indian divers could not survive under water for a whole hour since the small volume of air contained in the alveoli of the lungs is not renewed. Once these particles of ointment are consumed, these animals would suffocate.

On the other hand, renewal of air and exchange between air contained in the alveoli of the lungs and external air are absolutely necessary to maintain life. This is not due to the repair of some ointment in the air but to a completely different mechanical cause. We must find out whether and how the danger of suffocation can be obviated.

I observe that divers usually live immersed without notable injury during one minute with the small volume of air breathed in during one inspiration. After thirty successive inspirations, they remain under water for half an hour, although with an interruption. It seems unbelievable that they cannot live for half an hour with the head enclosed in a closed container which contains two or three times as much air as is needed for thirty inspirations, if the necessary precautions are taken of cooling and purifying the air. Consequenlty, if the air in the container was renewed about every half hour or more often if necessary, by opening the container, the divers could live longer under water.

Proposition CCXXII

Construction of a machine with which men can breathe
and live under water for several hours.
Tab. XIV, Fig. 8.

A container made of brass or tin BMHC with a diameter of two feet can contain the head A of a man, like a helmet. Its neck BC adheres to the shoulders, the vertex and the upper aspect of the chest. A waterproof coat of goatskin is tied tightly to the brass neck BC by strings. The man thus equipped if immersed can live under water for several hours breathing freely the air enclosed in the brass container BMHC as long as this air is renewed in due time, as explained below. To solve the two problems mentioned above, a three-foot-long curved brass pipe IQKL is connected to a skin purse K attached at its lower and curved extremity. The two orifices I and L of this pipe are inside the container surrounding the head. Through the anterior one I, air must be blown. The other L opens close to the vertex.

This device has two advantages. Firstly, air expired and blown into the long pipe IQKL is cooled by the water outside as is the air in the big container. Secondly, when we blow air through the mouth tightly holding the orifice I of the pipe this air is breathed out only through the mouth and not through the nose. When expiration and rejection of spirits occur by blowing air into this long and contorted pipe, the droplets of vapour must adhere to and condense on the internal walls of this pipe, as occurs in the top of a still, and flow into the purse K. Therefore, the air coming out of the other orifice L of the pipe, close to the vertex, is cooled, purified and dried. The nose and mouth thus do not inhale this warm and damp air just exhaled but air which has been cooled and purified. The man thus can keep breathing conveniently at least for half an hour without risking suffocation.

The quantity of water exhaled by the mouth as vapour during 24 hours does not exceed one pound, as observed by Santorio. This quantity can be received by the purse easily. The smoke and soot which perspire through the pores of the head and face must not bother you. Children who in winter sleep with the head beneath their blankets are not suffocated by these emanations. Therefore, divers in a similar close place should not be inconvenienced by them either.

The man can live only if the air contained in the brass container is renewed. At the top of the container there thus must be two brass pipes N and O closed by valves or taps to provide the man with air in an emergency when he arrives at the end of his reserve. After the valves have been opened, used air can be blown out through one pipe PMO whereas new air is entered through the other N. Then after closing the valves, the diver can be immersed again.

The goatskin coat must be shaped like the human body and its limbs, and be easily connected with gloves and shoes to enable the diver to move comfortably.

Moreover, the brass container which covers the head must comprise an anterior window 2, 3 closed by a glass fixed with a glue made of flour of quick-lime and white of eggs. The man can see through this window that which is in the water and on the bottom.

The weight of the brass container or the lightness of the contained air are no obstacles. The man with his equipment can be balanced approximately with water by adding leads or by increasing the volume of air in the container. Consequently, with this artefact a man obviously can breathe and live for a long period of time under water.

Proposition CCXXIII

Equipped with the artefact thus described, men immersed in the depths of the water can at choice move and remain immobile as fishes do. Tab. XIV, Fig. 8.

Nobody in his senses would say that the work of Nature is not as simple, necessary and economical as possible. Therefore, machines imagined by men, if they try to equal the operations of Nature and simulate them, undoubtedly achieve the same desired end. Since we want to move and remain immobile under water like fishes, we shall succeed if we adopt the same mechanism which Nature uses in fishes[18]. To be able to use the same machine when under water we must carry a big syringe RS attached to a belt D like a sword and containing a cubic foot of air. The extremity S is closed by welding. The piston T is tight. It can be pulled up or pushed down by a toothed rack XV, a worm-screw and a crank Y. Thus the air contained in the syringe can be condensed or rarefied.

If the man AF dressed with a goatskin, covered by the brass container BGHC, carrying the syringe RS containing air, has a specific weight less than that of water, part of his helmet MC will emerge above the surface of the water.

[18] Prop. 123

Addition of some pieces of lead increases the specific weight of the man and makes it about equal to that of water and only a small part of the vertex G emerges. Pushing the piston T down towards S condenses the air TS inside the syringe. The segment TΩ which was full of air initially fills with water. The syringe with its piston thus occupies less space in water and the volume of the man with his equipment including the syringe also occupies less space than previously. Consequently, his specific weight is increased and, in a first stage, equilibrium with water is restored[19]. Then the man remains immobile anywhere in the depths of the water. If then the piston T is pushed further, the air inside the syringe is compressed more, and more water enters the syringe. The specific weight of the man becomes greater than that of the water and the man slowly falls to the bottom. Inversely, pulling the piston T towards R rarefies the elastic air inside and expels water from the cavity TR of the syringe. This again makes the specific weight of the man less than that of water. As a consequence, he rises until part of the top GM of the container emerges.

The man thus must not necessarily move over the bottom of the water as do crabs. If he wants he can swim with flippers attached to his hands an feet like a frog.

Proposition CCXXIV

Building and using a submarine.
Tab. XIV, Fig. 9.

We showed that men can live for a short time in a closed space, breathing the air contained in this space, if this air is renewed using the artefact described in proposition 223. It thus will not be difficult to build a boat completely closed like a bedroom, which can remain immobile in the depths of the water, as fishes do. If we wish, we can even move this boat upwards, downwards and sideways.

The artefact will be similar to the previous one. The boat by occupying in the water a space equal, larger or smaller can, as fishes do, remain immobile under water or descend to the bottom or rise towards the surface. The bottom EF of the boat ACEG will be perforated at N N N and goatskin bottles ON, ON, etc. inside the boat will be attached upside down to the orifices N in the bottom by small nails or bound tightly round the protruding mouths of the orifices so that water flowing through the holes can fill the goatskin bottles but will not be able to enter the boat through the interstices between the nails. When all the goatskin bottles ON, ON inside the boat are full of water, the boat occupies less space in the water than previously and displaces less water than its own volume. Its specific weight is thus greater than that of water. Therefore, the boat goes down to the bottom like a stone. But if the goatskin bottles are compressed by a lever PO or some other means and the water is expelled from the

[19] In Arch. Auct. de Insid. Fluv. Prop. 1

boat through the holes N N, the boat will occupy more space in water than previously. Equilibrium is attained and the boat remains immobile under water. If then the boat is made lighter than the water it will rise.

Oars X X can be adapted to such a boat, emerging through lateral apertures V, V covered by double goatskins fixed with pegs and tightly tied to the oars, so as to prevent water from entering the boat. These oars, acting as legs, will be able to move the boat under water by pushing on the sandy bottom. We can make the boat lighter than water whenever it must go through some higher passage.

To ensure horizontal movement the oars must end with flexible flippers ZZ which, like the feet of frogs and geese, can expand when pushing the water backwards, and tighten and fold when the oars are moved forwards.

But perhaps it would be easier to move the boat, not with side oars, but with one flexible and resilient webbed oar set at the stern. As a result of its vibration, the boat will be able to proceed more conveniently through the water as fishes do propelled by their tail.

End of Part I

F I N I S

Part II

On the Internal Motions of Animals and Their Immediate Causes

Carlo Giovanni of the regular clerics of Jesus, general superior of the pious school of the Mother of God.

Benevolent reader, greetings.

The most famous Giovanni Alfonso Borelli left us nearly two years ago. The second part of his work *On the Movement of Animals* keeps him present in our memory and in the memory of learned people. What more active and more durable life than that which, trusted to scientific books, is transmitted from hand to hand and repeated by learned mouths without fear? Illusions are dissipated by the winds. What is expounded in this second part was suggested by the author in the first part. He investigates the internal movements, the structure of the muscles, the pulsation of the heart, the blood circulation through the vessels of the whole body in a perpetual movement. He analyses the cause of breathing, the organs of respiration and their function. He explains the spirits and nervous fluids to which he attributes the principal motions of the animals. He describes the digestion, the chyle and how nutrition is carried out with discharge of different excrements through several canals. He skilfully discusses sleeping and waking, the causes and periods of fever, and other good or bad modifications of life and health. He explained all these matters by such an easy theory, by so likely and obvious a mechanism that his compatriots said that he learned the tortuous paths of Nature and the wonderful assemblies of the animals and plants from Nature herself in some cave, like Numa Pompilius used to be taught by Egeria. Indeed he devoted all his energy to show that Mother Nature is active, anxious to avoid superfluous detours, foreseeing, arranging and managing everything simply. Do not fear difficulty because the author involves mathematics. One has not to be a swimmer from Delos not to drown in deep sea. As Socrates judged another book on nature written by Heraclites, everything is so clear that without knowing mathematics you will understand the demonstrations and recognize the evidence. The remainder of mathematics is clear for a physicist and a perspicacious anatomist. Hippocrates suggested to his son Thessalus thoroughly to study this part of mathematics not only for the beauty of life but also for the practice of medicine. He claimed that he applied arithmetic to the periods of the diseases and geometry to the position of the limbs and to the understanding of dislocations. The famous Borelli did not restrict them to these fields. He attempted to expand them to the conveniences of the rest of life and health. He was perhaps the first to use this method. Therefore, when reading this work, you take care of a double life, an intellectual life which generally is enriched by innovations and a natural life for the maintenance of which, healthy or ill, you will perceive as well as a physician, what to avoid and what to do.

You owe this part also, dear Reader, to the benevolence of the heroic Queen Christina who has the noble and high qualities inherent of Kings, is good to the poor, protecting the beseeching. She royally applies these qualities to them when they are alive. She has learned to awaken with inexhaustible generosity the ashes of learned men and to save them from oblivion.

Other works of the author remain which he either discussed in the High Academy of the Queen or elaborated in other times. Not to lose that which such a Man left, we consider editing and publishing these works in a book for their public utility. In the meantime use the present work well. Greetings. In Rome, from the Pious School near St. Pantaleo. December the 22nd, 1681.

Chapter I

On the mechanism of muscle contraction.

After dealing with the magnitude of the motive force which Nature uses to contract the muscles and overcome resistances, the causes of muscular contraction must be explained.

The main and essential action of a muscle is to contract. Contraction can be carried out in different ways. We thus must examine those which are possible and suited to explain the phenomenona observed in the muscles of the animals.

Proposition I

The rope AGCB (Fig. 1, Tab. XV) is fixed by a nail X at A, pulled straight downwards by a weight Z and pulled upwards towards A in the direction BCA by forces of equal magnitude applied at intermediate points CDEF. I claim that the whole length CB remains straight and tense, without puckering and only the upper segment CGA of the rope is loose and incurved; the sum of the forces CDEF is equal to the weight Z.

All the forces CDEF, equal to each other, pull the weight Z in the same direction BCA and none is opposed to another. They are equal in magnitude and velocity and are all directed towards A. Therefore, the segment CDB remains straight and tight. Puckering requires a movement upwards of parts of the rope and a movement downwards of other parts. Before tension, the rope was straight and point C was at D. Therefore, after upwards displacement of point C the length AGC of the rope is equal to the distance AD and thus longer than the distance AC. This is the reason why the rope AGC is incurved. The segment AGC is loose since it is no longer pulled from the nail X as it was initially. Finally, the weight Z is no longer held by the fixity of the nail X but only by the forces CDEF, equal to each other, which impart the same velocity to the different segments. Therefore, the sum of these forces is equal to the weight Z, etc.

Proposition II

Muscular contraction does not consist of simple tension of the fibres similar to that exerted on a rope raising a weight.
Tab. XV, Fig. 2.

The muscle ACB is attached at X in the extremity of the bone XT and at Z in the other bone articulated at T. I claim that it cannot be pulled upwards like a rope. Let us imagine that all the fibres are pulled upwards towards A by equal forces CEFD so that the point B is displaced to D, the point D to F, F to E and E to C. Finally, the proximal part CA of the muscle must remain loose and incurved[1]. This results from the fact that the intermediate segments are not puckered. By hypothesis, they all move towards A and are not displaced by opposite movements without which puckering and contraction cannot occur. But this is impossible and contrary to experience since, in vivisection, any part AC of the muscle remains hard, tense and straight. Moreover, we also observe that the extremity X moves as well downwards towards B if the bone Z remains fixed. Then all the parts CEFD would move downwards towards B with the same velocity and the distal segment DB would become loose and incurved. This is impossible. All the segments at first moved towards A and then would move the other way.

As a further consequence, the length of the muscle would not contract like a pulled rope. This is not what we see. We indeed observe that, if the loose and elongated muscle AB shortens, the same occurs for the segment AD. This is not one movement straight towards one of the extremities A or B but a true contraction, C moving towards E, and E towards C, and so on.

Proposition III

A fibre or a contractile spring with an extremity attached to a nail cannot lift a weight hanging from the other extremity higher and contract more than to its initial length when no weight was hanging.
Tab. XV, Fig. 3.

The fibre or the contractile spring AB is attached at A by a nail X. Its normal length when extended is AB. I claim that, if some weight Z is suspended from B, this weight cannot be lifted by the contraction of the machine so that the extremity B arrives closer to A, such as to D. By its nature, the machine AB requires an extension of the line AB. It can be lengthened or shortened only by an external force. If this force pulls the extremity B towards C, the initial distance AB becomes AC. If the extremity B is pushed towards A, the normal length AB is shortened and becomes AD. The weight Z is an external force which overwhelms the force of the machine AB since it succeeds in increasing its length, pulling away from A, displacing B to C, i.e. elongating the spring up to the length AC. If the contraction force of the spring cannot overcome a force equal to that with which the machine without suspended weight Z resists contraction, it will even less be able to pull upwards a resistance twice as great, i.e. the resistance of the machine itself together with that of the suspended weight Z.

[1] Prop. 1

Proposition IV

If the suspended weight Z is raised above D, beyond the normal extension of the machine AB, I claim that this is carried out by an additional force equal to four times the suspended weight.
Tab. XV, Fig. 3.

RS is the magnitude of the force exerted by the fibre or spring AB to contract after being stretched to AC by an external force and to expand if violently contracted to AD. The force EF contracts the spring AB and reduces this spring to the compressed position AGD. The force exerted by the rope acts against the rigidity and resilience of the spring, i.e. against the force RS. The force RS acting as would a weight hanging from D, opposes the force with which the rope EG tends to contract the spring AGD. This spring AGD is firmly attached at X. Therefore, the force ST exerted by the rope EF is twice the resistance RS due to the rigidity of the spring, as results from proposition 31. Then a weight Z equal to RS is suspended from D. The rope AEG acts with an additional force TY against the tension due to the weight Z. The rope is attached to the nail X. Consequently (proposition 31, in part I), the force TY exerted by the rope AEG is equal to twice the weight Z or RS. Therefore, the sum of the two forces ST and TY exerted by the same rope AEG is equal to four times the resistance RS or the weight Z. Q.E.D.

Proposition V

When considerable weights are suspended, the muscle is not contracted by the own force of the machines the fibres of which contract like a tight spring.

The machines are such that they can contract only after they have been stretched and the stretching force has stopped acting. Therefore, as long as it is not stretched, a muscular chain will not contract but will keep its normal length. Secondly, when stretched by some weight, its equilibrium is maintained and the chain remains immobile as long as the weight is suspended. Therefore, it is impossible that the muscular fibres contract by the simple force of the machine, at the will of the animal. Contraction can occur only when some external force is exerted by either a rope or another instrument. This external force must be able to overcome twice the weight together with the force with which the machine opposes contraction. That is four times the suspended weight[2]. This must be multiplied by the number of small machines in the muscular chain. The force of the machine thus is unable to raise the suspended weight since on the contrary it prevents such raising by opposing the contraction of the machine itself.

---

[2] Prop. 4

Moreover, the force with which the machines of the fibres contract by themselves is equal to the opposite force exerted by the antagonistic muscle. These forces being equal, balance each other and the muscles remain immobile as if no forces were exerted. This is similar to the weight and counterweight on scales. But a balanced force cannot raise a weight of several thousand pounds. Therefore, the vital action of a muscle does not result from the contraction force of the fibrous machines. On the other hand, this contraction force of the muscles is small and weak. Consequently, the muscles cannot exert by themselves the huge force necessary to raise heavy weights.

There is another argument to demonstrate the same point. The more the muscles and their fibres contract, the looser their small machines become. A machine exerts its greatest force when violently stretched and lengthened, like the tensed strings of a zither. When they are shortened they become loose and exert no force. Therefore, the force of a muscle and of its machines is reduced after its contraction and cannot sustain considerable weights.

It can be deduced that the contraction force of the muscular fibres maintains the muscle tense, without any puckering, in a favourable position to enable it to be contracted very violently by an additional external force and to carry out its function.

Moreover, if the muscles contracted by themselves, i.e. if contraction was carried out by the machines which constitute the fibres and not by an external force, whenever the extremities of the muscle would be brought as close together as possible, as occurs for the masseter muscle when the jaws are closed, then the masseter would have to contract spontaneously. This is not what is observed and the masseter muscle remains loose and soft until it is ordered to contract by the will. Therefore, contraction and tightening are not spontaneous for the muscle and do not result from the structure of the machines of the fibres but from another cause completely different.

Nevertheless it could be said that the magnitude of the force exerted by the muscle results from the small machines which constitute the fibres. This force is not exerted continuously because the contraction of the machines is impeded until permitted by the will. Similarly, when the spring of a ballista is tightened, shooting of the ballista is prevented by a cotter-pin or by a tie. As soon as the cotter-pin is removed or the tie unknotted a very violent shot takes place and the arrow is projected. Therefore, we shall show that:

Proposition VI

The considerable contraction force of the muscular machines is not at rest and does not interrupt its operation because it is impeded by an obstacle, neither does it act when this obstacle is ordered by the will to be removed.
Tab. XV, Fig. 4.

We assume that this great force exerted by the muscles by order of the will is normal and inborn in the muscles so that they are always able and ready to carry out this operation. When they are inactive, it is not because of a lack of power but because their action is impeded by some external cause. Similarly, the force of gravity always acts on the bodies. If the latter are immobile it is because their descent is prevented by the resistance of the ground or the force of a subjacent hand. It must be demonstrated that this assumption is impossible and erroneous.

To enable us to verify the hypothesis of the proposition, according to which the muscular chain AD made of machines and attached to a nail A can exert considerable contraction force, this chain must at first be pulled by some external force such as that exerted by the hand M which elongates the chain from D to B. In the normal and balanced position AD it has no force. The force M, as long as it maintains the chain tense, opposes by its permanent tension the contraction force of the small machines of the chain. We must consider that the muscle AB is pulled continuously by the force M and that it raises the suspended weight R only when the will or wish prevails and the force M stops acting and releases the chain so that the latter can contract. All the time during life that the muscle is inactive, the force M must be present, elongate the chain and maintain it lengthened by its persisting action. Such is the hypothesis.

I consider that the obstacle or its force M must be such as to prevent completely the force and contraction of the muscular machines. Therefore[3], the force of the obstacle must be half that of the machines since they balance each other. But the contraction force of the small machines is more than the weight of two hundred pounds which the muscles lift. Consequently, the force of the obstacle is equal to hundred pounds.

Actually, this hindering force is internal and exerted by the animal itself. It acts during the whole of its life with few short interruptions when heavy weights are carried by order of the will. The animal would be forced to considerable and permanent effort during its whole life not to its advantage or to lift weights but only, by a continuous conflict, to impede the action of the machines of the muscles. It would work without doing anything, just to keep the animal immobile. This ridiculous and inconsiderate action seems to be completely absurd and contrary to the most cunning economy with which the actions of the animal are carried out.

Moreover, when a spring is tensed, the fatigue is not perceived by the material of the spring but by the hands which are forced to maintain the tension. Similarly, considerable fatigue would be experienced by the internal force and by the organ of the animal preventing the action of the muscle by maintaining its machines stretched. Therefore, during our whole life, even when we rest and sleep, we should be subjected to permanent and nonsensical work and fatigue. On the contrary, as the return of a spring to its normal length is easy and appropriate to its nature, normal retraction of the small machines of the muscles which raise considerable weights, should be neither hard nor fatiguing but rather pleasant, as rest after work and violent exercise, and return to a normal state are most agreeable. But this is not what occurs and we become tired when

[3] Prop. 31 Part I

we carry heavy weights and not when we rest. Therefore, the contraction force of the muscles is not impeded when we are at rest and heavy weights are not lifted by the force of the machine when the obstacle is removed.

Besides, if some cause impeded the contraction force of the muscle and this cause acted continuously against the contraction faculty, as a weight always tends to go down in balanced scales although it actually does not move, this would destroy and paralyze the contraction faculty. If, on one hand, the muscle remained tense, swollen and hard even when it does not act by pulling a resistance because the machines pull and contract as much as they can and the obstacles permit; and if, on the other hand, that which could be carried out more economically was achieved by a long and laborious action, two causes should be assumed which would be present not together but successively. One would be the existence of a contraction force of the machines in the muscles exactly when heavy weights are carried by order of the will and then the force of the obstacle would not act. On the contrary, when the contraction force would not act, the force of the obstacle ought to be present all the time during which the muscles are inactive.

If Nature does not want anything else than to carry weights at given times and this can be carried out more economically just by contracting the muscles, what would be the advantage of a useless additional mechanism impeding this contraction? Such a mechanism would be useless, only ensuring rest and inactivity of the muscles.

Moreover, the contraction force of the muscle cannot be cancelled by absence of wish, i.e. by nothing. Therefore, in the absence of wish the impeding cause would be present which would cancel the contraction force. Afterwards, when desire surges this obstacle would be cancelled. This is a multiplication of useless operations to no avail. Consequently, without this obstacle one cause is sufficient to contract the muscles by order of the will.

Proposition VII

Muscles are not tensed by a vital movement and do not harden because their elements remain immobile.

Some people believe that water consists of some twigs which, since they are tortuous like eels, flow between clusters of the same fibrils and make water fluid. When they become rigid as a result of cold and they all remain stiff, they acquire the hardness of ice. They also think that immobility alone of the components is a sufficient cause of the consistency and hardness of all solid bodies. Since muscles when acting become tense and hard, they do not hesitate to assert that the will by ordering immobility in the muscles provokes in them tension and rigidity.

This concept appears to be erroneous as we showed in the book on natural movements resulting from gravity[4]. In the present instance, the fact that the

[4] Prop. 143

muscular fibres would become tense and hard as a result of the immobility of their elements would contradict the given explanation.

We assume that the muscle contracts to carry weights. This contraction must result not from immobility of the elements of the muscle but from their moving closer together. This movement can be carried out in the muscles only by exerting a considerable force sufficient to lift the suspended weights. As long as the elements of the muscle are moving they remain soft by hypothesis and no hardness appears so far. Therefore, the considerable force by which heavy weights are raised and the contraction of the muscle are exerted while the elements of the muscle are still soft and before they become hard and immobile. Consequently, immobility is not the cause of the elevation of the weights and of the contraction of the muscle.

Afterwards, the weights are kept suspended while the muscle remains hard. But that does not mean that the energy resulting from the immobility of the elements of the muscle is the cause of this suspension. This action results from an effort and a continuous impulse by which the elements of the muscle, wet and thus not hard, are contracted by some motive and laborious force. This can be deduced from the subsequent tiredness when the effort of keeping the weights suspended results in trembling.

Proposition VIII

Muscles do not contract by drying or by an alteration similar to that which occurs in burning hair.

There are other ways in Nature for filaments to contract and shorten. We observe that hair, nervous and tendinous fibres, intestines, skin and all membranes pucker and constrict when exposed to fire. Muscular fibres, when roasted and broiled, obviously contract. Hence, some people suspected that the muscles of animal could be contracted voluntarily in a similar way. We shall show that this is very unlikely and absurd. Firstly, these operations are carried out by the fire which burns and dessicates to the point of extracting their fluid and slippery elements from the glutinous substances mentioned above. As a result of the lack of these elements, the remaining resistant parts unite and hold firmly together. Hence, their volume diminishes and they contract as occurs for earth and countless similar bodies. But in our instance, this operation does not take place. Indeed, neither dessication nor combustion occurs in muscles when they contract violently. People say that dryness and combustion are not requested for nerve and tendon shortening; only alteration of their temperature is enough; muscles shorten considerably in boiling water although they are still soaked.

I reply that actually the internal parts of boiled nerves and tendons are brought closer together and become harder and denser as a result of the expulsion of the juices which filled the interstices between their fibres. This expulsion is caused by the heat of the surrounding water.

This operation cannot apply to the contraction of muscles, as can be suggested in several ways.

Firstly, as a result of an excess of heat in roasting or boiling, the tendons and membranes contract much more than the muscles themselves. When in action during life, only the muscular fibres tighten considerably and contract but tendons and membranes do not contract at all.

Secondly, when muscles contract quickly their temperature is not above normal and is not sufficient to roast or boil. In fishes and reptiles movements are even carried out in the cold.

Thirdly, hardening nerves by roasting or boiling requires much heat which modifies their structure and character so that they cannot resume their previous softness and looseness. This is not true for the muscles of animals since they contract and then suddenly loosen. It thus appears that the structure and character of the muscles are not altered. Their contraction thus does not compare with the alterations of nervous fibres broiled by the fire or puckered by boiling. Moreover, the heart and muscles of reptiles, when dissected and excised, move, contract and relax for a long time without heat, boiling or broiling, without persistent alteration and commutation of their material substance, but under the action of some motive force resulting from pain. Such a heart thus excised, after having been immobile and quiet, and apparently dead, if it is stung with a needle or irritated by bitter juice, suddenly revives and resumes its pulsations.

Finally, it must be seen whether muscles can contract by puckering their fibres. The following lemmas are presented to introduce this analysis.

Proposition IX

A rope AB (Fig. 5, Tab. XV) attached to a nail X is tensed by a weight Z hanging from B. Its mid-part is twisted and folded in one plane by forces R and S so that the rope remains tightened by the four tensions. The lines of action DO, IC of the forces R and S are parallel and bisect the alternate angles of the rope. I claim that the ratio of the sum of the forces S and R to the weight Z is equal to the ratio of four times the sine of the angle DGO, complementary angle of half the angle GDF formed by the rope, to the trigonometrical radius.

The rope DC is divided in two at F. The line GFE is drawn at right angles to DO, CI. Force S pulling the ropes DG, DF in the direction ODS acts against two resistances, the weight Z and the force which opposes pulling the point F towards D. The lines of action DG, DF are equally inclined to the height OD of the triangle. The point of intersection D is mobile along OD. Therefore[5], force Z is equal to the force which in F pulls the rope DF towards F. These forces balance force S. Thus[6], the ratio of force S to the sum of the two

[5] Prop. 69 Part I

[6] Prop. 69 Part I

equal and opposite forces, Z and its counterpart acting at F, is equal to the ratio of twice the height OD to the sum of the two equal lengths DG, FD. $S/(Z + F) = 2OD/(DG + FD)$. For the same reason, the ratio of force R to the sum of the two equal and opposite forces, the nail X and the force pulling the rope CE towards F, is equal to the ratio of twice the height IC to twice FC or the ratio of twice DO to twice DF. Therefore, the ratio of the sum of the two forces R and S to the sum of the four equal resistances (the weight Z, the nail X, the force pulling the rope DF towards C and the force pulling the rope CF towards D) is equal to the ratio of four times the height DO to four times the inclined rope DF. The four equal forces Z, X and the two opposite forces pulling the rope at F are equal to four times the force Z. Therefore, the ratio of the sum of the two forces S and R to the resistance Z is equal to the ratio of four times the sine DO to the trigonometrical radius DF, etc.

A fourth of the sum of the forces S and R or half the force S supports the weight Z. The other half of S balances half R. The other half of R balances the force exerted by the nail X. And so, the rope remains tense.

In the present instance, the angle GDF is 51.5°. Four times the sine DO is equal to the trigonometrical radius DG. Therefore, the sum of the two forces R and S is equal to the weight Z. If the angle mentioned above is more obtuse, the sum of the forces R and S is less than Z. If it is more acute, the sum of the forces R and S is more than Z.

Proposition X

With the same data (Fig. 6, Tab. XV), if the rope is folded so that DFC is parallel to BG and AE, and to the lines of action DS and CR, the sum of the forces S and R is equal to four times the weight Z.

The opposite directions RC, SD and the ropes AC, CD and BD are parallel. The angles and triangles GDF, FCE thus disappear. Consequently, the lines DO, DG, DF, CF, CI, CE are equal. Four times the sine DO is equal to DG. Therefore, the ratio of the sum of the forces S and R to Z is equal to 4/1.

Comment

The rope can be folded differently by turning its mid-part DFC (Fig. 7, Tab. XV) round the rod MN. Then the two opposite forces SR twist the rod and form a spiral GDFCE. The sum of the two forces S and R is equal to four times the weight Z also.

Proposition XI

The data are the same. If the mid-part of the rope is folded perfectly several times at CDEF (Fig. 8, Tab. XV) by the forces SR, HI, KL, MN, I claim that the ratio of the sum of all the folding forces to the resistance Z is equal to the ratio of four times the number of folds, CDEF, to one fold.

The rope AB is pulled straight by the weight Z. At the first fold C, the sum of the two opposite forces S and R was shown[7] to be equal to four times the resistance Z. But these forces do not do anything other than shorten the rope. They act in the same way as if the rope was attached by some tie at C, attached to the nail X and pulled directly by the weight Z from C to B. Then two additional opposite forces HI provoke a second fold D. Since the upper part DCA of the rope remains tense despite the intermediate fold C, similarly the lower part DB is tensed by the weight Z. The sum of the two opposite forces H, I thus is equal to four times the resistance Z. The sum of the four forces S, R, H, I thus is equal to eight times the resistance Z. The same can be said of the other folds E and F. Thus four times the resistance Z must be multiplied by the number of folds, etc.

The whole length of the folded rope is pulled upwards and downwards by the opposite forces so that half the forces pull the folds upwards and the other half downwards.

Proposition XII

Muscular contraction does not occur by puckering of the fibres as appears in earthworms.

Those who judge immediately from appearances without careful examination are often deceived, not that experiments are useless to acquire knowledge but experiments either are not exact as a result of some imperfection of the senses which are unable to discern minimal differences between things, or observation without a critical mind is often misleading.

This occurs when looking at earthworms. They seem to move by puckering, by incurving their long body into small alternate and successive folds. It is admitted that movement of animals does not occur without the action of muscles. One thus could think that the fibres of all the muscles contract by similar puckering, like the spires of a snake.

Actually, mechanical theory and common sense itself sufficiently show the absurdity of puckering of the muscular fibres. We are, therefore, convinced that muscles contract by puckering neither in the earthworms nor in other animals.

As shown in the three previous lemmas, two forces S and R are necessary to pucker the rope GDCE completely (Fig. 6, Tab. XV). They pull the rope in

[7] In the previous comment

opposite directions, S upwards and R downwards. Their sum is equal to four times the suspended weight Z. When the rope is folded several times, as many pairs of forces pulling in opposite directions are required as there are folds. The ratio of the sum of all these folding forces to the resistance Z is equal to the ratio of four times the number of folds to one fold. For example, if there are one hundred folds and the weight Z is one pound, the sum of the equal contraction forces will be 400 pounds. Such an expenditure of forces results from the useless opposite tension upwards and downwards without which the rope could not contract. But this is not the end of the absurdity. I also consider that no animal motive force can pull the resistance without taking support or being attached somewhere. Such a force cannot move like wind, as was shown. A machine or a fibre SO or RG necessarily must be attached at a fixed extremity, for example to a nail S, or tied at R. If the fibres must contract by folding at a, b and c, d (Fig. 9, Tab. XV) opposite forces f, g, h, i again must be involved. They are also equal to four times the resistance and must be multiplied by the number of folds in the transverse ropes SD, RC (Fig. 6, Tab. XV). Finally, these opposite folding forces do not move like wind but contract transverse ropes by additional folds. Therefore, the sum of these forces is equal to four times the resistances, multiplied by the number of additional folds and so on, ad infinitum. This thus is absurd. Consequently, the muscular fibres do not contract by puckering but by shortening of their whole length.

Proposition XIII

Mechanism enabling earthworms to crawl. How the puckering which we observe in earthworms actually is the effect rather than the cause of the contraction of their muscles.
Tab. XV, Fig. 10.

An earthworm is a long cylinder AE of soft and membranous substance surrounded by short muscular fibres extending over the length of the earthworm and linked like MK, LI, KH, IG, etc. Two movements are observed in earthworms. Firstly, the body ABCD is supported at D and the posterior part DE shortens, the tail E moving towards D. The skin DE puckers with multiple folds. Then, while C is supported by the ground, the next segment DC contracts in the same way with conspicuous folds. As a result of this shortening the posterior segment is swollen as much as required by the reduction in length. The segment CB is shortened in the same way. Secondly, while the belly BC is supported, the head A is suspended, seems to lengthen somewhat and is moved forwards. Then the head A takes support on the ground and B moves towards A, C moves towards B and so on.

It must be shown that this forwards movement of the earthworm does not result from the visible folds which appear in its skin. If the skin and the internal fleshy membranes produced the contraction of the earthworm by puckering, the forces opposite to each other which would produce the folding would have to be multiplied immensely. This would be absurd, as was shown[8]. On the other hand, the same operation can be carried out very easily and quickly by a contraction of the muscular fibres, like in all animals. Therefore, the movement of earthworms is carried out not by this absurd puckering but by muscular contraction which Nature exerts everywhere.

This is how the operation is carried out. The muscular fibre EG is fixed at G which is supported by the ground. The tail E is attached to its extremity, like a weight. Contraction of the muscles shortens the fibre EG and moves the tail E towards G. The same happens for F which is moved towards H. The skin and internal membranes do not shorten but always retain their initial length. They necessarily undergo some puckering as occurs with muscles of bigger and more perfect animals. Particularly, the membranous envelope of the diaphragm of a rabbit puckers into small folds when the muscular fibres contract. Similarly, the segment ED is pulled towards C, CD towards B and BC towards A.

Erection and displacement forwards of the head and anterior part MA, if they were bony or hard, could be carried out easily by the muscular fibres like the long cartilaginous tongue of the woodpecker AB in Fig. 11. The action of this tongue is so nice and cunning that I do not resist describing it in a short digression.

Digression

The osteo-cartilaginous tongue ABE (Fig. 11, Tab. XV) can be hidden in the bill. This tongue ABE with its bony root is twice as long as the bill DO. When the tongue is retracted, its two osteo-cartilaginous appendices BHE are folded up round the head GM. Two muscles CFE are inserted in the extremity of the mandible C and end in the extremities E of two ossicles BHE. Contraction of the muscles CFE pulls the osteo-cartilaginous roots BHE (which are the appendices of the tongue) towards the extremity C of the bill. Thus, the whole tongue AB surges from the bill, its most posterior part BE remaining inside the bill.

Nature provided the bird with four muscles to draw back into the bill at the quickest this tongue covered by ants. Two of them EGMO are reflected above the head up to O. They retract the two appendices EB of the cartilaginous ossicles.

The two other muscles BHIK, NPQR draw back the roots of the tongue and its principal appendix B straight into the cavity of the bill DO down to its deepest L. The tongue is long, almost as long as three finger breadths. The length of the muscles had to be more than twice as much. This is as much as the straight fibres of the muscles can shorten. To accommodate such long muscles in the small space provided by the head and neck of the bird and operate them conveniently, Nature cleverly led two muscles EGMO round the head and attached them to the upper root O of the bill and turned the other two BHIK in a spiral round a cartilaginous pulley IKPQR.

[8] Prop. 12

But such a movement can take place all the more if the tongue is osseous. When the tongue is fleshy and soft as in quadrupeds and man, we also see it surging from the mouth. It seems possible to deduce the mechanism which enables this from artificial contrivances. We see that a cylindrical leather balloon inflated with air becomes straight and surges. This occurs much more easily if the inside of the balloon is spongy because, when air is pumped violently into it, all the spaces must be filled as if it were by wedges until they are dilated as much as possible. Hence, the folds of the ball disappear, its walls are stretched maximally and the balloon itself becomes straight, hard and is moved forwards. There are examples in Nature in which tendinous and soft masses become turgescent, hard and move forwards as if they were balloons. The penis of horses, like a sponge, is swollen by blood and other spirituous juices and becomes turgescent. Its volume is increased in the three dimensions. However, it increases little in width and considerably in length. The lengthening of the penis of the horse results from its turgescence and hardness which can depend on the structure of the cavities which exist all along its length. Moreover, all the length must be surrounded by circular fasciae which oppose increase in width. Therefore, the cavities easily inflated in length can move forwards its huge mass.

The head of the earthworm, that of the snail and the horns of the latter can be moved forwards by the same mechanism. Consequently, the movement of the earthworm can be produced by muscular fibres in the same manner as muscles contract in other animals. The folds in the external skin thus are the effect and not the cause of the movement of the animal. As a conclusion, the muscles, by order of the will, in no way contract by puckering.

Proposition XIV

Muscles do not contract by condensing the length of their fibres and bringing closer together their extremities but their hardness and tightening result from swelling.
Tab. XV, Fig. 12.

When I see that a red-hot metal wire, when it cools, shortens and pulls upwards a suspended weight, I understand that this can occur only if the minimal particles of metal ordered longitudinally are brought closer together. In this instance it must be admitted that the particles of the wire move to each other by opposite displacements. If muscular contraction occurred in the same way, the opposite forces would be either internal and natural, such as the force of gravity, or external. Muscular forces are no proper and internal forces because the muscles do not act by themselves. They need an impulse of the will or animal faculty despatched by spirits or some similar body. This, however, without any support, like wind, would be forced to push any minimal particle of the muscular fibres in opposite directions and each particle would have to exert a force equal to four times the resistance of the suspended weight (as was deduced from proposition 11). Such operation seems to be most absurd and ununderstandable. Therefore, we must not reject the easy and perceptible solutions which Nature uses everywhere. This easy way consists of using the structure of a chain (which was discussed in part I, proposition 114) the links or rhombs of which are widened by wedges and can contract the fibrous chain easily.

As far as the second point is concerned, I consider the swelling of bodies the cavities of which are filled by some fluid or solid substance, such as sponges, goatskin bottles and other containers of the the same kind. Their volume obviously increases in the three dimensions and they become tense and hard. But careful examination does not show any increase in volume of the muscles. The length of a muscle contracts and shortens but its width and diameter actually do not seem to increase. They retain the same size. It is true that muscles become very hard and tense. But many people deny that this must provoke swelling. However, against this point of view, I claim that hardness combining with contraction is an obvious argument for swelling. Firstly, the volume of flesh contained in the muscle when elongated and flaccid is also contained in the same muscle when contracted since both correspond to the same quantity of muscular tissue. Two cylinders AB and DE (Fig. 12, Tab. XV) of inequal heights have the same volume since the excess of length of one CB in relation to the other HE is exactly compensated by a difference in diameter, the latter DH being wider than the former AC[9]. Therefore, if the shortening is equal to a third of the length, the base or the width reciprocally must increase by a third.

But, if this is true, why is this swelling not obvious? It is possible that, after having been loose, when they become tense, the fibres are more united and tighten together. They become thicker as a result of their puckering like the filaments of a rope which, when wet, are more closely tied together, become harder and increase in width. They thus increase in mass without enlarging their external surface. They are said to increase in mass not only because they become wider externally but also because their internal mass becomes wider and their internal cavities and interstices are filled. The density of a body indeed is not measured by the width of this body including its empty cavities but only by its solid parts. When the countless pores of a sponge or leavened bread or pumice-stone are filled by sand or some hardening juice, their mass increases although they retain their previous apparent dimensions.

Who thinks that the mass of the fibres can shorten without twisting or puckering like earthworms? In this puckering, the fibres necessarily must swell and become thicker.

Swelling and tightening of the muscles result not only from puckering and shortening of their fibres but also from an actual increase of the mass of the muscle and from the addition of a new substance. This is shown experimentally in the muscle of the heart. Its big cavities are filled by blood when the heart is loose and inactive. When the heart is tense and active, the blood is expelled and the cavities are filled by the fleshy substance alone. This can be felt. If a finger is introduced into an incision in the apex of the heart of a live

[9] Eucl. 6.32

animal, during each pulsation the finger undergoes a violent constriction and compression as if it were in a press During the pulsation of the heart, its external surface is not increased because of the hardness and contraction of its circular fibres. Therefore, the heart does not seem to increase in volume. However, its internal cavities are filled and closed so that, during pulsation, the volume of the fleshy mass of the heart is twice what it was between the pulsations. We shall demonstrate this. The same occurs in the fleshy ventricles of the birds.

Moreover, one can feel the increase and swelling of the masseter muscle by touching it, every time the jaws compress the opposite teeth violently, like a press.

Proposition XV

Muscles contract with considerable force because their fibres are swollen by an additional substance as if it were by wedges.

It was shown that hardness and tightness acquired by the muscle when acting can only result from swelling of its fibres. It is also obvious that turgescence and swelling do not occur without contraction and shortening of these fibres. It must be shown that these phenomena can be explained only if the cavities of the muscular fibres are swollen by another substance as if it were by wedges. From what has been said it appears that contraction of the muscle cannot result simply from puckering or condensation of the fibres produced either by the force of a machine or by an impulse of spirits. If this was true, the substance of the muscle would not increase but would keep the same volume. After contraction its width would increase as much as its length shortens. This obviously does not happen since the fleshy substance of the muscle of the heart during pulsation acquires a volume twice bigger than when loose. The external aspect is not altered but the cavities are filled by fleshy substance after the blood which was contained therein has been expelled. However, such increase of the volume of the muscle cannot be understood without addition of some new substance or without rarefaction of the cardiac muscle due to dilatation of its pores. The latter seems impossible since empty cavities would result in softness and loosening rather than in hardening of the muscle. It remains thus the addition of a new material substance which either fills the pores or dilates them by its force of percussion. Such a new substance does not lengthen but rather shortens the fibres. Therefore, it swells the fibres much more laterally and increases their thickness. Such swelling would be impossible if the particles of the new substance were not driven into the pores of the fibres like wedges or if they did not expand these cavities by their force of percussion. The action of percussion actually is comparable to the force and action exerted by a wedge. The proposition thus is verified.

Some people claim that Nature dares, and can, use something beyond the laws of mechanics. They speak as if the laws of mechanics were not necessary and as if Nature could carry out impossibilities against the laws of necessity prescribed by Divine Wisdom.

Others do not fear to assert that we must not measure the treasures of Nature with our modest capacity. In the same way as we pull weights, impart different movements to a sword and fidget, so animal spirits can move muscles. But these Gentle Men, if they discussed mathematics and philosophy as well as they are expert in anatomical dissections, would perceive that all the movements of the sword are composed of circular displacements each of which directly results from simple contraction of muscles which tighten and swell. They would thus understand that the force of the muscles is not due to jumps and prances but simply to an impulse resulting in wedging in and contraction of muscles.

Chapter II

On the untrue causes of contraction of muscles given by others.

Although Nature is admirable in all her operations, everybody is very surprised when considering the huge force and energy exerted by muscles and trying to understand the deep causes of, and the instruments and contrivances by which Nature carries out such operation. The ugliness of falsehood is more easily perceived by human imbecility than the beauty of truth. Therefore, we must first reject erroneous opinions. On the one hand, this is a part of science which must not be held in contempt. On the other hand, rejecting errors leads us more easily in our research of truth.

Proposition XVI

Muscles are not moved directly by an immaterial natural faculty.

We shall disregard the arguments which are usually put forward against immaterial natural faculties such as proposed by Galen. I only point out that muscles do not act without contracting. This contraction is powerful only as a result of the hardness and swelling acquired by the muscle. Such swelling cannot be conceived without the arrival and intromission of another body. But the tridimensional mass of the muscle cannot be increased and swollen by an immaterial faculty which has no dimension like an indivisible point.

But some people rather listen to the voice of Aristotle than to that of Nature. These do not fear to claim that muscles can swell as a result of the rarefaction of their elements. They think that this rarefaction can be achieved by an immaterial faculty by way of a simple movement. They also think that heavy or projected bodies are moved directly by an impetus, some immaterial being. I have done enough for them, I think, by showing that the impetus is nothing else than the movement itself which occurs in a mobile object[1]. Impetus thus is not an

[1] De vi percussionis Chap. 17

entity different from movement which continuously exerts an active force by driving heavy or projected bodies. Impetus is rather the persisting effect of the projecting cause itself. It is the movement carried out by the mobile object. Consequently, it is not an immaterial faculty subsisting by itself which succeeds in moving bodies normally and directly.

Against this theory one can mention the excellent experiment of a very learned modern author who observed the following: if a light chicken feather is added to a wooden javelin and this javelin is then thrown with great impetus, the course of the javelin is slowed by the feather; if the feather is pulled away, as often happens, the javelin freed of its impediment moves more quickly through the air. Similarly, a man can run more quickly after dropping the load which he was carrying. From this it thus seems possible to deduce that the cause of the accelerated movement of this thrown javelin results from the great effort of the impetus present in the javelin. This impetus was powerful but could not impart the same velocity because of the impediment of the feather.

To this objection I reply that actually the movement of the javelin is neither slowed by the feather nor accelerated after the feather has been pulled away. Its velocity is equal although it does not seem so. When carrying the feather posteriorly, the javelin rotates on itself like a boat which is forced by the activation of a stern-oar to describe a tortuous course. The sideways deviations, however, are so small that they can remain hidden and the movement of the javelin appears to be slowed only although it is as quick as when it is straight after the feather has been discarded. Consequently, the cause of the movement of the javelin is not the entity of an immaterial impetus, present in the javelin and acting continuously. The cause is the initial impulse given by the one who threw the javelin. Therefore, the persisting movement of the javelin results from this initial percussion carried out by the throwing body.

Moreover, either this rarefaction of the muscle occurs without any addition and then the full muscle would acquire a greater mass without addition of new substance, which is impossible; or the muscle rarefies by addition of many empty cavities and then, inside the yielding pores of a soft substance such as flesh, empty spaces would be able to persist during muscular contraction, which is impossible, not because Nature abhors vacuum but because fluid and soft substances combine and retract as a result of their weight. Consequently, swelling of the muscles by any simple immaterial faculty is impossible.

Proposition XVII

Neither spirit nor gas such as air can contract muscles.

Some people assert that animal spirits, very small bodies like gas, are the motive cause of muscles. But there are countless objections which they cannot refute. This spirituous gas must move the muscles by entering their cavities like wind or inflate the muscles by its abundance and plenitude or rather contract the muscles by enlarging their pores by ebullition and some rarefaction. These operations appear to be impossible because the action of a muscle is a pure contraction of its length. The extremities are brought closer together by moving in opposite directions with swelling and turgescence in width. Such action cannot be conceived without a wedging action since these spirits are driven into the pores of the muscles like iron wedges and they contract the fibres with so much force that the muscles can raise heavy weights. Such an operation, however, requires firstly wedges, secondly hardness of these wedges, thirdly a motive force to drive them in very violently. As far as the first prerequisite is concerned, I do not see whence so much "spirituous gas" would come and in what cavity these spirits would be hidden. If, however, they are assumed to be present, they will have no hardness since they are looser and softer than sponge, wool or cotton. Air enclosed in a water-skin resists compression from a weight acting on it as much as if air were a mass of iron. But this phenomenon is utterly different from the action of a wedge or a muscle since it does not result in lifting the supported weight but only in resisting its compression. To exert the action of a wedge, it would be necessary to be able to insufflate and drive in or to produce additional air in the pores themselves. Hence, this most difficult question reappears: what causes this huge motive force which pushes an additional huge quantity of spirituous gas into the pores of the muscle? The difficulty consists in that violent projective impetus cannot be imparted to bodies of very small density like cotton or air. Indeed we observe that such light substances, even when projected by the force of the fire of a war machine, suddenly fall and are scattered. Similarly, air pushed by any wind cannot be driven into the gaping pores of a sponge so as to swell its volume and make it turgid. To say that this additional quantity of spirituous gas is generated by order of the will inside the pores of the muscle when the muscle contracts, and disappears or is destroyed when the muscle relaxes, is like a dream. On the other hand, ebullition does not seem possible in muscles. This is obvious. Neither can heat be produced then in the muscles beyond the usual temperature. Even more: fishes are cold. Finally, this joke is contradicted by the following common experiment. Muscles are incised longitudinally in a live animal and immersed in water in which they are violently agitated because of the pain. With such abundant and violent ebullition of this spirituous gas thus provoked in the muscles, countless bubbles of air should leave the wound and rise in the water, as occurs with red-hot iron. But this does not appear at all. Consequently, muscles are not swollen and moved by material spirits.

Proposition XVIII

Muscles cannot contract by swelling of their pores by juice or blood.

There are distinguished authors who believe that muscles are swollen by blood sent copiously from the arteries and which cannot flow away through the veins. This occurs in the penis. The action and effect are different in these organs. The penis is not only enlarged but also lengthened whereas muscle is shortened. However, since both are irrigated by blood and become red, their turgescence and hardening require the filling of their cavities by the same blood.

Actually, since the spongy substance of the penis is expansible in all directions, the penis can increase in the three dimensions. But the fibres of the muscles are so resistant and firm that they cannot be elongated, no more than the strings of a zither. Therefore, when their pores are filled by the flow of blood, they expand only in width and not in length. The muscles thus must shorten as a result of the puckering of their fibres. But it is easy to demonstrate the falsity of this opinion by showing the difficulty of this operation. To lift a big rock iron wedges are not enough. Above all, considerable motive forces must be found to move the rock by hammering in the wedges. Similarly, for muscles to lift big weights it is easy to have an abundance of droplets of blood which can be driven into the pores of the muscular fibres like wedges. But where shall we find the motive forces sufficient to elevate considerable weights by driving small bloody wedges violently into these pores? Such huge motive force as is necessary either must be proper to the blood or results from an external impulse. It cannot be proper to the blood.

Firstly, the force of blood appears to be very weak. This is surmised from its slight wavy movement which is so numb and weak as hardly to overcome the weight of a straw. This movement can rather be considered as a fermentative movement than as a displacement. Therefore, if the muscles were contracted by an inflow of blood then a minimal motive force would overcome a force hundreds and even thousands of times greater than itself, directly and without any machine. This is obviously impossible. Consequently, muscles do not swell and contract because of an inflow of blood.

Secondly, if blood was supplied with a motive force sufficient to lift such heavy weights by swelling muscles, it would certainly be able to overcome smaller resistance by its own force. Indeed it would be able to push itself through the arteries and to flow back through the veins, carrying out its own circulation. But this does not occur. Nature is compelled to rely on the impulse of the heart to distribute the blood through the arteries and she needs the help of valvules and the compression of the muscles to bring back the blood through the veins to the heart.

Thirdly, as shown by experience, compression by the muscles of the abdomen, by those which are involved in respiration and by those of the limbs combine with the peristalsis of the veins to propel the blood through the veins. The absence of such impulse would result in venous stasis. In other words, the effect would produce its cause: compression by the muscles which is thought of as the effect would produce the movement of the blood which was supposed to be the cause of the contraction of these same muscles.

Fourthly, the blood ascends with difficulty through the veins of the legs and arms because of its weight. It flows upwards so slowly that the veins are often varicose and protrude through the skin. But, when the legs and arms are raised, their veins immediately collapse and the skin becomes white and flat because the blood can flow downwards easily. The motive force which is supposed to belong to the blood thus is so weak that it cannot overcome the weight of the blood itself. Therefore, this force cannot possibly raise weights which are hundreds or thousands of times its magnitude, by swelling the muscles. Consequently, the muscles cannot swell and contract alone by the force of the blood.

This is confirmed, moreover, by the difficulty and impossibility of the operation. Blood is not completely fluid. For its most part, it consists of fat, glutinous and fibrous elements. If very fluid water oozes in narrow pores and canals not suddenly but gradually, it will be much more difficult for blood with its more fatty and more glutinous consistency to enter and leave these narrow pores rapidly in abundance. Therefore, the glutei and quadriceps muscles, if they can be filled and swollen by a great mass of several pounds of blood to be brought from here and there, would swell and deflate not instantly but in the course of one or two seconds.

I remember having laid a naked man on a table. The table was supported on the edge of a prismatic piece of wood corresponding to the middle of the buttock where the centre of gravity of this man lay. Then the calf and quadriceps muscles were contracted, thus swelling, to stretch the legs. This should have disturbed the equilibrium, giving advantage to the feet at the expense of the head, by moving from the whole body to the legs the great quantity of blood required to swell the distal muscles. However, this did not occur.

Moreover, if the muscles are overabundantly swollen by blood, the exits from the muscle must first be closed or restricted. Afterwards one would expect that, since as much blood is retrieved from the arteries as was required to swell the muscles, some time would be necessary to evacuate this mass of blood into the adjacent very small veins. But this is in contradiction with experience since muscles, even the vasti muscles do not take a long time to swell and to deflate. They do it at once.

It must be added that a prolonged action of a muscle would impede the blood flow through the vessels penetrating into this muscle. Swelling of the flesh of the muscle, by its mass, necessarily compresses and squeezes the small internal vessels. As long as they are closed the blood cannot flow through. But this does not occur since blood never circulates more abundantly and energetically than when the muscles work for continuous movement.

Finally, muscles contract not only while blood is transfused to them from the arteries but even when they are very altered and this blood flow is missing, after the arteries have been resected. The muscle of the heart of a turtle, excised and immersed in water, pulses during two or three hours without one drop of blood. Therefore, muscles do not contract because they are swollen by a mass of blood, which they are not, but for a completely different cause.

Proposition XIX

Muscles cannot be contracted by the impulse imparted to the blood by the motive force of the heart.

The heart appears as the primum movens of the animal system. Therefore, those who think that all the other movements of the animal depend on the heart are worth indulgence. We indeed see that the heart pulses continuously and powerfully sends blood through the arteries to all the parts of the animal.

This suggested to some people that the muscles could be contracted by the motive force of the heart which would quickly send and drive the blood into the pores of the muscles and thus swell them as if it were with wedges. Others thought that the heart is not the only and principal cause of but only a help to the contraction of the muscles. We shall demonstrate that both assertions are false and impossible.

Without any doubt the heart is one of the main muscles of the animal. It is made of fibres convoluted into spirals, tendinous and fleshy of the same nature, consistency and structure as the fibres of all the other muscles. Its fibres seem to swell, harden and contract like those of the other muscles. Since Nature does not like mutiplying causes and instruments nor variety, the heart must swell and contract in the same way and for the same causes as all the other muscles.

If we thus show that the muscle of the heart cannot contract and pulse as a result of the impulse of the blood sent into it, this will plainly demonstrate that the muscles are not swollen by this same motive force but are moved by a completely different cause.

As appears in anatomy, the heart is provided with two coronary arteries which send blood into its own muscular mass. They originate from the aorta before the latter leaves the pericardium. They have their own valvules preventing backflow of the blood. There is a mass of blood in the flesh of the heart itself. This blood comes from the arteries, is absorbed by the coronary veins and is carried back to the right ventricle in a peculiar circulation, as Harvey says.

The immediate motive force which moves the blood through all the arteries is the energetic contraction and pulsation of the heart which, like a press, squeezes out the blood contained in the left ventricle into the aorta, hence into the coronary arteries, and hence into the fleshy substance of the heart. Thus intromission of the blood into the pores of the heart muscle is the effect of the contraction and pulsation of the heart itself. Since an effect cannot produce its own cause, it is impossible that the intromission of the blood into the pores of the heart by a force similar to that of a wedge, contracts the muscle itself. Consequently, the heart muscle must be contracted by a cause completely different. Therefore, the other muscles of the animal cannot be swollen by the blood either.

To clarify the problem, I consider that, at rest, the heart is turgid. Not only its ventricles but also its vessels, pores and interstices between the fibres are saturated by blood like a sponge. During pulsation the whole mass of the heart is contracted and hardened by the swollen fibres. Therefore, the blood must be expelled like by a press, from the ventricles, the vessels and all the spongy substance of the heart. Arrival of new blood from the coronaries into the pores of the heart is impeded since, during the pulsation, the heart retains its hardness. This cannot be understood without constriction of the cavities. This contraction of the heart is so energetic that not only it spills all the blood contained in it but pushes this blood to the very extremities of the animal. Therefore, the blood forced out and sent into the coronary arteries has no greater force than the cause of its propulsion, i.e. the contraction of the heart. Thus, a smaller motive force which is the effect cannot by pushing overcome the greater energy of the impulse which is the cause. It cannot open the lumen of the vessels and the pores so tightly closed. Consequently, during the pulsation of the heart, the blood will not be able to flow from the coronary arteries and be driven into the heart. But, if there is no blood, there can be neither swelling nor filling. Thus it is impossible that the blood coming from the coronaries carries out the swelling by which the heart hardens during its pulsation.

From this demonstration it appears that the coronary arteries work differently from all the other arteries. The other arteries push and transmit the blood at the same time and by the same action as the pulsation of the heart. But the coronaries transmit the blood into the mass of the heart between the pulsations when the heart is relaxed and its vessels and pores are dilated like a sponge to receive the blood. During the pulsation of the heart the parts of the coronary arteries, which are in the pericardium before reaching the heart, fill with blood almost up to rupture. Return of the blood is prevented by their valvules and downstream the vessels are constricted by the contraction of the heart. The force, with which the transverse fibres of these arteries attempt to free themselves from this violent tension (like the tensed strings of a zither), after the pulsation, suddenly propels the blood which was retained and pushes it into the ramifications of these vessels scattered in the whole mass of the heart. During the next pulsation of the heart, the blood contained in these vessels is prevented from flowing back by the valvules and by the push of the arrival of new blood. Therefore, it must move through the open lumen of the coronary veins and, hence, into the right ventricle.

The muscular substance of the heart thus is filled and swollen by the blood from the coronaries, not when the heart pulses but rather when it relaxes. Consequently, the heart is contracted and hardened not by the blood but by another cause.

We thus shall demonstrate again that the blood similarly sent by the arteries into the other muscles of the limbs can neither contract these muscles nor even help this contraction.

The same motive force by which the heart pulses also expresses the blood through the whole body as if it was by a press. The blood expelled by a force of this magnitude, like a stone of fruit squeezed between two fingers, swells the muscles with which the limbs raise heavy weights. Actually, forces equal to each other and the same energy produced by the same organ cannot, directly and without any machine, overcome resistances considerably different from each other.

We suppose that a motive force which could be equal to a resistance of 100 pounds is required to drive the blood through the whole body of the animal. We also suppose that this motive force must not only drive the blood but altogether support directly a weight of 1000 pounds. Nobody in his senses will deny that such an action requires a motive force ten times greater.

We shall consider the twofold action of the heart. Firstly, when the muscles of the limbs are inactive contraction of the heart only pushes the blood through all the body of the animal. Secondly, contraction of the heart pushes the blood and at the same time drives it into the bigger muscles of the animal with enough energy to lift a weight of a thousand pounds. The heart seems to use ten times less of its effort and motive force in the former action than in the latter. This is obviously false. The heart actually beats at the same quiet pace and with equally laborious energy in the first instance as in the second. This is felt in vivisection. A finger introduced through an incision in the heart is always squeezed with the same force, like by a press, whether the muscles of the animal are acting or whether they are at rest. Consequently, the muscles of the limbs cannot be swollen and contracted by blood sent by the force of the heart pulsation.

Another confirmation consists in that the heart never pulses more strongly and never pushes the blood more energetically than during high fever. Thus then all the muscles ought to be energetically activated up to prance. Actually not only are they not activated more violently but on the contrary they are very quiet. The proposition thus is demonstrated.

Proposition XX

Investigation of the cause of the contraction of a wet rope.

Some people will perhaps suspect that the turgescence and hardness which constitute the contraction of the muscles may be due to the inflow of blood in these muscles in a way similar and for the same cause as wet ropes quickly suck in water. They are swollen by water with as much force as if it were by wedges and, not only contract, but even exert a force sufficient to lift the heaviest weights. Therefore, this opinion must be examined.

The cause of the contraction of a wet rope must be sought for.

Air pressure cannot be the cause of the contraction of a wet rope. This results from what was demonstrated in the book on natural movements resulting from gravity[2]. Indeed, we showed that water cannot be driven into the smallest cavities of pipes, sponges or into the interstices between the filaments of a rope by the weight or the pressure of ambient air.

Although water rises pushed by its own weight in capillary tubes and in sponges, this cause does not seem to be sufficient to contract a rope with a heavy weight suspended. This can be demonstrated easily. The contorted fibres of a rope form a reticular bundle of chains the links of which are dilated by water wedges. This confirms what was demonstrated above[3]. The rhomb AEBD (Fig. 13, Tab. XV) is suspended from A and pulled by a weight Z attached to B or to K. A force however small can dilate the ropes AEB and ADB and thus raise a weight Z however big[4]. A hanging weight Z of more than 100000 pounds can be lifted by a force of one pound which widens the small ropes until the angle EAD formed by the ropes is increased by two seconds of

[2] Chap. 8 Prop. 183
[3] Chap. 16 Part I
[4] Prop. 99 Part I

an arc. Then the weight Z is raised and rises for a minimal distance. This distance is smaller than 1/100000 of the length AEB of the rope.

This contraction is multiplied by the number of rhombs in the chain AK. However, the same weight Z is hanging from one rhomb AB as it is from all the countless others BC, CK, etc.[5] and the dilatation is equal for all. In any of the rhombs the dilatation ED is proportional to the shortening of half the height AR. Therefore, the dilating forces are multiplied not to support a bigger weight but to raise it higher. If the shortening of the chain AGK is 1/100000000 of its length, since the angles of all the rhombs are equally widened by all the forces, this widening will not be more than two seconds of an arc.

But experience shows that a wet rope the diameter of which is equal to the fourth of the breadth of the ring finger shortens by much more than 1/16 of its length. This requires a very large angular opening[6] of the ropes, appproximately by 41°. The force dilating the ropes then is more than one and a half times the resistance Z which weighed not less than 16 pounds.

In our experiment the volume of water soaking a sheet or a cylinder of smaller height, as measured from the shortest fibres constituting the rhombs of the rope, must have been less than two drops. However, the energy of the weight of so little water was enough to raise 16 pounds. In the table given above[7] it appears that these two drops should have raised about eleven pounds which is false. It is impossible for ropes to be dilated up to an angle of 41°, to be shortened by a 1/16 of their length and to carry a weight of 16 pounds by the minimal weight of two drops of water. To carry out this work about 11 pounds of water are required, as results from calculation.

On the other hand, the motive cause of the contraction of the rope cannot be anything else than the series of drops which swell the rope, as wedges would do, by being driven into the spaces delineated by the filaments contorted into spirals which compose the rope. Such intromission is possible only if the ways closed or restricted are violently opened by dilatation of the interstices which were virtual because of either the consistency and hardness of the rope or the pull of the suspended weight. Both force the fibres very close together.

The magnitude of the force necessary to drive the wedges into the pores of the rope can be suggested by the importance of the resistance which they overcome. The pores of very hard wood like boxwood, walnutwood and similar ones are very tight, rigid and difficult to dilate because of their hardness. However, we observe that the fibres of these woods, when wet, obviously become dilated and move away from each other. They swell and occupy more space in width. This can occur only if the smallest water molecules are driven violently into the pores of the boxwood. Therefore, we are compelled to admit that the droplets of water are able to exert this considerable force. Otherwise such operation would not take place.

It is thus worth-while to consider the mechanism of this admirable operation.

---

[5] Prop. 99 Part I

[6] Prop. 99 Part I

[7] Prop. 99 Part I

Water obviously is driven by its weight into any pores empty or containing some very rarefied air since, according to Archimedes[8], fluids with a lower pressure are expelled by other fluids with a higher pressure and give way to the heavier ones. In these areas where pores are so wide and voluminous these pores thus must necessarily absorb the molecules of water. But it is difficult to understand how the minimal weight of the droplets, by drilling the closed pores of wood or of a rope can expand them and separate the fibres tightly united by the hardness of the wood or of the rope and by the tension of a suspended weight.

I think that this problem can hardly be solved except by the theory expounded in our book on the force of percussion[9].

Minute droplets of water can be driven into pores of wood or of a rope, which are bigger than these droplets. But this is impossible without displacement and a displacement cannot take place without an impetus. Consequently, when molecules of water are carried into wider pores of the rope, they do not act inertly simply by their weight but are subjected to a movement and an impetus against the resistance of the tightness of the fibres of the rope and the suspended weight. Actually, the small force of the droplets of water subjected to a movement and an impetus can overcome any immobile resistance whatever its magnitude. This was demonstrated[10]. Consequently, a flow of minute droplets, by the energy of their movement and percussion, can be driven into the pores with the force of a wedge, or into the interstices between the fibres of a rope and dilate these with a force sufficient to overcome the resistance of a suspended weight, heavy and immobile.

Proposition XXI

Muscles are not inflated and contracted by the blood violently injected or by air pressure or by the weight of the blood in the same way as wet ropes contract.

We showed[11] that water cannot be driven into capillary tubes, in sponges and filters by air pressure. Using the same experiments and arguments, it appears that blood cannot be forced into the pores of the muscles by the pressure of ambient air.

It is possible to prove this even more obviously by the following experiment carried out by the most zealous Boyle and also in our Accademia Del Cimento. Many animals, shut up in the vacuum of a Torricelli tube, move and are agitated for a long time before they die. In this space almost deprived of air or where air is very rare there is no air pressure to drive blood into the muscles and swell and contract them. Therefore, the violent convulsive movements

[8] De Insiden. Fluido.
[9] Prop. 190
[10] De vi percuss. Prop. 190
[11] De Motion. natur. à gravitate pendent. Chap. 8

which are actually observed should not take place. It must be concluded that blood is not driven into the muscles and does not contract them by air pressure.

Furthermore, the theory reported in the preceding proposition which applies so well to the contraction of a wet rope is not sufficient to explain muscular contraction. We shall demonstrate this.

Firstly, a rope swells and shortens when all its internal parts are wet. It loosens and lengthens when completely dry. Inversely, the muscle loosens, softens and lengthens when maximally irrigated by blood, like a sponge. It hardens, tightens and contracts when the blood is expelled totally or for its most part. This appears in the heart and also in the other muscles which become paler when contracted and are red when relaxed. A muscle cut by a knife when loose bleeds abundantly contrary to what happens when the muscle is swollen and hard.

This is confirmed by the example of the penis which is tensed and made erect by an afflux of blood, as is the rope by water. When turgid, it is red, contrary to the muscle. Consequently, swelling and tension of the muscle do not result from violent intromission of blood into its substance but from another cause completely different from that of the swelling of a rope by water.

Secondly, blood is continuously sent into the muscles by the arteries which are always pulsating willy-nilly. Therefore, if the muscles, like a wet rope, were swollen and contracted by the blood driven by the force of the arterial pulse, either they would always remain contracted as occurs in convulsions or they would pulse like the heart.

Somebody may object that there are some valvules in the muscles which impede the blood entering the pores of the muscles except when they are opened by order of the will[12]. But such a subterfuge does not make sense for several reasons. The motive force which must drive blood into the pores of the muscles like wedges, is so great that it can lift considerable weights. This force acts against the resistance of the valvules before the entry, during all the time when the muscles are not acting. Then the muscles act against the resistance of the limbs and suspended weight. Consequently, Nature is overworked continuously and uselessly throughout the lifetime to enable the muscles, when they swell for a short time, to support the limbs and suspended weights. Nothing more stupid can be imagined. Moreover, the force exerted by the valvules ought to be equal to the impulse of the arterial blood and should also be exerted all the time, even when the muscles are inactive. This would be useless and nonsensical.

Moreover, if the force of blood swells the muscles as wedges would do and as water swells a rope, there remains the greatest problem of how to loosen and deflate: all these small wedges of blood inside the pores of the muscular fibres must be removed. But this is most difficult and can occur neither instantaneously nor without a force equal to that with which these wedges had been driven in.

---

[12] From Prop. 6

Thirdly, since the face of a man blushes when he is angry or ashamed and grows pale when he is afraid, it is commonly thought that blood is moved by emotions and directly subjected to the will. I cannot endorse this opinion. The blush and afflux of blood to the cheeks directly result from vigorous and frequent activation of the heart in anger and shame. This in turn entails the diffusion of more blood through the arteries than can be absorbed by the veins of the face. Hence, the capillary veins are full and turgescent and the skin blushes. Inversely, in fear the pulse slows down and less blood flows through the arteries than returns through the veins. Hence, the empty capillary veins entail a pale colour. Will by itself and directly cannot produce blushing or growing pale. Consequently, blood cannot be sent to a well-determined muscle and retrieved from another by a direct action by order of the will. It is rather moved, according to Nature's requirements, by normal organs which work willy-nilly without our knowing and even when we sleep. But muscular swelling and contraction are directly subjected to our will which is unable to move the blood. Consequently, the muscles do not act because of blood driven into them like wedges.

Fourthly, a rope cannot be soaked suddenly but only progressively. The particles of liquid cannot freely move through the narrow pores of the rope and penetrate into its deepest parts. They arrive there slowly, as taught by experience. In contrast, muscular contraction is very quick, like a lightning. Consequently, this contraction cannot result from a displacement of blood penetrating into the pores of the muscles.

One may consider the example of a filter or of a sponge: when dry they can be swollen by water only after some time; if wet they suck water very quickly up to turgescence. But the structure and mode of action of a rope is very different from those of a sponge which is very loose. After being squeezed and its water content expelled, a sponge expands again and its pores fill with air. Therefore, a sponge can suck surrounding water easily and rapidly since the wet pores do not slow down the entry of water. But water cannot be squeezed out from a hard rope and the pores of the rope cannot fill with air afterwards since the rope undergoes neither dramatic reduction in volume nor expansion. Consequently, the expanded pores of a wet rope cannot be filled by air and afterwards absorb water again, as a sponge does. Although soaking a rope would not require much time, drying would be very difficult and would require a long time. As would be expected such drying would occur by evaporation of the water. No artificial means that I know has been imagined so far to dry a rope instantaneously. On the contrary, a muscle empties itself and deflates at once. Consequently, muscle does not contract by intromission of blood in the same way as a rope is swollen and shortened by water.

Finally, the heart of a turtle removed and divided into parts, and the muscles of eels or snakes, immersed in water, move and contort for hours although blood neither flows into the pores of these muscles nor can be provided since there is not a drop of blood available. These muscles cannot either be swollen by some internal fluid, like ropes, since they are equally wet when they contract and when they relax. Consequently, there is no way the muscles can contract, like wet ropes, by an afflux of blood.

Chapter III

On the likely causes of contraction of the muscles.

From what was said above, we are forced to admit that muscular contraction depends on causes very different from an afflux of blood driven either by the motive force of blood itself, or by air pressure, or by the impulse of the heart, or by the weight and movement of blood itself. Muscles do not contract in the same manner as wet ropes do. We shall attempt to investigate as best as possible the actual cause of this admirable operation and the organs and mechanisms through which it is carried out.

Proposition XXII

Contraction of the muscles requires two causes. One is present in the muscles themselves, the other is external to the muscles.

With few exceptions, the muscles contract when we wish, by order of the will and brain which is the seat of sensation and movement. The orders of the will are transmitted by the nerves as everybody says, convinced by experience. This in itself refutes any action of an immaterial faculty or of airy spirits. Thus, some material substance must be transmitted through the nerves to the muscles or some commotion must be communicated by this substance which can at once carry out a very powerful contraction.

Since swelling, hardening and contraction do not occur in the nerves themselves which transmit the motive faculty, but outside the nerves, in the muscles, the substance or faculty transmitted by the nerves is not sufficient by itself to carry out the contraction. Something else must be added which is found in the muscles themselves or is provided abundantly there. Mixing of these substances results in something which, like a fermentation or an ebullition, produces this instantaneous contraction of the muscles.

Such an operation is possible as appears from countless experiments which are observed in chemistry: concentrated spirit of vitriol poured into oil of tartar, all acid spirits mixed with fixed salts suddenly boil by fermentation. Consequently, in muscles some similar mixing may occur. This entails sudden fermentation and ebullition which fill and expand the pores of the muscles, resulting in turgescence and contraction.

Not only is such a mechanism possible but it must be admitted since it explains precisely the phenomena which is neither possible nor easier otherwise and since Nature never relinquishes usual, easy and direct ways of action. Therefore, we can conclude with a high degree of probability that the operation carried out by Nature in muscles is not different.

Proposition XXIII

Analysis of the structure of the fibres of the nerves
and of their force and function.

It remains to investigate what is sent through the nerves, with what force and how it is pushed, and through which canals.

Firstly, to be logical, the structure of the fibres of the nerves must be analysed. The nerve appears as fasciculate, or capillary, composed by several thin fibres bound by some membranous envelope. Any fibre can be hollow as are reeds and blood vessels, although to the naked eye they seem to be solid and full. But who can see the pores and small canals in our skin? Who can see the lumen of the veins in fleas or intracutaneous small worms? Their veins and ducts are so narrow that we can hardly imagine them. However, they transmit blood to feed the animalcule. If it is not impossible for the nervous fibres to be hollow tubes, we can more readily admit that they are tubules full of some spongy and wet substance similar to the marrow of elder, giant fennel or sugar cane. This is confirmed by the fact that the nervous fibres are soft, flexible, slippery and always wet. They accept and absorb wet nutrient and they secrete some juice. All these properties require spongy pores irrigated by some juice. This is similar to being filled by some spongy and wet substance.

The spongy cavities of these nervous fibres thus are conceived as being always soaked and filled up to turgescence by some juice or spirit transmitted from the brain. In a bowel full of water and closed at both ends, impulse at an extremity compressed and slightly percussed is instantly transmitted to the other extremity of the turgid bowel. The adjacent elements of the liquid are aligned in a long row. By pushing and shaking each other, they transmit the movement to the extremity of the bowel. Similarly, as a result of some slight compression, jolt or irritation at the origins of the canals of the nervous fibres which are in the brain itself, these fibres thus shaken and activated must secrete some drops of this juice which swells their internal spongy substance, into the fleshy mass of the muscles.

Proposition XXIV

Nervous juice can be secreted into the muscles by the will.

It seems unquestionable that animal spirit is a fluid, very delicate, extremely pure and mobile substance. We cannot conceive that an act of will and of wish of our sensitive faculty occurs when the animal spirits are completely immobile and numb. But it seems necessary that the spirits be activated in the brain by some motion as is required by their mobile character.

Therefore, it is possible that the cerebral juices or spirits thus activated displace or shake the origins of the fibres of some nerve by a movement of concussion or by stinging mordant (which they may be able to provoke). They thus irritate and titillate the nerve. The structure and temper of the nerves are very delicate and sensitive, as experience shows. When touching the inside of the nostrils or of the ears with a light straw, the nerves are shaken so violently along their whole length that they provoke convulsive sneezing and coughing. Therefore, it is not surprising that a slight commotion or irritation of the nerves in the brain provokes some convulsive shaking through their whole length, resulting in the secretion of some droplets of this juice which swells the canals of the nervous fibres.

The distal orifices of these nervous fibres are scattered everywhere in the mass of the muscle. Although they are open, the spongy structure itself with which the fibres are provided plays the role of valvules. Indeed droplets hanging from a wet sponge do not flow out. A shaking force is required to express them.

This may be the cause why the nervous juice is secreted and instilled in all the mass of the muscle by order of the will.

Proposition XXV

It can be believed that the usual ability of the animal spirits to stimulate determined nerves in the brain is not innate but acquired by exercise and experience.

I confess not to understand the mechanism by which the movements of the spirits thus activated in the brain by order of the will, direct these spirits to titillate certain well-determined muscles. If I wish to extend the hand, the spirits are not directed to the nerves of the feet or of the chest but are carried to those which correspond to the hand and titillate them.

Attempting to provide an explanation, I think that not all the acts of will are carried out by the same motions of the spirits. In any act of will the spirits do not move to the same parts, in the same way, with the same rhythm and velocity but in very different ways, so that there are as many movements of the spirits in the brain as there are acts of will. Hence, in any act of will, spirits are directed and carried to a determined area of the brain where the nerves are located which are aimed at carrying out this act of will.

Such internal movements do not result from natural and blind necessity as the fall of heavy objects does. They may occur as a result of practice unconsciously acquired by frequent unconscious repetition of the acts. This appears in countless experiments. The fingers of zither players touch and pull the different strings with incredible velocity and art without thinking but by some habit. Conscious reflexion is certainly not necessary in such action. If zither players wish to think and take care of moving their fingers according to the rules of art, they hesitate and are confused rather than they emit proper harmonic sounds.

Similarly, it is possible that, taught during childhood by trial and error, we acquired the habit of sending spirits to the nerves of the legs when we decided to walk and of acting in the same way for the other movements. Many observations show that this opinion is not completely absurd. I indeed often observed the staggering of children and the efforts which ants exerted to push up grains of barley on uneven slopes. After many vain attempts they avoided the useless motions and finally retained that which led to their goal. I am convinced that initially the animal spirits, when for example they want to move the hands, try and experiment countless movements because of their inexperience, sending spirits to the feet or to the tongue or anywhere else. After these attempts to no avail they discard the vain and useless efforts and finally retain the movement which has been approved by experience. Such habit is acquired in childhood, learned without reasoning, from practice and not by study, in such a way that afterwards we unconsciously retain the acquired knowledge of where to move the spirits in the brain to carry out certain motions of the limbs.

Proposition XXVI

Blood is the other element which is involved in muscular contraction.

As said above, the nervous juice by itself cannot produce effervescence and swelling in the muscles. If it rarefied by itself without any adjuvant and mixing, and produced swelling, it would entail turgescence and effervescence in the nerves themselves where the motive cause, the order of the will, is and acts and where the substance able to rarefy, i.e. the nervous juice, is abundant, but not in the muscles where the juice is instilled scarcely and drop by drop and where mixing with other juices would rather weaken the force of this nervous juice. It does not seem possible that the obstacles be removed in the muscles. The narrowness of the place in nerves more considerable than in muscles would rather increase the percussive force of rarefaction as appears in blunderbusses and in mines full of condensed air. It is not the narrowness of the nerves either which can prevent effervescence. In closed spaces no flame can burn. Since the muscles are closed everywhere as are the nerves, swelling of the muscles cannot result from true ignition. This is obvious. Something else thus must be found in the muscles which, combined with the nervous juice sent by order of the will can produce instantaneous rarefaction and swelling.

Which other element can be found in the muscles, susceptible of being subject and matter to fermentation and ebullition of the muscles?

In the muscles nothing else can be found than fibres, lymph and blood which is abundantly brought in by the arteries and fills all the fleshy mass of the muscles. We know that the fibres and flesh, stimulated by their own contraction force, are unable to contract so powerfully by order of the will when stung by nervous juice. This was demonstrated above[1]. The same is true for the lymph or the blood pushed by animal spirits or by ambient air or by the compression of the heart or by its own weight. These factors cannot contract the muscle as water does for a rope[2].

---

[1] Chap. 1
[2] Chap. 2

The only remaining possibility is that fermentation and ebullition result from mixing of nervous juice with either lymph or blood, similarly to what we observe in chemical reactions. It does not seem possible to explain this operation otherwise.

Proposition XXVII

Necessity and mechanism of ebullition and swelling in the muscles.

Undoubtedly there is no way to understand swelling and deflating of the muscles which instantaneously occur by order of the will. We should have thought that to be impossible if chemical experiments had not suggested that similar operations are carried out by Nature everywhere. Mixing acid solutions and vinegar with salines of another kind either fixed or alkaline provokes ebullition and reaction. For our purpose the experiment recorded by Willis is interesting. If you pour into warm blood spirit of wine, of hartshorn powder, of soot, of vitriol or some other liquor, above all spirituous or saline, this provokes surprising ebullition and effervescence.

Without any doubt, the juice sent into the nerves by the brain is extremely spirituous, saline and volatile since it is the instrument of sensibility and movement and because its sweet and pleasant taste obviously points to its saline nature.

Blood is abundantly provided with alkaline salts. This appears from tasting, from the reaction and union of urine poured in blood and from chemical experiment. Consequently, pouring nervous juice into warm blood must necessarily result in effervescence and ebullition.

Fibres and flesh of the muscles are continuously soaked by blood, like a sponge. Consequently, these spirits instilled with some convulsion by an order of the will, can be sent from the nervous canals full of spirituous juice and poured into the blood contained in the muscles. They thus can provoke effervescence and ebullition in the fibres almost instantly, either in the spongy canals of the muscles or in their interstices. This results in swelling, hardness and contraction of the muscles.

The mechanism of such ebullition and effervescence is not different from that of common fermentation. In this no new motive forces are created but the forces which were impeded can exert their natural properties. Particles of a combined substance which could move by themselves if they were not impeded by their binding with thicker and more earthy elements, after being freed of their ties and restored to their own individuality, are able to exert their own mobility.

We observe indeed that percussion of rocks by steel breaks the structure of the stone so that sparks which were retained in this texture can get out. The same happens with wood when rubbed, scraped or compressed.

Similarly, plain water by dissolving the union of chalk enables sparks which were hidden in the cavities of the chalk to get out freely. This results in ebullition.

Similarly, salts in solution penetrate the pores of corals and metals. By scraping away with their edges like rasps and dissolving these hard bodies they enable the exit of sparks which were hidden in their pores.

Similarly, the same salts in solution, by dissolving other fixed salts of tartar, permit the free exit of sparks and other mobile particles, which provokes hot ebullition.

On the contrary, oil of vitriol by dissolving a combination of ammonia salts permits the exit of cold mobile particles but not of igneous particles of which ammonia is deprived. This provokes cold ebullition.

In the same way spirit of wine, spirit of hartshorn and finally nervous juice which is extremely spirituous and saline, by dissolving a combination of alkalified salts with which the blood contained in the muscles is abundantly impregnated, are likely causes for mobile saline particles to be able to exert freely their mobility, once their jails have been opened. They thus could produce this ebullition which swells the pores instantaneously.

The cause and mechanism by which nervous juice is instilled in the muscles with a convulsive force by an order of the will and produces their instantaneous swelling, are deduced from what was said above. Contraction will continue as long as the cause of the bursting is present, i.e. the instillation of nervous juice. When it stops, the turgescence of the muscles disappears, as light disappears when the flame which continuously renews it is removed.

Proposition XXVIII

Refutation of the objections which can be raised against the theory thus presented.

The position of those who admit their ignorance of the causes of natural matters has always appeared to me as more commendable than that of those who take the liberty of being daring in Philosophy. Both, however, are wrong. Neither must we accept hypotheses which do not conform to the behaviour of Nature and to common sense, nor must we be deterred by any objection before having examined its importance carefully. Therefore, I cannot praise those who deny that muscular contraction can result from effervescence of the spirituous juice flowing from the nerves and mixing with the saline of the blood inside the pores of the muscles because of the following objections.

Firstly, such ebullition is imaginary and not confirmed by the senses.

Secondly, muscles when contracting do not increase in volume.

Thirdly, effervescence due to chemical reactions persists for a long time and does not stop instantaneously. But muscular contraction is quicker than lightning, stops instantaneously, is repeated as many times as we wish and persists as long as we wish.

Fourthly, movement persists for a long time in muscles of beheaded turtles or insects and of frogs, eels and vipers after removal of their heart. Therefore, such movement is not caused by an outflow of spirits originating in the brain and impelled in the blood by the heart.

Fifthly, it appears that movement of a muscle can result directly from a motive fibre by a mechanism which we ignore since the heart muscle of a turtle and the muscles of the legs of a frog can pulse for one or two hours and then revive and pulse anew when stung by a needle or touched by some corrosive fluid.

I think that all these objections can be refuted.

I reply to the first objection that, if we do not perceive the ebullition which occurs in the pores of the fibres, this ebullition can be deduced from its necessary effect. This effect is the swelling and hardening of the mass of the muscle. They must necessarily result from violent wedging by another body, as was shown[3]. Similarly, because vessels of chemicals, eggs and chestnuts make cracking sounds and split in a big fire, one deduces that huge ebullition takes place in their inside, although this does not appear to the eye.

The reply to the second objection has been given above[4]. The muscles actually swell and increase in volume. This is demonstrated by the hardness and obvious increase of the muscle of the heart.

To the third objection I say that effervescence can be instantaneous in muscles as it is in salines of chemical reactions. For better understanding of this phenomenon I point out that the reason for which ebullition is achieved quickly and stops instantaneously in a muscle is not that the fluids of this fermentation are separated from each other and stored in different receptacles, once expansion is completed (as some people thought). The reason is that they are dissolved and consumed, either both or one of them. Similarly, in gun powder all the components, carbon, sulphur and nitre are ignited and consumed together. They take fire and are extinguished instantaneously. But the flame could persist if more nitre powder was added again and again. Then the interruptions would be so short as to be imperceptible. In the same way, a burning log when rotated describes an uninterrupted circle of fire.

Actually the effervescence of salines is different from the ignition of nitre powder since these fluids do not combine suddenly by small quantities as do powders of nitre, sulphur and carbon exactly mixed. When new unions of these fluids occur, successively in different places, ebullition continues for some time. If, by some artifice, countless particles of saline spirit fell like dew on the other saline fluid, effervescence could be instantaneous, these fluids being consumed rapidly. This effervescence could persist for some time if the rain of dew was repeated more often.

It seems possible to verify that in muscles. The multiple nervous filaments, like the roots of a tree, originate from the nerve which arrives into the muscle and are scattered between the muscular fibres. From their orifices minute droplets of spirits can be poured simultaneously after convulsive irritation. All these droplets can provoke instantaneous effervescence with the blood which is present in the muscles. This effervescence is similar to the ignition of nitre powder. Such ebullition can stop instantaneously by its nature, once the spirituous droplets are burned out. It can also continue as long as new and continuously

[3] Prop. 15

[4] Prop. 15

repeated convulsive irritations can produce new pouring of spirits which in turn provokes new effervescence by combination with blood. This results in swelling of the pores of the muscle and continuous powerful contraction of the muscle can be achieved.

To the fourth objection I answer that it is not surprising that, for a short time, after being beheaded and their heart excised, turtles can move and contract their muscles, frogs can jump and vipers crawl since the spirituous juices sent by the brain into the nerves so far and the remnants of blood in the pores of the muscles can carry out posthumous effervescence when the nerves are irritated by stinging, as previously they were shaken when stimulated not only in the brain, but even in their filaments.

But you ask how a few droplets of spirituous juice which remained in a nerve are sufficient to carry out so much ebullition for hours. I answer that in the same way odoriferous substance flows out of some grains of musk or civet perfume, i.e. material particles dispersed in air disseminate odoriferous smokes in a thousand rooms, courts, streets and places. Such emanation persists several months without any sensible diminution of the volume or weight of the musk and civet perfume. In the same way the very spirituous juices present in divided nerves can be released by new convulsive irritations, subdivided into the thinnest oozing (i.e. in particles as minute as the odoriferous granules dispersed in air). Combined with the remnants of blood in the pores of the fibres, they can provoke the effervescence and swelling of the muscles and their contraction. These can be repeated until all these fluids are burned out or hardened or the muscular fibres are dried out. Experience clearly supports this point of view. The heart of a turtle, after being excised and immersed in water, pulses for a long time until its external envelope dries and puckers. When moistened again it is revivified and pulses anew. Finally, after one or two hours, when it seemed completely dead, stung by a needle or irritated by some corrosive liquid, it pulses twice or thrice. I conclude that the sensitive faculty not completely extinct in this heart, when hurt by the corrosive stinging of the nerve can express the very last droplets of nervous juice, with great strain, by contorting the nerve itself. You are not surprised that the motive and sensitive faculty can remain in a heart excised and in its nerves. This is always claimed by the Aristotelian Philosophy which assumes a divisible soul in reptiles: the excised tail of a snake moves because it retains part of the soul. Certainly, if this part did not perceive the pain, it would not contort and become activated in the same way as it was shaken previously when united to the living animal. This slight stinging by a needle is not so dessicating as to result in a contortion similar to that which occurs in wet wooden boards when one of their faces is exposed to a fire. Neither has the needle the force to broil, as skins and hair are burned by fire. Neither does the steel of the needle induce alterations similar to those produced by corrosive fluids. This stinging evokes only a weak movement in the fibres of the heart. This movement, which is slow and weak, cannot provoke in the heart an impetus and a shock greater than it is itself. Moreover, the weak impulse of the needle cannot move all the other parts of the heart which are not touched by the needle and which are not bound together very firmly. The nerve thus is shaken on its own, irritated by the stinging in the same way as what occurs in the living animal. Consequently, a nerve, even excised, perceives pain.

The fifth objection is refuted by what has been said. Those who claim that the considerable energy with which muscles carry heavy weights by order of the will, is proper to the motive muscular fibre, let themselves be deceived by the obscurity of their words and overlook the impossibility of their own assertion. We demonstrated that the contraction force of the machine itself constituted by the fibres is very weak[5]. It cannot contract by itself to lift heavy weights. Such contraction must result from an external cause, different from the material force of the machine and which violently contracts this machine. These authors actually confuse two forces. They think to have solved the problem by using the word motive fibre which would exert so much force by itself. "We do not care, they say, how this occurs".

This is as if somebody inquiring about the cause of the ticking of the hours by the clock which is a contrivance and its envelope both made of metal elements, was answered that the ticking results from metallic engines without caring about how this works. I praise the sincerity of those who plainly recognize their ignorance of what is hidden. But the effort of those who endeavour to investigate the causes, at least probable, of the effects of Nature is not blameworthy. It is indeed something to progress a little if what is beyond is not known. That is the reason why I do not like this hasty method of discussion. We attempt to go further and to find out whether the metallic elements are the actual causes or whether they are the instruments of these movements, whether they actually provoke the movements or whether they are driven by some other external force. Moreover, if we cannot see, we at least attempt to guess the configuration of the wheels, their connections and arrangement. We try to know whether the wheels are moved by a steel machine, or by a suspended weight or by the force of the wind or by flowing water and by what mechanism their motions are carried out with such regularity and order. If we could not find the same explanations for the muscles, we at least attempted to investigate the forces. Firstly, I point out that the fibres should not have been called motive fibres since these fibres are not machines which move themselves. They behave purely passively and, by themselves, are inert. They move only when an impulse arrives sent through the nerves by the will. Secondly, although we do not see the organic structure of the fibres, we perceive from obvious signs that they are similar to a contractile chain made of links which can contract by intromission of wedges. Swelling, tension and contraction cannot be explained by any other structure and by any other mechanism than that which we described above since the operations of Nature are easy, simple and follow the mechanical laws which are the laws of necessity.

[5] Prop. 7 Part I

Proposition XXIX

Mechanism by which the weak ebullition which occurs in the muscles is able to exert a huge force.

From our demonstrations[6] the force and energy of the muscles appear as huge and inexpressible. If all the machines by which the muscles act are considered, the force exerted by Nature is thousands and thousands of times greater than the resistance. In contradistinction, the ebullition occurring in the muscles and which we have considered above seems to be weak. The mechanism by which the small energy of the fermentation can exert an immense force thus must be explained.

Firstly, the example of a wet rope shows that such an action is possible and actually occurs. The rope contracts and raises heavy weights for no other cause than the small weight of the particles of water which, when driven in the fibres, like wedges, provoke the contraction of the rope. This, however, results from the huge number of wedges acting together and from the energy of the movement and the percussive impetus with which these countless granules driven into the pores and the interstices of the fibres can overcome any immobile resistance, as was demonstrated[7].

The structure of the muscle and its contraction are not completely different from the composition and operation of a wet rope. Both are made of countless fibres with frequent small ties. They thus present with oblong rather than round- shaped pores which are dilated beyond 40° by intromission of corpuscles as by so many wedges which, multiplied immensely, act together with the force of the movement and the energy of percussion against an inert and immobile resistance.

The only difference lies in the rapidity of the operations. In a rope the watery wedges are driven in slowly and get out during a much longer time. In the muscles the wedges provoke inflation and deflation instantaneously.

The reason for this difference is that the watery wedges in a rope retain their solid volume, come from outside and are driven into the pores of the rope whereas in muscles the wedges do not come from outside, they are generated in the interstices of the fibres when the muscles are inflated and they are destroyed and dissipated when the muscles deflate. Therefore, they do not retain the same solid volume. They acquire it by rarefaction and then lose it by condensation as appears in the ebullitions and explosions thus mentioned, in which very mobile particles dilate spherically and form clusters of small light and hollow bubbles separated by much empty space. Therefore, they easily gather and can be reduced to their small previous volume instantaneously. The mechanism of action of the muscles thus can be deduced, I think. All the substance of the muscle is inflated by the swelling and bursting of countless sparkling vesicles dispersed throughout all the mass of the muscle and acting as the

---

[6] Chap. 17 Part I

[7] Lib. De vi percuss. Prop. 190

watery wedges in a rope. As a result of the strength and inextensibility of the fibres, dilatation and swelling must necessarily entail contraction and shortening of the length of the muscle. The vesicles are multiplied immensely throughout all the muscle and act by their force of percussion, as said above. This results in this huge power by which heavy weights can be raised a considerable distance, over more than six-tenths of the length of the muscle. This was demonstrated above[8].

Chapter IV

On internal movements in animals and first on blood circulation.

The movements of the muscles depend on the will. But there are other movements which seem to be completely natural such as the tireless movement of the heart. This can be investigated more easily in its effect, the blood circulation which it produces. Therefore, we shall first discuss blood circulation.

Proposition XXX

Modern authors demonstrated that there is obviously a blood circulation.

It is clear that a triple movement can be observed in the blood of living animals. One is the movement which stirs the components of the blood according to their innate tendency and the requirement of their nature. This movement is appropriately called fermentative movement. Another is the movement due to the vibration of the particles of the blood which are rarefied and condensed by the force of a machine as will appear below. The third movement is the displacement of the blood from the heart through the aorta and the other arteries to all the parts of the body, into the spongy substance of the flesh, viscera and glands. Hence the blood is absorbed by the capillary veins and collected into the big veins. This is similar to the streams which are increased by the sources, the rivulets and torrents which they receive in their course. All the veins end into the big vena cava and finally discharge into the right ventricle of the heart. From there all the mass of the blood is carried through the pulmonary artery to the lungs where it is distributed. It is then collected by the pulmonary vein and brought back to the left ventricle of the heart from which it flows in the aorta again. This movement of the blood is called CIRCULATION. It was discovered partly by Cesalpino and afterwards, not long ago, demonstrated by Harvey so obviously that nobody can question it any longer.

---

[8] Prop. 20

It is useless to repeat what has been discovered and described with admirable clarity by others concerning this movement of the blood. I thus will consider only what has been omitted or inadequately explained by others concerning the causes, the mechanics and the circumstances of this movement.

Proposition XXXI

How the movement of the blood is continuous although it seems to be interrupted in the heart.

The impulse of the blood results from compression in the heart and is not continuous. It is regularly interrupted by short stops. As a consequence, the blood does not leave the heart in a continuous flow as do streams and rivulets from springs. It flows by jerks but regularly. The movement of the blood thus must not be considered as continuous. Between two successive pulsations there is a stop with absence of movement. Rather than a true steady movement it seems that it can be designated as an interrupted and stirred-up movement.

Nevertheless, I do not hesitate to claim that the blood circulates through the body of the animal in a continuous and uninterrupted movement. Although the heart does not pour blood into the arteries during its diastoles, the blood does not stop and remain completely immobile and stagnant in the arteries, viscera, flesh and veins when the heart is at rest. The blood keeps moving but with varying velocity. This is obvious in the arteries. The blood which filled the arteries keeps flowing out until they are completely empty, when the afflux of blood from the heart has stopped after the heart has been removed in frogs, or if breathing is impeded as at death's door, or in strangulation, or if the diaphragm cannot move after the thorax has been opened. This results from the fact that the arteries themselves are constricted by contraction of their circular fibres. They are also compressed by the contraction or the peristaltic movement of the muscles of all the body. Hence comes the old tale that arteries do not contain blood but vital spirits the dissipation of which leaves the arteries empty in corpses.

Afterwards blood continuously flows in the veins, not only when they are filled by the blood coming from the arteries and driven by the pulsation of the heart but even when the heart is at rest. This is suggested by the fact that then the blood flows through the vena cava to refill the right ventricle of the heart. We also observe that blood continuously flows from a vein divided in a wound as if it were from small fountains, not only when the heart pulses but also when it is at rest.

Proposition XXXII

Mechanism of the continuous flow of blood through the veins.

Blood certainly flows through the veins. This is proved by experience. But the cause and the mechanism of such movement are not so obvious as those of the impulse in the arteries. Indeed, there are no anastomoses between the capillary veins and the smallest arteries. The blood thus cannot be sent from the arteries directly into the veins since these vessels are separated from each other. We believe that there is some communication between the ends of the arteries and the capillary veins across the intermediate spongy substance of the flesh of the viscera or across the cancellous tissue of the bones, like in the pores of pumice-stone. However, we do not perceive which motive force can drive the blood into the capillary veins. Firstly, the impulse which sends blood into the arteries during the systole of the heart must progressively weaken and finally be exhausted in the narrow extremities of the vessels and intermediate pores. Secondly, the orifices of the small veins cannot always remain dilated and open since their consistency is neither hard nor bony but rather membranous, soft and slippery. Therefore, they can easily be closed and impede entry of new blood. Thirdly, we cannot resort to compression of viscera and flesh to explain the passage of the blood. Indeed, we observe that blood is sucked by the capillary veins not only when muscles contract and exert compression but even when they are at rest and completely relaxed. Consequently, blood cannot be expressed by compression of the muscles which does not exist in Nature.

This is confirmed by the fact that capillary veins are present and contain blood in the brain, in bone marrow and even in the hardest substance of bones which is not compressible. But blood circulation cannot be denied since we observe that all the blood flowing from the arteries is absorbed and carried back to the heart by the veins. This does not result from an attractive force. Such an action is impossible as we demonstrated[1]. We are forced to say that blood is absorbed by the capillary veins by the same mechanism by which capillary tubes, sponges, filters, ropes and all porous objects absorb contiguous water and become wet[2]. This is nothing else than the weight of the fluid itself, which is increased by the impetus of its own movement and by the impulse transmitted to it by an external force. Blood, like water, has some weight. When the motive force of the weight meets the open venous capillaries (the membranous connections can never be so tight and closed that no access persists, as appears in the pores of a rope), its motive energy must overcome the resistance of the narrowness of the orifices. Therefore[3], blood is driven into the capillary veins in a manner similar to what occurs in a filter.

Collected in the small veins by the same impetus by which it was driven in (since such intromission is impossible without movement), the blood can be pushed a little further by its own impulse and by an external force. This force is

[1] De motion natur. à gravitate pend. Chap. 6
[2] De motion. natur. à gravitate pend. Prop. 185
[3] De vi percuss. Prop. 90

provided by the next quantity of blood which pushes. Similarly, we observe that water sucked in by a filter falls drop by drop from the inclined superior rim of the filter. Afterwards, the confluence of several small veins gives a larger vein in which the movement of the blood is slowed down. The initial impulse is weakened so that blood needs help to be pushed further. This is above all the force with which the circular fibres of the veins contract, the innate peristaltic force of the fibres which contract the vessels, helped by the compression of the ambient and inspired air, by the weight and elasticity of this air, by the contraction of the muscles and the movements of the different viscera and liquids which flow through the body of the animal, as we shall see later.

Proposition XXXIII

How and why blood progresses in veins.
Tab. XVI, Fig. 1.

Membranous valvules are appended to the internal walls of the veins, as in HIKL. They are half bags or hollow vesicles AONMP and BONQR adhering to the walls of the veins. Most often a couple of opposite valvules are found at the same level, touching at NO. Their convex bottom MN, QN is directed towards the beginning of the capillary veins beyond HL, from which blood arrives. Their concavity PO, RO is orientated towards the heart in the direction IK. It remains to demonstrate that blood must proceed towards the heart as a consequence of the structure and position of these valvules. The segment HMQL full of blood is constricted one part after another by its circular fibres and by the surrounding muscles and viscera so that its walls S, T come closer together towards V. Then the vein loses its cylindrical shape and is transformed into two funnels HVL, MVQ the capacity of which is less than that of the cylinder. Therefore, the blood which was contained in the spaces VHS and VLT is expelled through the orifice HL. The remaining volume of blood contained in the spaces VSM, VQT is ejected through the orifice MQ towards IK. Thus, as a consequence of the compression of the walls of the vein, the blood is expelled in equal quantities to opposite sides. This would occur if there were no valvules. Because valvules or membranous bags are appended to the opposite walls MP, QR of the vein, the blood pushed by the compression exerted at ST must pass through the slit NO. The yielding liquid contained in the valvules is squeezed and expelled by the arriving blood. The sides of the valvules NO are separated from each other, leaving the way open through which blood arriving from MSTQ can be driven in and pass beyond AB. After the blood has passed the borders of the valvules PO, RO, the slit NO must necessarily close spontaneously. Indeed the blood must fill the valvules as a result of its volume and fluidity. The soft sides of the valvules are stretched until they adapt exactly to each other and must close the slit NO tightly.

The vein is not constricted simultaneously everywhere but successively at one place after the other. After the blood has passed into ABCD beyond the valvules, the walls AD, BC become constricted while ST is not. Because of the tight closure of the slit NO, the volume of blood contained in the spaces EAG, FBG cannot flow back towards AB since it meets the obstacle AOB formed by the valvules filled by blood. The blood thus is forced, by reaction, to move towards DC like sport javelins meeting a wall. Under the same compression, the blood which was contained in the spaces EDG, FCG is propelled beyond DC. In the compressed areas thus a double volume of blood is expelled during the same time through the same opening DC. But when a double volume of fluid is sent through the same orifice during the same time, its velocity must be twice greater. Thus, as a consequence of the valvules, compression of the veins propels the blood towards the heart by doubling its velocity. The blood does not progress in a continuous flow but with phases of slowing down and inequal velocities.

Proposition XXXIV

Normal composition of the blood is maintained by its circulation.

As shown by experience, wherever blood stagnates either inside or outside the animal body, its red part quickly coagulates and separates from its serum or albumen. Thereby its structure is dissolved and corrupted. On the contrary, as long as blood keeps moving inside the vessels of a live animal it maintains its normal composition and temper for the mechanical reasons explained below.

Mixing of all the components of the blood can be maintained only by continuous activation and shaking. This is carried out in the vessels at first by the heart which pushes the blood through the arteries with a strong impetus. Afterwards, in the veins where the impulse of the heart is missing, the activation is promoted during the passage of the blood by constriction of the circular fibres and viscera, by compression by the inspired air and by the muscles. But these compressions would not be sufficient if there were no valvules appended to the veins, which enable blood progress and beating.

It is worth considering the anastomoses of capillary veins and the distribution of the valvules. There are not many in some veins and many segments of veins are deprived of any. The anastomoses and the absence of valvules, however, seem also to have a function. Because the structure of the veins is loose and soft they can be constricted and contracted by their own circular fibres. Hence, the lowest part of a vein can be distended abnormally by blood forced back by regurgitation in great quantity and with great impetus from segments of the veins deprived of valvules, whereas its highest part is constricted. And vice versa when the constriction is loosened, blood is activated and beaten by a reflex regression, although imperceptible, and thus can maintain its adequate composition. Arteries do not need valvules, except the semilunar valvules of the heart, because the strong impetus with which blood is ejected through the arteries is sufficient to beat and mix it, wedging the white fluid particles into the red glutinous particles, as we shall show below.

Proposition XXXV

Most important effects resulting from the velocity of blood circulation.

We indicated the usefulness of blood circulation for the blood itself. We now must consider why Nature is forced to circulate blood with such velocity that it flows through the arteries, is led in all the parts of the body and carries out its whole circuit in little more than three minutes, as appears from the observations of Harvey. This has several significant consequences.

Firstly, during each pulsation of the heart a great quantity of blood flows through and is ejected by the narrowest arteries. The greater the quantity of liquid flowing from these arteries the quicker it moves, as was demonstrated by B. Castelli[4]. Therefore, the blood is sent into the spongy pores of the flesh and viscera, like full and quick torrents.

Secondly, the energy of the blow which the blood projected by the heart strikes in the distal parts depends on the velocity and the volume of blood thus pushed, as we demonstrated[5]. As a consequence, the empty and closed pores of the flesh and viscera are opened by the percussive impetus of the blood. They are filled and saturated by this abundant inflow. Therefore, the numb, useless and excremental elements present there are forced out and expelled through the pores of the skin or through the vessels destined to expulsion and transportation.

Moreover, parts of the animal are repaired since, as a result of the shape of their small pores, adequate particles of blood, congruent with these pores, are driven in violently like wedges and conglomerate, as stones of different shapes compose a mosaic work. Nature, using this quick wedging of blood, not only carries out repair of parts but also clears the blood itself of its excrements in the viscera of the animal.

Thirdly, thanks to its velocity and the violence of its impetus, the blood succeeds in scraping off and carrying with it countless other particles such as spirituous, saline and fermentative particles which are deposited in the brain, viscera and glands and restore and improve the disturbed composition of the blood.

The action of Nature is better discovered by observing that the scraping thus described does not occur in large vessels but after blood has flowed from the extremities of the capillaries where very narrow canals conveying nervous spirituous and fermentative juices and disseminated in viscera and glands like the roots of a tree, end. This seems to occur so that the spirituous and fermentative particles combine with some molecules of the blood and mix. Hence, the blood which repeatedly passes acquires a different consistency and nature.

All these advantages require continuous and very quick movement of the blood. Therefore, it is likely to these ends that Nature designed the very quick circulation of blood.

---

4 Lib. de aquae fluxu.

5 De vi percuss. coroll. Prop. 28

Proposition XXXVI

Why the blood must repeat very often its passage through the same ways.

It is clear that a stream keeps its identity. The parts which are gone are replaced by the arrival of new waters and the latter are replaced in turn. To achieve this, a huge quantity of water is necessary or the water which has flowed by must be returned to the source. Then the stream is maintained by a circulation.

Nature proceeds similarly. Blood circulates very quickly through all the body of the animal to the ends described above and this circulation is perpetual. A sufficient volume of blood had to be provided to maintain the flow of blood in the animal: 18-20 pounds of blood are normally found in man. This blood circulates quickly, in three or four minutes. Therefore, not to interrupt the flow, the circuit of the mass of blood must be repeated.

Besides this necessity, there are admirable functions and economies achieved by Nature in repeating the circuit of blood through the same ways frequently. If blood did not repeat its circuit frequently it could not get rid of the excrements of urine in the kidneys, the chyle could not be mixed with the blood in the heart nor the blood be purified in the liver, and other operations necessary to the economy of the animal could not be carried out. All this will be explained in its place below, not to disturb the order.

Chapter V

On the heart and its pulsation.

Knowing the effect produced by the pulsation of the heart, we can more easily investigate its cause which is the movement of the heart, its motive force and its mechanical action (Tab. XVI, Figs. 2–4).

Proposition XXXVII

Structure of the heart.

The heart appears not as a parenchyma but as a muscle of exactly the same nature as the other muscles of the limbs. After the fleshy fibres have been made turgid and thicker by boiling, it is even clearer that the heart is composed of strong fleshy fibres of the same prismatic shape, the same colour, consistency and strength as the fibres of the muscles of the limbs. In the same way they are inextensible, resist stretching and contract, tighten, swell and become hard during pulsation of the heart. However, the fibrous make-up of the heart is different from that of the other muscles, in that its flesh is firm, hard, uniform and rubicund. It is not loose and soft. Its prismatic columns are not separated by membranes and countless tendinous fibres as are the fibres of the other muscles. Moreover, the arrangement and configuration of the fibres of the heart are very different from those of the other muscles. The fibres of the heart are not straight and parallel to each other but curved and coiled. They are amazingly intricate. Their texture is not similar to that of a wicker-basket, as Vesalius believed, but they are arranged in a more sophisticated way. Immediately underneath the pericardium, from the basis of the heart and from the circular tendinous orifices of the vena cava and pulmonary auricle, and from the origins of the aorta and pulmonary arteries, a layer of fleshy fibres spreads. These fibres are parallel to each other and they are orientated from the basis straight to the apex of the heart where they are diversely curved and intertwined, reflecting towards the cavities of the ventricles. This layer is followed by other layers of fibres, oblique and descending in spirals. They are more and more inclined and orientated also towards the apex of the heart. Before reaching the apex, they intersect and cover each other and other ordinary fibres. Hence they are deviated to the inside and partly reflected in oblique and transverse spirals like scythes, towards the base of the heart. Part of them seem to compose internal columns to which the strings of the tricuspid and mitral valvules are attached and others are intertwined transversally and form the cavity of the right ventricle.

I was the first to see this admirable structure in Pisa in the presence of the famous Malpighi in 1657. Afterwards, more recently, others made the same observation. Finally, the famous Lower and Lorenzo Bellini investigated the exact texture of the heart by undoing the binding of the fibres as complex as a ball of wool. I missed observing that, after boiling, the fibres are easily separated before the intertwined bendings are undone.

Countless ramifications of nerves originating from the two vagus nerves are scattered between the fleshy fibres of all the heart. Moreover, two coronary arteries originate from the aorta before it leaves the pericardium. They are provided with valvules which prevent backflow of the blood, as was observed by the famous Bartholin. They carry blood to the muscular flesh of the heart, not to the ventricles. Hence the blood passes through the coronary veins which meander also through the flesh of the heart and after a peculiar circulation discharge in the right ventricle.

Two additional hollow and round muscles adhere to the heart. They are called auricles. They act as the door-keepers of the heart. After the blood has flowed from the vena cava and pulmonary vein into the ventricles, they close tightly and prevent backflow of the blood. They consist of fleshy fibres which form muscular columns similar to those which bulge into the ventricles.

Besides the auricles, even the trunk of the vena cava close to the heart is of muscular nature. It consists of rubicund circular fleshy fibres which help the constriction of the venous trunk, like a goatskin bottle or a bowel full of water from which water is expelled by compression of the hands and prevented to flow back.

Proposition XXXVIII

Description of the action of the heart muscle.

When the heart contracts like all the other muscles, all its flesh swells, straightens, tightens and hardens with a maximum impetus. This results from the structure of the heart. The fibres of its columns and their bundles are not attached to firm bones or tendons, as is usual for the other muscles, but they originate and end in the heart itself which is suspended and unstable. The fibres are firmly anchored into it. Consequently, turgescence of the fibres of the heart is not aimed by Nature at pulling and bringing their extremities closer together. In contracting the fibres swell and decrease the cavity. In so doing they squeeze out the blood contained in it, like a press, as stones of fruit are projected by pinching. This results from the fact that at every pulsation and tightening of the heart the blood which was contained in the ventricles is ejected in the arteries with a maximum violence as if it was propelled by the piston of a syringe. This appears more obvious after having introduced a finger into an incision in a live heart, as observed by Bellini and Lower. Then, during every pulsation the finger is squeezed, like by pliers or by a press, and is retained by the flesh of the heart swollen and hardened. Although the external surface of the heart is not increased during a pulsation, all its fleshy substance actually swells and increases.

After these remarks, some lemmas must be presented which are useful to understand the mechanical action of the heart muscle.

Proposition XXXIX

In a series of arithmetical proportions A, B, C, D, E, the smallest of which is A, the terms closer to the smallest one decrease proportionally more than those which are more rêmote from it.
For example $B/A > E/D$.
Tab. XVI, Fig. 5.

In a series of arithmetical proportions, the differences between two adjacent terms are equal. Thus, $B-A=E-D$. Since $A < D$, $(B-A)/A > (E-D)/D$[1]. And by combining: $B/A > E/D$ and so on.

[1] Ex Pr. 2 lib. 3 Elem. Eucl. Auctoris

Proposition XL

In the same circle AB (Fig. 6, Tab. XVI) two concentric areas ABC and DCF are designated. Their widths AD and DE are equal. In these areas two series of an equal number of circles are drawn. Their common centre is K. These circles are in arithmetical proportions and their arithmetical averages are GL and HM (G and H divide the lengths AD and DE in two equal parts). I claim that the ratio of the sum of all the circumferences of the series ABC to the sum of all the circumferences of the series DCF is equal to the ratio of the circumference GL to the circumference HM.

In the series ABC, the double of the circumference GC of the mathematical average, multiplied by half the number of terms of this series, equals the sum of the circumferences of the area ABC. This was demonstrated by mathematicians in the past. Similarly, the double of the circumference HM multiplied by half the number of terms of the series DCF is equal to the sum of the circumferences of the area DCF. There is an equal number of circumferences in both series. Consequently, the ratio of twice GL multiplied by half the number of terms in the series ABC, to twice HM multiplied in the same proportion is equal to the ratio of the circumference GL to the circumference HM. And, therefore, the ratio of the circumference GL to the circumference HM is equal to the ratio of the sum of the circumferences of the area ABC to the sum of the circumferences of the area DCF.

Proposition XLI

Two spirals AMD and DOE (Fig. 7, Tab. XVI) following each other immediately are in the same ratio as the circumferences GL, HP of the arithmetical averages of the circumferences of the circles delineating these spirals.

In an Archimedean spiral[2] two movements, one along the straight line AK and the other about the centre are carried out equally during the same time. Thus, the velocities decreasing in continuous arithmetical proportion, with which the point A moves along the spiral AMD are equal to the velocities with which all the points of the straight line AD successively rotate about the centre K. These velocities are measured by all the circumferences described by these points in arithmetical proportions. Similarly, all the velocities with which point D moves over the spiral DOE are the same as the velocities with which the points of the straight line DE rotate. These velocities are measured by the arithmetically proportional circumferences described by these points. Therefore, the ratio of the velocities with which point A describes the spiral ABD to the velocities with which point D describes the spiral DOE is equal to the ratio of the sum of the

[2] De Spiral. desin. 2. in Archim. Auct.

circumferences decreasing arithmetically in the area ABD to the sum of the circumferences in the area DCE or equal to the ratio of the average circumference GML to the circumference HOP[3]. But these circumferences are described in the same time as required by the genesis of the spiral. Therefore, the ratio of the spiral AMD to the spiral DOE is equal to the ratio of the sum of the velocities of the course of point A over its spiral to the sum of the velocities of the course of point D over its spiral. Consequently, the ratio of the spiral AMD to the spiral DOE is equal to the ratio of the arithmetical average circumference GL to the circumference HP.

Proposition XLII

Any number of spirals KNE, EOD, DMA (Fig. 7, Tab. XVI) follow each other in the same plane and the smallest is KNE. I claim that those closer to the smallest decrease proportionally more than those more remote from the smallest. For example, the ratio of DOE to ENK is greater than that of AMD to DOE.

The circles GM, HO, RN are drawn. They are the arithmetical averages between the extreme circles enclosing each spiral. In the spirals which follow each other regularly the straight lines AD, DE, EK are equal and are divided in two parts at G, H, R. The differences GH, HR thus are equal and, therefore, the differences between the radii GK, HK, RK are equal. The circumferences are proportional to their radii[4]. The circumferences GM, HO, RN of the circles thus are in arithmetical proportion. Therefore, those which are closer to the smallest circle decrease proportionally more than those which are more remote[5], i.e. the ratio of the circumference of the circle HO to that of the circle RN is greater than the ratio of the circumference GM to the circumference HO. The spirals are proportional to the averages of the circumferences which circumscribe them[6]. Therefore, the ratio of the spiral DOE to the smallest spiral ENK is larger than the ratio of the spiral AMD to the spiral DOE. Q.E.D.

Proposition XLIII

A ball of wool AB (Fig. 8, Tab. XVI) is made of threads AB, CD, EF of even thickness and of the same consistency, concentrically or spirally wound round an inflated bladder so that there is a central cavity HK. I claim that, if the cavity HK is filled by soaking the threads LM, NO, etc. (Fig. 9, Tab. XVI), the internal threads NO, PQ necessarily must pucker, the external threads LM remaining taut.

[3] Prop. 40
[4] Prop. 14 lib. 5 Eucl. Rest. Auct.
[5] Prop. 39
[6] Prop. 41

The ball of wool is supposed to be cut along its axis. When dry, the circular layer AE, FB (Fig. 8) is made of concentric or spiral adjacent threads. When the ball is soaked, the circle LM (Fig. 9) is filled by its swollen but contracted threads NO, PQ. The threads are assumed to be of the same thickness and consistency. They are all similarly porous. Therefore, water must swell them equally and proportionally. Their shortening must be proportional to their length when dry in the area which they occupied. The spirals AB, CD, EF (Fig. 8) which were more remote from the centre before being soaked shortened in smaller proportion[7] than the spirals LM, NO, PQ (Fig. 9) brought closer to the centre by the swelling which results from being soaked. Therefore, the spirals LM, NO, PQ close to the centre are longer than the spiral space in which they are contained. They thus must pucker to be able to remain there.

Proposition XLIV

Mechanical process by which the cavity of the ball of wool is filled when the ball becomes soaked.
Tab. XVI, Fig. 9.

The crust LM of the ball of wool is formed by the external threads and resists stretching because on the one hand the threads cannot be elongated as a result of their spiral contortion and on the other hand they are swollen by the water which penetrates their pores. They contract with such a force that, like the iron circles of a barrel, they resist the force of the internal parts which are swollen. Although the external threads are swollen and thickened, the consistency and tense hardness of the convex surface formed by the threads impede swelling and expansion of their convexity. Therefore, swelling of the internal threads must increase either the external surface or the depths of the threads. But the external surface cannot increase because the other threads intertwined with those of the surface impede this expansion by their contraction. The alternative consists of protruding in the depths towards the centre of the ball. All the adjacent layers NO, PQ of threads thus must move somewhat towards the centre g of the cavity in the ball and start to fill in this cavity hk.

The underlying layer NO of threads equally soaked also swells and shortens. Its convex circumference will not remain as tense as the first rigid crust was. Its concavity will move much further into the depths towards the centre. Because its threads shorten proportionally to the lengths they had before being closer to the centre — although less than their vicinity to the centre would require — they are too long for the narrow space. Consequently, the concave surface of the threads NO puckers and the underlying layers PQ, yielding to the more superficial ones, are forced much further towards the centre g than previously. This fills even more the cavity. The same occurs with the third, the fourth and the other layers which are swollen always more proportionally, pucker and move much further towards the centre like wedges until the cavity hk of the ball is filled completely. Such was the proposition.

---

[7] Prop. 41

Proposition XLV

The data are the same. I claim that all the internal spirals of the ball of wool must pucker unevenly, the folds increasing the more, the closer they are to the centre of the ball.

The internal spirals of the ball of wool become closer to the centre for two reasons. Firstly, they are pushed towards the centre by the swelling and contraction of the external threads. Secondly, their own swelling brings them further and further towards the centre. They contract less and less proportionally than required by the place which becomes smaller and smaller. Therefore, the folds must be larger and larger the closer they are to the centre of the ball.

Proposition XLVI

The data are the same. I claim that the external shape of the ball of wool when soaked must neither increase nor decrease.
Tab. XVI, Fig. 9.

Firstly, it cannot expand since the threads LM of the ball are assumed to be inextensible either because of their strength or because of their spiral contortion similar to that of ropes. But, when soaked, ropes do not lengthen. They rather shorten with great force. Thus, the external threads of the ball cannot be elongated and, consequently, the external surface of the ball cannot increase.

Secondly, the same external threads of the ball cannot contract although they try with a huge effort, because the mass of the internal threads NO, PQ swells and hardens, thus impeding contraction of the external threads LM. The external threads thus maintain their initial length. Consequently, the external shape of the ball of wool does not decrease. Neither can it increase. Thus, it does not change and keeps the same volume and shape as previously.

Proposition XLVII

A hollow ball of wool ABR (Fig. 10, Tab. XVI) is made not of one but of several threads attached to a ring or to a fixed radius of the ball AETR and to the surface of the cavity. The threads are contorted concentrically or spirally. If the cavity is filled by swelling of the threads resulting from their soaking, the internal threads M, O, Q (Fig. 11, Tab. XVI) must pucker unevenly, the folds increasing more and more towards the centre. The external shape remains unaltered.

A rope is made of hemp and consists either of one complex spiral thread or of the same thread divided into several parts also with a spiral texture. The external spirals tightly surround the internal ones and cannot be displaced. We observe that, when soaked, ropes swell equally, contract and raise equal weights.

Similarly, balls made of one or several threads ABR (Fig. 10) must swell in the same way and their cavities must be filled M, O, Q (Fig. 11). The origins and ends of the spirals are maintained firmly in the same places. Either they are linked and continue one into the other as in Fig. 8, or they are attached to rings or other firm places of the ball as in Fig. 10. In both instances the threads are the same with the same shape, arrangement, density. By their tension the external threads spirally tighten and surround the internal ones. They must be swollen by water in the same way and pucker in the same proportion. Puckering is the more pronounced the closer to the centre. The volume and size of their external shape remain the same.

Proposition XLVIII

When the pores of the fibres of the heart muscle swell as a result of internal rarefaction, the cavity of the heart must be filled by its flesh without any alteration of the external shape.

The heart is a hollow ball made of countless spongy fibres, very strong and inextensible in length. They are attached to the four annular tendinous orifices of the heart. Their ends are attached to the surface of the cavity of the heart and are convoluted and intertwined in spirals. All these fibres swell as a result of internal rarefaction, as the threads of a rope or of a ball of wool become turgid when soaked[8]. As a ball made of threads must be transformed, the cavity of the heart must be filled when the internal fibres are unevenly puckered, swollen, and tautened. Their puckering is the more pronounced the closer they are to the centre of the heart. The external shape remains, neither increasing nor decreasing.

Proposition XLIX

Constriction of the ventricles of the heart cannot result from the contraction force of its fibres.

If everybody agrees that the characteristic action of all muscles is the contraction of their fibres, it also seems to be agreed that the action of the heart muscle itself is also the contraction of its fibres. The heart is not attached to limbs which it would move by its contraction. Its contraction must only squeeze its cavity. It must be seen whether the cavity of the heart can be reduced by simple contraction of the fibres. Firstly, if the heart is similar to a ball made of threads convoluted in spirals, the fibres when they contract cannot exert a force by pulling their extremities towards their centre. This results from what was said above[9]. Actually they rather relax by puckering. As a weight cannot be elevated

[8] Prop. 47
[9] Prop. 47 and Prop. 48

by a loose rope, the opposite walls of the cavity of the heart cannot be pulled violently and meet by the action of loose and puckered fibres.

Secondly, we assume that the spiral fibres of the heart do not circumscribe tortuously the heart down to its cavity but rather extend straight into the ventricles where they collect into bundles which compose several small cylinders or ropes. It thus may be thought that simple contraction of the fibres could shorten these cylinders totally and so the opposite walls of the ventricles of the heart could meet.

Actually this seems impossible. If this occurred, one would have to admit a puckering of all the surface of the cavity of the heart between the bases of these small columns or ropes. Countless interwoven fibres would come closer together and also pucker, i.e. they would have to become looser. Therefore, they would not act by their contraction. This is against the hypothesis.

Moreover, there are no small cylinders in the right ventricle of the heart, or very few of them. Consequently, such mechanism does not take place.

Furthermore, total shortening of these small cylinders or ropes would be possible only if the lengths of the spirals ending in the columns themselves ran between other fibres, like in sheaths, or round pulleys. This is contradicted by autopsy. Actually, the fibres of the heart interwoven in different directions are so tightly attached to each other that they cannot carry out different and opposite movements between the other fibres which surround them.

Finally, tension of the heart muscle cannot be explained by simple contraction of its fibres. The volume of the muscles of the limbs which actually contract appears rather to decrease. The volume of the heart increases since its cavities are filled by fleshy substance and its external shape does not decrease. Consequently, the heart is not tautened by the contraction of its fibres.

Proposition L

The cavities of the heart are squeezed not as a result
of shortening of the length of the ventricles
but because the lateral walls come closer together.
Tab. XVII, Fig. 1.

This appears from the position itself and from the configuration of the ventricles of the heart ABC, also from its action. The left ventricle FG, opened from the apex to the top, extends over the whole length of the heart from the base AB to the apex C which ends thin and pointed in the wall GC. Since the external shape of the heart does not shorten during pulsation, the length FG of the cavity does not decrease, i.e. the base of the cavity does not come closer to the apex.

Moreover, the base AB and the apex C of the heart cannot at all swell and thicken because the base lacks fibres and flesh, being all occupied by the four large orifices D, E of the veins and arteries. The wall GC of the apex is very thin. Therefore, the cavity cannot be filled by swelling of fibres of the base and apex of the heart since such fibres are absent. Consequently, it must be filled by swelling of the very thick lateral walls.

Finally, if a finger is introduced into a ventricle of the heart of a live animal through an incision, constriction of the lateral parts is perceived but no meeting of the apex and base.

Proposition LI

Mechanism of the action of the heart.

It thus appears that the cavity of the heart cannot shorten for mechanical reasons. The heart can contract only if its fibres contract. Its parts which are deprived of fibres thus cannot contract. But among the multitude of fibres which go down from the base of the heart less than a thousandth reaches the apex. (If they all met there, one fibre superimposed upon another, the apex would be very thick. This does not occur since the wall is very thin there.) All the other spiral fibres are intertwined and reflected into the heart transversely before they reach the apex. They thus cannot pull the apex of the heart towards its base and, consequently, cannot shorten the length of the cavity of the heart.

The lateral walls of the ventricles are squeezed for the following reason. All the almost countless fibres of the heart turn round the sides of the heart obliquely and transversely. They form several layers superimposed upon each other like membranes. When the fibres of any layer which are adjacent to each other laterally or which lie at the surface swell, they forcibly must push each other sideways. They thus expel each other from their proper location and move each other further, always towards the base and the apex. Therefore, the distance between the base and the apex should rather increase. Actually the other external, obliquely transverse, surrounding fibres intersect them and prevent their elongation and protrusion as we shall show below. Consequently, the fibres must swell internally towards the cavities and so the inflated walls of the cavities must meet.

Moreover, the septum of the heart consists mostly of transverse fibres which swell and cannot shorten. Its width increases considerably. The external walls also must swell towards the internal cavities. The opposite walls thus must meet whereas the base remains remote from the apex.

Proposition LII

The heart acts by constricting its ventricles, compressing and expelling the blood which it contains, like a press. This does not result from twisting of the spiral fibres of the heart but from swelling and tautening of these fibres. Tab. XVII, Fig. 2.

Without any doubt, the volume of the heart considerably increases during its pulsation but its external shape a, n, c, o, b and its dimensions neither increase nor decrease. The fleshy substance of the heart is NAMO when not contracted.

Almost half the total volume ABC comprised in the external shape then is filled by blood. When the two ventricles FG, HI are full of blood in men their cavities are almost equal to the volume of the fleshy substance when not swollen NDMO. During the systole, the blood is expelled and both ventricles fg, hi are filled by the swollen fleshy substance n, a, m, o of the heart itself. Therefore, the external shape is not changed but the volume of the heart during contraction becomes about twice what it was before the pulsation. Actually such increase of the fleshy volume is not possible without considerable swelling and turgescence of its fibres. Consequently, the fibres of the heart become twice thicker during pulsation. This being established, I consider that the external fibres of the heart exert their maximum force, not by shortening, but by resisting stretching like the iron circles of a barrel, so that the perimeter of the heart is not increased and the fibres swell inside the cavity, as was explained for a ball of wool[10]. Similarly, the internal fibres when they swell and thicken, exert their maximum force in forming turgid, hard and tense folds which act as wedges. Not only do these folds fill the cavity but also they expel the blood contained there by an energetic compression in a movement similar to that of stones of fruit projected by pinching.

The tiny columns of flesh connecting the opposite walls of the cavities of the ventricles thicken, tauten and simultaneously shorten. Their fibres swell and pucker, so that they fill the cavity. They exert their greatest force by their tautening and swelling so that they act as wedges. They exert no force by their contraction to bring together the opposite walls of the cavity because they are loose as a result of the puckering of the length of their fibres. Any way they could not join the opposite walls exactly together: these muscular columns indeed could not shorten totally since muscles do not normally contract by more than a third of their length. However, it cannot be denied that the columns, like ropes, are used to maintain the arrangement of the internal parts of the heart and to prevent an exaggerated distension which could be provoked by too much blood injected from the veins.

The fleshy prominences which protrude into the ventricles and to which the membranous filaments of the triangular valvules of the heart are attached act not only by swelling but also by protruding like the tongue and the penis.

All this applies to the left ventricle and to the auricles. But in the right ventricle there are not so many columns. Constriction there results from thickening of the external wall n, by swelling and shortening of its fibres so that the concave internal wall by swelling becomes plane and the cavity is reduced. Turgescence and swelling of the fibres of the septum m or fairly thick middle wall also contribute to the filling of the right ventricle. Its convex surface protrudes further and fills the cavity. The space thus is filled and the walls touch each other along a lenticular curve.

All this occurs according to the plan of Nature which at first fills and expands the pores of the spongy fibres by rarefaction, as explained above. This has two consequences. In the muscles of the limbs this turgescence of the pores of the fibres produces a secondary effect: shortening of the muscle and strong

---

[10] Prop. 47

pull of the limb. But in the heart Nature uses swelling and thickening of the fibres to fill the ventricles of the heart. The mechanism, however, is the same and consists of a force acting as a wedge which dilates the pores of the fibres.

The blood is not expelled from the heart by spiral twisting of its fibres as twisting of soaked linen clothes expels the water contained in them (Fig. 3, Tab. XVII). Some people have expressed this opinion recently. It is true that expelling the liquid from the heart, as well as from a soaked linen cloth when twisted, results from the constriction of the cavities and pores which were full of liquid. But I deny that this constriction in the heart and the expulsion of blood result from the same cause, the same instruments and the same mechanism as the twisting of a soaked linen cloth. Before being twisted, the fibres of the linen cloth AB, CD were loose and thus offered many interstices which could be filled by water. When the cloth is twisted, its fibres are forced to form several turns ac, bd about the bundle of linen which retains approximately the same length. Therefore, the fibres must not only be elongated in long turns, but also be tautened and thinned. Consequently, the sides of the fibres are smoothed by disappearance of the folds and touch each other, the interstices disappear and the water which they contained must be expelled. Actually, in the heart filling of the ventricles is very different: the volume of the substance of the heart does not decrease; it rather increases and doubles. The fibres of the heart are not elongated but rather contracted as required by the nature of muscles. These fibres do not touch each other and their interstices are not filled as a result of violent tension and expansion but by the swelling of the fibres. The folds are not removed and smoothed by tension. On the contrary, the fibres pucker as a result of the turgescence of their pores. The cavities of a linen cloth are suppressed by twisting the fibres and water is expelled. But the cavities of the heart are not filled by a similar spiral twisting of the fibres of the heart since their spirals during tension do not twist nor become tighter and more oblique. On the contrary, they become less oblique. Consequently, such twisting of the fibres does not occur in Nature and thus cannot result in filling of the ventricles of the heart and expulsion of the blood.

The spiral arrangement of the fibres of the heart, however, is not useless. It is involved in a firm binding of the walls so that the aspect and shape of the heart are maintained. Nature successfully convoluted the fibres spherically from the external tendinous orifices of the vessels to the internal columns. These fibres are intertwined and woven in an admirable arrangement.

Proposition LIII

Cause of the pulsation of the apex of the heart on the left side of the chest during systole.

The heart is suspended from its ligaments in the middle of the chest, like a pendulum. During each systole its apex is brought into contact with the chest and strikes it. This must result from a dilatation of the heart due either to its displacement or to its erection or to bending of its apex. Experiment refutes the first two explanations and thus confirms the last. It remains to investigate the mechanism of this effect which seems to result above all from the arrangement of the fibres of the heart. A curved bowel ligated with a thread and partly filled by water is pulled straight by its weight. If the water is pushed by a piston, the bowel bends as required by its shape. Its hanging lower end protrudes and strikes the hand which is in the vicinity. This obviously results from the curve-shaped skin of the bowel which is longer over its convexity and shorter in its concavity. In the left part of the ventricle of the heart also, the left wall is shorter, less fleshy and less curved than the two walls of the right ventricle. Hence, during the systole, the apex of the heart must rise towards the left part of the chest and can strike it with a force proportional to the violence of the erection. This can also be due to or helped by the erection of the heart which lies obliquely or by the position and arrangement of the fibres which are spirally convoluted from the right aspect of the base of the heart to the left aspect of the apex. When the fibres swell a bundle of fibres can be distorted and protrude anteriorly towards the left and striking can occur.

Proposition LIV

Action of the auricles and valvules of the heart.

The end of the vena cava in most animals is muscular over its whole circumference so that it contracts like a sphincter. There is no such support at the extremity of the pulmonary vein. Both these veins end in two muscular auricles which are hollow. They are like bags appended to the sides of the heart. They do not differ much from these tortuous roundabouts which usually are added to the sides of the doors of jails so that the prisoners cannot easily run away.

The structure of the auricles is similar to that of the left ventricle in its hollow part. The auricles are made of intertwining fleshy fibres which coalesce inside into several columns connecting the sides of these bags. They are followed by three membranes in the right ventricle and two in the left. These membranes are very strong. They are triangular. Their bases are implanted tightly over all the internal perimeter of the tendon of the orifice of the heart. The surfaces of the membranes branch out inside the ventricles in several tendinous ropes. These are connected to the extremities of the protuberances which arise in this region.

The function of these admirable structures must be described. Firstly, the fleshy part of the vena cava is not strong enough not to be disrupted by the afflux of blood. Rather by its constriction it propels the blood in the oblique cavity of the auricle which it makes turgid. This is helped by peristaltic constriction of all the vena cava and by compression of the muscles and viscera of the whole body. As a result, blood flows in a whirlpool through the orifice of the auricle, fills its cavity and flows into the right ventricle. Blood is forced from the pulmonary vein into the left auricle and then into the left ventricle for the same reason.

The two auricles are constricted simultaneously like presses by the contraction and compression of the muscles in an order such that at first, their orifice being closed, they prevent the blood from flowing back into the veins from which it came. Then acting with the force of a press they drive the blood into the ventricles until these are turgid. Their action is followed by the compression of the heart by which the blood itself fills and swells the tricuspid and mitral valvules, thus closing the orifices of the veins and preventing backflow of the blood. Consequently, the blood must be expelled into the pulmonary and aorta arteries. All these facts were demonstrated by many experiments carried out by Harvey and others.

Proposition LV

Analysis of the mechanism of these operations.

After the description of the phenomena, it remains to analyse their mechanism. Firstly, the auricles of the heart are round and hollow muscles made of fleshy fibres convoluted in spirals, intersecting each other. They end in columnar bundles and in depressions in a structure similar to that of the ventricle. The auricles thus must also act like a press by the same mechanism as the heart does. They must be inflated, tautened and constricted by wedges driven into the pores of their fibres and thus expel the blood which they contain.

Secondly, it will be shown that constriction of the auricles must precede contraction of the ventricles, although they seem to move simultaneously. If it is not so, the auricles and ventricles contract together at the same time or the heart contracts before its auricles. In the former instance the triangular valvules are useless before the heart has contracted, and after this contraction is completed: indeed closing of the valvules would be useless when the blood cannot flow back. Before the heart is contracted the blood cannot flow back since it is not yet in the ventricles and thus cannot be pushed by the systole. After contraction of the heart, the expelled blood which is no longer in the heart and in the arteries, can regress even less. When the ventricles contract, the orifices of the veins must necessarily be tightly closed by the triangular valvules so that the blood is pushed forward into the arteries and not backwards. If the auricles contracted at the same time they would spill the blood which they contain into the ventricles. They would thus open the closing achieved by the triangular valvules because their arrangement is such that they are opened by the arrival of blood. Therefore, at the same time blood would be pushed into the heart by the auricles and repelled by the heart. The opposite efforts and movements would cancel each other and thus would be in vain. Moreover, when two exits are open in the same ventricle, the compressive force of the heart is divided into two equal parts which push half the blood backwards and half forwards into the arteries. Nature would provoke half the effect by a stupid double effort. Besides, the triangular valvules would be completely useless since they would be forced to open exactly when they should close. Consequently, it seems impossible that the auricles and ventricles contract at the same time.

If we assume that the ventricles contract first and the auricles afterwards, this would be even more absurd. Half the blood contained in the ventricle would flow back and be driven into the open auricle and hence into the vein.

Consequently, contraction of the auricle must precede and be followed immediately by contraction of the ventricle. Then all the operations proceed at best and economically. The auricle being compressed, firstly backflow of the blood into the vein is prevented. Secondly, blood is expressed from the auricle into the cavity of the heart. Thirdly, the orifice of the heart is closed by the contracted auricle. Fourthly, the cavity of the heart being dilated by the mass of blood, the curtains formed by the triangular valvules are drawn like doors, as we show below. Then, fifthly, the heart contracts and all the mass of blood contained in the ventricles, which cannot flow back because of the double closing, is forced to leave through the open door of the arteries.

Thirdly, the action of the right auricle is different from that of the left because the blood from the big pulmonary vein must rush very quickly into the left ventricle as a result of its weight and of the compression of the lungs acting like inflated bellows. Consequently, for the small auricle to be sufficient to transmit the blood, the flow into the left ventricle must be very quick. On the other hand, the orifice of the heart can be closed exactly by the small auricle together with the mitral valvules.

In the right auricle the slowness of the blood flow must be compensated by the large size of the canal. Moreover, the right auricle must not only close the orifice of the heart but also send rapidly into the right ventricle the blood which flows in slowly. Therefore, the muscle and the cavity of the right auricle had to be larger and stronger than the muscle of the left auricle.

Fourthly, the triangular valvules close the orifice of the heart exactly, in the following way. These membranous triangular valvules are implanted at their bases, like flags, on part of the circular tendinous ring of the orifice of the heart. Their other sides are attached by several tendinous ligaments to fleshy protuberances which are present in the opposite part of the cavity of the ventricle, as a sail spread out against the rays of the sun is maintained taut by several ropes. Hence, when blood flows in, the cavities of the ventricles are dilated and these membranes which were closed are pulled transversely through all the space of the ventricles. That is why the apices and the sides of these triangular valvules, pulled by intertwined ropes, must unite and, being connected, form one conical surface which is larger than the plane of the orifice corresponding to the circular base of this cone. This is followed by a systole during which the internal walls of the ventricles meet. Then these ropes are loosened and meet together. Therefore, the faces of the triangular membranes must make folds and take a sinuous shape because their surface is not decreased since the membranes do not contract. Hence, the mass of blood pushes back these membranes and bends them as the sails of a ship bend when filled by the wind. Afterwards, these curved sails are moved towards the tendinous orifices of the heart. They are much larger than these orifices. Therefore, they must occlude exactly these orifices before the systole is completed. The blood contained in the cavities of the ventricles thus must be expelled completely from the ventricles through the arterial orifices where an exit is open, by a steady contraction of the heart until the walls have met everywhere.

Fifthly, if closure of the auricles did not precede, the membranous valvules could not resist the violence with which the blood compressed by the heart would stretch and tear them. Therefore, foreseeing Nature provided a vault, i.e. the floor of flesh formed by the contracted auricle, to close with a double door the large orifice of the heart.

Consequently, the action of the auricle is longer than the systole of the heart since it begins before the heart contracts and ends when the systole is completed.

Sixthly, the arterial orifices of the heart being also open do not need such a device to prevent backflow of the blood into the heart. After the blood has left the heart and most of it has been expelled from the proximal extremities of the arteries, it cannot be pushed back, either because the muscles do not press it with as much force as the heart does, or because it has already left the arteries. Therefore, weaker valvules are sufficient because the force which anyway can be exerted by the arteries is small. Three semilunar valvules much weaker than the triangular valvules, without auricles, are sufficient to prevent regurgitation of the blood expelled from the heart.

Then the following lemmas are useful to analyse the force of the motive faculty of the heart.

Proposition LVI

If a rope turned round a compressible sphere or cylinder contracts uniformly over its length, the ratio of the force of contraction to the resistance of the sphere or cylinder is equal to the ratio of the radius of the sphere or cylinder to the circumference described by the rope.

Tab. XVII, Fig. 4.

The longest circumference of a compressible cylinder or sphere or its cross section RSO is ligated by a rope the axis of which is CEDH. The length of the rope is measured by its axis which divides its diameter in two. The length of the rope is contracted uniformly so that all its particles shorten in the same proportion. I claim that the ratio of the force which contracts the rope to the force with which the sphere or cylinder RSO resists constriction is equal to the ratio of the radius AC of the circle to its circumference CEH.

The extremity B of the rope BE is attached firmly at R and the extremity C of the rope HC is pulled by a force M from C to M. Then the circular circumference CHDEB must diminish more and more, forming smaller and smaller circles until, when the rope is completely pulled, the circumference of the cavity ROS is brought into contact with the centre A. During such tension the size of the enclosed sphere or cylinder is more and more reduced until the sphere or cylinder disappears and its centre is brought into contact with the rope at R beneath the nail at B. The ratio of the circumference CEH to the radius CA is equal to that of the circumference IN to the radius IA[11]. The differences between the homologous sizes thus are in the same ratio. The ratio of the difference between the circumference CEH and the circumference IN, i.e. the length CM of the pulled rope to CI or to the shortening RA of the radius, i.e. to the movement of contraction of the enclosed sphere or cylinder, is equal to the ratio of the circumference CEH to its radius AC.

This being said, I mention that the force with which the sphere or cylinder resists constriction is multiple. It consists of as many parts as there are physical points in the circumference of the cylinder RSO. These forces are exerted along the radii of the circle, pushing from the centre to the periphery when they are dilating and, inversely, regressing towards the centre when the radii are compressed. The singular points thus travel the same distance along equal radii during the same time. Consequently, their velocity is the same. Therefore, the whole force with which the sphere or cylinder resists constriction has the same velocity as the movement of one point of the circumference. Thus the velocity with which all the force resists the constriction of the cylinder is measured by the movement along one single radius. On the other hand, the velocity with which the force M pulls the rope is measured by the length MC of the pulled rope or by the circumference BEDH. Consequently, wherever the forces are in equilibrium, they are in inverse proportion to the velocities with which they can be displaced during the same time. Thus the ratio of the force M to the force with which the cylinder resists constriction is equal to the ratio of the movement through CA to the movement through CM or to the ratio of the radius AB to the whole circumference BEDH.

The same movement results if all the small particles of the rope contract coming closer to each other as by pulling the rope by the length CM equal to the circumference BEDH. Consequently, the force contracting the rope carries out a movement equal to the circumference BEDH. The resistance is the movement of the compressed cylinder along the radius BA. The ratio of the force contracting the rope to the resistance of the compressed cylinder thus is equal to the ratio of the radius BA of the circle to its circumference BEDH.

Proposition LVII

Two pairs of scales FB and BH (Fig. 5, Tab. XVII) are connected at B. Their fulcra are C, D. The arms CB, BD are equal. FC is longer than DH. The weight V suspended from the adjacent extremities B is balanced by two counterweights R and T. The weights V and R are given as are the ratios of the arms CB/CF and BD/DH. The weight T must be found. If the weights R and T are given, the weight V can be found.

CB/CF = R/P in which P is a part of V. The weights P and R thus balance each other. By hypothesis, the whole weight V remains balanced by the two weights R and T. Therefore, the remaining part Q of the weight V balances

[11] Lib. 5 Prop. 14. Eucl. Rest. Auct.

the weight T. Thus, $Q/T = DH/DB$ and the weight T is known. Secondly, $R/T = DH/DG$. At the extremity G a weight S equal to R is suspended. Thus, the moments of S and T are equal. Actually, S is balanced by the same weight Q. Therefore, $S/Q = BD/DG$ and
$PQ/S = (FC + DG)/BD = (FC + DG)/CB$ and $S/T = DH/DG$. Thus the ratio PQ/T is composed of the ratios $(FC + DG)/BD$ and DH/DG. These two ratios form the ratio of the rectangle $(FC + DG)$ DH to the rectangle $DB \cdot DG$. But $PQ/R = PQ/S = (FC + DG)/BD$. If DH is given a common length, this is equal to $(FC + DG)\,DH/(BD \cdot DH)$. Therefore,
$PQ/(R + T) = V/(R + T) = (FC + DG)\,DH/BD\,(DG + DH)$. This enables one to find V which was sought for.

Proposition LVIII

The ratio of the force constricting a goatskin bottle full of water to the resistance of the water expelled through an adjunct pipe is equal to the ratio of the cross section of the goatskin bottle to the cross section of the pipe.
Tab. XVII, Fig. 6.

The goatskin bottle ABCD is assumed to be cylindrical. A pipe GH also cylindrical is adjunct. The water contained in the goatskin bottle is pushed from AD to BC by a piston LM. The water AC thus expelled rises from G to H. The force pushing the piston is P and R is the resistance of the water to expulsion. I claim that the ratio of the forces P/R is equal to the ratio of the surface of the circle AD to that of the circle IG. The same volume of water moved during the same time successively fills two equal cylinders AC and IH. Therefore[12], their lengths and the surfaces of their cross sections are in inverse proportion. The velocity of the expelled water or of the resistance R is measured by its displacement or by the distance GH. The velocity of the compressed water or of the movement of the piston or of the force P is measured by the distance AB. Consequently, the ratio of the force P to the resistance R[13] is equal to the ratio of the velocity GH of R to the velocity AB of P, or to the ratio of the surface of the circle AD to that of the circle IG.

Comment

If the length GH is nine times that AB and if the force exerted by the water IH is equal to one pound, the force pushing the piston is equal to nine pounds.

[12] Lib. 6 Prop. 32. Eucl. Rest. Auct.
[13] According to mechanics

Proposition LIX

If additional water is pushed by a piston from a bigger pipe into a smaller containing water, the ratio of the force pushing the piston to the resistance of the initial volume of the water thus pushed further into the pipe is equal to the ratio of the surface of the orifice of the bigger pipe to that of the smaller.
Tab. XVII, Fig. 7.

There is a volume of water GH in a pipe. A piston is pushed to BC and the water rises from H to O. P is the force pushing the piston and R the resistance exerted by all the volume of water GO. I claim that the ratio P/R is equal to the ratio of the circle AD to the circle IG. The volume of water GK is equal to AC. The volume of water GO represents the water present initially and the water introduced. GH is the volume present initially. Therefore, the volume of water HO is equal to the volume AC expelled or to the volume GK. Adding KH gives KO which is equal to the volume present initially GH. The water GH is displaced from H to O during the same time as the water expelled moves over an equal distance GK since the cross section of the pipe is even. Therefore, all the water is moved with the same velocity as that of the expelled water rising by GK. The ratio of the force P pushing all the water GO to the resistance R is the inverse ratio of the velocity of the water GO (which is the same as the velocity of GK) to the velocity with which the force P pushes the piston. But the ratio of the velocity GK to the velocity AB is equal to that of the circle AD to the circle IG[14]. Consequently, the ratio of force P to the resistance of all the water is equal to the ratio of the cross section of the larger pipe AD to the cross section of the smaller pipe IG.

Comment

If there are 57 ounces of water present initially KO and 3 ounces are introduced, there are in all five pounds of water GO. This water is moved with a velocity IK which in the preceding proposition was assumed to be nine times the velocity AB. Therefore, the force pushing the piston is nine times the resistance of five pounds. The force exerted thus is equal to 45 pounds.

It must be noted that either the water sent into the pipe resists the impulse by its weight as in any erect position, or in recumbent position it resists by its inertia since it slows down the impetus of the body which pushes depending on the magnitude of the motive force, as we will show[15].

[14] Prop. 48
[15] De vi percuss. Prop. 27 and Coroll. Prop. 28

Proposition LX

If the water contained in the tube AC (Fig. 8, Tab. XVII) is pushed by a piston LN into the adjoining soft pipe EH which is completely squeezed by the weight S, the ratio of the force PQ pushing the piston to the sum of the resistance R of the water to the impulse and the weight S which impedes the dilatation of the pipe is equal to the ratio of the rectangle EFHG formed by the length and the dilatation of the filled pipe, to the square of the displacement AB of the piston.

The impulse of the piston LN provokes two effects simultaneously: straight flow of water over the length of the pipe EF and lateral expansion EG of the pipe by the volume of water acting as a wedge. Water resists straight impulse by its inertia[16] since the magnitude of the impetus which is given to the water is subtracted from the propelling force. Moreover, the force of the bandage circularly resists the lateral impulse as well as the weights S in so far as they impede expansion of the tube. Finally, when the dilatation is completed the moments of the opposite forces balance each other. The ratio of the component P of the pushing force to the resistance R of the water which flows straight is equal to the ratio of the velocity of the movement through EF to the velocity through AB since the force P pushing the piston is moved during the same time. Similarly, the ratio of the remainder Q of the force to the resistance S with which water and the weight constrict the pipe EH and impede its dilatation is equal to the ratio of the velocity exerted by the weight over the width EG to impede expansion of the pipe, to the velocity with which the force Q pushes the piston through AB. These two ratios combine into the ratio of the rectangle FE · EG to the square AB · AB. Therefore, the ratio of the force PQ to the resistance RS is equal to the ratio of the rectangle FE · EG to the square AB · AB.
$PQ/RS = (FE \cdot EG)/(AB \cdot AB)$

For example, if the length of the pipe EF is one hundred including AB which is ten, if the width is two, if there is one pound of water GEFH and if the incumbent weight is nine, the rectangle GE · EF then is 200 and the square AB · AB is 100. Therefore, the ratio of the force PQ pushing the piston to the resistance RS is 200/100 = 2.

[16] De vi percuss. Chap. 10 Prop. 27

Proposition LXI

The conditions are the same. The soft pipe EH (Fig. 9, Tab. XVII) is full of water but not turgid. Surrounded by bandages and squeezed by a weight S, its dilatation is resisted. The ratio of the force PQ pushing the piston to the sum of the resistance R of the water present initially and of the water added, to the force S with which the weight and the bandages oppose dilatation of the pipe, is equal to the ratio of the rectangle GKZO formed by the straight movement of the added water and by the dilatation of the pipe, to the square of AB formed by the movement of the piston.

To drive the water ABCD into the pipe EGHF, this pipe which is suppposed to be full of water must dilate to KM to be able to receive the additional water. It must then be observed by what motions the water is introduced and how the water present initially expands the pipe. The water GEFH because of the contiguity of its parts constitutes a cylindrical column. Its circular base GE cannot be compressed and pushed towards H without all the length of the column GF being also pushed since the posterior particles push those immediately in front like the fibres of wood are pushed straight ahead. But this column is soft and the walls GH are not firm but can be stretched. Consequently, the fibres GH, EF of water pushed straight are forced to deviate and are brought to occupy the space MHGK.

By these two movements, straight and transverse, all the water present initially and the water introduced must move forwards, pushed by the force of the piston so that the water GEIO fills the space of dilatation ZOHM. The additional water pushed by the piston fills the initial space GEIO and the space of dilatation KGOZ. The velocity with which all the water moves straight through the length GH of the pipe is measured by the displacement of the additional water EIZK through the space EI since the water present initially moves with the same velocity as that of the water introduced additionally, as said above[17]. But the velocity with which all the water simultaneously moves transversely to fill the expanded width is measured by half the difference between the diameters EK and EG. The supported weight and the bandages surrounding the pipe also move and give way with the same velocity with which the water expands in width. Consequently, the component P of the force pushing the piston acts with a velocity AB against the force of all the water resisting with the velocity EI. The other component Q of the pushing force acts with the same velocity AB against the force of the weight and bandages which oppose the dilatation with the velocity GK. Therefore, the ratio of the total force PQ to the sum of the resistances of all the water, the weight and the bandages is equal to the ratio of the rectangle GKZO to the square AB · AB.

For example, AB is 5, AD is 6, EI is 10, the width EG is 2, the magnitude of the expansion of the pipe is 1 and the length EF of the pipe is 30. Then the

[17] Prop. 59

volume of water AG, or EZ or GM is 30 ounces. The water present initially is 60 ounces. All the water EKMF is 90 ounces. The ratio of the sum of the resistance of all the water EM and the force exerted by the supported weight and the bandages, to the force with which the piston is pushed is equal to the ratio of the square of AB = 25 to the rectangle GKZO = 10.

In this hypothesis, the ratio of the force driving the piston to the total resistance is 2/5. Then we will show that:

Proposition LXII

If additional water is sent by a piston into a rigid pipe full of water, a volume of water equal to the volume which was injected will flow out of the orifice of the pipe.

A rigid and inextensible pipe cannot contain more water than the cylinder of water by which its cavity is supposed to be filled. Therefore, if we want to push additional water into the pipe, since there is no interpenetration of the bodies, this additional water acts as a wedge of wood and expels through the orifice a volume of water equal to that which is injected by the piston.

Proposition LXIII

If a soft and dilatable pipe which cannot retract, is full of water but not turgid and as much additional water is driven into it by a piston as is sufficient to fill the dilatation of the pipe, the additional water will not flow rapidly out of the narrow orifice of the pipe.

The force exerted by the piston acting as a wedge to push the water fills the soft pipe and makes it turgid. The additional space in the dilatable pipe is thus filled by the injected water. The pipe is expanded as much as it can be. It will resist further dilatation as if it was rigid. We suppose that the progress of the piston ends in this situation. Consequently, water does not spontaneously flow rapidly from the pipe because of its weight or would not flow out at all from the orifice of the pipe if the latter was erect. If the pipe was recumbent, water would flow out slowly as a consequence of the narrowness of the orifice. Neither is it violently ejected because the previous impulse of the piston cannot do more than filling the soft pipe and, once the pipe is filled, we supposed that the piston is not pushed further. Neither can water be expelled by a contraction force, like a spring, which the pipe is supposed to lack. There is thus no way for the water to flow rapidly out of the pipe. Such was the proposition.

Proposition LXIV

If more water is injected by the piston into the pipe than enough to fill the dilatation of the pipe, the water in excess of what is necessary to fill the dilated space will be expelled.

The conditions are the same. The volume of water AC (Fig. 9, Tab. XVII) is larger than the dilatable space GM in the pipe. Therefore, after the space KH is filled by the portion VC of the volume of water AC, the pipe GM is turgid and cannot dilate more, as if it was rigid. Then according to proposition 62, the excess AX of water pushed by the piston must be ejected through the orifice FM.

Proposition LXV

The conditions are the same. If the pipe filled by the piston until being turgid expels so much water as to deflate and remain soft and with the same size as it was initially, the pipe is not evacuated by the impulsive force of the piston but by another external cause.

The pipe can return from its turgescence to its initial soft consistency only if as much water leaves it as what had been introduced by the piston. The volume of water which resulted in the turgescence of the pipe can neither leave spontaneously, as was shown, nor be expelled by the force of the piston because, when the piston pushes, water flows to the more adjacent places which resist less and are easily stretched, like the soft walls of the pipe. Therefore, the action of the piston fills and dilates the pipe but does not expel rapidly the water through its narrow orifice. Similarly, when water is pushed with a piston into a bowel perforated by a needle, only a few droplets leave through the holes whereas the bowel is quickly filled and dilated. Only after the bowel is turgid, does water begin to get out more abundantly because then the walls of the bowel, as if they had become rigid, cannot be dilated further and water, as if it was pushed by a wedge, can flow out quickly.

Moreover, water cannot get out through the small orifice of the pipe immediately after the impulse of the piston since then this impulse has stopped. Thus, in no way is water expelled from the pipe by the impulse of the piston. Therefore, the water which ensured turgescence of the pipe must necessarily be expelled by another cause. This cause can be the compression exerted by an incumbent weight or the force of a machine or bandages which surround and constrict the pipe.

After these preliminary lemmas, I come to the matter of the force of the heart. And first of all:

Proposition LXVI

Description of the method by which the motive force of the heart muscle can be investigated.

The simplest way of analysing a motive force consists of knowing its most notable effect. This is the weight which the force raises. The force equals the resistance if they are directly opposed. On scales with equal arms, the force balances the counterforce. One does not prevail on the other. If there is a machine which amounts to a lever, the forces are inversely proportional to their lever arms or to the velocities with which they can be moved. We examined the forces exerted by the muscles of the limbs using this method. Actually this cannot be applied to the heart which was destined by Nature, not to pull heavy limbs with the weights they carry, but to express a small volume of blood contained in its ventricles, by its contraction, like a press. A weight does not always oppose an expulsive force. When we push a rock downwards the rock does not resist our impulse by its weight. When we pull a boat floating horizontally this boat does not resist by exerting its weight, for reasons of equilibrium. Moreover, a force can take advantage of the resilience of a machine, such as a spring or something similar the resistance of which often cannot be compared with a weight. Such is the action of the heart. Therefore, we cannot appply to the heart the method of measuring the force of the muscle by a suspended weight as we have done so far.

In this matter we cannot perceive the cause from the effect. We are thus forced to guess the cause itself from some other sign from which we find the magnitude of the effect. Such a sign is the similarity and analogy of the heart muscle with other muscles of the same animal. We said indeed that all the smallest fibres or the small machines of the different muscles in the same animal have the same strength and, when healthy, exert the same motive force in the same time. The equal masses of two muscles contain an equal number of small fibres. Consequently, if we know the total motive force of one of the two equal muscles, we can surmise the total force of the other. The fleshy volume of the heart is about equal to that of one temporalis and one masseter muscle. We know the total motive force of these muscles. We thus can assert that the motive force of the heart is probably equal to that exerted by these two muscles which close the mandible.

Proposition LXVII

The motive force of the fibres of the heart muscle by themselves would be able to raise a weight heavier than 3000 pounds.

We demonstrated previously that the weight which is carried by the whole muscle or by a whole bundle of fleshy fibres can also be lifted (although less high) by one layer of as many small fibres as there are fibres in the bundle[18].

[18] Prop. 120 Part I

Therefore, it is not enough to know by experience that the four muscles which close the mandible raise a weight of 300 pounds without being helped by any lever[19] and that one temporalis muscle with one masseter muscle supports a weight of 150 pounds. The effort actually is carried out by as many small fibres or by as many muscles as there are complete fibres comprised in these two muscles. Any whole fibre in these muscles is not shorter than two finger breadths (some longer fibres being compensated by some shorter). In the space of one finger breadth of any fibre there are more than twenty small machines[20] or small pores aligned like the links of a chain. We assume, however, that there are no more than ten small fibres. Therefore, the length of the fibres which compose the two muscles contain more than twenty small fibres. Since a weight of 150 pounds is carried by one layer of small machines of these muscles, to obtain the whole force which Nature exerts in these muscles, the force developed by one layer, i.e. that sufficient to carry 150 pounds, must be multiplied by twenty. Consequently, the force exerted by Nature to swell all the small pores of these two muscles amounts to more than 3000 pounds although its moment is no more than 150 pounds.

But the fibres of the heart act by swelling not to contract but to swell and to fill the cavities of the heart by their volume, and express and push the blood contained in the ventricles, acting as a wedge, like stones of fruit are projected by being squeezed between two fingers. The cortex of the heart thus can be compared to a small weapon called the blunderbuss. The inflated internal fibres play the role of the gun powder which is ignited and rarefied. The blood expelled from the ventricles can be compared with the shots projected by the blunderbuss.

These operations are suggested by the structure of the cortex made of fibres convoluted in spirals which resist stretching as if they were bandages or iron circles round the barrel of the heart. The barrel of a blunderbuss, however, is different from the cortex of the heart in that the former only resists stretching by its strength. The fibres of the heart also resist stretching by their strength and by the contraction force of their linked external face. But, moreover, the concavity of these fibres puckered by swollen folds exerts the same action as a wedge since, foaming phials being turgid, it develops a force similar to that of exploding gun powder. Any particle of gun powder by its rarefaction propels the part of the bullet which corresponds to this grain and all the particles when ignited together push as many parts of the bullet as there are particles. Similarly, the singular small fibres or small machines of the heart, by their sudden and simultaneous swelling, push the equal particles of the bullet of blood which is ejected from the heart. Thus the total resistance of the blood is made of many parts each of which is balanced by the force exerted by one small fibre of the heart. Hence, the absolute force of the heart is equal to its moment, i.e. equal to the absolute resistance of the ejected blood.

[19] Prop. 87 Part I

[20] Prop. 15 Part I

If any small fibre of the heart exerts a force equal to that exerted by any small machine of the temporalis or masseter muscle, as we suggested above[21], and if the number of small fibres contained in these two muscles is equal to that of the small fibres of the heart, since the fleshy mass of the heart is equal to the fleshy mass of these two muscles, then we can deduce that the force exerted by the sum of all the small fibres of the heart if they pushed the arm of a pair of scales divided in two at its centre can be more than 3000 pounds.

It must then be shown that the heart by its contraction overcomes a resistance greater than the force exerted by the heart muscle itself, or by its small fibres or small machines; in other words that the absolute force of the heart muscle is smaller than its moment. To this end, it must be demonstrated firstly that:

Proposition LXVIII

All the arteries of an animal cannot remain empty of blood after a pulsation is completed.

It appears that the total cross section of all the veins is larger than that of all the arteries, perhaps four times larger. All the mass of blood flows in both. This normally is not more than 18 or 20 pounds in an adult person. The veins appear always to be full of blood because of the transparence of these vessels. One could doubt, however, whether the arteries are always filled with blood or whether they let blood flow through only during their pulsation and remain empty afterwards. If they were bloodless when at rest, then, when exposed, they would appear constricted and flat like ropes or ribbons. This is contradicted by autopsy. They retain their wide and round shape and their cavity is full of liquid. The fact that their cavity is full can be palpated by the fingers. When the veins are exposed, if the arteries remained bloodless after pulsation of the heart, the cavity of the veins would increase three times in diameter as a result of the addition of a volume of five pounds of blood. This does not occur and the veins do not increase at all.

Moreover, in animals whose arteries are visible through the skin, like frogs, the livid colour suggests that they are full of blood. The arteries thus do not remain empty after pulsation but they are filled by at least a quarter of all the blood and, in men, they contain about five pounds.

When they pulse, the arteries are very turgid and not more than three ounces of blood are provided by the heart. Consequently, if the arteries were empty before pulsation, the three ounces of blood sent by the heart would fill a space twenty times greater than this volume of blood. This is impossible without rarefaction which certainly does not occur. Therefore, these five pounds of blood would come from somewhere else, either from the heart or from the veins by a backflow. But both are impossible.

[21] Prop. 120 and Prop. 127 Part I

It must be added that the three ounces of blood sent by the heart cannot fill the space of more than a half foot of the arteries close to the heart. Therefore, if they were empty before pulsation the further arteries would remain empty even after the pulsation. They thus would not pulsate simultaneoulsy with the heart. Since they are swollen by blood, heart and arteries pulse together.

Finally, experience shows that during pulsation blood flows from any artery which has been incised, whether this artery be large or small. The arteries keep ejecting blood like fountains. This would not be possible if all the arteries were not full of blood in their whole length with such a turgescence that blood would rush and flow out because of the exiguity of space. After the pulsation, blood is neither replaced nor reset into the arteries except for a small quantity of three ounces which by its volume could not fill half the capacity of the arteries. Therefore, if about five pounds of blood did not remain, the arteries could not be made turgid again.

I know that some people say that a small part of blood rarefied as a result of the heat of the heart can fill the cavities of the heart and of the arteries. Such imagination is contradicted by experience itself as others have noticed.

Corollary

From the previous proposition it is deduced that, after pulsation, the arteries remain filled by about a quarter of all the blood of the animal. This is about five pounds. The volume of blood sent into the arteries by the heart represents one twentieth of the blood contained in the arteries. Three ounces of blood ejected by the heart into the arteries fill not more than half a foot of the space inside the arteries close to the heart. This is three or four times the length of the ventricles.

Proposition LXIX

The movement of the blood in the arteries is three times quicker than that of the heart pushing the blood.

All these movements are carried out simultaneously: swelling of the heart, reduction of its cavities by the internal mass of flesh, expulsion of the blood which filled the ventricles, flowing of this blood into the arteries and progression of the mass of blood which was in the arteries previously and which is pushed by the blood coming from the heart.

The first three of these operations occur in the heart with the same velocity: the fibres of the heart swell across the width of the ventricles by the same movement as they constrict the ventricles and eject the blood which they contained. The last two operations are also carried out with the same velocity: the blood pushed into the arteries by the heart occupies exactly the same space as that by which the volume of blood which was previously in the arteries must move forwards since one part follows the other in the same passage.

Actually, if the movement of constriction of the heart is compared with the progression of the blood in the arteries, their velocities are certainly not equal since the former is carried out over a distance equal to the length of the ventricles which does not exceed three finger breadths and the distance travelled by the blood in the arteries is equal to half a foot. The latter distance is three times the former and the movements are carried out during the same time. Consequently, the movement of the blood in the arteries is three times quicker than in the heart.

I mention that the movement of the blood in the arteries is always the same. Either three ounces of blood sent by the heart fill the dilated space in the arteries, or some part of it is expelled from the arteries after repletion. In both instances, as much blood present previously in the arteries is moved forwards as the space occupied by three ounces of newly introduced blood. These three ounces of blood travel more than half a foot.

Proposition LXX

The force exerted by the heart in contracting its muscle is smaller than the resistance exerted by the blood in the ventricles in a ratio of two to three.

By comparing the simple action of the heart compressing the blood contained in its ventricles with the resistance of the blood to compression, it appears that such compression is carried out by the external fibres of the cortex which contract and by the internal fibres which pucker and swell acting as a wedge.

The external fibres and the convex aspect of the next fibres act as a rope convoluted about a sphere or a cylinder. Therefore, the ratio of the force contracting the external fibres of the heart to the resistance of the compressed blood is equal to the ratio of the radius of the circle to its circumference[22]. The force thus is six times smaller than the resistance. But the deeper fibres of the heart and the concave parts of the external fibres act as a wedge or a piston by swelling and puckering. They push the blood contained in the heart straight from its periphery to its centre. In this action the moments are equal for the same radii, and impulse and repulse occur at the same time. The force of the internal fibres thus is equal to the resistance of the blood which they compress. But the external fibres of the heart which contract are less than perhaps a tenth of the whole mass of the heart. Therefore, the force of the internal fibres is about nine-tenths of the force of the heart. The resistance of the corresponding blood is also about nine-tenths of this same force. But if the force of the external fibres was one, the resistance of the blood was six. Therefore, if all the force of the heart is ten, the resistance of all the blood compressed in the ventricles will be fifteen, i.e. a ratio of two to three.

[22] Prop. 56

Proposition LXXI

The compressive force exerted on the blood contained in the ventricles is equal to the force with which the same blood acts as a piston pushing blood in the arteries. The ratio of this force to the resistance of the blood is 1/40 approximately.
Tab. XVII, Fig. 9.

In the diagram illustrating proposition 61, ABCD is the cavity of the ventricles filled by a volume of three ounces of blood. Its width AB and its height are both of three finger breadths. The arterial ramifications must be as wide as the main trunk GZ. Longer ramifications compensating for shorter ones, we can consider that all the arteries constitute one trunk corresponding to the arterial cylindrical pipe GF. During the diastole, the pipe GF remains soft and loose although it is filled by five pounds or 60 ounces of blood. During the systole, three more ounces of blood are injected. The pipe EM swells and is filled up to turgescence by 63 ounces of blood. The ratio of the cylinder GF to the cylinder EM of equal length, and hence the ratio of the base EG to the base EK, is equal to 60/63. The ratio of the diameters EG/EK is equal to that of their square roots or about 7.7459/7.9373. By division, the ratio of the lines EG/GK is approximately 7.7459/0.4086 or about 40/1.

The action of a piston is not different from that of the heart as far as their effect is concerned. Both in the same time with the same velocity reduce the cavity of the heart by moving to the wall BC the opposite wall AD or the base of the piston. The liquid contained in the ventricles resists this reduction. The liquid thus resists both compressions with the same strength. The force exerted by the piston is equal to the force with which the liquid contained in the pipe AC resists reduction. We can undoubtedly assume that the force of the piston replaces the resistance with which the blood resists compression of the heart. Consequently, both the force which would push the piston into the heart and the force exerted by the blood in the ventricles to resist compression are in the same ratio to the force with which the blood present initially in the arteries and the additional blood, together with the other obstacles, resist the impulse. Therefore, the ratio of the force with which the blood is compressed in the ventricles or the force with which the same blood compressed by a piston pushes blood into the arteries, to the resistance of all the blood in the arteries and of the fasciae which impede dilatation of the arteries is equal to the ratio of the rectangle KG · GO (the length GO of which is 9 finger breadths and the width GK of which is equal to 1/40 of a finger breadth) or of its content GE to the square AB · AB (the side of which is 9 finger breadths)[23] or the movement of the piston and the reduction of the ventricles. This ratio is 1/40 approximately.

If some part of the three ounces of blood were expelled outside the arteries by the pulsation of the heart, the space GM corresponding to the dilatation of

[23] Prop. 61

the artery would be filled by less than three ounces. Therefore, the base GH would increase by much less than a fortieth of its initial size. The ratio of the rectangle KG · GO to the square AB · AB thus would be smaller than previously. The force of a piston of the heart would be much less than a fortieth of the resistance of the blood in the arteries and of the constricting fasciae.

Proposition LXXII

The ratio of the motive force which contracts the heart muscle to the total force with which the blood resists expulsion into the arteries is equal to 1/60.

The ratio of the absolute force which contracts the heart muscle to the total force with which blood resists expulsion into the arteries is the product of two ratios. One is the ratio of the force contracting the heart muscle to the force with which the blood compressed in the ventricles pushes blood into the arteries like a piston. This ratio is 2/3 [24]. The other is the ratio of the force of the blood in the compressed ventricles to the force with which blood in the arteries resists expulsion. This ratio is 1/40 [25]. Multiplying the two ratios 2/3 and 3/120 gives a ratio of 2/120 or 1/60. Such was the proposition.

Proposition LXXIII

The moment of the motive force of the heart muscle overcomes the resistance of all the blood in the arteries and of the fasciae which oppose their dilatation. This resistance is more than 180000 pounds.

The absolute force exerted by the heart muscle in swelling all its pores is sufficient to raise directly, without any machine, a weight of more than 3000 pounds[26]. But the ratio of this force to its moment or to the force which opposes the movement of blood in the arteries is 1/60 [27]. Therefore, the absolute force which opposes movement of the blood in the arteries and which is overcome by the force of the heart is more than 180000 pounds. Q.E.D.

To make more obvious the mechanism of such a wonder, the cause of this huge resistance which is overcome by the heart must be investigated. And firstly we must refute the common assertion that blood is expelled through the orifices of the arteries, flowing out as from a fountain by no other cause than the impulse of the heart. We will show that:

---

[24] Prop. 70
[25] Prop. 71
[26] Prop. 67
[27] Prop. 72

Proposition LXXIV

During any pulsation of the heart, either all the mass of three ounces of blood sent by the heart or a portion of it is expelled outside the arteries by a motive force different from the piston action of the heart.

The arteries are soft and dilatable pipes full of blood but not up to turgescence, as said above[28]. During their pulsation an additional volume of blood is sent into them by the contraction of the heart, as by a piston. This volume is sufficient to fill their dilatation or may be in excess. In the latter instance, the surplus is ejected outside the arteries by the pulsation of the heart. After the pulsation is completed, the turgid arteries return to their previous soft consistency and dimension. Therefore, the volume of blood which had been driven in by the piston of the heart, or its remainder, must leave the arteries. But the blood introduced cannot leave spontaneously because of the very narrow orifices of the cavities of the arteries since it lacks a proper motive force. Neither is it expelled by the impulse of the piston of the heart, except for a part, because the arteries are not hard, made of iron, but soft. The force of the heart by its direct action dilates the arteries. This dilatation is rather the cause of retention than of expulsion of the blood. Consequently, all the blood or the additional blood is expelled by another and different cause.

Proposition LXXV

The cause of the expulsion of the blood from the arteries through their orifices is a compression and a constriction of the arteries by the force of a machine which contracts the fibres of the vessels and viscera.

Obviously, when a soft and extensible pipe such as is the artery, is full of blood or any other liquid up to turgescence, the liquid cannot get out except if it leaves spontaneously or if it is ejected by some force. The first alternative is possible only if the liquid is moved forwards by some innate motive force. This motive force is either the weight, when it can be exerted, or rarefaction of the liquid which requires more space than the cavities of the vessels and then the liquid is forced to leave these narrow vessels.

Blood does not leave the arteries as a result of the motive force of its weight because the heavy liquid does not fall where there is no declivity and it is unable to ascend. But the arteries of the head, arms and legs when these are raised are not declivous and, nevertheless, blood flows out of their orifices, pushed upwards with maximum impetus. Consequently, blood does not flow from the arteries as a result of some innate force of gravity.

Neither is there any rarefaction since we observe that arterial blood is as warm, as fluid and as voluminous as venous blood. But blood is ejected from the arteries by some external force. This force or impulse is produced by the

[28] Prop. 68

piston action of the heart or by the compression of the walls of the artery by a weight or by the constriction of all these walls by the force of a machine.

Firstly, the impulse of the heart does not expel all the volume of blood sent by the heart outside the arteries but, at most, part of it. This was demonstrated[29].

Secondly, the blood does not leave the arteries as a result of compression by a weight. Denuded arteries are not compressed and, nevertheless, they expel blood. Some ramifications of arteries enter solid bones. There, because of the rigidity and hardness of the holes, turgid arteries cannot be compressed by the weight of the bones. Moreover, the arteries which are suspended and adhere to overlying bones or tendons cannot be compressed by these bones or tendons since they can neither move nor dilate. Neither can they be compressed by the underlying tissues which by their weight tend to move away from the arteries rather than to compress them. Similarly, all the arterial branches close to the skin in supine position support the minimal weight of skin and subcutaneous fat. Such small compression would be unable to exert sufficient force as required to provoke a very quick outflow of blood. Although the weight of flesh and viscera resting on arteries in some places compresses them and contributes to the expression of blood, this force is weak except in the lungs and it is not exerted everywhere. Therefore, it does not constitute the general cause of this considerable expression of blood.

The last way of expressing blood, by contracting the transverse fibres, not only is possible but is very usual in animals. Food is swallowed by the oesophagus only if the latter is constricted by contraction of its circular fibres (as a stone of fruit which is ejected when pinched). Food thus is swallowed by repetitive compression and pushing. The faeces are moved in the intestines in the same way. Similarly, other liquids are propelled in the viscera of animals by peristaltic movement.

Moreover, apart from the easy and obvious mechanical ways of expelling liquids from soft and turgid pipes thus described, no other is observed in the operations of Nature or of engineering. Nature regularly takes the simplest ways and dislikes roundabouts and changes. Therefore, Nature expels blood from the arteries by one of the ways thus described. But all of them, except the last one, were shown to be impracticable. The last one consequently must be retained.

But we will proceed to firmer and more convincing arguments. Pulsation of the heart is followed by two effects: filling of the arteries by the blood sent into them and outflow of blood from the same arteries. These two operations cannot occur simultaneoulsy since the former is accompanied by dilatation of the arteries and the latter by their constriction. Such opposite movements cannot take place together. It is thus necessary that filling of the arteries precedes and their constriction and evacuation follow.

Filling and turgescence of the arteries precede. They are not possible without violent stretching of their transverse fibres. But we know that all fibres of the vessels, like those of muscles, intestines, tendons, membranes and skin, resist

[29] Prop. 74

stretching and have the strength of contracting like a stretched spring. We observe that all fibres in their normal surroundings undergo some violent tension: if divided, they contract and shorten. This would not happen if these fibres were balanced between contraction and looseness, as the string of a bow, when untied, does not undergo contraction or stretching.

But if all fibres in their normal state undergo some stretching, when the arteries are full of blood, their cavities are larger and their transverse fibres must be elongated much more and undergo much more violent stretching. This dilatation of the arteries is followed by constriction which is impossible without shortening of the circular fibres. Such contraction is characteristic of these fibres. Therefore, after the violent stretching due to turgescence, arteries necessarily must exert this force of a machine which they possess. Consequently, this force acts as a rope which would surround and constrict the arteries and violently expels blood from their orifices.

Proposition LXXVI

If the force exerted by the heart not only produces the maximum dilatation of the arteries but additionally expels from these arteries part of the blood sent into them, this force overcomes a resistance of more than 135 000 pounds.

I suppose that the dilatation of the arteries is not sufficient to receive all the volume of three ounces of blood sent by the heart. Then the surplus must be expelled from the arteries by the piston action of the heart. I claim that the force exerted by the heart must overcome a resistance of more than 135 000 pounds. The resistance which the heart must overcome only to fill the arteries with blood up to turgescence is not less than 180 000 pounds. This was shown. But for the blood sent by the heart to leave the arteries, another force is necessary which by compressing and constricting the arteries expresses the blood which made them turgid. Blood leaves the arteries only if it simultaneously dilates the narrow paths inside the muscles and the tissues of the viscera through which it must pass and which were constricted by fasciae and by their mechanical force. Therefore, the force which expresses and ejects the blood from the arteries must overcome this additional resistance. When expansion of the arteries has not a capacity of three ounces of blood, the surplus must be expelled from the arteries by the same piston impulse of the heart. Consequently, in such instance, the heart must exert a force greater than if it had only to dilate the arteries.

The resistance to the impulse of the blood which is exerted to open paths between the fleshes and inside the viscera is considerable since the blood must drill its way through pores in solid parts of the animal body, with great impetus. Not only these pores are narrow but they have different shapes like a sieve. A blood particle of the same shape, i.e. of the same type, is driven violently through any hole, like a wedge, so that tissues of a type are fed by particles of the same type. Moreover, these particles, once introduced, like nails, expel useless and excremental particles from the pores of these solid parts. Hence this perspiration is called imperceptible. For this reason and for other ends which we ignore, Nature is forced to exert such a great motive force to circulate blood continuously.

If in one pulsation alone the heart overcomes such resistance, greater than 135 000 pounds, since the heart pulses at least a thousand times an hour, in an hour it overcomes a resistance of 135 000 000 pounds and in a day a resistance of more than 3 140 000 000 pounds*.

Such a great force is undoubtedly surprising and would be completely unbelievable if there was no energy of percussion which by its nature can overcome any finite static resistance.

Chapter VI

On the causes of the movement of the heart.

The opinion that muscles are different from the heart prevailed among the ancient authors so much that they believed that their substance, structure and mode of action were completely different. They also thought that they were moved by different motive causes. The muscles would be moved by the feeling and appraising animal faculty or by the will, the heart by some blind vital faculty. Stimulated by the experiments of modern authors, we seem to be able to analyse something more reliable. First we will show that:

Proposition LXXVII

The direct motive cause of the heart is not different from that which moves the muscles of the limbs.

The heart is a muscle of the same nature as the other muscles. Both are made of the same components, bundles of fleshy fibres and of tendinous substance, contractile, of the same shape, arranged and bound in the same way, similarly penetrated by nervous ramifications, similarly irrigated by blood from arteries. As far as their action is concerned, the fibres of the heart as well as those of the muscles of the limbs act by their swelling and straightening. Going as far as their direct motive cause and their way of action, from what was said it seems impossible for the heart to be swollen and moved by some immaterial faculty, or by spirits, or by blood violently injected. We showed that the muscles of the limbs could not possibly be moved by these same causes either. It remains that, as all muscles contract by swelling of the vesicles of their pores, the direct cause of tension of the heart is the swelling of the vesicles of its pores resulting from fermentative ebullition of elements of tartar of the blood by some spirit-

---

* 135 000 000 × 24 = 3 240 000 000!

uous juice instilled from the orifices of the nerves. Nature uses to carry out similar operations in the same way and with the same faculties and instruments. She does not like variety. Therefore, the same arguments pertain for the heart as those discussed for the action of the muscles.

Proposition LXXVIII

The first and indirect cause of the motion of the heart seems to be different from that of the movement of the muscles of the limbs.

It is obvious that among all the muscles of the limbs we move those which we wish and when we wish, we keep them acting as long as we want and suddenly stop their action at will. But the heart muscle seems to be on its own, not subjected to the will. It moves all the time willy-nilly, like a water mill, even during sleep. Moreover, the heart cannot either continue its action for some duration of time or stop it at will. By some blind necessity, it carries out very energetic and momentary pulsations alternating with pauses and stops, also temporary. This persistent action is never interrupted as long as the animal is alive and healthy.

By general agreement the movement of the muscles of the limbs was called voluntary and that of the heart natural. An act of will consists of a choice and movement by which we tend to apprehend what is good and avoid what we know as evil. This does not seem to apply at all to the movement of the heart. In this movement there is no sensation or apprehension, no choice of what is good or avoidance of pain. Since the very beginning in the incubated egg this motion of the heart takes place by a natural or rather an instinctive and blind necessity, when we cannot conceive any perception and apprehension of good or evil to exist, in this very small initial egg. Even in a heart removed from an animal, still warm and put in water, pulsation persists for several hours after all connection with the brain has been severed by division of the nerves and no sensation or wish can initiate the movement of the heart. Consequently, one can think with good reason that the first and indirect cause of the movement of the heart is different from that which moves the muscles of the limbs.

Proposition LXXIX

The movement of the heart can result from an organic necessity, as an automaton is moved.

As results from what was said[1], the direct cause of the movement of the heart is the fermentative ebullition of the juice of tartar of the blood provoked by mixing with spirituous juice instilled from the nerves. But the nerves are shaken in the brain by another cause different from the will, spontaneously without any

[1] Prop. 77

rousing. Nervous juice is instilled into the heart by itself. And, therefore, pulsations occur. Since the primary cause is unknown and imperceptible, if we want to be led to the action of the heart by an analogy with other actions of Nature (as required by a correct philosophic method), we must refer to motive forces which, by some natural and inborn energy and by some necessity, act spontaneously and always repeatedly such as fire, the fall of weights, the flow of water or others of the same kind. These occur always and continuously by their nature. They never stop except if their action is impeded or their motive faculty is alternately extinguished or weakened or suppressed and then revivified or resumed after synchronic interruptions like the flame of a candle which at one time is present and at another time absent or a combustible object successively brought closer to the flame and retrieved. Or finally, the agent and the patient continuing, the operation is carried out by way of a valvule or a machine which now forbids and then allows to carry out the operation with the interruptions thus mentioned.

Nature presents examples of such operations. Clouds are continuously moved by the wind. Water surface and flags are moved in an undulating way as if they were hit by repetitive synchronous blows. We also observe shivering and oscillation of rods and weights hanging from ropes. Above all a clock seems to represent the action of the heart in life. The clock is made of cog-wheels by which the force of a weight acting permanently moves a hand which indicates time. The pendulum oscillates with repetitive falls downwards like waves, in the same manner as the heart moves.

All these examples, if carefully considered, cannot apply to our case. It is impossible to simulate, to suppress or to restore the animal motive faculty of the spirits. Neither can the orifices of the nerves be successively present and absent since, as long as the animal is alive, spirits and nerves are present in the brain. Although it does not seem impossible to set valvules in the origins of the nerves, apart from the fact that they do not appear, it would remain to enquire why such valvules would close and open, at synchronous short intervals. This is as difficult as for the spirits to move and stop in the nerves directly by themselves.

The same difficulty forbids to imagine in the orifices of the nerves of the brain pendulums or cog-wheels or something similar which would produce oscillatory movement. Not only are such machines not found in the brain but they seem to be opposed to the simplicity in which Nature delights in her operations.

Another organic structure must be sought for, which not only is possible and simple but which is also observed here and there in natural operations and is sufficient to explain the pulsation of the heart such as described above.

This, if I am right, may be as follows. Although sponges, filters or capillary glass tubes are full of liquid permanently, water does not flow out from their lower edges or orifices continuously like from a fountain but drop by drop, separated by regular stops. The cause of this seems to be the excessive narrowness of the canals which impedes free flowing of the liquid because the particles or molecules of liquid are not completely smooth but are nappy and viscous. Therefore, they are forced to proceed slowly through the cavities of the surface which are also rough, rubbing their asperities. Hence they are forced to flow not straight but sinuously and interruptedly except if the droplets are pushed by an excess of compressed liquid. The same is experienced more obviously with viscous liquids the molecules of which impede each other mutually, even if they flow in the open and are not impeded by asperities of the canals. This occurs with sputum spilled by the mouth.

This being established and confirmed by experience, one must think that in the brain spirituous liquid, somewhat viscous, always touches and moistens the orifices of the nerves which go to the heart and the narrow cavities of which are full of spongy substance and are soaked by the same viscous spirituous juice. Because of the narrowness of the cavities this juice can flow out uneasily except if forced by excessive turgescence or by bitterness or if the fibres are stimulated by stings. The juice or spirit is expressed drop by drop into the flesh of the heart as a result of peristaltic puckering. Its mixing with the humour of tartar of the blood results in ebullition and explosion. Since the ejection of the first droplets is momentary and intermittent, beating of the heart occurs and then stops until some liquid is collected for the second time and provokes turgescence of the nerves and second ejection of drops by peristaltic constriction, and so on. In the brain there is always plenty of spirituous juice of the same temper and the narrow spongy arrangement of the nerves also impedes its outflow in the same way. Therefore, the periods of rest are equal in duration as long as the nervous juice retains the same degree of fluidity or is not pushed by some greater violence.

If it happens that the cerebral juice becomes more fluid or more bitter or is violently shaken as a result of unusual fermentation, as perhaps occurs in fevers, or without fermentation simply because of accelerated movement as in anger, then there is also an outflow drop by drop but the intervals between the ejections of the drops are shorter and more frequent. Then repletion and turgescence of the nerves are quicker as are their discharges, as we shall mention.

This is not contradicted by the fact that after division of the nerves and excision of the heart, its pulsation continues for some time since the cavities of the nerves of the heart can remain full of spirituous juice. This juice can be instilled further for some short time. Afterwards, the heart when stung by a needle or irritated by some gnawing juice can, by peristaltic constriction, express some droplets from its nervous fibres or some remnants of spirituous juice still present there.

But it may seem strange that repetitive pulsation does not continuously occur in all the muscles of the limbs as it does in the heart since the orifices of the nerves going to the muscles of the limbs are always in contact with, and moistened by spirituous juice, as occurs for the nerves of the heart.

This objection can be met by remembering, as we suggested elsewhere, that in the brain the orifices and canals of the nerves going to the limbs can be shaped so that spirituous juice cannot be expressed out of them into the muscles if they are not shaken convulsively by a motion of the spirits. This depends on the will. The orifices of the nerves of the heart can be more open and their spongy canals more accessible so that the fullness itself of these spongy canals produces an outflow of droplets even without irritation and without convulsive movement. Hence it is possible that continuous moistening of the orifices of the nerves going to the muscles of the limbs does not produce pulsation similar to that which we observe in the heart.

This discussion, if I am right, sufficiently explains that the movement of the heart can occur by a natural instinct or by organic necessity in the same way as an automaton moves.

It is not superfluous to consider whether there are some reasons to question if movement of the heart can result not from pure mechanical necessity but from the same faculty of the soul by which all the other muscles are moved.

Proposition LXXX

The movement of the heart may be caused by the animal cognitive faculty but unconsciously as a result of acquired habit.

From the following reasons it can be surmised that pulsation of the heart results from the same animal faculty which appears in the sensation of pain and in the lust for pleasure. This faculty transmits spirits to certain parts of the brain to shake the roots of well-determined nerves which go to a certain muscle rather than to another.

The sensitive faculty of the soul which to a good purpose or by fear of evil sends spirit or instruments to move those muscles and not these is the same which provokes a most energetic or a very weak pulsation of the heart depending on whether some unexpected good happens or some terrible evil is imminent. Both changes in the pulsation result from fear and belief which are the cognitive faculties of the soul. Thus, this movement of the heart results from sensation and lust, not really from some unknown organic necessity. One cannot believe that the heart is moved now by one cause and then by another. Everybody agrees that even the usual motions of the heart are not due to some mechanical necessity, like in automatons, but to the same animal power of sensation and lust which carries out that big pulsation in joy and this small in fear.

Nothing prevents the usual motions of the heart to occur without our assent, even against our will. Several other motions of the limbs which undoubtedly depend on the will occur without notice and sometimes even against our will, by acquired habit. For example, the muscles moving the lower leg certainly depend on the will. However, when we stand on one foot, to keep our body erect and balanced, antagonistic muscles alternate their action. Now they pull, then they are loose like the rigging in a bad ship. This oscillation stops when the centre of gravity is on the vertical across the plantar sole. Even the ignorants contract the muscle on the side opposite to the inclination to prevent falling. It may happen that somebody does not want to move this muscle because it is injured and painful. The muscle, however, pulls against his will to avoid falling. So efficient is the habit acquired from childhood.

The muscles of the eyelids which carry out voluntary movements also close the eyes by habit not only while we ignore it but even against our will if a friend brings his hand near, although we know he does not mean any harm.

Consequently, it is not impossible that an action which is carried out from mere habit and which we did not notice to have wished, which we even thought we did not wish, be called voluntary. Such a habit is acquired only after many and frequent actions ordered by the will have been carried out. This repetition, by training the spirit, develops a certain ability and trains the organic instruments, making them quicker to react. It seems that the power of habit consists in this.

In the small embryo in the egg recently discovered by Malpighi, not only all the organs admirably built by Nature like an automaton are present but also humours, spirits and the sleeping animal faculty. Awakened by the external heat, this faculty perceives, by way of the nerves, some painful sultriness in the heart turgid with blood. As a reaction it attempts to eliminate this inconvenience by contracting the tiny heart muscle. The first pulsations thus are carried out from a preliminary sensation of lust, however with laborious effort. Similarly, when a musician starts being trained playing a zither, the first and complicated motions of his fingers are carried out by order of his will but not without difficulty and pain in the fingers. After many repetitions, however, when he has acquired ability and habit, the same movements can be carried out not only easily but even without thinking or even taking notice. Therefore, it does not seem stupid or ridiculous to wonder whether pulsation of the heart could not occur without being noticed, from mere habit, by the same animal faculty which moves at will all the muscles of the limbs, rather than from blind mechanical necessity.

Neither is it a wonder that the heart of a turtle, when excised, pulses in a dish for several hours. The muscles of the limbs of all animals and even more so the muscles of the back of snakes are undoubtedly the instruments of voluntary movements. However, when excised they move for a long time without receiving orders from the will through the nerves from the brain which has been separated. Consequently, the motive cause of the movement of the heart may be the same animal faculty which moves all the muscles in a live animal.

Muscles used for voluntary movement continue to move for a long time after being separated from the animal because the instruments and possibility of action which the will of the animal used remain in them: fibres, nerves and juices which they contain. When irritated by some stimulus, these elements are organized for the movement by which nervous juices are instilled into the heart. Then fermentative bursting and swelling follow by mechanical necessity. Such movements of excised muscles cannot occur by themselves or from a cause different from that which moved the muscles in the live animal. The movement thus occurs as a result of this unconscious cause.

It can be concluded that the reflex action by which the sensitive soul takes note of its will does not seem necessary to its willing it.

Chapter VII

On breathing.

We discussed at length the motions of the muscles of the limbs which undoubtedly are voluntary and the pulsation of the heart which is considered as spontaneous and natural. We must now deal with mixed movements which, although being spontaneous, can also be controlled by the will. Breathing uses to be counted among these. We shall discuss it first since it is obviously connected with the pulsation of the heart. This movement is not well known. So far its actual motive causes, its instruments and its mechanism are controversial. Even its primary function is not clearly perceived. Therefore, we shall proceed methodically, starting from facts.

Proposition LXXXI

Description of the facts observed in breathing.

Air is sent into the chest through the mouth and nose repeatedly. It does not remain there as food and beverage do in the stomach but soon after its entry it is rejected from the chest through the mouth and the nose. The intervals of time between two respirations are not as frequent as the pulsations of the heart and are not synchronous with the latter. They occur at different times. Moreover, respiration can be altered arbitrarily as far as frequency, duration and depth of breathing are concerned. But once outside the womb we have started to breathe, we are unable to stop completely. We are forced to absorb some air day and night and to reject it in a continuous alternation. Interruption of breathing entails suffocation and death.

The substance which can be breathed by terrestrial animals and birds can be nothing else than air. Inhaling water or any liquid causes immediate suffocation. And air must be light. If it is condensed significantly either by compression or by some other means, like in mines, or if it is very rarefied, the animal also suffocates.

If the elements which move during breathing are analysed, it appears that air is driven into the lungs through the trachea until they are inflated like goatskin bottles or balloons, and during expiration they deflate. We experimented by dividing an intercostal muscle. During any inspiration a lobule of the lung extrudes through the aperture until it is inflated. If touched with a finger this lobule appears to be hard. During expiration it becomes flaccid and soft.

When dissecting live animals we observe that during each inspiration the cavity of the chest becomes larger and more spherical, the ribs are dilated and rise with the sternum towards the clavicles, the intercostal spaces become smaller and the diaphragm is depressed downwards.

The expansion of the thorax resulting from inspiration of air can be measured using the following artifice. I took a 52-finger breadth-long cylindrical glass tube. The diameter of its cavity was a little more than a quarter of a finger breadth. When full of liquid it contained 14 cubic finger breadths. The orifice of the tube was immersed in a bowl full of soap-suds. I took a small quantity of viscous water to fill the lower extremity of the tube as if it were a diaphragm. I pinched my nose and applying my mouth to the tube I inhaled as much air as my lungs are used to breathe in during one quiet inspiration. The drop of viscous water rose but did not attain my mouth. Hence I deduced that the volume of air which I breathed in was less than 14 cubic finger breadths. But let us assume it was 15 cubic finger breadths. Since the diameter of my thorax is less than 15 finger breadths, the volume of my thorax, approximately that of a compressed sphere, is 3375 cubic finger breadths. After inspiration 15 additional cubic finger breadths of air are introduced and the volume of the thorax attains 3390 cubic finger breadths, the cubic root of which is approximately 15.02. Consequently, the introduction of inspired air increased the diameter of my thorax by not more than a fiftieth of a finger breadth. Hence I understood that during quiet respiration, the movement of the thorax must be little apparent since we can notice only some slight rising of the sternum towards the neck. When the abdomen is open, however, we observe that during each inspiration, the diaphragm contracts and moves downwards, flattening its surface which previously undulated towards the centre of the thorax. This is what happens during quiet inspiration. During deep inspiration, however, the thorax dilates much more and inhales twice as much air as during normal breathing.

After the description of these facts, we must find out by which force breathing occurs. And firstly:

Proposition LXXXII

Air and lungs are not the actual causes of respiration.
They are only involved passively in the process.

It is clear that respiration results from a double movement: that which fills and inflates the lungs with air and that which empties them. Each of these two movements comprises two parts. During inspiration air is introduced and the lungs are dilated to receive this air. During expiration the lungs are constricted and air leaves them.

Firstly, air is not the effective cause of these movements because air has no motive force beyond its weight and elasticity. These cannot exert any action since fluid air is balanced in the atmosphere like water in the ocean and is uniformly compressed. As long as it remains immobile it thus cannot dilate the lungs violently enough to fill them and afterwards constrict the lungs so as to leave them. Moreover, respiration is carried out by an animal faculty according to the requirements of Nature and when the will prevails. Therefore, it is not carried out by the external force of air which would provoke breathing continuously even against our will.

Secondly, the lungs are not the effective cause of respiration since they are not made of fibres like the muscles and thus cannot move by themselves. They can only be pushed passively by an external cause. Therefore, they can neither dilate to inhale air nor constrict to expel it. Moreover, there is no attractive force in Nature. Consequently, breathing must be attributed to some other cause since air and lungs are involved in the respiration merely passively.

Proposition LXXXIII

The cause of inspiration is the force of the muscles which expands the cavity of the chest, and the weight and elasticity of air.

During inspiration air is sent into the cavity of the chest. But air can be driven into a space filled by some other substance only if this space is enlarged. Consequently, to allow for inspiration, the cavity of the chest must at first be expanded. The parts of the animal body cannot be moved without contraction of muscles. Therefore, the movement of the thorax which enlarges its cavity necessarily must be carried out by the muscles which are involved in the dilatation of the chest.

But after the thorax has been dilated, it remains to find out by what mechanism and by what force air is sent into the chest. Firstly, air is not inhaled or attracted by the lungs themselves since, as mentioned elsewhere, there is no attractive force in Nature[1]. Secondly, it is also demonstrated that air does not arrive as a result of fear of vacuum[2]. The remaining possibility is thus that air flows into the dilated cavity of the thorax by its weight and elasticity in the same way as it spontaneously falls into a well and fills it pushed by its weight. As a whole, all this operation occurs as when a balloon including a goatskin bottle and having with this a common orifice, is pulled with the hands and thus dilated. The goatskin of the bottle adhering to the envelope of the balloon must expand also. Then ambient air arrives by its weight to fill the cavity inside the goatskin bottle contained in the balloon, provided that the orifice of the balloon is filled by the bottle and air cannot get between the envelope of the balloon and the bottle. The thorax is like the balloon and contains the soft lungs which adhere to the concavity of its walls like the goatskin bottle. The orifice of the thorax is the tracheal tube which ramifies into the lungs. Therefore, when the balloon formed by the thorax is dilated by the force of the muscles, air spontaneously rushes in by its weight and elasticity to fill the lungs.

---

[1] De motion. natur. à grav. pend. Chap. 6

[2] De motion. natur. à grav. pend. Chap. 6

Proposition LXXXIV

The movement of inspiration is carried out by the intercostal muscles and the diaphragm acting simultaneously.
Tab. XVIII, Fig. 2.

It is usual to distinguish respiration as normal, quiet and spontaneous; violent and deep; and very violent. Quiet and spontaneous respiration occurs mainly during sleep and rest, by the diaphragm alone with the intercostal muscles or at least by half of them. Very violent respiration is carried out by the same muscles and all the muscles of the chest and abdomen. As far as quiet respiration is concerned, when the abdomen of a live animal is open, one can feel and see that during each inspiration, the diaphragm which was loose and undulating, protruding towards the heart, moves downwards, becomes less concave and tends to become almost a flat disc. In a guinea-pig, I saw that the diaphragm was lowered towards the abdomen with such violence, as a result of pain, that it became convex downwards. However, spontaneous inspiration does not occur without the help and action of the intercostal muscles. Actually, on ourselves and on sleeping people we see and feel that the ribs dilate and the sternum rises towards the neck. This movement of the thorax cannot result from a contraction of the fibres of the diaphragm since when they contract the circumference of the diaphragm decreases and is pulled centripetally. Consequently, the floating ribs which are appended to the other ribs and the sternum are pulled downwards by the diaphragm. Therefore, far from producing an elevation of the ribs, they impede such elevation and constrict the circumference of the thorax. Inspiration thus could not occur. Consequently, to carry out some inspiration the diaphragm and intercostal muscles necessarily must cooperate.

It also seems impossible to me that the two kinds of fibres which intersect each other in the intercostal muscles have different functions, i.e. that all the parallel fibres EF, GH, IK inclined to the left be used to the dilatation of the thorax and inspiration, and those LM, NO, PQ inclined to the right constrict the thorax and produce expiration.

The normal and proper action of muscular fibres is to contract. Therefore, all the fibres attached to two adjacent ribs DC, AB, when they act, must necessarily shorten. But, when the tightened fibres shorten, the ribs AB, CD to which the fibres are attached must be brought closer together. They thus must produce a reduction of the thorax. Consequently, when they contract, the fibres LM, NO, PQ cannot possibly produce an expansion of the thorax. Such an expansion can occur only by a separation and displacement of the ribs from each other. Such an increase of the interval between the ribs could not occur without elongation of the fibres which connect them. Then the muscular fibres LM, NO, PQ would act by becoming elongated which is against the nature of muscles.

Moreover, if one kind only of intercostal oblique parallel fibres had to dilate the thorax, the ribs would be twisted and distorted. If there are two adjacent ribs AB, CD parallel to each other at the surface of the cylinder ACDB, it is obvious that the oblique fibre EF, when shortening, pulls its extremity F towards C. At the limit, the fibre will become parallel to the spine AC. But the part CF of the rib is longer than the part AE. The segment of rib CF thus must bend and extrude from the surface of the cylinder in which the rib AE lies. The same applies to the other fibres GH, IK, etc. Hence, all the surface of the cylindrical chest would be folded and deformed, which does not occur.

Consequently, all the decussating fibres which connect two adjacent ribs provoke the same effect. They bring these ribs closer together. This occurs by the same mechanism as when a suspended weight is pulled vertically upwards by oblique ropes inclined to opposite sides, as shown above.

I dare not assert that the help of chest muscles like the serrati is necessary for deep inspiration. The two serrati majores muscles are attached to the bases of the scapulae. Vesling thus thinks, not without reason, that these muscles pull the shoulder downwards rather than pulling the ribs upwards. It is likely that the two postero-superior serrati muscles maintain erect the three cervical vertebrae to which they are attached. Finally, as admitted by the anatomists, the two postero-inferior serrati muscles dilate the lower part of the thorax. Thus they are not used for inspiration. Only the triangular muscle remains which can be counted among the intercostal muscles.

The other muscles of the chest and thorax are not involved in the inspiration. Anatomists agree on this point. Everybody also says that the abdominal muscles can only constrict the lower part of the abdomen. This pushes the abdominal viscera upwards. These muscles thus impede lowering of the diaphragm and dilatation of the thorax. Together with the diaphragm incurved upwards, the viscera are driven inside the chest like a wedge and thus produce expiration rather than inspiration. We experience that on ourselves by feeling. In violent coughing the abdomen is constricted by its muscles not when we inhale air abundantly but afterwards, during the subsequent violent expiration and expulsion of air from the chest. Then sputum is scraped from the walls of the bronchi by the rapid flow of expelled air and is spilled through the mouth.

All the other muscles having been excluded, only the intercostal muscles and diaphragm remain to carry out inspiration.

I am not of the opinion commonly held that the intercostal muscles and diaphgram are insufficient to carry out a very energetic inspiration and that the pectoralis muscles which move the arms must contribute to it. I indeed observe that the muscles of the arm which carry a small weight of one ounce can also overcome the resistance of a weight of a hundred pounds. The muscles which move the mandible also exert little force when the teeth gnaw soft food and considerable force when grinding hard bones. The difference does not result from the number of muscles involved but only from the unequal motive faculty which moves these muscles. Similarly, there is no need for several hammers to split a stone. One is enough if the stone is hammered with sufficient energy.

It thus can be concluded that inspiration, whether it be quiet and normal or violent, is carried out only by the intercostal muscles and the diaphragm acting simultaneously.

To understand the mechanism of respiration some lemmas must be presented.

Proposition LXXXV

The two extremities A and C (Fig. 3, Tab. XVIII) of a semi-elliptic spring ABC are fixed to the wall FH and the plane ABC is inclined to the plane of the wall FH. The spring is pulled upwards towards the perpendicular from B to E. I claim that the concavity of the spring is further from the wall at E than at B and at the end of the pull the spring spontaneously will return to its previous inclination.

BO and EN are drawn perpendicular to the plane of the wall. The sagittae DB and DE of the spring are equal. The angle BDO is smaller than the angle EDN and the angles at O and N are both equal to 90°. Therefore, the perpendicular EN is longer than BO. Consequently, the spring is further from the wall in position AEC than it was in its previous position ABC.

Secondly, the extremities A and C of the spring ABC are fixed to the wall FH as if they were rigid nails. Therefore, after the pull, if the spring is no longer retained in the position AEC by some force, it must by its resilience spring back to its previous inclination BDO. Q.E.D.

Proposition LXXXVI

An extremity A of the spring (Fig. 3, Tab. XVIII) is fixed to the immobile wall IF and the other to the mobile rod GC. The concavity B of the spring is pulled in the direction BM forming an acute angle MBD. I claim that the extremity C moves from A and that C is raised towards M.

If the apex B of the spring ABC is pushed towards D or if B is pulled by a rope towards D, the apex must be depressed and the curvature diminished. If the apex B is not pulled directly towards D but obliquely in the direction BM forming the acute angle MBD, the apex B is also depressed and brought closer to D. But B can be brought closer to D only if the segments BC and BA are more curved than previously. The segments resist this bending by their tension and by the force with which they attempt to expand. Therefore, the extremities A and C of the spring are forced away from each other. Since the extremity A is fixed to the wall and C is mobile along the rod CG, C moves away from A. As the oblique pull is carried out obliquely in the direction BM, from B to M, the mobile extremity comes closer to M. Hence, the extremity C, with all the spring, rises towards M in a circular movement about the centre of rotation A. Q.E.D.

Proposition LXXXVII

The extremities A, D, H of several semi-elliptic springs (Fig. 4, Tab. XVIII) are attached to the fixed column PS and the extremities CEI to the mobile rod QR. The planes of the springs are parallel and form acute angles with the plane PR drawn across the extremities of the springs. The apices B, O, M of the springs are pulled in the direction MB parallel to the plane PR in the obtuse angles, i.e. from M towards Z. I claim that the semi-cylindrical concavity ABCIML is expanded and, after the pull has ceased, spontaneously returns to its previous narrower shape.

Each of the springs, such as ABC, fixed obliquely to the wall PS and to the mobile rod QR, is pulled towards L in the direction MB which forms the acute angle LBX. The more the spring is pulled[3], the more its apex B recedes from the plane PR. All the springs always remain parallel. Therefore, the apices B, O, M of all the springs recede from the plane PR further than previously. They thus enlarge the semi-cylindrical space and increase its capacity.

After the pull has ceased, each of the springs spontaneously returns to its previous more inclined position as a result of its resilience[4]. Therefore, all the springs which are parallel, by their resilience, narrow the space which had been increased and occupy their previous more restricted position.

Proposition LXXXVIII

The conditions are the same. Other springs AFC, DNF, HGI are attached additionally in the same way and order on the opposite side of the same plane PR (Fig. 5, Tab. XVIII). I claim that the cylindrical cavity ABMGFC expands and, after the pull has ceased, it returns to its previous smaller capacity.

The cavity is expanded when the springs on one side are pulled upwards and constricts spontaneously after the pull has ceased[5]. Consequently, the springs on both sides will do the same. Q.E.D.

[3] Prop. 85
[4] Prop. 85
[5] Prop. 85

Proposition LXXXIX

The conditions are the same. I claim that on each side the circumferences of the springs come closer together either very little or not at all. They move away from the fixed column PS and the rod IC significantly rises towards Q (Fig. 5, Tab. XVIII).

Firstly, if the adjacent and parallel springs ABC and DOE (Fig. 5, Tab. XVIII) are equal and have the same structure, and if the intervals AD, CE between the points of fixation are equal, the space delineated by these two springs is a parallelogram, although curved. But, in a parallelogram, whether rectangular or not, the opposite sides are always equal. Therefore, the distance BO between two adjacent springs is always the same, whether the springs are depressed and inclined or whether they are pulled upwards in their erect position to the plane PR.

If the spring ABC (Fig. 5, Tab. XVIII) is shorter than the next one DOE, the space which they delineate is a trapezium. The more a trapezium a, d, o, b (Fig. 6, Tab. XVIII) from being oblique becomes perpendicular, such as in position a, d, o, b, the more the interval OB decreases since the lines bz and BZ drawn parallel to a, d are also equal as a consequence of the parallelograms. Therefore, in the triangles BZO and bzo, the two sides bz and BZ are equal, the two sides oz and OZ are equal and the obtuse angle bzo is larger than the right angle BZO. The base bo thus is longer than the base BO. The opposite would be true if the spring ABC (Fig. 5) was larger than the spring DOE.

Secondly, all the springs are pulled upwards. But their extremities ADH (Fig. 5, Tab. XVIII) being fixed to the wall cannot be pulled whereas the extremities CEI are fixed to a mobile rod. Therefore, the extremities CEI of the springs can be raised with the rod QR.

Thirdly, the apices of all the springs BOMFNG (Fig. 5) are pulled obliquely and centripetally. Therefore, their opposite extremities move away from each other[6]: the distances between C and A, E and D, I and H increase. The extremities ADH being fixed to the wall PS are forced to remain at the same place. Consequently, the extremities CEI together with the mobile rod QR are moved away from the fixed wall. Such was the proposition.

After this preliminary we shall easily describe the mechanism of inspiration. And first

Proposition XC

When the intercostal muscles contract together with the diaphragm, the cavity of the chest must expand and air must be inspired.

The bony ribs ABC, DOE, AFC, etc. (Fig. 5, Tab. XVIII) not only are curved but present with the shape and consistency of a ballista. They are hard and, if bent by considerable force, they rebound like springs.

[6] Prop. 85

The posterior extremities ADH of the ribs are solidly attached to the vertebrae which constitute the spine PS. Their anterior and cartilaginous extremities CEI are attached less firmly to the sternum QR which is mobile. The ribs thus can easily be bent, suspended and expanded together with the sternum.

Twelve pairs of ribs are disposed on both sides forming two series bent downwards and equally inclined to the plane DR drawn across the spine and the sternum.

Finally, there are decussated intercostal muscles which when contracting pull the circumferences BOMFNG of the ribs upwards, towards the neck, in the directions MB, GF forming acute angles with the planes of the ribs. Consequently, the cylindrical cavity of the chest is increased[7] and the distance BF between the two sides is lengthened. Moreover, the sternum CL significantly rises together with the ribs, towards the neck Q[8]. When rising, the sternum recedes from the stable spine DS.

Consequently, the cavity of the chest expands in two directions: transversely from right N to left O and postero- anteriorly from D to E. The height GI of the chest also increases, as we shall show.

The circumference of the diaphragm HMIG is attached to the vertebrae at H, to the bony extremities of the ribs, to the cartilaginous extremities of the lower ribs over all the periphery M and G and to the sternum at I. The centre of the diaphragm hangs from the mediastinum and pericardium. Contraction of this muscle thus has two effects. First, the height of the thorax is stretched and elongated towards the abdomen RS by tension on the flexible cartilaginous extremities of the ribs MIG. This overcompensates the small ascent of the thorax towards the neck Q. Second, the diaphragm itself which was undulating and protruding upwards is depressed and flattens. Therefore, the bulging which the diaphragm formed into the chest is suppressed by its depression and flattening. Hence the thorax takes the shape of a sphere or a half egg, the apex of which is the neck and the base the plane of the diaphragm which corresponds to the section of the egg. Such a depression of the diaphragm must necessarily increase the capacity of the thorax. Previously the width and depth of the chest had been increased by the intercostal muscles. Therefore, the intercostal muscles and the diaphragm necessarily must dilate the cavity of the chest. Air must be driven into the empty chest by its weight and elasticity. This is how inspiration occurs. Q.E.D.

Proposition XCI

Inspiration is not completed by simple arrival of air into the chest expanded by the intercostal muscles and the diaphragm. Additional compression is required which is achieved by the resilience of the ribs.

After the chest is dilated ambient air spontaneously flows into its cavity. It is necessary for this air filling the chest to retain the softness and rareness of

[7] Prop. 87
[8] Prop. 89

external air, to ensure balance between external and internal air so that they do not exert any motive force. Actually to dilate the other narrow canals in the lungs which are closed by the collapse of their soft walls, some motive force is necessary and much more force is required to insufflate with sufficient impetus the air received in the bronchi into these narrow canals where it plays the role of wedges. When we dilate bellows or a goatskin bottle with our hands, ambient air spontaneously fills their cavity and remains in them completely immobile. If we want to blow on a fire or produce music from a pipe connected to the bellows, an opposite action is necessary, that is a compression of the bellows to express the air which they contain. In the same way the force which compresses the thorax and violently pushes the air contained in it cannot be the force of the intercostal muscles and diaphragm. These muscles can only dilate the thorax but not constrict it. Neither can this force be exerted by the pectoralis muscles which are not involved in the respiration. The energetic impulse given to the air present in the lungs thus results from a cause other than the force exerted by muscles of the chest. If I am right, this cause can be investigated as follows.

The elements of bellows are thought of as being maintained together by compression by a steel machine such as a press. Then, if we want to dilate the bellows with our hands we must overcome the resistance of the machine by stretching this machine. When we stop pulling with our hands there is a strong rebound of the machine which returns to its previous position and thus expels air violently.

The thorax is like the bellows. It is maintained compressed everywhere by as many machines as there are ribs. During inspiration, the ribs are pulled and dilated violently together with the thoracic bellows by contraction of the intercostal muscles. Then, suddenly, when their pull stops, the machines of the ribs are necessarily relaxed and let to themselves. They thus resile and constrict the thorax with great impetus. Compression of the bellows is useless or of little use if the bellows are not closed by a valve, since air would flow back and would not be expressed through the pipe violently. Similarly, the epiglottis moved by its muscles must close the trachea. Then spontaneous constriction of the ribs condenses and simultaneously pushes the air contained in the bronchi. Thus, a double force, the impulse mentioned above and the elasticity of the compressed small machines which constitute air and attempt to dilate, drives air in all the recesses and diverticula of the lungs, inflates the alveoli of the latter and, in so doing, expresses and decants the blood contained in the collateral vessels.

Proposition XCII

During quiet and normal expiration air is not ejected by the force of some muscles but as a result of the quietness and absence of action of the intercostal muscles, of the relaxation of the diaphragm and of the opening of the epiglottis.

As mentioned above[9], the need of Nature is not satisfied by simple inspiration of air but requires very powerful compression which is provided by the resilience of the ribs. This compression is possible only if the trachea is tightly closed by the epiglottis. Therefore, during the very short time of this compression air cannot get out of the closed trachea. But it leaves immediately after the compression is finished. Indeed the orifice of the trachea can open immediately. Then air must be exhaled from the chest without being expelled by any muscle. If muscle acts by contracting, certainly the diaphragm does not act at all. When it relaxes and returns to its previous position, it undulates and protrudes into the thoracic cavity. As a result, air is forced to make place for the protruding diaphragm. It leaves through the mouth and nose.

Moreover, the air compressed in the chest, by its normal elasticity attempts to find space where to expand. Therefore, its gets out spontaneously through the mouth without being ejected by any force or action of muscles.

Proposition XCIII

Compression by the ribs and the abdominal muscles contribute to violent expiration.

As shown by experience, in coughing and in similar very energetic expirations, the thorax and abdomen constrict powerfully when air is expressed. The abdomen is certainly compressed by its muscles. The viscera are then pushed upwards with such violence that the diaphragm is propelled into the cavity of the chest. As a result of this protrusion air is expelled through the trachea. The thorax cannot be constricted by anything else than the resilience of the ribs.

When one is forced to cough, air is first inhaled profusely. Afterwards, the epiglottis being open, the abdominal muscles and all the resilience of the ribs suddenly are used by Nature to expel air in a rapid flow which can clean the bronchi by spilling sputum. This avoids suffocation and removes the inconvenience which irritates the nerves of the trachea. Consequently, in such instance, besides the causes described above, the abdominal muscles combine with the permanent resilience of the ribs.

[9] Prop. 92

Proposition XCIV

During expiration the lungs are not completely evacuated.
A considerable volume of air always remains in them.

The substance of the lungs is not solid and dense but contains many cavities which are full of air. After breathing out they remain dilated because the trachea and bronchi are made of many cartilaginous rings which cannot be constricted and thus constitute large open tubes. The remainder of the lungs is also made of countless alveoli in which the bronchi end and which cannot be constricted. Otherwise they would not fill the large thoracic cavity. These tubes and alveoli do not fill with blood or some other liquid after expiration. Consequently, they must be kept dilated by the air previously got in.

This can be observed in cadavers in which the lungs remain expanded so as to fill the thoracic cavity, and the bronchi and alveoli are obviously full of air.

In live animals also a profuse quantity of air remains in the lungs after normal expiration is completed. This is confirmed by the following experiment. After quiet inspiration, the volume of air which is easily breathed out in the summer usually amounts to 18 or 20 cubic finger breadths, as measured with the narrow tube described above. After quiet expiration has been completed, we can additionally exhale again the same quantity of air in the tube, i.e. 20 additional cubic finger breadths. This is made possible by strong abdominal compression as a result of which the diaphragm sinuously incurved compresses the lungs in the thorax like a wedge and expresses the air contained in them. This is the most conspicuous demonstration that, after easy expiration, the lungs were not completely devoid of air. This is further confirmed by the fact that after having blown in something we are forced to carry out a deep long inspiration to compensate for the loss of air which must remain in the lungs.

I said that we breathe out more than we breathe in during quiet respiration, not because we exhale more air, as some people think, but because inhaled air is rarefied by the warmth of the viscera and thus occupies more space. The addition of water vapours does not alter the substance of air: it does neither increase nor condense the quantity of small spiral machines since it is something extrinsic to air. This is confirmed by the fact that the difference between inhaled and exhaled air is much more in the winter than in the summer as far as the difference of temperature between condensed winter air and the viscera is greater than that of rarefied summer air and the viscera.

Proposition XCV

Differences in the structure of the thorax
and breathing in different animals.
Tab. XVIII, Fig. 7.

It is clear that all the animals which live in air inhale in their chest and exhale air. But most of the fishes, instead of air, inhale and exhale water. Therefore, the fishes seem to breathe water.

The structure of the thorax and lungs is also different. Terrestrial animals, except for turtles and others of the same kind, have light bony ribs shaped almost in the same way, ending in a cartilaginous process and attached to the sternum. Birds have no floating ribs but either the ribs CB, FE, IH or their appendices BA, ED, HG are osseous. They form flexible angles at B, E, H inclined upwards towards the neck. These appendices are attached to a huge bony sternum, the breast bone RS, at A, D, G by angles also flexible. There are also decussated intercostal muscles. Their fibres when they contract necessarily pull the angle B towards the neck K, the angle E towards B and H towards E. As a result, the ribs are straightened, the distance betweeen the sternum ADG and the spine CFI is increased and the capacity of the thorax augmented. Consequently, ambient air fills this space.

Birds have no fleshy but a membranous diaphragm which extends at different places to the lower abdomen. This diaphragm is a separation but does not contribute to inspiration as it does in terrestrial animals, because it lacks muscular fibres.

Moreover, birds have perforated lungs which adhere to the ribs. The vessels of their lungs can constrict as well to enable blood to be decanted by the resilience of the ribs which compresses the air contained in the thorax, as said above.

Turtles have no light ribs but are covered by one continuous thoracic bone like a vault. Hence they cannot bend and have no diaphragm. The lungs are replaced by two oblong vesicles containing blood vessels. They are not alternately filled by and emptied of air but remain always inflated. Air is only partly renewed in them. When new air is needed external muscles adhering to the skin carry out its absorption. These muscles form an internal cavity when they do not contract. When their fibres contract, the muscles flatten together with the external skin. This enlarges the internal cavity of the chest and air fills this space.

The vesicles and the air which they contain are repeatedly compressed as if it were by a press, by round and sphincterial fibres which surround them and by internal muscles. As a result, air squeezes and beats the vessels contained in the vesicles by its elasticity. And so the blood is decanted from the arterial to the venous canals.

In insects respiration occurs in much the same way or in an analogous way. This is surmised from the observations of Malpighi. I do not know, however, whether the glowing of the glow-worms results from inspiration of air or from systole of the heart which is near the tail. I think that glowing appears when the external skin of the thorax is distended by the filling of the latter, i.e. when puckering which induces opacity is flattened.

In fishes gills open below their throat. In these gills there are several curved combs made of countless rubicund hanging threads which include vessels. These vessels transmit the blood received from the heart. The combs replace the lungs. Water is absorbed profusely and is compressed by the gills. This decants the blood by squeezing and beating the vessels as if it was with a press.

Amphibious animals, as are all the cetaceans, have lungs of the same structure as the terrestrial animals. They breathe air except when they are immersed. Then the air retained in the bronchi and lungs is squeezed repeatedly by the muscles of the chest and decants the blood. I noted in dolphins an epiglottis different from that of terrestrial animals and which is considerably elongated perpendicularly towards the palate[10] so that the cleft which is in the mouth, above water, predominates. Otherwise respiration seems to be carried out by the ribs and muscles, without a diaphragm.

Chapter VIII

On the primary function of respiration.

Nature usually achieves several advantages from one action and from the same instrument. This appears particularly true in respiration. The greatest usefulness offered by breathing is conservation of life and this can be considered as primary. We thus must see how this important advantage is attained and what mechanics are implied in its achievement. Firstly, erroneous opinions must be refuted and replaced by more likely theories.

Proposition XCVI

Respiration was not instituted to cool and ventilate the flame and heat of the heart.

According to the Ancients, respiration contributes to maintaining the vital heat. This opinion originated in some similarity of words and in the analogy between the action of the heart and the fire in a kitchen. The Ancients thought that a vital flame resides in the heart. The flame glows in a stirred up wood fire. Fire in a closed space immediately dies out. It is vivified and blazes when ventilated by air. Similarly, in the absence of respiration animal life dies out. Therefore, the opinion prevailed that this imaginary vital flame residing in the heart needs air ventilation and dies out in the absence of respiration. This seemed to be confirmed by the fact that those who remain too long in a very hot bath die like a flame which dwindles and dies out in a closed space without ventilation nor cooling. Hence the lung used to be called the fan and ventilator of the heart.

This opinion prevailed among the Ancients since they self-assuredly claimed that burning heat resides in the heart and needs continuous cooling by inhaled air.

[10] Prop. 92

Such a huge flame would be ignited stupidly in the heart so that it would need being continuously repressed and extinguished. The goal of moderating the vital heat could be attained with greatest economy by stimulating a smaller flame.

Firstly, I consider that no fire or flame can live in the heart because it would be suffocated and extinguished in this small and enclosed place. The chest comprises the mediastinum which surrounds the pericardium. The pericardium contains the heart the ventricles of which are surrounded everywhere by fleshy walls and valvules. Thus it is impossible that fire ignited in the ventricles lives even if air could be insufflated repeatedly into them. Fire would have to be exposed in the open so that a continuous afflux of air could vivify it, as said elsewhere. Confronted with this impossibility, modern philosophers are forced to deny a fire and flame visible in the heart. But they admit only a boiling and intense igneous heat which would burn the hand if it could be touched. Using one's senses shows that this also is false. After having incised the chest and heart of a live animal and introduced a finger into the heart we found no burning heat but warmth as moderate as that of the other viscera of the same animal. To determine the temperature of the heart accurately, at Pisa, I took care of opening the chest of a live stag and immediately let a thermometer be introduced through the wound into the left ventricle of the heart. I saw that the highest temperature of the heart did not exceed 40° which is the usual temperature in the sun in the summer in our country. Afterwards, using similar thermometers, I measured the temperature of the liver, lungs and intestines in the same stag alive. It appeared that the heart has the same temperature as the other viscera. Therefore, the heart is not the principal fire-place which must be cooled and ventilated because of its excessive heat.

Moreover, that air which permanently remains in the bronchi and lungs is as warm as the heart. This can be felt. Therefore, even if this warm air could be sent into the heart, it would not cool it. The air which is inhaled does not progress beyond half the trachea and seldom fills the trachea. Consequently, it can cool only the trachea itself which it attains but not the heart which is very much remote since warm air, blood and pericardium liquid which are all as warm as the heart are interposed between the arriving air and the heart. Actually, the warm air compressed and condensed contained in the bronchi is warmed up because the warmth distributed in the rare air is constricted into less space after this air is condensed. Therefore, it is warmed up by the inspired air rather than being cooled.

As far as ventilation is concerned, I claim that it is impossible for the fire-place of the heart to be vivified by breathing. To burn coal it is necessary that the wind directly touches the embers and stirs up their flame. When enclosed in a room embers cannot really be stirred up by the wind which blows along avenues. Similarly, a fire-place enclosed in the ventricles of the heart can be neither attained nor stirred up by the air inhaled into the bronchi which do not lead to the heart. And if a route was open so that inhaled air could be insufflated into the heart, this would not stir up but would rather suffocate the vital fire. Indeed, air is considerably compressed and condensed in the lungs by the resilience of the ribs, as said above, and a fire is suffocated and extinguished by air too condensed as it is by any liquid. This is taught by experience.

Expired air gets out warmer. This does not mean that it has acquired this tepid warmth from the heart after having cooled it. This reaction is actually carried out by the warmth of the lungs themselves which are in direct contact with the air and not by the heart which is remote. Neither should it be said that those who die after remaining in a warm bath for a long time die because of a lack of cooling of the heart: sometimes in the summer air is warmer than that of the bath but, nevertheless, it does not induce fainting and suffocation. Actually, in the bath there is a very different cause to the accident: vapours or profuse smokes or poisonous smokes which are usually exhaled by coal, mines and storepits of wheat.

To summarize, I notice how little this common theory is consistent. The heart is compared with the fire of a kitchen. The supporters of this theory say that the warmth of life is extinguished by lack of respiration because it cannot be cooled nor ventilated by air. They prove it by the fact that those who remain a long time in a hot bath suffocate. Actually the fire of a kitchen is extinguished neither by the hot air from an eolipyle nor by the hot air of a furnace. It is rather increased and stirred up. Thus, similarly, the fire-place of the heart which they suppose to be very hot should not be extinguished by the air of the bath. If they say that it is suffocated by a lack of ventilation, we know that a flame is ventilated as well by warm as by cold air. Inversely, too much wind extinguishes the flame.

Moreover, the heart is not ventilated by respiration because the fire-place present in its ventricles is not struck by the wind of respiration which is absent and by which it is not touched. Similarly, fire in a room is not stirred up by external wind.

If they say that fire can live without wind as long as it is not enclosed in a small space, I claim that the fire-place of the heart is very exiguous inside the narrow walls of its ventricles. Therefore, fire should rather be extinguished.

If finally they say that these two actions must be considered as combined, i.e. that breathing simultaneously cools and ventilates the fire-place of the heart, I reply that if ventilation prevents extinction of the fire, it prevents its diminution and I cannot understand extinction of the fire without its diminution. Therefore, ventilation by increasing the fire continuously, keeps it going. On the contrary cooling by its nature decreases excessive ardour of the fire. Consequently, the same action of breathing produces two opposite effects at the same time: it increases and decreases the ardour of the vital flame of the heart. This is impossible. Therefore, the function of breathing is neither cooling nor ventilation of the fire and flame of the heart.

Proposition XCVII

Expiration is not instituted to expel soot generated by the fire in the heart.

Our chimneys exhale smoke carrying this ugly soot which impregnates their walls. Seeing warm vapours produced by the mouth of animals and expelled as smoke in the winter, the ancient philosophers thought that smoke was emitted by the fire-place of the heart and soot was carried outside. Actually, such impure fire is not found in the heart but only moderate warmth. Thus this tale ends in smoke.

This moderate warmth is not particular to the heart but is rather general. There it is generated by the blood as occurs in the other viscera and flesh. Moreover, this moderate warmth cannot generate ugly and sooty smoke because warmth as such has no particular effect in the heart. It does neither liquefy nor burn nor boil the blood.

Neither is there any fermentation in the heart, because it is a muscle like the others. In its cavities fermentative juice neither exists nor is generated. There is only pure blood which does not stay there but is in continuous transit. Consequently, as in muscles there appears but little refuse, it seems incredible that so much refuse and soot be produced in the heart that they need a cloacum in the lungs through which they must be expelled.

I do not deny that some water vapour leaves the mouth together with the expired air. But this vapour does not come from the heart. It is secreted by glands in the trachea, the throat, the palate and the nose. It lines the surface of these cavities. Thus small droplets which form the vapour are scraped by the quick flow of exhaled air.

The reasons which convince me that the exhaled water vapour comes from glands in the trachea and in the throat rather than from arteries and veins of the lungs are the following. Firstly, Nature built its excretory workshop with the highest capacity in the kidneys to extract serosity from the blood. Therefore, it would be useless to carry out the same work in the lungs to eject a few drops of water transformed into vapour. Moreover, water by its nature most willingly absorbs the salts which are present in the blood and draws these salts from the blood. Hence, water which has stayed in the veins and is excreted in the urine or sweat is impregnated by very bitter nitrous salt of ammonia and salt of tartar. If water vapour left the veins of the lungs to be exhaled, this vapour would carry with it the same bitter salts which are present in the blood and would unnecessarily injure the lungs considerably by its bitterness and mordant. This is suggested by experience. Urine and sweat which no doubt come from the blood of the vessels are bitter and salty whereas the vapour exhaled during expiration is insipid and sweet like the secretions of glands.

Finally, the malodorous breath of some people does not mean that such corrupted smoke is emitted by the lungs. The stomach affected by drunkenness and by fetid food eructs these foul smells because the glands of the oesophagus, throat and palate are affected together with the stomach. Hence, clear air leaving the lungs and passing through a fetid place is contaminated by these foul vapours. This is confirmed by the fact that, when the disease of the stomach is cured, the foul smell disappears. On the contrary, when the lungs are affected, as long as they are not putrid, ulcerated or sphacelated, fetid smells are not exhaled.

From all this it can be concluded that during expiration, neither soot is ejected as it is from burning wood, nor refuse or rubbish of the heart and blood since the exhaled vapours not only are similar but are even clearer than those emitted by insensible perspiration.

Before going further, some lemmas must be presented.

Proposition XCVIII

Two aggregates of different corpuscules cannot be mixed without being activated by opposite movements such that the particles of one aggregate can be driven among the particles of the other.

As usual in mechanical sciences, the subject of any proposition must be isolated from variants and material circumstances or the conditions must be the same. Therefore, in our instance we assume two heaps of bird-seed, one of whitened seed, the other of black seed. They are round, equal in shape and in weight. If the two heaps are put together contiguously in the two sides of the same vessel or bag and not moved, they will never mix so as to let any white seed touch different black seeds. If the two aggregates are shaken and the seeds are separated from each other or if the black seeds move circularly away from the white ones with the same velocity, one also observes that they can never mix. If the black seeds are immobile or move slowly and all the white seeds move more quickly towards the black ones, the white seeds must be driven like wedges between the black seeds which can admit the white seeds only if they separate from each other. Thus, at first black granules are moved sideways. Then, by a circular reflex they return to the places let free by the white seeds. After repeated meetings and separations, and mutual penetrations, shaking is carried out like an ebullition. Then it is possible that nowhere black seeds remain without touching white seeds.

Proposition XCIX

The conditions are the same. A mixture can be undone by the shaking which provoked it unless another cause maintains union.

In our example the bird-seeds are equally round and smooth. Therefore, there is no firm, resistant and lasting adhesion between them. Because of their smoothness they easily glide on each other. This smoothness can separate the seeds as well as it achieved their union previously. As a consequence of their casual and whirling meetings by which the white seeds could be mixed, they could also be separated from each other either in their totality or in small clusters completely haphazardly.

The combination once started, to maintain it and the contact between the different white and black seeds, these seeds must be connected by some glue or some tie or some small hooks so that they cannot separate from each other. This is what happens in a mixture of flour and water where the particles of water after meeting the rugged particles of flour are sucked greedily by them and enter in their pores and recesses so that they can no longer be separated. The same occurs in cement and in similar mixtures. The more they are activated, shaken and beaten, the more the particles of water are subdivided and driven among the small rugged particles. As a consequence, the small rugged and consistent clods are subdivided and their mixture with the small particles of water results in uniform softness.

Remark

In the same way smoke ejected from a chimney, or droplets of water or small particles of earth dust are mixed with particles of immobile air to give fog.

Filth and excrements similarly muddy water and wine. In all these mixtures one observes opposite and whirling movements and driving of particles like wedges, without which such mixtures are impossible.

Proposition C

If part of a bag is occupied by white seeds and the bag is not completely filled by additional black seeds, both seeds can be mixed by repeated compression and squeezing of the bag.
Tab. XVIII, Fig. 8.

In the bag ABDE the part ALFCMB is full of white bird-seeds and the remainder FEDC of the bag is incompletely filled by black seeds so that it remains slack and can be constricted and dilated. The extremity ALFCMB is compressed so as to occupy the smaller volume AGFCHB. It was full of white seeds previously. Therefore, the seeds which were contained in the spaces ALFG and BMCH must be expelled. Since the bag is slack, they must be driven into the cavity FEDC. The bag will expand at I and K as much as it was constricted at G and H. But the black seeds can move sideways to I, K only if the white seeds expelled from L, M are driven like wedges between the black seeds. This interposition of white seeds results in a mixture since white and black seeds are in mutual contact by their sides. Consequently, repeated and mutiple compressions on both sides can result in general mixing of white and black seeds. This will persist if the contacts once achieved cannot be undone because of asperities or glue, as said above.

Proposition CI

If the end of the bag FEDC (Fig. 8, Tab. XVIII) is rigid and hard, I claim that the white balls contained in ABCF can never be mixed with the black ones, whatever the number of times they are compressed.

If this is not true, they will mix in this area if possible. But mixing can occur only if each white seed is driven between a pair of black seeds so that black and white seeds alternate. The white seeds touch black ones laterally. They thus must occupy a width twice as large as FC or ED. Consequently, the mixed seeds will dilate the hard pipe FEDC which is not dilatable. This is against the hypothesis. Therefore, mixing cannot occur.

Proposition CII

The data are the same. If the bag (Fig. 8, Tab. XVIII) is open at ED and continues straight ahead, I claim that the white balls compressed in ABCF do not mix with the black ones present in FCDE.

The white balls compressed in the space ABCF are pushed straight from FC towards ED. They cannot retrocede because of the closed extremity AB. Therefore, the black balls contained in FEDC are pushed by the white ones in the same direction towards ED. They do not resist this impulse since they can freely get out through the orifice ED. On the other hand sideways displacement is resisted firmly by the solid walls FE and DC. The black balls thus must comply with the straight impulse and not move sideways away from each other. But without sideways displacement and mutual separation of the black balls, the white ones cannot be driven in between them and thus cannot mix with them. When the first black balls escape with the same velocity with which the next white ones push them, the white ones cannot be driven between them like wedges. Consequently, mixing is impossible also for that reason.

Proposition CIII

If there are two liquids, one white AG and the other black GD, in the narrow and soft canal AC (Fig. 9, Tab. XVIII) and if one extremity DC is open and the other AB is closed, I claim that the liquids will not mix as a result of squeezing and compressing the canal.

The liquids resolve into hard and imaginary particles, as we showed in the book on natural movements resulting from gravity. Therefore, all that which was demonstrated for black and white bird-seeds can be applied to the particles composing the liquids, as far as their mixing is concerned.

Firstly, if the whole length of the pipe is compressed simultaneously and is reduced to AEFD, the white liquid ABGO is too voluminous to be contained in the restricted pipe AEIO with the same length. The surplus will move forwards as far as DF since it cannot flow back because of the closed extremity AB. But the place was initially filled by the black liquid which cannot stay there together with an equal volume of white liquid. Therefore, one of the liquids or parts of both must leave and move forwards towards KL where the way is open. But the white liquid cannot get out while the black remains. The latter would have to flow back from D towards O. Such backflow would result from the fact that the passage way is open, and would be produced by an opposite impulse due to the white liquid from O towards D, which is impossible. It would also be impossible if part of the black liquid remained in the space OF. Therefore, all the space AEFD must be filled by white liquid alone and the black liquid flows forwards to fill the open and free space DFLK. Consequently, compression of the tube AC does not mix the two liquids, the white AG and the black GD.

Secondly, if only the part AG of the pipe is compressed, the white liquid cylinder MPGO must move straight forwards. It will expel the black liquid cylinder which does not resist the straight impulse since the passage way DC is open. But its lateral expansion is resisted by the walls OD, OG. Therefore, the black liquid GODC must necessarily flow towards KL where a passage way is open which permits its escape. Actually when two adjacent liquids move with the same velocity to the same places and they are not activated by opposite movements and beaten repeatedly, it is impossible that the particles of one be driven like wedges between any pairs of particles of the other. Consequently, they cannot mix.

Proposition CIV

In wide canals with an open end some disturbance and mixing of parts of the liquids can occur but not so in narrow small canals. Tab. XVIII, Fig. 9.

The theory presented above with very narrow pipes with an open end is confirmed by the same experiment but is subject to some variation in wide canals. In the same figure, AC is a very narrow glass tube. Part DCGO is filled by red wine, part OMPG by white wine and part ABPM by water coloured blue. Whether the three liquids are shaken or pushed by some spirit or by a piston, they never mix nor leave their respective positions. When moved they progress in the same order, separated from each other as if they were solid wood cylinders.

But if the canal AC is very wide and liquids similarly coloured are placed in the same positions, they do not remain always well separated but become murky and mix.

The cause of this discrepancy is that the median parts close to the axis of the liquid cylinder can move with a greater velocity than the côllateral layers which are impeded by the asperities of the vessel. This unequal displacement of the different parts results in swirling of the liquid and differences in activation. Hence, the minimal particles of liquids acquire different positions and situations. Consequently, they mix partially. It is different in narrow tubes in which the particles composing the liquids are unable to swirl because of the exiguity of the place: they impede each other by mutual friction of their asperities and by friction against the rugged vessels. Therefore, the particles of liquids move with the same velocity and keep the same relative position. They thus cannot mix.

Proposition CV

The conical and soft funnel ABCD (Fig. 10, Tab. XVIII) is sudivided into several other funnels DE, FG, HM and into other thinner branches. Its base AB is closed. Fluids or white and black seeds are sent separately into the large funnel and they are compressed and shaken.
I claim that they will not mix.

The seeds or the liquids are adjacent and arranged along the length of the large funnel. Then under compression white seeds alone, at the exclusion of the black ones, must be driven into the branch DE placed in their alignment. Similarly, the black seeds which touch the orifice F of the funnel FG will fill the latter and white seeds will enter the small canal HM. No seed of a different colour can be admitted in any of them. Consequently, however considerably the small canals are compressed and squeezed, the white seeds present in DE or in HM and thus contained in the intact walls never can mix with the black seeds present in the separated small canal FG.

Secondly, layers of white and black seeds are superimposed. The white layer adjacent to the orifices of the small canals D, F, H (Fig. 11, Tab. XVIII), when compressed, will enter in all the pipes DM, FN and HO since it cannot retrocede because of the closed extremity AB and the push by the black seeds. Therefore, the white seeds must move forwards where a passage way is open. The black seeds cannot mix with the white ones which precede because these are pushed towards the open orifices and they cannot wander sideways to make place for the black seeds acting as wedges, as a result of the rigidity of the walls of the small canals and of their constriction carried out during the compression of the initial funnel.

After all the layer of white seeds has been expelled, the next layer of black seeds arrives at the orifices D, F, H. Under compression, the white seeds which initially were distributed in the spaces DM, EN, HG are pushed forwards by the second layer of black seeds ready to enter for the reasons mentioned above. Therefore, white seeds fill the spaces MQ, NR, OS and black seeds the spaces MD, FM, HO left by the former. Aferwards white seeds follow the black seeds. In the narrow spaces thus described it is impossible for black seeds to mix with white ones, as said above. Consequently, the black seeds always remain separated from the white ones. Therefore, in such progress, mixing never occurs.

Proposition CVI

If mixed white and black seeds are put in the same ramified funnel,
I claim that repeated squeezing can separate the white from the black seeds.

Mixed seeds contained in the large funnel can be driven into the first ramifications only after being subdivided into several heaps put in different places. The ramification is divided into several branches or smaller pipes. Therefore, any heap of mixed seeds must again be subdivided into other smaller heaps and these into even smaller ones until one arrives at all the narrowest pipes in any of which a small number of seeds can be contained. The straight series of seeds are alternately white and black. Finally, the narrowest pipes are also compressed, by hypothesis. Therefore, the seeds which they contain must be separated because the walls of the soft tubes, when compressed, meet together and thus separate the adjacent seeds. Hence, the seeds of any straight series are completely separated. All the mixing is undone since no black seed touches any white one. Firstly, they are in distinct tubes. Secondly, in the same tube, the seeds are separated from each other.

After this discussion, we understand that from the orifices of the extremities of the most tiny tubes black seeds can be discharged at times distinct from those at which white seeds are expelled. Consequently, all contacts are dissolved and no mixing remains.

Proposition CVII

Two different liquids mixed in the same funnel can be subdivided into very small particles by squeezing and sometimes the mixing can even be undone.

The liquids are composed of minimal hard and shaped granules. Their heaps made of white and black molecules which touch each other are not different from the aggregate of bird-seeds described above, except for their size. Therefore, if the smallest tubes of the funnel mentioned above could be narrowed in the ratio of the size of the granules composing the liquid to the size of the bird-seeds, then the mixture of granules of fluid would be undone by the same mechanism as the mixture of bird-seeds. Actually there are countless examples of such narrow tubes in Nature. The pores of the skin are narrower than the smallest blood particles: blood cannot get out through them whereas sweat can. There are other examples in other parts of the animal, in plants, in metals. Consequently, mixing of liquids can be undone in such funnels.

Famous anatomists believed that respiration was instituted to mix the principal constituents of blood in the lungs (i.e. serum, red elements, chyle and lymph) so that the smallest part of an element would be touched by and interspersed with the smallest parts of the other elements. They thought that this was achieved by repeated compression by the alveoli inflated by air. Therefore, as is my custom and to save words, for the sake of truth, I will explain what makes this assertion questionable.

Proposition CVIII

It is impossible that the heterogeneous components of the blood mix completely in the lungs however squeezed they are.

It is assumed that blood undergoes significant alteration during its circulation through all the body of the animal. On the one hand blood is deprived of most of its best substance part of which is used to feed the parts of the animal and part of which is dissipated in moving the muscles. On the other hand, during its circulation blood absorbs and carries away many other particles, partly refuse, partly inorganized chemicals. Finally, additional contributions of chyle from the thoracic ducts and from lypmhatic vessels flow in the right ventricle of the heart. These elements do not reach the proper degree of mixing in the right ventricle. This is common sense and apparent to the eye: in dogs ripped open some hours after feeding, the right ventricle is usually full of white liquid due to the chyle eructed by the thoracic ducts. Some authors conclude that mixing of the blood must occur in the lungs. Actually complete mixing is possible only if every minimal particle of one kind is in contact with minimal particles of other kinds. Such mixing of the minimal particles thus should occur in the lungs.

But without opposite and swirling movements and without repeated beating, the particles of one kind cannot be driven like wedges between the other particles. Consequently, such opposite movements and beating of the blood should occur in the lungs. But this is possible only in a large place like a bag or a goatskin bottle closed everywhere, as was mentioned[1]. The cavity of the lungs thus should be such, which it is not. The lungs actually are composed of countless ramified and very narrow canals like the roots of a tree. Therefore, it is impossible that the heterogeneous particles of blood mix intimately in the lungs.

Moreover, different liquids cannot mix in narrow and soft canals closed at one end and open at the other, even if they are repeatedly squeezed and beaten. This also was mentioned above[2] and demonstrated by the sensible experiment with very thin glass tubes. Therefore, mixing of the heterogeneous liquids which compose the blood cannot occur in the lungs made of a huge multitude of small canals.

Finally, the vessels of the lungs are similar to the ramified funnels described above and in them no mixing can result from squeezing, as said above[3]. Therefore, mixing cannot be achieved in the lungs. Consequently, although the movement of breathing squeezes and beats the blood vessels of the lungs, this does not imply that the heterogeneous elements of the blood must be mixed. Neither is the structure of the lungs such that intimate mixing of the components of the blood can be carried out in them.

Proposition CIX

Respiration and the movement of the lungs undo an improper and confused mixing of the blood by dividing it into its smallest elements.

After completing its circular course through the body of the animal, blood considerably altered is brought back by the veins to the right ventricle. There it is

[1] Prop. 100
[2] Prop. 103
[3] Prop. 105

further altered by addition of chyle. It is thus no pure blood which is driven by the heart into the pulmonary artery but rather a turbid mixture of different liquids. These liquids not only are altered in their qualities and consistency so that some can be very fluid and others dense and coagulated but they are not even distributed where they should be. This turbidity and improper arrangement cannot be brought to perfect composition and union without initial dissolution and destruction of the texture of bad combination and even worse composition. Consequently, Nature is forced to separate and subdivide down to the minimal the elements of the blood which were wrongly heaped up so as finally to compose a useful mixture in a proper order and best arrangement.

The solution brought forward by wise Nature is suggested by the structure itself of the lungs and by their movement during breathing. These are unable to mix blood and can achieve nothing else than breaking down and subdividing the blood into its smallest elements, as was shown[4]. The lung contains several ramified funnels which end in soft capillary tubes. The different liquids contained in these capillary tubes must necessarily be separated from each other and subdivided into their smallest elements by squeezing and compression, as mentioned above. As a result of such division, the curdled glutinous elements can become fluid again and mix with the others. Consequently, far from being used to mix the blood, squeezing and compression in the lungs undo prior mixing by dividing the blood into minimal particles. This is their direct action.

Proposition CX

The life of the animal does not depend completely and absolutely on respiration in so far as the latter contributes to blood circulation.

The opinion prevails that respiration is absolutely necessary to maintain life since we observe that perfect animals, once they have begun to breathe, cannot stop breathing for a significant period of time without suffocating and dying.

Respiration transmits blood from the right to the left ventricle of the heart. Without this passage, the blood cannot carry out its circulation. And we see that as long as circulation of the blood continues, life is maintained. Some people conclude that the main function of respiration is to move the blood and thus to maintain life.

There are some objections against this too absolute assertion. Firstly, animal life does not seem to consist only of the action of the heart, of its movement and of blood circulation: in hysteria and in some acute conditions, as I personally observed, the pulse of the heart and arteries uses to stop for several hours as does respiration without the movements of the limbs being impeded. Do not say that some vermicular movement of the heart persists although hardly and this can maintain life. When the heart of a frog is excised from the chest, life and movement persist. Frogs jump and escape without heart. Then, however,

[4] Prop. 107

blood cannot circulate. The same is observed in snakes. It thus appears that animal life does not consist of blood circulation. It can fortuitously happen that, when what is actually the true and proper cause of animal life fails, blood circulation also fails and errors can be committed by taking for the cause what is not the cause.

If indeed interrupting blood circulation does not provoke death by itself, then interruption of respiration which contributes to blood circulation interrupts this circulation but will not by itself provoke suffocation and death.

Moreover, in moribunds we observe that breathing stops before pulsation of the heart. Therefore, respiration is not the cause of the movement of the heart, as some people say. Consequently, if animal life cannot be understood without vital motions, one must say that such movement is not only that of the blood and of the organs which move the blood, primarily that of the heart and secondarily the respiration of the lungs. But some other different movement of something else is the primary movement on which life depends.

Proposition CXI

The main function of respiration is not the transfusion of blood from the right to the left ventricle of the heart.

If respiration had been instituted only to transfer blood from the right to the left ventricle, Nature obviously would have committed an error against her own law according to which she pursues economy everywhere. Without such an important apparatus burdened by so many risks, she could let the blood seep through a simple canal, like in embryos, or carry out this percolation by one single ventricle, like in fishes.

Do not object that the blood had to transit and seep through the narrow pipes of the lungs and needed the strong compression of the pulmonary press: such laborious action by way of the respiration indeed would be to no avail. Firstly, compression could be achieved in the lungs without alveoli inflated by air, more easily and more completely by some muscle as usually occurs in other viscera and glands. Secondly, we observe that without the compression of a press, nutritive juices are introduced through the very thin canals of the roots of trees. The blood contained in the capillary veins in the brain and bones also is properly sucked and percolated in the same way as liquids enter sponges, filters and capillary tubes without any external compression. Consequently, this compression of the pulmonary press so laborious and loaded with danger is superfluous. Since such nonsense cannot be attributed to Nature, one must say that the main function of breathing is not the transfusion and percolation of blood from the right to the left ventricle of the heart.

Proposition CXII

Air inhaled during breathing is the most important cause of the life of the animals.

Everybody knows that the life of animals depends on several causes, on the integrity of the whole body and its principal elements such as the head, heart, arteries, veins, stomach, intestines and viscera, also on the humours such as the blood mass and others, on the motions of these humours such as circulation of the blood, spirituous juice and chyle. Since solid as well as liquid elements are continuously dissipated, food and drink are necessary to repair and restore them. Such reparations, however, would not be possible without instruments and workshops in which they would be prepared, and without the presence of external adjuvant causes such as warmth, cold, humidity and others of the same kind. All these factors take part in life. None of them seems to be the most important and primary for life since, even if they fail, life does not disappear totally and immediately but goes on imperfectly for some short time. For example insects and reptiles when their body is divided in two, do not perish immediately. They still move for several hours. Similarly, beheaded flies fly. Frogs jump and escape after their heart has been removed. Eels and snakes move during a whole day and thus exert their vital movement after their arteries have been dissected, after they have been emptied of their blood, without their stomach and completely disemboweled. In the same way, when their blood circulation and pulsation of their heart are failing, sick men sometimes survive for one or two hours. But we observe that frogs and insects survive after their heart has been removed and there is thus no blood circulation. Also when they are deprived of food, drink, sleep, external warmth, etc., animals live, although uneasily, for some days. Consequently, none of the factors mentioned above is the most important and primary cause of the life of the animals.

All these factors being excluded, it only remains to examine the force of the air which the animal absorbs either by breathing or in any other way. We easily demonstrate that this air is the principal factor in producing life. As long as they breathe, animals live. But, if air is prevented from entering the lungs, they immediately suffocate and perish. Closing of the mouth and nose, strangulation of the throat by a rope or obturation by water or by a catarrh, are followed by the death of the animal.

The experiment which must obviously prove this assertion is the temporary removal of air in the pump of Boyle or, even better, in the vacuum of Torricelli achieved by using quicksilver. All animals shut up in such a machine fall moribund immediately but, if air is carefully returned, the same animals come back to life.

Conversely, flies and other insects, frogs, lizards and snakes which lived for some time and moved in the presence of air, after their bodies had been divided, beheaded, without heart and blood, immediately fall and agonize after being placed in such a vacuum, i.e. without air.

More surprising, fishes which do not breathe and spend their life immersed in water cannot live if they are completely deprived of air. Gassendi claims to have observed that fishes perish in frozen stagnant water if the ice is not drilled. After the ice is drilled they rush to absorb air with such avidity that they cluster with open mouth in the vicinity of air. Similarly, small fishes which swim in glass vessels live for a long time when the orifice of the vessel is open. When it is closed they soon perish.

Such avidity for air in fishes does not result from the emptiness of the bladder which balances them in water: in dead fishes beneath ice, the bladders were found to be as inflated as in other fishes alive. Moreover, we observed in the Accademia Del Cimento a fish the bladder of which had been disrupted and emptied in the vacuum of Torricelli. This fish was removed quickly before it died and was placed again in a fishpound. It lived there for a long time afterwards. During about a month it crawled on the bottom of the pond like a snake and could not raise itself or become erect because, as a result of a lack of air in its bladder, its specific weight was higher than that of water. Since this fish lived without an inflated bladder, it is concluded that fishes need air for another important necessity. This other necessity does not seem to be anything else than the desire to maintain life since shut up and completely deprived of air fishes die.

This is also confirmed by other experiments. Authors mention that ill people nearly moribund were returned to life by sneezing. This movement is nothing else than energetic driving of air into the lungs.

Moreover, the famous Thruston mentions to the Royal Society of London that D. Croone strangled a chicken with his hands until it seemed dead. Air was insufflated into the trachea and the chicken came back to life.

Similarly, the famous Needham strangled a dog with a rope until the heart stopped pulsing. Then the abdomen was quickly opened, a tube was adapted to the receptaculum chyli described by Pecquet and air was energetically insufflated. As a result, the blood and the movement of the heart started again and the animal came back to life.

All these facts sufficiently suggest that the most important and principal cause of the vital movement of animals is the air absorbed by respiration.

Proposition CXIII

By respiration of air the particles of the blood can be mixed.

It thus appears that the function of respiration is neither cooling of the superfluous heat of the heart, nor ventilation of a vital flame, nor mixing of the heterogeneous elements of the blood which would be achieved by squeezing of the inflated alveoli of the lungs, nor transmission alone of blood from the right to the lelft ventricle to enable circulation. On the other hand, we feel that such a considerable apparatus as are the vessels and organs of the lung was made to some great purpose. We will attempt, if possible, tentatively to expose this purpose.

Firstly, air insufflated into the bronchi through a tube does not appear to enter the pulmonary vein. Despite this experiment famous authors self-assuredly assert that inspired air mixes with blood in the lungs. But, if I am right, they do not seem to base their opinion on solid reasons. They say that the walls of the vessels of the lung are perforated by narrow and inconspicuous holes like the pores of the skin. Inspired air could be driven into the veins through these holes. But this is the controversial point. Although we must admit these pores, it is not certain that air can pass through them. We observe that spirit of wine passes through the pores of a pig bladder but neither water nor air. Oil and water penetrate through the pores of the intestines and skin but not air. From this experience we deduced elsewhere that the particles composing air are more branchy and larger than the particles of water. It is obvious that air does not pass through the pores of veins more than it penetrates through the porous membranes thus mentioned. If air is insufflated through a tube into any vein, it does not leave the vein.

These authors say that chyle is extracted from food loaded with air and carries particles of air in the blood. This can be believed. But this suggests that air is not sent from the lungs into the blood vessels. There is a royal way to mix air and blood through the chyle. Therefore, it is inferred that air is not introduced through the impenetrable paths of the vessels of the lung. It thus appears that the famous authors did not prove entry and mixing of inspired air with the blood in the lungs.

If I may express my opinion, I consider that countless particles of air are mixed with any liquid. This appears in the experiment of Torricelli. If there is water above the quicksilver, when vacuum is created in the upper part of the tube, many bubbles of air rise through the water. They were not visible previously because of their exiguity but develop in huge bubbles when the atmospheric pressure is removed. This is confirmed in ice in which many bubbles of air appear which were invisible in water, as said elsewhere. Above all it must be noted that a smaller quantity of air is found in frozen water in a closed vessel than in stagnant water frozen in the open. In frozen enclosed water very few bubbles of air can be seen whereas they are many and bigger in water frozen in the open. The reason for this difference is that, as air easily scrapes water particles from the surface of water, so small spiral particles of air are driven into the water. This is suggested by the fact that wet linen spread in the shadow dries even in the winter and, if wind blows, it dries very quickly. This is certainly due to the fact that the particles of air activated by the wind scrape small grains of water and by the violence of the wind many particles of air must be driven into the water which retains them and from which they cannot escape.

Therefore, water beaten in air becomes foamy. As water retains big bubbles of air, others smaller and some invisible because of their smallness remain in the water since they cannot be retrieved, as shown by experience.

In the recesses and alveoli of the lungs there is always some aqueous and serous juice instilled there to make the lungs slippery and flexible. This serosity beaten by the wind of inspired air will no doubt be transformed into foam. Hence, this water is impregnated by particles of air. Since water can easily exude from and penetrate into the pores of the veins, it must carry with it the particles of air thus included and mix them with the blood. This effect seems to be unquestionable. Thus although air alone cannot pass through the pores of the skin or of the veins, it can pass mixed in water, subdivided into mimimal particles as we see minimal particles of air transpire through the pores of wood or of the skin with smoke and sweat. But larger quantities of air even insufflated by a piston cannot get out through these pores because the ramifications of the particles of air are intertwined and linked and thus are impeded to pass through narrow pores, as shown elsewhere.

This is, if I am right, an expedient way to transmit air into the veins of the lung and an easy way to mix minimal particles of air with the other small components of blood.

Proposition CXIV

The particles of air mixed with the blood do neither increase the fluidity of blood nor produce effervescence in the heart by their elasticity or their nitrous nature.

These famous authors who claim that inspired air mixes with blood attempt to explain the function and mode of action of this air added to the blood. Firstly, they say that the particles of air mixed with blood increase the fluidity of the latter. I cannot agree with that opinion. If air, thanks to its great rareness and fluidity, increased the fluidity of water, no doubt the more air would be mixed with water, the more water would become fluid. Then foam would be more fluid and would flow more easily than pure water. But this is not what occurs.

Fluidity results from the subdivision of the bodies into minute particles the surfaces of which are smooth, the shapes less rugged and angular, and the weights equal, as we showed elsewhere. But air does not seem to be more fluid than water or mercury because air is made of branchy and spiral particles of larger shape and circumference than the particles of water since water and mercury can pass through the narrow pores of skin whereas air cannot. Therefore, air mixed with water will rather decrease the fluidity of the latter. Air is not more fluid than water because it is rarer and lighter. Indeed oil is less fluid than water although it is lighter and rarer.

Moreover, a heap of inflated bladders is less fluid than one of wooden balls, the other conditions being equal, i.e. if they are equally smooth and slippery, because the surfaces of the bladders coapt easily as a result of the softness of their envelope and the compressibility of the contained air. Larger surfaces of contact stick together. But the wooden balls, as a result of their hardness, have almost points of contact and, therefore, can easily roll over each other. It is different for the bladders which, as a result of larger contacts, stick together like pieces of mosaic works. Therefore, they move over each other only with difficulty. Similarly, the particles of air which are spiral and yielding not only are united by larger contact surfaces but are even attached to each other. It is, therefore, difficult for one spiral to move over another and to flow. This is different for water and other liquids which are made of hundreds of rigid and hard particles. Therefore, their contacts are minimal and hence their fluidity is greater.

From this I conclude that air, because of its consistency, when mixed with blood, rather diminishes the fluidity of the blood.

Some authors also claim that air induces rarefaction of blood by its elasticity. This would result in some effervescence and fermentation which would give blood its perfection and stimulate the movement of the heart. Others do not hesitate to say that air has a nitrous nature which promotes the movement of the blood when activated by warmth. If I am right, all these opinions are expressed self-assuredly rather than being based on solid arguments. Air mixed with blood would induce rarefaction by its elasticity. This would occur either because air transmutes the substance itself of water by conferring to it a rarer consistency or because included bubbles of air become bigger. But neither alternative can occur as can be seen in a thin glass tube in which a mass of water occupies the same volume whether or not air is added. Actually, particles of air included in water are rather compressed and condensed by the weight of the incumbent water.

In order to perceive more clearly the mechanism of this operation, imagine several inflated vesicles to be mixed in a heap of sand. The grains of sand no doubt touch each other as they did previously. The elasticity of the vesicles thus does not rarefy the sand and the vesicles are rather compressed by the weight of the incumbent sand and become denser. Therefore, the composite of sand and vesicles is not rarefied. Similarly, air by its elasticity will not rarefy the water in which it is contained.

If the authors thus mentioned actually wish that air, by its elasticity, disrupts and separates the particles of blood which conglutinated or coagulated in veins and viscera and that air fluidifies blood in the lungs by its swirling movement, it appears that this adjuvant is useless since there exists a very powerful cause which can separate the conglutinated parts, reduce them into minute elements and thus make the blood very fluid. This cause is the expression through the capillary tubes of the vessels achieved by the press of the lung, inflated and compressed by the resilience of the ribs.

The assertion of the nitrous nature of air does not seem to be more true whether this nature results from the mixing of earthy particles which contaminate the air or whether it is part of the proper intrinsic nature of air. In the first instance indeed, air itself contains not only inflammable nitrous atoms but also cold particles inducing condensation. Therefore, it does not induce rarefaction of the blood. In the second instance, the nitrous nature of air and its force of rarefaction are a pure material condition of its machines for which warmth can be as beneficial as it can be harmful, i.e. the machines can now be condensed and then rarefied by warmth. We indeed observe that the strength and rigidity of a metal spring are weakened and softened by fire. But the same fire strengthens and hardens pieces of wood and javelins.

Moreover, if the particles of air actually were nitrous they would burn and crackle when in contact with the flame of a candle. But this does not happen.

I also notice that the proper substance of air is not rarefied by the action of warmth but rather by mixing with exhalations or particles of fire as if it were by activated wedges. The spiral elements of air caught by these particles are rather compressed and condensed as we suggested elsewhere.

Finally, there is no heat nor any ebullition in the left ventricle of the heart. The warmth there is not more than what is found in the right ventricle. We experienced this by introducing fingers through an incision into the heart of a live animal. Therefore, this nitrous nature of air introduced by recent authors is completely false. Consequently, the actions of air mixed with the blood during breathing are completely different from what has been thought so far.

Proposition CXV

Mechanism of the continuous oscillatory movement which air mixed with blood can produce.

Compressible and resilient machines have two important properties. Firstly they can remain immobile at a certain degree of dilatation only if the force which compresses and squeezes the machine is equal to the force with which the machine normally attempts to dilate. Since equality is indivisible, if one of the forces is greater or smaller than the other, equilibrium is disturbed and the smaller force yields to the greater as occurs in scales. The second property has nothing to do with a balance: in a balance once equilibrium is altered, the lighter scale rises and then remains high and does not come back down. It is different for the machine. If it is compressed beyond equilibrium, it does not remain constricted but is dilated spontaneously by its resilience. In carrying out this movement the mobile acquires a new impetus and this impetus by its nature must produce its effect which is a further displacement of the mobile, as we demonstrated. Similarly, a pendulum when coming down towards the vertical acquires force by its movement, by which force it is raised again from its lowest position to a height almost equal to that from which it went down. The same occurs with a flexible rod, with sea waves and with countless other phenomena. Consequently, while the movement due to the resilience and dilatation of the machine is carried out, the machine unquestionably acquires new impetus which moves it beyond the limits fixed by its natural expansion. Therefore, the machine necessarily comes back to its normal compressed shape. But during the movement which it carries out when constricting, the machine also acquires another impetus which cannot remain unused so that necessarily it constricts the machine more than normal and so it goes on. Therefore, the machine is forced to pursue its oscillatory movement like the waves and the pendulums.

But the described to and fro movement slowly weakens as a result of external impediments so that the oscillations continuously diminish and finally stop. If we want to produce such oscillations without interruption, we must neutralize the weakening of the oscillations by new external impulses so that the movement is maintained by the addition of forces.

I consider that the minute particles of air are spiral machines which can be compressed by an external force and then spontaneously resile like springs, as appears in a pneumatic blunderbuss. Such aerial small machines mixed with blood necessarily undergo some compression either by the weight or by the compression of surrounding viscera, vessels, flesh, membranes and skin which are parts of the animal. If this compression was uniform and always of the same degree, the aerial small machines also would maintain their state of compression unchanged as required by equilibrium. But, the liquid and solid elements of the animal are activated and moved in a thousand ways, now by internal, then by external causes so that the aerial small machines which are mixed with the blood do not remain balanced in one state of compression. By the effect of external shaking they must start the oscillatory movement described above. This movement can never stop as long as the animal is alive because new percussions are never lacking, from breathing, from pulsing of the heart and arteries, from the movement of the muscles and from a thousand other internal and external causes which can set up again this trembling movement of the small machines of air. Therefore, the particles of blood must always be shaken by a peculiar movement. They must be activated by this oscillatory movement of the small machines of air included in the blood.

Proposition CXVI

Introduction of air and its mixing with blood by respiration produce and maintain life in animals.

An automaton seems to present some ressemblance with animals since both are mobile organic bodies which comply with the laws of mechanics and both are moved by natural faculties. But let us see whether it is possible to investigate the properties of natural things in some way by using the knowledge of artificial things.

A clock is made of several cog-wheels artificially connected which are moved according to a certain law by a suspended weight to mark the course of the sun and moon and to carry out other movements. If the motive force of the suspended weight acted freely by rotating all the wheels immediately like spinning tops, it would complete the course of the clock and thus would not conform to the movement of the sun and moon. To obviate this inconvenience, it is usual to add a weight or an oscillatory pendulum which by carrying out to and fro movements of equal duration according to the laws of mechanics, regulates the power of the motive cause and the movement of all the cog-wheels so that their action corresponds to the course of the sun and moon.

I also observe that the life of the animals or their vital operations consist of continuous and uninterrupted movement. The limbs and all solid, fluid and spirituous elements are activated when the body is in movement or is displaced from one place to another, when it eats, digests, makes chyle and transforms food into blood, when it feeds and repairs lost elements, when it expresses sensations. In summary, nothing remains stable in an animal as long as it is alive.

All these vital movements are carried out with certain and well-determined velocities, rhythms and periods. If they were carried out with much greater velocity, they would no longer be vital and life would not be preserved.

Besides the organs which are built with the utmost perfection in the animal, undoubtedly there is also a motive cause which gives the impetus. This is the spirit which is inside and, spreading through all the limbs, moves the mass of the animal body.

This spirit actually would impart frantic movement to the organs of the animal as a result of its very mobile nature and thus the animal would not carry out the vital actions as required by Nature. Therefore, a regulating machine must be added in the animal (which is an automaton of Nature) like in a clock. This regulator by its mechanism refrains the motive force so that it does not infringe the laws instituted by the Divine Architect. This machine seems to be similar to the regulating pendulum of a clock. It must also regulate by its oscillatory force the movement of blood and spirits to prevent these from flowing frantically and foolishly.

The clock stops as if it was suffocated when the movement of the pendulum is interrupted. Similarly, animal life dies out when the oscillatory movement is stopped or impeded. Hence we discover the great mystery of the necessity of air in animals: as long as the animal is alive, particles of air must be driven into the blood continuously because the small machines of air sent into the blood must carry out this oscillatory movement like a pendulum, as said above. The particles of blood contiguous to these aerial machines necessarily must be activated by the same oscillatory movement. As a result of this primary movement, all the elements of the animal move at a regular rhythm in the same way as the cog-wheels of a clock are pushed and regulated by the oscillations of the pendulum. The activation of the spirits or of the nervous juice and of the muscles may result from this movement, at least partly. Blood is carried in a continuous flow to all the parts of the body of the animal, like a stream. It carries with it the cause and stimulus of the movement which is the origin of life.

The presence of air and its mixing are observed also in fishes and plants which present some sort of respiration. There is general agreement that more perfect animals carry out better and more divine movements than do plants and imperfect animals, the more so the more the movements inside themselves are lively and powerful. Perfect animals thus had to be provided with more and nobler organs so that such excellent life could be exerted. Hence, in perfect animals, air must be renewed by quick-acting and continuous additions by way of the respiration. On the other hand, life is always endangered since it consists of such motions which cannot be maintained without air being provided.

Proposition CXVII

Why the foetus begins to breathe after birth and continues breathing.

In the embryo in the womb of the mother the blood is transmitted from the right to the left ventricle through the foramen ovale and through the ductus arteriosus. In this passage air cannot be mixed with blood and there is no respiration. However, vital movement can be maintained in the foetus because continuous intromission and mixing of air with blood are ensured by the lungs of the mother. These indeed send blood continuously enriched by particles of air into the placenta. Hence it is absorbed by the umbilical veins, carried to the heart of the embryo and circulated. By this artifice, Nature maintains the life of the embryo in utero without that the embryo breathes.

But, at birth, air must be used up for the reasons given by the famous Thruston, as we will show. As a result of pain suffered when leaving the uterus of the mother, the embryo is stimulated to move all its muscles without any other purpose than reacting to the stimulus of pain and the inexperience of the motive spirits. During this frantic general commotion it is impossible that the diaphragm and intercostal muscles refrain from contracting. But they cannot contract without expanding the cavity of the chest. This in turn forces ambient air to flow in and to fill the bronchi and the pulmonary alveoli of Malpighi. Afterwards, the ribs moving back and resiling, the blood which is contained in the vessels of the lungs is expressed into the left ventricle like a torrent. As a result, a new and large route is opened to the blood through the lungs. Therefore, the narrow and transverse foramen ovale stops being used. The flow of blood closes its membranous operculum. Then, once the vessels of the lung are evacuated, the large and straight route of the pulmonary artery is open through which the blood can be eructed from the right ventricle more easily. It cannot flow back because of the semilunar valvules. The blood thus neither needs nor can any longer use the narrow and more difficult oblique passages. Therefore, both being neglected progressively coalesce and close. In these conditions, the blood is forced repeatedly to follow the same route and to continue for the same reasons and mainly because of the considerable practical advantages that the animal acquires with the lungs. It is stimulated and attracted by these advantages. Respiration not only enables muscles and above all those of the abdomen to contract, as a result of which faeces are discharged, viscera and vessels open and blood can circulate; but also particles of air driven into the blood in the lungs impart to the blood the oscillatory movement which maintains life. As a consequence, the animal acquires vivacity and alertness. Therefore, this small animal is stimulated by a natural need ànd is attracted by the advantages to continue breathing throughout its lifetime.

Proposition CXVIII

Solution of the problem of Harvey.

The famous Harvey proposed a problem which, according to distinguished authors, has not been solved satisfactorily so far. When the foetus has left the membranes, after he has inhaled air once, he cannot survive a while without air but, deprived of it, dies immediately. Why?

If I am right, it seems possible to find the answer in what was said above. Indeed life is possible without respiration in utero because breathing of the mother makes up for this lack. Particles of air are continuously and abundantly mixed with the blood of the foetus. These particles impart vivifying mobility to the mother's blood, as said above. This vivified blood is led through the umbilical vessels to the heart of the embryo. This has the same effect as if the foetus itself breathed. The small heart receives the blood impregnated by aerial particles by which the primary oscillatory movement of life can be exerted. Afterwards all the mass of blood can circulate from the right to the left ventricle through two canals as if it passed through the lungs by the effect of respiration. The impulse given by the pulsation of the heart muscle occurs equally well in the foetus. Thus it is obvious that the embryo can live in utero.

After birth, the animal is no longer fed by the blood prepared previously in the lungs of the mother. It must make and compose blood in its own workshops. Thus the animal must achieve the absolutely necessary addition of air in itself. There is no other way than that in which this addition was achieved in the lungs of the mother. The embryo thus can use its own lungs which are completed and ready to work. Life cannot continue without the contribution of air which can be called the salt of life and without blood circulation which is promoted by breathing. Therefore, as long as the animal which has left the uterus lives it breathes. This seems to be the reason why the animal cannot live for a moment without breathing.

This is to confirm a traditional theory.

Proposition CXIX

Explanation of why air retained in the lungs for a long time induces the same anxiety, suffocation, and finally death as does impeded breathing.

Absence of inspiration is harmful. This is easily understood. In these conditions the puckered vessels of the lungs do not expand and unfold, and the alveoli of Malpighi do not fill. Therefore, the vessels cannot be opened as if it were by wedges and the blood cannot be transmitted from the right to the left ventricle. This is harmful and finally deadly. It seems normal that filling the bronchi and inflating the alveoli with air kept for a long time eliminates this inconvenience since the vessels are unfolded and air wedged in promotes the passage of blood. Experience, however, teaches that the animal suffocates not less from prolonged and continuous presence of air in the lungs than from its absence.

Lack of respiration is pure deprivation followed by lack of blood circulation and lack of vital movement. These are equally deprivations. But it seems that this unpleasant feeling of anxiety and suffocation cannot be produced by nothing. It is rather produced by a positive cause inducing a painful solution of continuity. The lacks thus mentioned, therefore, should not provoke such unpleasant anxiety. Consequently, it is worth-while to explain this effect.

Firstly, I consider that the inconvenience of itching seems to originate from tickling and slight friction of small nerves, not from laceration or solution of their continuity. The slightest friction with a straw or with the wattle of a chicken feather introduced in the nostrils or in the hair of the skin causes the most unpleasant itching. The same is provoked by water repeatedly injected on the face and in the mouth. These frictions or tickling do not cause solution of the continuity. Actually the remedy for such inconvenience is scratching of the skin with the nails.

A crumb of bread or a drop of drink inadvertently entered in the throat causes such unpleasant itching that the animal keeps coughing in an effort to expel and spill it. In heavy colds the sputum which tickles the internal wall of the trachea provokes coughing by which the animal attempts to get rid of the inconvenience.

Permanent filling of the lungs by air can provoke the same inconvenience as its lack. When respiration is hindered, a great quantity of breathed air retained for a long time, by its volume, can distend, irritate and prick the internal wall of the trachea and its alveoli, like catarrhous humours.

Pain can also result from a lack of air because the pulmonary arteries and their ramifications full of blood up to turgescence can entail the same pricking and tickling by compressing the membranes and vesicles of the trachea. This can result in unpleasant itching and anxiety with a feeling of fear produced by the lack of circulation of the blood. Continuous abundance of air impedes the arrival of blood in the pulmonary vessels which are compressed by the turgescence of the alveoli. Lack of air impedes the emptying of the vessels and the expression of blood. These inconveniences are complemented by a third one due to the interruption and absence of mixing of particles of air in the blood. As long as the movement of the thorax and breathing are interrupted, whether the lungs are empty or full of air, the alveoli are neither emptied nor squeezed by the resilience of the ribs. Hence, the aerial particles are neither expressed nor driven into the veins. Finally, a fourth inconvenience is added: the vain effort of the muscles which, normally irritated by the stimulus of a need, attempt to restore breathing. All these causes produce an unpleasant feeling of anxiety and of painful suffocation.

Death results from the interruption and stop of the vital movements, i.e. from an absence of the trembling and oscillatory movement of blood and spirits and from the absence of blood circulation on which life depends.

This death, however, is progressive. Everybody agrees that the feeling of destructive pain in the nerves and brain, seat of the spirits, induces great disturbance of the movements. Sensations are confused, disturbed and soon progressively blunted. Simultaneously they produce a kind of frantic convulsions. These finally stop with death when the vital movements are totally extinct.

Proposition CXX

Why and how the animals suddenly fall and die in rarefied air or in vacuum.

Animals enclosed in the rarefied air of a pump of Boyle and in the vacuum of Torricelli suffer suffocation and anxiety and suddenly fall moribund. This was mentioned above. We now have to investigate the mechanism of this suffering.

Firstly, the opinions of others have to be considered. They claim that the symptoms thus described occur because in such instance blood stops circulating as a result of an excess of air or of the rareness of a very delicate ethereal substance and also because, as a consequence of a lack of air, the fluidity of the blood is not maintained. But these opinions are sufficiently refuted by what was said above. Life does not die out in the absence of respiration and pulse in hysteria and other conditions, or at least the patients do not die immediately as occurs in vacuum. Moreover, frogs and insects live and jump for some time without a heart and divided into several segments. But, even when they are intact, they fall moribund in vacuum as suddenly as if they were struck by a thunderbolt.

Consequently, they do not perish as a result of the lack of blood circulation or of less fluidity of their blood since the blood was found as fluid in animals dead in vacuum as in others who died in the open.

Let us see whether we can offer a more likely explanation. Unpleasant itching, pain and anxious suffocation which occur in vacuum must be much more severe than what we described in the preceding chapter, as we shall explain soon. But, however unpleasant and acute they are, they must not floor the animal like a thunderbolt by blunting the senses and inducing convulsions and by killing the animal quickly. Actually we observe that crucified persons remain alive for a long time, their senses are not blunted and they do not present convulsions. Therefore, some other very powerful cause produces these severe lesions. In an attempt to find this cause, I observe that the particles of air left in water and in the bladder of fishes expand considerably and achieve turgescence in vacuum since there all atmospheric pressure is removed. Therefore, normal elasticity of air, thus let free, can expand by resilience according to its natural tendency.

The aerial particles contained in humours, membranes and parenchyma of the viscera of the animal must exert their elasticity in vacuum and dilate (This is obvious: in vacuum frogs are deformed by swelling, their eyes protrude and, when put back in air, they immediately deflate and return to their previous size.) In vacuum all the pores of the flesh, vessels, nerves and viscera become turgescent, swollen by the particles of air which they contain. All the soft tissues of the animal are inflated and turgescent since they are full of countless vesicles swollen by rarefied air. As a consequence, all the vessels and the small canals of the nerves are compressed violently as if it were by a press. Therefore, the vessels being obstructed, all motions of the humours and spirits must be interrupted. Hence, all feelings and movements in the animal are subjected to considerable disturbances and interruptions: the senses are blunted and frantic convulsive movements are carried out. Finally when all the ways are occluded and the animal movements are stopped, death must follow.

However, this fight, squeezing and wedging in of the turgid vesicles of air are necessary to produce pain and unpleasant anxiety from the beginning. Moreover, such itching feeling of painful anxiety is necessary to stimulate the nerves which end in the chest. By natural instinct and from acquired habit, the animal attempts at contracting the diaphragm and intercostal muscles. These muscles contract with great difficulty and at the price of a considerable effort because their motive force is decreased and because they miss the help of air as we shall show. The animal becomes tired by its vain effort and thus its anxiety and deprivation increase. Animals actually exert such useless effort in rarefied air. This appears particularly in birds who often inhale with open bill and attempt to catch some air.

These are the two important causes which produce the most unpleasant feeling of anxiety in the first place.

The painful anxiety combined with the disturbance and diminution of the vital movements of the blood and spirits induces painful suffocation with blunting of the senses and convulsions. Soon complete interruption and stop of the vital movements lead to death. Such suffering occurs immediately because air suddenly disappears in the machine and at the same instant the vessels and paths of the liquids and spirits through which feelings and movements, i.e. life, are exerted, are compressed.

Proposition CXXI

Why dyspnoea occurs during powerful and laborious effort of the muscles and how it is provoked.

Distinguished anatomists observed that the nerves of the diaphragm have their origin in the brachial nerves. Henceforth they thought that this was so that the diaphragm also is involved in laborious movements of the arms and is forced to dyspnoea. But it is easy to see that somebody who climbs a slope with the hands tied is dyspnoeic also.

We mentioned above that a voluntary action like any sensitive operation of the soul cannot be carried out without movement of spirituous juice and that muscles cannot be stimulated to move without the nerve roots which innervate these muscles being shaken in the brain by some convulsion by a motion of the spirits through which the will exerts its action. Consequently, for several muscles to pull powerfully and energetically, as in running, jumping and dancing, the spirits in the brain must be shaken with the same vehemence and power as if they were boiling. The more the spirits are shaken the more the muscles are stimulated. The spirits violently shaken by a frantic impetus in the brain and wrestling also contact the orifices of the nerves which innervate the heart. Therefore, this is not different from anger. The heart must pulse powerfully and violently since the orifices of its nerves in the brain are always open and can absorb some of the spirits as a result of any extraordinary commotion of the latter. Consequently, every time the muscles of the limbs contract powerfully the heart must also pulse more quickly and more strongly.

Then since the heart pulses more quickly and powerfully, the blood must accordingly flow more rapidly through the vessels as a result of the impulse of the heart. As a consequence of the compression of the heart, when the muscles contract, the left ventricle remains empty and the right ventricle together with the pulmonary artery is filled by blood in excess. Because of the turgescence of the heart, its fibres are stretched exaggeratedly and almost break. This results in pain and anxiety. This cannot be helped except by breathing which promotes the passage of blood through the pulmonary small canals.

The passage and decanting of blood through the thinnest pulmonary pipes cannot be achieved either by inspiration alone or by expiration alone but by an alternation of both, i.e. by the momentary resilience of the ribs which follows inspiration and precedes expiration. Hence, every time the lungs are considerably loaded by blood as a result of assiduous and frequent flowings in, they also need frequent discharges. Therefore, frequent alternations of inspiration and expiration are required and breathing becomes dyspnoeic.

Moreover, frequent pulsation of the heart requires frequent respiration for another reason. When the pulse is accelerated the blood flows quickly through the heart like a stream and if particles of air are not provided frequently by respiration the quality of the blood is not uniform because air is not mixed with all its particles. As a result, blood perhaps would be unable to carry out the effervescent rarefaction in the muscles, above all when movements are so powerful and frequent.

Consequently, when muscles contract laboriously, respiration must become frequent or dyspnoeic.

After the movement has been completed and the muscles rest, dyspnoea persists for some time. This seems to result from the impetus given to the spirituous liquids with the blood and to the organs by the previous powerful movement. Similarly, the waves of the sea remain agitated after the storm has quieted down.

We experience this fact in ourselves. Every time we rotate our body about our feet in a vertiginous movement or when we are on board a ship buffeted by the wind, even after the agitation has ceased and we are at rest, the walls seem to turn and we vacillate so that we cannot remain erect. We undergo all that against our will. Consequently, these symptoms must result from internal motions of the spirits which retain the impetus of a previous movement.

This thus appears to be the cause why laborious motions of the muscles are accompanied and followed by dyspnoea.

Proposition CXXII

Why one holds his breath during laborious and difficult effort of the muscles.

We know by experience that when carrying heavy weights or exerting considerable effort we are forced by Nature to retain the inspired air in the chest. It is deduced from this that we hold our breath for some advantage and that this has some important usefulness. To carry out very powerful and continuous movements, muscles require spirituous juice and blood not only abundant but also transmitted in a continuous flow. It seems that this can be achieved very easily if the big arteries and nerves are compressed by an external force as if it were by a press. This strong and continued compression expresses spirituous juice and blood into the muscles which must exert such great effort.

Such compression obviously can be exerted by the inspired air since the aorta and the most important nerves extend in all the cavity of the chest and all the abdomen. Thus air compressed by the thorax as long as breath is held squeezes like a press the vessels and all the viscera depressed by the diaphragm pushed downwards by this air filling the chest. Simultaneous contraction of the abdominal muscles takes part in this action by compressing the branches of the remainder of the aorta and the ramifications of the nerves. Henceforth, the arteries and nerves send spirit and blood in a powerful flow, like fountains, into the acting muscles. They carry into these muscles the substances for effervescence which makes possible great and continued effort. This seems to me to explain the necessity of holding one's breath during laborious motions of the muscles.

Proposition CXXIII

Why muscular effort in rarefied air provokes dyspnoea.

Among the wonders which I observed at the top of mount Etna in 1671, there was an unexpected effect resulting from the excessive rarefaction of air. Up there indeed the least movement to climb among the rocks produced such tiredness that even young strong fellows and peasants were often forced to sit in order to rest and to restore their strength by frequent and dyspnoeic respiration. I also noticed that this inconvenience was caused neither by excessive thinness of this air nor by some alteration of its quality. Indeed when sitting or riding we breathed the same air and, however, could breathe without dyspnoea as it is usual at sea level. I gave a solution of this problem in the supplement to my *Meteorology of the Eruptions of the Etna*. But, after further tought, I can no longer endorse this solution and I came to a more likely one.

Experience shows that powerful and laborious movement produces dyspnoea. We described the cause and the mechanism of this effect. Consequently, if we demonstrate that moderate motions to climb in steep places cannot be carried out in rarefied air witout considerable work and fatigue, it will be clear that dyspnoea is produced by effort.

It thus remains to show that a movement of the muscles which is carried out in air of normal density with tolerable difficulty and without undue fatigue, cannot be carried out in rarefied air without considerable work and tiredness. This is true, we will show it.

An action may be more laborious in two ways: (1) If the resistance is increased whereas the motive force is neither altered nor increased; (2) if the motive force is decreased whereas the resistance is not diminished. The force exerted by one hand to carry two pounds is greater than to carry one pound. The work to carry two pounds with one hand is greater than with two hands. Consequently, for the same effort it is easier to carry a weight of two pounds with two hands than a weight of one pound with one hand.

Air retained in the chest compresses the spirit and blood vessels by its elasticity and thus helps the effort of the muscles, as mentioned above. Therefore, rarefied air (although it is compressed by the thorax as is dense air), empties the vessels less completely and thus helps less the effort of the muscles. Thus, in rarefied air the motive force of the muscles is decreased by lack of the help which results from the elasticity of condensed air. Consequently, in rarefied air the same resistance is overcome by a decreased motive force with more work and fatigue. This explains the proposition.

Excessive rarefaction of air induces immediate deadly fall but condensation and excessive density of air also suffocate and kill animals. This must be discussed. And firstly we must see:

Proposition CXXIV

Why and how water sent into the lungs suffocates the animals.

When the mouth and nose are closed and the throat is constricted, the reciprocal movement of entry and exit of air must be interrupted. Then the effects on the passage of the blood stop and blood circulation on which life depends is impeded.

But it does not seem possible to impede this reciprocal movement in animals who live immersed in water. Water can indeed be sent into and out of the bronchi by the movement of the thorax and the blood can be expressed by the fluidity of the water. Thus circulation can be continued as occurs in fishes. Since they do not present subsequent suffocation at all, it is worth-while investigating the reason for this phenomenon.

Firstly, if a drop of water is introduced into the trachea or if catarrhal mucus provokes such unpleasant convulsive coughing, it is not surprising that filling the trachea with water induces much more severe unpleasant suffocation.

This unpleasant feeling occurs in the trachea and not in the small bronchi since mucus does not provoke coughing as long as it is in the depths of the cavity of the lung. Only when it arrives in larger bronchi does it seem to achieve unpleasant titillation. Therefore, water similarly would neither provoke coughing nor induce suffocation if it stagnated in the small bronchi.

But the error of such reasoning is obvious. In animals immersed in water not only the small bronchi but also the trachea must be full of water. Therefore, considerable irritation and very unpleasant convulsions must result. These must be the more painful the more the effort of Nature at eliminating the inconvenience by coughing is prevented. This effort is prevented because water being much heavier and denser than air cannot be evacuated through the mouth as a consequence of the weak compression of the thorax impeded by the weakening of the diaphragm and the pressure of the water which remains in the cavities of the lungs.

Moreover, stagnation of water in the smallest bronchi, by its weight and density, permanently compresses the vessels full of blood adhering to the bronchi and thus impedes the passage of the blood. Thus, there is interruption of blood circulation without which life cannot continue.

Additionally, there is a lack of frequently repeated driving of particles of air into the blood and the vital oscillatory movement described above thus stops also.

It is true that in animals immersed in water life does not stop immediately as occurs in the vacuum of Torricelli. Actually water can be driven only slowly and with difficulty in the smallest pipes and in the alveoli of the lungs as we experimented with glass tubes. Therefore, air which remains a long time in these spaces can be expanded and constricted repeatedly by the movement of the thorax. This enables sparse and difficult passage and decanting of blood, almost by seeping. The same occurs in suffocation and strangulation.

The course of events is as follows: as soon as the throat is constricted, obturated or filled the most severe and painful anxiety begins which activates and shakes the spirits in extraordinary ways. The muscles react with convulsive movements. Hence, feelings are disturbed and finally become numb. The animals no longer perceive pain although convulsions continue. This was experienced by some people who by chance were saved from suffocation. Finally, after some minutes, the animals die.

Proposition CXXV

Very likely causes of suffocation usually resulting in different manners from thick or too condensed air.

It is obvious that pure and clear air becomes very rarefied or condensed. Condensation results from compression, like in pneumatic blunderbusses, or from the weight of a huge mass of some incumbent air, as mentioned. Sometimes air can be made denser by addition of heterogeneous particles, earthy, aqueous, oleaginous, igneous, saline and countless others.

Animals immediately suffocate in the air of a closed room full of coal smoke, in emanations of vapours of mines of metal and also in the dense air of graves in which corpses or wheat have been kept and in the lowest layer of air in the cave of the lake of Agnani near Pozzuoli. This also occurs in the air usually found in deep wells. All this confirms the suggestion that air when too dense induces suffocation no less than water.

This is true. But I suspect that the effect of suffocation is often erroneously attributed to something which is not its actual cause. Indeed air in the cave of Agnani, in the mines and graves thus mentioned is dense and full of heterogeneous effluvia so that it puts out a candle as does water. At first glance, it seems that there animals die in the same way as they are suffocated by water. But if it was true, they could not be brought back to life by other means than those by which men immersed in water and half drowned are treated. These are suspended by the feet and water is rejected through their mouth, flowing out by its own weight. Dogs half suffocated in the cave of the lake of Agnani are not resuscitated this way. On the contrary, thrown in the lake they immediately come back to life or, retired from the cave and lying on the ground, they recover gradually.

It appears that in this operation the dense and heavy air, like water, cannot be ejected from the lungs by its flowing upwards if the mouth is not facing down and, therefore, the animals are not brought back to life. Thus, some poisonous gas is carried to the brain by the nerves of the nose and throat and there by malignant contagion disturbs the quality of the spirits. As a consequence, the dogs immediately fall moribund. These poisonous particles must be very heavy so that they cannot rise as high as the respirable air in the upper part of the cave.

Similarly, smoke of burning coal does not condense the air of the room so as to provoke suffocation by its density, like water, since it does not put out a fire as does water or the air in the cave of Agnani. It must rather be said that the sulphurous emanations of coal and the smokes prick and contaminate the brain through the nose and induce unconsciousness and finally death. It is true that the air of the room, coming from burning coal, does not kill by its thickness and density since on the contrary it is commonly thought to be rarefied by too much heat. Actually, the abundant igneous particles in the air of a Turkish bath, without smoke of coal, induce anxiety and fainting.

Neither does air moistened considerably by water vapours kill animals. Actually air which remains in the bronchi and pulmonary alveoli is never dry but always wet. It is moistened by exudation of serum which is transformed into vapour by the warmth inside the animal. This is demonstrated by the water vapours which are always present in expired air.

When southern and rainy winds induce clumsiness and apathy, I think that this results from putrid exhalations carried by the winds. These exhalations disturb and blunt the brain or they obture the pores of the skin and thus impede insensible perspiration.

I will not deny, however, that pure air compressed at a maximum, like in a play-ball, would be totally unsuited for breathing. The smallest bronchi and the thin alveoli of Malpighi, if compressed by the great elasticity of such air, would be stretched and torn. This would lead to very unpleasant anxiety. Passage and circulation of blood also would be impeded because expiration would be very difficult as a consequence of the excessive resistance of ambient air.

The immediate function of respiration thus consists of driving and mixing air with blood to maintain the vital movement. After expounding this immediate function it remains to describe another important function which leads to the perfect preservation of the animal. This is the making and formation of blood from which the spirits of the animal are generated, the solid parts are fed and the other operations and motions are carried out.

Proposition CXXVI

What others said on the origin of the qualities.

It is obvious that composite bodies belonging to the sphere of the moon can be transmuted from one state into another, losing their initial structure and faculties and acquiring new ones different from the first. The newly acquired structure, since one ignores in what it consists precisely, is usually described in general terms relating to the aspect rather than to the essence of the thing with such words as act and entelechy or perfection. Nobody denies, however, that such transformation or improvement is acquired by mixing different simple or less composite bodies. The authors also agree with Aristotle that intimate mixing requires that the elements susceptible to be mixed be divided into minute particles so that the different particles unite by mutual contact. This general theory usually is not questioned. Only the origin of the qualities and active faculties which appear in the results of mixings is discussed. Some people think that they are created, passing from not being to being. Others believe that the forces of Nature do not create anything new but that only what was hidden and bound as if it were with ties, appears. Once their ties are dissolved and broken, these substances can act and work in different ways according to the different character and modification which the situation, the position, the order and the shape of the particles allow and require.

This seems to be confirmed not only by correct reasoning but also by countless experiments. Different masks of colours result from pure alteration of the surface. The motive forces which are latent in bodies become active when the ties which retained them are dissolved. Afterwards mixed with others of different characters and modified by a different structure of the mixing, they can exert qualities of a surprising variety.

Chemicals resulting from mixing, fermentation, separation and sublimation of some liquid and solid particles not only present different colours but also bitter, sweet and other flavours. Sometimes active forces are made inert, sometimes they are brought to life wonderfully so that they can act powerfully.

Proposition CXXVII

Instruments and mechanisms by which mixings can be carried out in plants and animals.

To allow for a perfect mixing which gives a composite body of a determined kind, i.e. to obtain such structure and such faculties of action, it is at first necessary to divide the components into minute elements. For solid bodies this is achieved by squeezing and grinding, for soft or liquid substances by driving them into very thin tubes. In this way not only are they subdivided into their smallest elements but, if perhaps some of their parts are conglutinated and curdled, their agglutination will be dissolved by the violent wedging into small and narrow canals. If then these slippery bodies are not homogeneous, the particles of one kind can be separated from particles of a different nature by the particular instruments which art and Nature use now and then. One of these is a sieve in which bird-seeds are separated from barley-corns. Nature also emits sweat through the sieve of the skin of the animal, leaving the blood with which sweat was mixed. Similarly, the roots and vessels of plants, perforated by holes of different shapes, draw from the earth bitter juice in the colocynth, sweet juice in the sugar-cane, acid juice in the apples of Marseille, oily juice in olives, and select different other juices in other plants. Consequently, in this way, if the orifices of the sieve or of the small tubes have the appropriate shape to draw particles of different shapes, the different liquids will be collected in distinct tubes the shapes of which conform to the shapes of the particles of liquid.

Such an instrumental arrangement could not only resolve composite bodies into minute elements and separate elements of one nature from others of another nature, but could also, like chemical resolutions, separate oily particles or spirits which by nature are mobile by dissolving the ties which retained them so that they could freely exert their motive character. Moreover, in the passage through the orifices of the tubes soft and liquid particles can acquire a shape similar to that of the prototypes for which the holes were drilled in the tubes. This applies to lead, wax, and clay, for example. We use the same artifice to create round, corrugated and hollow shapes corresponding to the shape of the holes through which the mass of substance was expressed.

Once this preliminary disposition is achieved, it remains to discover by which instruments, by which artifice and mechanism the particles are recombined in due order to acquire the new conformation and character.

Firstly, among the artefacts I consider the work of a craftsman in mosaics. In such work, stones of different shapes and colours are artfully and orderly adapted and glued together in due places, positions and configurations so that they represent a man, a plant or another object. If these stones were very small and mixed in some liquid as are the pigments of painters, somebody could build an instrument which would by itself carry out the work of the craftsman without laborious arrangement by the hands of the latter. Imagine tubes full of pigments of different colours arranged like a hair net so that the orifices end in due places on the table to be painted. Then simple expression and ejection of the pigments would reproduce the figure of a man and of a plant as pretty as if they had been assembled in their due places by the hands of the craftsman.

I do not hesitate to claim that this artifice is usual and very familiar to Nature.

Every time I ponder that Nature cannot and will not produce and feed plants and develop their leaves and fruits without driving aqueous, aerial and saline particles through a sieve made of countless small holes which give access to the hair of roots, I wonder for what necessity and to what significant advantage Nature always uses roots and tubes. After long pondering on the subject, no other explanation remains than the arrangements thus described, the position and the shape of the particles. This shape is partly acquired by passage through appropriate holes of different shapes in the roots and partly results from a proper distribution of the particles which the texture of the roots conglutinate like a mosaic work. Thus, by some necessity ruled by the Divine Architectural Wisdom, the liquids filtered and carried in a plant induce the configuration and characteristics proper to this plant as we shall show.

This is obviously suggested by the fact that, from the same earth irrigated by the same water, countless different roots, as a result of their difference in shape and texture, give rise to plants different in magnitude, shape, colour, savour, and smell, with countless properties.

It is even more wonderful that the same liquid sucked from the roots of a plant if, on its way through the stem of this plant, it meets by chance roots of moss or of another grass or a twig of another plant parasite of the first and included in its cortex, acquires characteristics totally different from those it had previously.

After this discussion I revert to my subject.

Proposition CXXVIII

After having completed its circuit through the whole body of the animal, blood is carried back to the right ventricle of the heart, considerably altered and unfit for nutrition.

During muscular activity, as long as blood is violently driven from the arteries into the recesses and pores of the flesh, a great deal of its spirituous and aerial particles are dissipated and consumed. Nobody will deny that. It is suggested by the subsequent tiredness, diminution of strength and exhaustion. Moreover, the elements which give the impetus, with countless liquid particles of the blood are lost to feed the solid elements of the animal, bones, tendons, flesh, nerves and spirits. Hence, blood remains less fluid, numb and almost cadaverous, partly as a consequence of the loss mentioned above, partly as a result of the alteration of its structure when some elements have been removed and displaced from their adequate place and position. It can even happen that some of these elements which are prone to union and coagulation curdle. Actually, the red part of the blood coalesces in cold and its fluid and serous part condenses in warmth. Therefore, the quality and correct constitution of blood can be deformed and altered in surprising ways before the blood is brought back to the heart.

Then disordered addition of chyle and lymph in the heart and in the pulmonary artery disturbs the blood even more. Therefore, the blood altered in so many ways is unfit for the nutrition of the animal.

Proposition CXXIX

Mechanism of the making of blood in the pulmonary vein.

I finally arrive at the last and superb function of the lungs and respiration. Firstly, the making of blood in the lungs is not due to some ferment acting there. This results from the fact that no fermenting juices are kept in the lungs since the alveoli of Malpighi are full of air alone and the other vessels are always cleaned by a continuous flow of blood. Therefore, as the division of blood into minute elements can occur in the lungs without fermentation, similarly mixing and making of the blood occur without ferment. As mentioned above, the blood is carried back to the right ventricle considerably altered and deformed. From there it enters the pulmonary artery together with new chyle.

This confused combination is violently driven by the press of the lungs through countless small arterial canals which divide all these liquids into minute particles, separating those which were conglutinated and curdled. Then in the filtering process they receive aerial particles transmitted from the alveoli of Malpighi. At the same time, during their passage, they are forced to conform to certain shapes similar to those of the holes in the tubes. After this preparation which we discussed above, there are other thinner ramifications of the tubes which form a very abundant hair like the extremities of the roots of trees. Well-determined liquids are absorbed by these capillary tubes, for each tube those which correspond to the shape of the orifice of the vessel and can enter it. Soon these vessels which are differently bent and contorted like a net, as Willis observed, arrive at larger veins and end in suitable places to compose a structure similar to a mosaic work. There, they spill the droplets which they contain. In so doing they can make an exact and perfect mixture and the appropriate composition of the blood.

From there the portions of blood rejuvenated, vivified by spirits transmitted by the nerves and activated by the aerial machines are carried to the main trunk of the pulmonary vein which discharges them in the left ventricle. The blood then is ejaculated with a powerful force and distributed through the whole body of the animal.

It seems to me that this hypothesis can be assumed in such a difficult and hidden problem as is the function of respiration.

We shall pass to the conjectural investigation of the components of the blood and of its structure. And first to be led by an analogy from the better known to the lesser known:

Proposition CXXX

Examination of the structure of milk from its analysis.

It appears that milk is composed of two different liquids, one glutinous-fat or caseous and the other aqueous-serous. They can be mixed by dispersion of the particles of one into the liquid of the other, as minute droplets of wine scattered among particles of water compose a mixture or as bird-seeds are mixed with barley-corns.

We must accurately analyse how the caseous and aqueous elements are arranged to compose the liquid consistency of milk and by which mechanism the ferment of rennet dissolves this mixture so that the caseous elements are separated from the serum, unite and coagulate.

Familiar experiments show how this composite liquid can become white. Dust of marble, plaster or wheat dispersed in water creates a uniform white colouring. Similarly, white molecules of caseum can be mixed with water and give the milk its white colour. Separation of the dust mentioned above results from its higher specific weight: this dust is precipitated to the bottom of the vessel. This is not true for milk. The caseous elements are not precipitated but float, move transversely to unite and coagulate in floating curds. It must be concluded that all the elements of milk have the same specific weight as those of water.

In this instance I do not understand how the glutinous elements of milk present in water and separated from each other can unite and coagulate without imagining that they adhere to each other by some magnetic force, which nobody in his senses asserts.

Viscosity of the particles of milk is not enough since these particles, when floating in water separated from each other, can unite only if they approach each other by opposite movements. Then they form bigger molecules. After this union the bigger molecules are more distant from each other and it is more difficult for them to unite.

Neither can the caseous small grains of milk unite as a result of a considerable fermentative boiling. So many and so different opposite movements do not appear during coagulation of milk. On the contrary, we observe that milk coagulates suddenly without any disturbance. The particles of milk are rather scattered and divided by a great effervescence as occurs in any movement of a liquid into another liquid, such as we observe when it rains and when pouring red wine into water. The particles then are scattered like smoke or resemble the expansion of a cloud.

If we conceive the glutinous particles of milk not as round and light molecules but as branchy like tufts of cotton, then when dispersed in water and separate from or contiguous to each other they cannot unite and form a clot without some magnetic force. If the hairy extremities of the tufts are assumed to be intertwined they must by their nature lack a force of contraction. Otherwise they would not remain in their previous liquid state and caseous curds would always float in the serum. Since such coagulation does not occur without addition of rennet, this acid and bitter juice must impart a glutinous contraction force to the caseous particles. This does not seem to happen. The caseous particles of the milk by their nature are glutinous more than enough since they spontaneously coagulate although late.

Moreover, acid and bitter salts are effective not by their glutinous force but rather by their incisive and penetrating ability whenever we observe some viscous juices dissolving and liquefying like in sputum, white of egg and blood serum. These are made friable by acids.

Although vinegar seems to enhance coagulation of clear soup and this is another operation similar to milk coagulation, we do not know whether this happens by chance, by separating and expressing the liquid from the glutinous bodies, as leathers and tendons when boiled and dissolved coagulate into gelatin after separation from water and evaporation of the latter. It thus appears that acids and bitters enhance coagulation of gelatin and milk not by their own constricting action but by removing the obstacles which prevent the action of the constricting force of gluten. On the other hand we know that skin and tendons are very glutinous. To harden gelatin one adds whites of eggs which are very viscous.

Consequently, the structure of milk must be such as to explain the phenomena thus described. For its analysis, I observe that, for two heterogeneous liquids to be able to mix intimately, both must be divided into minute particles and any particle of one kind must touch a particle of the other kind, as bird-seeds and barley-corns are confused in a heap. If one of the liquids is glutinous it cannot be divided into small elements because its cohesion forbids total disjunction of its parts. They can rather expand in very thin layers or membranes between which droplets of the other and very fluid liquid are driven by the impetus of the beating. Forced into the pores of the first liquid, they can form a texture similar to foam which is a mixture of viscous water and air contained in the vesicles of water. In the same way, very fluid water mixed with a more viscous liquid like white of egg and dispersed in its pores can give rise to a new kind of foam loaded with water, by energetic beating. By the same mechanism the gluten of white of egg permits its stretching into membranous phials and opposes their rupture. Therefore, droplets of water can be forced and included into these phials.

This composition of the foam has not been imagined gratuitously but is confirmed by strong arguments. In the same way we can surmise that a foamy structure of milk may result from the mixing of its glutinous juice and its aqueous serum. Droplets of water scattered and included inside the resisting pores of caseum necessarily acquire a spherical shape because of the compression exerted from everywhere. On the contrary the walls of the small pores, as a result of the softness of gluten, can be stretched and expanded into very thin membranes which cannot be torn or easily disrupted because of their viscous resistance. Therefore, it seems possible that the composition of milk is similar to that of foam.

Such structure is very likely since it can easily explain the phenomena thus observed. We indeed understand that the very thin walls of the vesicles which compose the foam of the milk can be corroded, torn and disrupted in a moment by the acid and bitter juice of rennet. As a result, droplets of water flowing from the pores and collected float here and there. At the same time, glutinous skins contract spontaneously as a consequence of their own viscosity, unite and coagulate since they are no longer stretched by the water which previously filled their pores.

The same coagulation can occur in milk without rennet in the same way but later because the foam loaded with air is dissolved little by little. It seems to me that this can be surmised about the structure of milk.

Proposition CXXXI

Blood analysis achieved by distillation seems to be unreliable.

It is admitted that blood can be divided into its components only by chemical distillations. Firstly pure water is extracted from the blood in a still, then some spirits, salts and sulphur; at high temperature, burning spirits are pulled out and several fixed salts remain in the sediment.

This division appears to be suspect for several reasons. The action of fire indeed is but a violent and very quick motion and an outflow of igneous atoms like a very rapid wind. In its blind impetus it attempts at removing any obstacles which would impede its travel. These small atoms of fire do not care to divide bodies which they meet into their heterogeneous elements. Mixtures of very different substances if they cannot withstand the impulse of fire let themselves willingly rise and be expelled upwards as straws, fragments, cinders accumulated with countless other heterogeneous substances are ejected with the smoke from a chimney. The most burning flame, as a result of its velocity, ejects quite a number of intact and unburned grains of nitrate powder upwards from mortars themselves. Therefore, all bodies whatever their nature, whether mixed or separated, are equally easily pushed upwards and sublimated by a fire if their specific weights are equal or if they are equally disjoined from the earthy and composite particles. In other words, all these bodies are equally movable. But if the heaviest and indestructible heterogeneous particles of a mixture, like metals and glass, are subjected to fire, then igneous wedges driven inside these particles succeed in activating them for a while but do neither stretch them nor send them at a distance. However, it may happen by chance that some separation results from fire because the less heavy particles, connected by looser ties, of smaller mass and larger size, can be expelled by the fire together with the smoke, leaving the tenacious heaviest particles with the sediment.

Some elements are separated in the condensing chamber of the still and others remain at the bottom. From this fact one cannot infer that their nature is different. Only these particles of the same salt which are small or have a large shape or a situation easily movable are raised by the force of fire. These are called volatile salts. The particles which are thicker, intertwined with each other and with particles of earth remain at the bottom of the still. They are called fixed salts.

Surprisingly the particles of sulphur and of fire which are extremely inflammable and volatile cannot be separated from the particles of earth but remain caught in their recesses.

Moreover, the evaporated, burned, calcinated and distilled substances are altered considerably by the action of fire because many igneous particles have mixed with them and remain in the pores of these bodies as the famous Boyle and others noticed.

Therefore, chemical analysis of the blood is unreliable. Firstly, in such analysis, the substances which compose the blood not only cannot be obtained pure but they are even changed by expulsion of aqueous elements and addition of igneous particles. Secondly, the proper structure, arrangement and composition of the components of the blood, or rather the mixed composition which they presented previously, does not remain the same and can no longer be recognized because it is dissolved by the action of the fire and its configuration and structure are destroyed.

These chemicals, salt, sulphur and mercury are very general principles from which not only the blood but all mixtures are made of. Consequently, from this chemical work no particular composition of the blood is deduced and particularly what the components of the blood are, its nature, the properties, faculties and structures of its components.

However, I am not the one who denies that wonderful operations are sometimes carried out by chemical research. But I deny that chemical reactions are always and everywhere useful, above all in our instance concerning the analysis of blood.

Proposition CXXXII

Resolution of blood into its components and structure of its composition.

As shown by experience, two components of the blood above all spontaneously separate from each other when blood is extravasated and is left quiet. Albumen is separated immediately from the red elements in the same way as the serum of milk is separated from its curdled caseous part. The red part of the blood also coagulates spontaneously, surrounded by albumen which retains its fluidity so that it looks exactly like the serum of milk.

The two parts thus mentioned of the milk as well as of the blood are unaltered and pure. This appears from the fact that they separate from each other spontaneously and immediately without perceptible action of any external cause, without fermentative ebullition, without addition or subtraction of any material substance.

Both of the primary parts are further resolved into two other different substances. The albumen consists of some viscous liquid very similar to the white of egg. Both retain their transparency as long as they are fluid. Both coagulate by the action of fire. They become opaque and white. They have very similar flavour and smell. The albumen of blood, however, is impregnated by copious aqueous serum containing salts of ammonia, of tartar and others. These salts are easily collected by evaporation in a still.

The primary red part consists of some very glutinous slippery substance and some red juice which when intertwined together spontaneously coagulate in curds. After repeated washings with water following coagulation, a glutinous part appears collected in white fibres or in reticulated membranes similar to those of tendons. This is clear in the blood of doves. The curdled red part cleaned by water is reduced into a red powder which remains at the bottom of the water.

Besides these five components, many particles of air are mixed with the blood in the lungs, together with spirituous juices provided by the nerves. This results from what was said above.

The blood made of the heterogeneous elements thus described, as long as it moves in the vessels of a live animal, is foamy like milk. This seems very likely to me. I indeed could not imagine any other structure which would explain all the phenomena. As there is liquid serum in milk, very liquid serous albumen is found in blood, so similar to water that all the ancient authors believed that the albumen which is separated from the blood was urine until D. Dumel, Professor at Padua, recently discovered this error of the senses.

As there is a coagulable juice in the caseum of milk, a viscous and glutinous juice is found in blood. After coagulation this juice changes into fibres or reticulated membranes. These fibres so condensed do not pre-exist in the vessels of a live animal. This is suggested by the fact that these fibres and white membranes of the blood, when inspected with a microscope, are thicker than the capillary blood vessels and, therefore, could flow neither in nor out of these vessels since they would obstruct their narrow orifices at least with the length of the filaments and the width of the membranes. Thus one must conclude that the glutinous albumen of the blood retains its slippery and liquid consistency as long as it moves in a live animal.

As a foamy conformation results from mixing of gluten and water in milk, there is general agreement that, in the blood of a live animal, the very liquid white serum is divided into minute globules by the continuous squeezing and beating which occur during circulation. These minute globules are scattered all over the glutinous liquid which, because of its tenacious viscosity, can be stretched into thin membranes but not divided into separate globules. As a consequence, the vesicles have a foamy structure. They are full of albuminous juice, copiously diluted by many particles of water mixed with this juice.

A very likely rupture of the caseous vesicles in milk leads to their coagulation and the outflow of particles of water. In extravasated blood also the membranous vesicles can spontaneously contract and break as a result of the thinness of their walls the glutinous substance of which quickly acquires consistency. Thus, the foamy structure of blood is dissolved as occurs in vulgar air foam. Hence, the serous granules flow out and unite. Soon the viscous vesicles deprived of the serous wedges which stretched them can freely come closer together and coagulate as required by the nature of gluten. So the curdled red part of blood floats in its serous albumen. Therefore, it seems that the coagulation of milk and the separation from its serum can be explained only if the structure of milk is foamy. It is also likely that the structure of blood is foamy. This hypothesis precisely explains all the phenomena.

Order would require a discussion on the secondary functions of respiration. Everybody knows that the voice and thus speaking are produced by air expired from the lungs. Excrements are discharged by holding one's breath and compressing the abdomen. Inspired air is used for sneezing, coughing, yawning, spitting, vomiting. Many famous authors wrote abundantly about these phenomena. It would not be sensible to repeat what others have said. Little can be added which was not said by others hitherto or on which I disagree with them. What could be added was mentioned incidentally above or will be said below. I thus pass to another subject.

Chapter IX

On cleaning of the blood in the kidneys.

After dealing at length with the making of blood in the lungs, we shall discuss the manner in which blood is purified in other viscera. Firstly we shall describe the function of the kidneys.

Proposition CXXXIII

Kidneys do not contribute directly to the life of the animal.

Everybody knows that blood is purified in the kidneys where it is deprived of an excess of serum with excremental elements of salt and tartar. It thus appears that the kidneys are not organs which directly contribute to the preservation of life as do the heart and lungs which produce and maintain the vital movement of blood.

This removal of serum from the blood no doubt does not seem so necessary that, if it is neglected for some short time, the blood becomes completely useless. Indeed the greatest part of the blood mass, nine-tenths of it, does not pass through the kidneys during one circuit of all the mass of blood through the body. Therefore, the nine-tenths of the blood carry with them the urinary excrements to the head, arms, thorax, liver and spleen. If this soiled blood was harmful and unsuited for nutrition Nature would act stupidly by bringing it to the principal and noblest seat of the spirits, i.e. to the brain, so that there it would feed the spirits with dirty and muddy juice and would soil other principal parts, thorax and viscera with urinary excrements. Since such nonsense is incredible it must be admitted that urinary humour if not in excess is useful in the blood. If there is too much and for too long a time it imparts to the blood excessive fluidity and bitter fermentation. A little more serum in the blood is useful and perhaps necessary for the brain and the upper part of the abdomen but it is harmful for the muscles and the other parts of the body. On the one hand, diluted and more fluid blood can enter and flow more easily through narrow vessels. It thus can scrape them and carry away with it alkalified salts of which there is abundance in the former parts. On the other hand, Nature provides the brain, viscera and chest with many excretory vessels, salivary ducts and lymphatic vessels so that they absorb this abundant serum and they filter it after it has been used.

Proposition CXXXIV

Necessity of drinking much.

Nobody would divert into his own home a source so abundant that it would not only exceed the domestic needs but also require considerable work and expenses permanently to dry the rooms and court. It would have been worthwhile to send into the house a source which would not exceed the needs. Nature thus seems to be very wrong in filling the animals with too much drink which they are to reject permanently with considerable inconvenience and danger.

Since it is wrong to attribute our ignorance to Nature, we shall attempt to investigate the utility and advantages which Nature retrieves from the absorption, retention and expression of the aqueous serous humour by the kidneys. Firstly, water is drunk copiously so that, mixed with dry food in the stomach, it helps its maceration and fermentation. This soon composes the chylous juice by extracting particles of food and making them liquid. This constitutes the vehicle of the chyle and nutritive juice as it is of blood and spirituous juice, and also excrements. A heap of dry sand cannot be driven in and flow through narrow canals before being made slippery by addition of much water. Similarly, the solid molecules of food, chyle, blood and other liquids of the animal could not be moved and flow through the narrow canals of the body of the animal because of their roughness and the irregularities and angles of their various shapes. They thus had to acquire slipperiness and fluidity by addition of water. Nature achieves this fluidity by maintaining liquid in the stomach, intestines, chyle and blood. If this liquid, after entering, immediately left the body and thus was eliminated, continuous blood circulation which maintains life would be interrupted and stop.

With the permanence of aqueous serum in the blood, flesh and viscera, Nature achieves another obvious advantage. Alkalified salts and salts of tartar are absorbed by this serum. Indeed the fixed salts which agglutinate tenaciously and which are as if they were welded in the fibres and pores of the flesh and viscera cannot be removed by any other means than washing out and soaking. We experience that in meat and fishes preserved in salt since the structure and arrangement of the particles of water are so appropriate to absorb and retain tenaciously the molecules of salt in their pores that they are forced mechanically to catch the salts avidly and to retain them, as we showed previously. To allow for absorbing drinking water more easily and conveniently, Nature devised a way to keep water drunk during a period of rest but warmed in the viscera and activated by blood circulation so that, by repeated blows and beating, it penetrates all the diverticula in solid and liquid elements and thus can suck and absorb salts hidden everywhere.

After being loaded with salts, aqueous serum would be very harmful either because it would sting and prick the very sensitive membranes and nerves by its bitterness or because after long fermentation it would become harmfully corrupted, which is against Nature. It was thus necessary to eject such harmful excrement from the body of the animal. The mechanical separation occurs in the kidneys. Its mechanism will be described.

We demonstrated the necessity and extensively explained the usefulness of this excretion. It remains to investigate by what means liquid serum is separated from blood mechanically in the kidneys. For an accurate description, we must first consider the means used by Nature to separate two heterogeneous liquids which previously were mixed.

Proposition CXXXV

Separating action of fermentation.

One calls fermentation an internal movement of the elements of a composite body which are activated by their own motive force or by that of another additional body. This internal disturbance is followed by several different effects. Among them we shall consider only the dissolution of the texture of the composite body by which the heterogeneous elements can be parted from each other and then put in different separate places.

There are two kinds of mixing. Either the heterogeneous particles are simply in contact with each other or they are bound by tight links. In the former instance separation is achieved by one operation of displacement alone. In a heap of mixed bird-seeds and wheat grains, linking between the grains has not to be disrupted. Since they are simply in contact there is no uniting force. One has only to collect the bird-seeds, using either their own motive force or some external impulse, in a place separated from that where wheat grains are collected.

When minute particles of water from a cloud, floating in the air gradually collect into bigger drops, they fall by their own weight and thus separate from air. Similarly, minute bubbles of air or particles of oil, scattered in water, collect little by little and are pushed upwards to the surface of the water.

In a still, water is separated from salts and from composite plants by an external force. Fire transforms water into vapour in the boiler and cold condenses vapour into water in the condensing chamber.

Separation can also be carried out in another way, by an intrinsic force and by an external force. This is achieved either by direct action or by removing impediments. This is obvious in almost all chemical precipitations and sublimations in which the motive force of the fire or of an additional ferment dissolves the earthy and viscous ties in the pores and recesses of plants and minerals. Hence, their prisons being disrupted and dissolved, the igneous elements or spirits and other mobile parts which were included can freely exert their motive character and separate.

Sulphurous, oleous liquids or salts divided into minute particles can remain floating in water in two ways: either, as a consequence of the smallness of their mass the excess of specific weight of one liquid is smaller than the resistance which links mutually the hairs of the liquids as said above; or the particles of oil linked with heavier salts form molecules of the same specific weight as water. In this way mixtures can be combined from wax and sand, from oil and salts of lye into soap. They float everywhere in water. To separate oil or salts from water the equality of the specific weights must be altered either by increasing or decreasing the weight of the liquid milieu or that of the floating bodies. This can be achieved in many ways, by adding or subtracting igneous particles, or by rarefying the salines, or by condensing the liquid milieu or the floating bodies themselves, or by dissolving the ties between the floating elements. These actions are carried out by some internal movement or by fermentation. Then the mixed elements are separated spontaneously from the liquid by their own weight and regain their places.

For the same reason, minimal particles of metal corroded by salts of aqua fortis like by rasps, float in the water as a result of their smallness, as mentioned above. Addition of another liquid less heavy than aqua fortis can increase the fluidity of the latter or decrease its specific weight. Metallic powder then precipitates to the bottom of the water as a consequence of its excess weight. On the contrary, if the solvent is more saline than oil of tartar, its particles by their impulse can carry with them the floating and balanced particles of metal in their movement downwards.

Separation of the serum from the milk seems to be somewhat different. Acid ferment, by corroding and tearing the glutinous skins of the foamy vesicles of milk, enables the small grains of aqueous serum to get out. They were imprisoned in the vesicles and so water was accumulated round the caseous curds.

It can be deduced from all this that simple action of ferment by itself can dissolve by internal movement a tight and firm combination of heterogeneous elements as if they were not united by ties other than simple contact. However, to separate all the particles of one kind from those of another kind and collect them in different places, another motive cause is required, different from ferment. This cause is the very different weights of the heterogeneous liquids which can be separated from each other by natural motions.

Proposition CXXXVI

Fermentative separation cannot occur in narrow tubes but well in large vessels.

Fermentation in fluids occurs with some boiling and rarefaction of the elements of the mixture, as experience shows. Such effervescence cannot occur without some fluid elements moving among other equally fluid elements. This is possible only if the latter make place for the former. The latter are forced by a swirling movement to take the place left by the former. Consequently, these circumvolutions must be repeated often in the process of fermentation.

Such opposite and swirling movement by heterogeneous fluids cannot occur in very narrow tubes. We experiment with thin glass tubes in which drops of water, red wine, quicksilver and air can never mix together by interchanging their previous positions in opposite and swirling movements whatever the number of times they are shaken and insufflated. This is due to the fact that the fluids are not made of immaterial molecules but of many shaped, hard and rugged particles united by simple contact. Therefore, they cannot be activated by opposite and swirling movements in very narrow tubes because their different adjacent shapes compose an agglomerate very similar to a mosaic work. Hence, one molecule cannot move above another in a very exiguous place in which they cannot lengthen to free the way for the passage of their rough extremities. This obstacle is even increased by the internal asperities of the tubes. Consequently, fermentation, if it consisted of simple opposite and swirling movement of the elements, could in no way occur in these narrow tubes.

Actually, fermentation requires overturning and rarefaction of the elements which need much more space as we observe in boiling. Thus if the initial simple swirling displacement could not occur in an exiguous place, fermentative movement is even more impossible which requires, besides local swirling, rarefaction and swelling.

And even if fermentation was not impossible in narrow tubes, it would be stupid and inappropriate. Therefore, one can hardly believe that wise Nature would carry it out. Not only swirling and rarefying movements would be very difficult to carry out in narrow places but such places would not be able to contain all the juice which must ferment if the tubes were not multiplied. Then see how difficult and laborious it is to carry the fermentative juice, distribute it and put it in so many distinct and separate small places and afterwards to undo the work by using more and repeated labour. Once the fermentation is completed, the heterogeneous juices would indeed have to be collected from so many separate small vessels and each of them replaced in its own large vessel. These complicated works, useless for attaining the goal of fermentation and separation, are avoided completely if, instead of narrow vessels, a huge vessel able to contain all the fermentable juice is used.

Proposition CXXXVII

Separation of heterogeneous liquids by fermentation cannot be carried out at once but only slowly unless the liquids are very inflammable and spirituous.

Many opposite and swirling movements of one liquid into another must be carried out to enable them to ferment, i.e. to boil, to rarefy, to subdivide and to separate. Firstly, the ferment juice must diffuse and spread in all the elements of the liquid mixture so as to touch them directly or indirectly by altering intermediate elements since there is no physical action without contact.

Secondly, after contact has been made, the mixed elements must be pushed by the force of the ferment to be able to separate and be disjoined from each other. Such separation, however, is impossible without opposite and swirling movements by boiling. It is thus carried out by rushing from one place to another.

Thirdly, after the separation of the heterogeneous elements another movement is required to collect in one place the separated elements which are homogeneous.

It is impossible for all these movements to occur at once like a thunderbolt. They can be carried out slowly during some length of time, as experience shows in maceration and fermentation of must and plants. These last not only hours but even whole days. The reason is that a liquid cannot move freely into another. Necessarily by mutual strikings and repellings, they impede and slow down each other, either as a consequence of the asperities of their shapes or by their difficulty at bending. Because of their fluidity the particles are spread like smoke which is displaced and moves at a very slow pace.

There is no slowness, however, with spirituous ferments and mixtures and with those remarkably reactive such as gun powder ignited by a flame. We do not deal with these now.

Proposition CXXXVIII

Separation of mixed heterogeneous humours can occur without any fermentation and be achieved mechanically by the configuration of the small vessels.

As shown above, fluidity of humours cannot be conceived without actual division of their masses into minute hard, firm and shaped particles united not by tenacious ties but by simple contact so that any of them can be activated, flow and revolve about those which remain immobile or move differently. There is no other way of explaining the facility with which liquids flow and mix together.

The different natures and properties of liquids no doubt depend on the different consistency, structure, configuration and movement of the molecules of which the liquids are made. The molecules of water are all homogeneous, similar to each other and different from those of oil and other liquids. Hence there is general agreement that the particles of these liquids are different in structure, magnitude and configuration.

If the molecules of two heterogeneous liquids are equally slippery so that they can mix by simple contact, then their mixture can be compared with mixed bird-seeds and barley-corns. An example is water and oil divided into minute particles. Mixed bird-seeds and barley-corns can be separated using a sieve. Similarly, water and oil but not air pass through the pores of leather and wood. Quicksilver penetrates through the pores of gold but not water, oil or air. The liquids thus mentioned can be separated (not less than grains) from the other different liquids with which they were mixed, as a result of the structure of the sieve, without any fermentation. As grains pass intact through a sieve, similarly water and oil pass pure and unaltered through the pores of wood or of leather.

It is for a mechanical reason that some liquids penetrate and not others: the shapes of their molecules correspond to the shapes of the small pores through which they can pass. Another liquid with molecules of discordant shapes does not pass.

Often mixed liquids are not equally slippery and their particles are not united by simple contact but are bound by tenacious ties such that two components make a third substance. The result from the mixing of several liquids is a mixed liquid the nature of which is different from that of its components. The texture of such a mixture cannot be dissolved so easily as when there is simple contact. Although such connection can be dissolved by fermentation, the simple mechanical structure of a sieve made of funnels like the roots of a tree, however, is sufficient by itself to replace fermentation and to separate the heterogeneous molecules. We shall demonstrate it.

To make it clearer, I notice that fermentation which dissolves firm connection between heterogeneous molecules does not occur without the presence of ferment, i.e. a motive cause. The mechanism consists of driving mobile particles between linked molecules which constitute clusters. These clusters are divided and separated by a wedging action. Therefore, the liquid to be dissolved necessarily must be contained in a large vessel, as mentioned above.

On the contrary, in the absence of ferment, if these heterogeneous conglutinated molecules, i.e. this same mixed liquid, are pushed violently into the narrow pores of a sieve or into non-dilatable capillary funnels, then the wedging action is exerted in a reversed way. Hard and pointed wedges tear and dilate soft and fragile pores. But liquid and soft wedges undergo squeezing and corrosion when passing through the narrow hard passages of a sieve or of funnels. The heterogeneous particles forming the clusters are like liquid wedges. However tightly connected and linked they are, they must be disjoined and separated from each other by the narrow passages through which they are forced with quick impetus since they cannot pass through the very narrow canals if they remain united.

This separation of the heterogeneous particles of a mixed liquid which results from the necessity of passing through narrow passages obviously replaces fermentation and is more economical. Whereas fermentation is slow, this disruption and separation can be carried out at once. This occurs the more easily the more violently the mixed liquid is driven through the funnels. Therefore, the advantage of separating heterogeneous juices can be obtained more quickly, with less instruments and less work, by simple passage through narrow vessels. Such was the proposition.

Proposition CXXXIX

In the kidneys serum is not separated from blood by fermentation.

We discussed at length the composition of the blood and the artifices by which heterogeneous liquids can be separated from each other. We now can more reliably describe how serum is separated from blood mechanically in the kidneys.

It is superfluous to refute this old attractive faculty with which some people dreamed that the kidneys attract the urinary humour by some magnetic virtue. We showed that there is no attraction in Nature and almost everybody finds such attraction ridiculous now.

Fermentation actually must not be carried out at an equal pace. Distinguished and famous authors stubbornly claim that urine cannot be separated from blood without some ferment acting in the glands of the kidneys. I shall bring forwards the reasons why I cannot agree with this opinion.

Firstly, urine, although mixed with all the components of the blood, for the most part is dispersed in the albumen of the blood and is contained in minute vesicles of albumen. These vesicles do not seem to be resistant while the blood circulates through the vessels of the live animal since it appears that the substance of albumen is so liquid that it is not distinguished from pure water. This has been recognized since antiquity. Hence, it is surmised that the particles of water are not connected and bound to the liquid albumen of the blood by tenacious and strong ties but they are only in mutual contact in the same way as particles of air or oil scattered in water are mixed with the latter by no stronger ties than contact. The action of ferment actually consists of dissolving tenacious ties connecting heterogeneous fluids. Therefore, where there is no such connection the action of ferment is useless. Consequently, it would be stupid to use ferment in the kidneys to separate the urine from the albumen of the blood which is already separated since they are bound only by contact. But, however tightly and tenaciously urine is connected with blood, severing and separation are achieved in the kidneys by another cause, without formation of ferment. Severing of the ties must be carried out mechanically in the kidneys as a result of the violent wedging of blood into the very narrow tubes of the arteries which end in the renal lumps. Therefore, Nature would act stupidly if she added ferment to no avail or at least to sweat in a long, difficult and nonsensical work in order to achieve a separation which she was able to achieve much more economically.

Moreover, if there was ferment in the glands of the kidneys (although it does not appear there) and if the mechanism described above was missing, I claim that fermentative separation of urine from blood would be completely impossible. Firstly, commotion, contortions, rarefaction and boiling which are needed for fermentation would be impossible in the very narrow pores of the kidney glands.

Secondly, the renal glands ought to have a huge capacity for the fermentation to be carried out conveniently. It is thus impossible that such fermentation occurs. Blood circulates very quickly and steadily like a torrent from the arteries into the glandular lumps of the kidneys. Therefore, blood does not stay in these glands but passes through them rapidly. Even if its course was not continuous, it would be carried out at least during one pulsation of the heart. But, since the arteries by their peristaltic movement express and push the blood during the heart diastole, blood flow through the kidney is steady and quick. There is thus no delay in the kidneys and, during such momentary passage, the activated movement of fermentation cannot occur, as was shown. Consequently, it is impossible that urine be separated from the blood in the kidneys by way of fermentation.

This is confirmed by the fact that immediately after copious drinking urine is excreted abundantly. I do not understand how such important and complicated fermentation could be achieved so quickly. If this is impossible in large vessels, as said above, it is even more difficult that this be carried out with so much care in the pores of the glands of the kidneys.

It seems to me that the fermentation which usually takes place in the stomach and in the recesses of glands misled some. These famous authors did not notice that even in sufficient space fermentation is achieved by long contact of the mixed juices with ferment and stagnation of these juices. But in the kidneys and in the other glands without obstacles, such fermentation is impossible because the fermentation cannot be carried out during the momentary passage of blood, like a stream.

Proposition CXL

Urine is separated from the blood in the kidneys mechanically as a result of the narrowness and configuration of the vessels.

The general mechanism used by Nature in the kidneys must now be described. Firstly, the cause of the separation of serum from the blood is the impulse and powerful movement of the blood through the milking arteries the existence of which is not questioned whereas ferment exists only in the opinion of some people.

Secondly, the narrowness of the capillary arteries and of the pores of the renal lumps is obvious. This narrowness and the momentary passage of blood through these canals forbid and oppose any fermentation but contribute admirably to the division of blood into its elements. The quicker and more energetically wedges of blood are pushed into the canals, the more easily they are broken into minute particles. Hence, the elements of urine separated from those of blood to which they were united by simple contact, flow into the pores and interstices of the renal glands. Then who will admit that the elements of blood are selected and separated from the elements of water and reset in separate receptacles by virtue of some magnetism or ferment, as if it were by some occult means? Certainly, not wanting to dream of miracles, we insist on saying that there are two kinds of holes like in a sieve. There are venous holes which absorb the particles of blood alone because of their appropriate configuration, and not those of water. There are also the proper vessels of the kidneys which have shapes suited to absorb the particles of water and not those of blood. This operation is carried out generally in plants by Nature, as mentioned above. Even some people who assert the presence of ferments in the kidneys, compelled by this truth, admit this last point.

Proposition CXLI

Mention and refutation of different opinions.

To confirm the theory which we explained above it is worth-while to refute the objections brought forward by others.

Firstly, a famous modern author says that precipitation of blood is provoked by the force of some rennet separating the serum from the blood as the serum is separated from the milk by some rennet. The explanation is: the substances which induce coagulation of milk are also diuretic; therefore, since by making blood flow they precipitate more, they induce an abundance of urine.

Who does not see that the obscurity of this reasoning originates from the familiar method of discussion of the chemists for whom it is enough to mention these metaphoric words of fusion and precipitation? They do not care about the means and mechanism by which such fusion and precipitation are achieved. Therefore, they ignore whether they can apply to the matter in discussion.

The author thus mentioned intended to demonstrate that urine is separated from blood in the kidneys by fermentation. To this end he assumed that diuretics induce prolonged flow of urine by way of fermentation by rennet in the kidneys. But how does he demonstrate this? Because, he says, the same diuretics which separate the serum from milk equally separate urine from blood. But what does he conclude from this? He says that diuretics separate serum from milk by inducing coagulation. This is true. But then do diuretics not separate urine from blood by coagulating blood? Certainly not. On the contrary, he concedes that diuretics induce profusion of urine by making blood flow or making it more fluid. According to this concession, diuretics separate urine from the blood not by way of the same fermentation which they exert in the milk but by a completely different action. Therefore, one cannot conclude that separation of urine is carried out by way of fermentation.

Moreover, ramification of the vessels in the kidneys, like the roots of a tree, would be superfluous, useless and even harmful. Indeed serum is separated from milk by way of rennet most easily in a huge vessel. On the contrary, such operation is almost impossible in very narrow tubes, without delayed and interrupted passage. Because of the narrowness of the vessels, ferment cannot be driven almost instantaneously into the mass of milk to tear its foamy vesicles and the elements of milk cannot move in these structures, the particles of water leaving the glutinous particles. Therefore, the foamy structure of milk cannot be dissolved. Hence the viscous elements would not coagulate nor the aqueous elements be collected.

He misuses the word rennet and contradicts himself by saying that precipitation of blood does not occur by coagulation like that of milk but in the same way as excrements are separated from liquids in must and urine and collect at the bottom of the vessel. But the impassable difficulty remains of the narrowness of the small canals of the kidneys. Separation and collection of the excrements cannot be achieved in these except perhaps after a long time, but not during the very quick passage of the blood through the kidneys.

Even if this operation was possible, some capillary canals would be required the orifices of which would be shaped so that they would absorb urine but not blood and some others which would absorb blood but not water. If on the one hand this must be admitted to be necessary and on the other hand such structure of the capillary vessels is sufficient by itself to select and separate the aqueous elements of urine, leaving the elements of blood (without any precipitation by fermentation), as easily as sweat flows through the pores of the skin, leaving the particles of blood with which it was mixed in the veins, then what would be the purpose of such an expense in a laborious and useless apparatus and in fermentation?

He claims that diuretics are of the same kind as what induces coagulation of milk. It appears that blood does not coagulate like milk. However, it must be recognized that diuresis is not known and has not been investigated. Physicians attribute it to different causes. Perhaps such a quantity of urine is produced elsewhere than in the kidneys and by another mechanism. Then what we have described would not be the cause of the effect. Perhaps what is said to possess a diuretic force dissolves and liquefies the pituitous concretions present in the pores of the flesh or opens the way so that aqueous juices can be filtered into the veins. Afterwards the structure of the vessels of the kidneys is able by itself to absorb and discharge urine.

Proposition CXLII

Answers to other objections.

Now I arrive at other objections written by a very famous anatomist against our opinion.

This anatomist self-assuredly finds "very obscure the way in which urine is separated in the kidneys. Although (he says) all what follows the work of the glands (i.e. the kidneys) is sensible, since, however, this small and simple structure of the canals of the glands is hidden, we can ponder this structure so much that we probably solve the problem. This machine must carry out the separation by its internal shape. Actually one doubts whether it corresponds to those which we use to human ends and which we depict as very similar. Although the structures of sponge, sieve, riddle and tubes appear to be analogous, it is however very difficult to assimilate the texture of the kidneys to that structure whatever the similarities. Nature is very fecund in her ways of action, machines which we ignored are found which nobody yet understands."

To discover whether there is obscurity in the subject itself or in the discussions, the state of the question must be presented. It is asked by what means and what instruments urine is separated from blood in the kidneys: is it by fermentation or rather by the sieve of the kidneys or by both? It seems that the author in this text attempts to refute the second opinion which he appears to assert distinctly and denies at the same time. He says indeed: "it is necessary that this machine (meaning the kidneys) achieves the separation by its internal configuration." Can the internal configuration of the kidneys be anything else than the shape of the canals? I ask him. He thus concedes what he attempts to refute when he recognizes the necessity of such an operation. But, he says, we do not know the exact and precise structure of the canals: are they round-shaped or angular? What does it matter? It is sufficient to know that separation of urine occurs by expression through the openings. Only this was questioned. Cloth, skin and flesh are porous like a sieve although it is not very apparent and humours pass through them. This is suggested by a most reliable sign: we see water and sweat penetrate through the skin and blood pass through the flesh and be absorbed by the veins. This results from the preconceived opinion that bodies cannot penetrate each other. But from the passage of water through a cloth or leather, one must necessarily conclude that cloth or leather are perforated by holes and that water has flowed through them.

As a result of the variety of shapes some holes can absorb this liquid but not that one. This is suggested by an obvious sign. Indeed we see that sweat passes through the pores of the skin but not blood with which sweat was initially mixed; water penetrates through the pores of wood but not air or mercury;

water and quicksilver pass through the skin of a bladder but not air; quicksilver is driven into the pores of gold but not water; and in the kidneys aqueous serum penetrates through the urinary tubes without blood whereas pure blood is absorbed into the veins. It thus can be concluded that such heterogeneous liquids are separated mechanically because of the structure of the orifices of the vessels comparable to a sieve. This results from the fact that liquids are not made of immaterial particles but of many shaped and hard corpuscules. These corpuscules can pass only through holes corresponding to their shape but not through holes which are shaped differently. Although the shapes of the liquid molecules and holes are invisible because of their smallness, and although secretion of the liquids does not appear, they must be admitted, however, since their existence has been suggested sufficiently above by a very obvious sign, the effect of the penetration.

But he insists: "Nature is very fecund in her ways of action and machines are discovered which were ignored and which cannot be understood".

It is very true that we ignore countless actions of wise Nature who wanted them to remain hidden. But for those which she showed openly, she is far from enjoying multiple and different machineries. On the contrary she operates with the simplest and easiest means and instruments. For example, plants live in one way, by sucking their nutritive juice from the ground through their roots. Animals feed in one way, by eating their food and digesting it in their stomach and intestines. No animal can live without a heart. No displacement can be carried out without fibres and muscles. Nature thus always carries out similar operations without a multiplicity and variety of instruments but using one mechanism only and the simplest machine as required by the circumstances. The habit of Nature is invariable. She always works with the simplest, economical and necessary means. On the other hand, as a consequence of their shape, the molecules of liquids can penetrate through the appropriate holes and not through holes which do not correspond to their shape. Aqueous serum is absorbed in the urinary tubes but not blood. It must be concluded that Nature acts in the kidneys according to this one law of necessity.

Do not say that the luxury of Nature appears in the multiplicity of flowers and plants. Yourself are rather mistaken or deceived since Nature always acts seriously. She never deviates from her ends and always carries out the shortest and simplest operation. If sometimes she seems to play by discarding her usual actions, such variation itself is due to some new material requirement which Nature is obliged to meet. Thus, Nature never plays spontaneously and does not like changes.

As a conclusion, every time the necessity of some operation is perceived so that nothing else can be done because of material or mechanical requirement, Nature no doubt acts according to these laws of necessity. Inversely, from the fact that Nature carries out some operation, we must assert that this action is the simplest and is carried out according to the material and mechanical laws of necessity and that it is impossible to do otherwise or better.

The particles of aqueous serum which are hard and shaped (as we showed) are absorbed and filter through the urinary tubes of the kidneys but the corpuscules of blood are excluded from these tubes. We thus must assert that the molecules of urine are absorbed in these vessels by material and mechanical necessity and it is impossible that particles of blood be admitted in these same vessels.

After discarding fermentation because it is difficult and impossible, can this necessity be anything else than the similarity of shape between the vessel and the molecules of serum, and the dissimilarity of the particles of blood? Certainly not, if we do not introduce in the discussion ambiguities and mysteries, thus deviating from the royal way of philosophy.

The author continues "I am amazed that so many and so different bodies are separated in these glands in the state of Nature (meaning in good health) since water leaves with sulphurous salines and similar particles, and in disease even the remnants of abscesses and sometimes the dirt of the whole body whereas the particles of blood are retained. There are many of these and those particles are perhaps of different shapes. And, however, I think Nature designed the smallest and simplest structure of these glands."

Firstly, experience shows that water avidly absorbs saline corpuscules and retains them tightly in its pores or interstices. The shape of the droplets of soft water is not altered by salts caught in its pores. This is suggested by the fact that water loaded with salts enters the same pores as soft water. Tasteless and salt water get through the pores of hard wood and flesh with equal facility so that the internal parts of a beam floating in the sea are impregnated by salts. Other dyes and pigments mixed with water do not penetrate wood in the same way. Hence it is not surprising that urine loaded with salts can penetrate through the urinary tubes contrary to blood. The particles of blood are separated from the aqueous serum as a result of the narrowness of the arterial funnels but not salts because the particles of blood are very different from water as far as structure and configuration are concerned. But salts are driven into the pores of water so precisely that they do not alter at all the external configuration of the droplets of water.

Not surprisingly, in disease the remnants of abscesses and other dirts of the body mixed with serum are urinated whereas the elements of blood are retained. Then the shapes of the renal tubes can be torn or dilated more than normally. As a consequence, other dirts of the body consisting of even more minute molecules can be accepted by the enlarged canals. These retain, for example, the same round configuration, very different from the branchy shapes of the molecules of blood and, as before, do not accept blood. Similarly, not only bird-seeds, hemp grains, linseeds but also sand and angular sawdust smaller than bird-seeds can get through the round-shaped expanded holes of a sieve but bigger and oblong barley-corns cannot enter the same holes even if these are somewhat enlarged.

But I wonder that this author who describes as very obscure the separation of serum in the kidneys and its filtering because the holes of the vessels cannot be seen although they are confirmed by their effect and by an obvious sign, then so self-assuredly says that he observed that urine became tinged with blood sometimes when blood was more fluid than normal and boiling by the force of an excessive movement or of a violent activation. The blood in the urine coagulates at the bottom of the vessel. Hence and how could he observe that blood is liquefied and boils as a result of a movement more violent than normal? Blood never boils, is activated or made more fluid, when flowing from a wound or in blood-letting or after running or after the body has been shaken. Only after too much drinking is blood usually more diluted. It is more likely that some capillary veins are torn by violent activation or shaking of the body. Then some drops of blood can tinge the urine in the renal pelvis or in the urethra. This is confirmed by the story of this author. He says indeed that "there was bleeding and swelling in the testicles and other parts." This plainly indicates rupture of vessels and not boiling of the blood.

The famous author also observes that, after the different elements have been excreted with urine, similar elements remain in the blood. Therefore, some parts of the same salts and excrements are absorbed in the urinary tubes. The others mixed with the blood are excluded. Seemingly it can be understood that this would not occur if the separation of urine in the kidneys resulted from the configuration of the vessels because these salts would be provided with the same and only shape of the tubes of the kidneys. Thus either all would be excreted with urine or all would remain in the blood.

This difficult point can be explained by the mechanism by which blood is cleaned in the kidneys.

Imagine that the water of a torrent is muddy and at the side of the torrent there is a well in which water can be cleaned by filtering through clay as through a sieve. Then the muds will not be separated from all the water of the torrent but only from that portion which was absorbed by the well. Similarly the blood of the torrent in the aorta is soiled by aqueous serum and salts. A small part of it is deviated and enters the milking renal arteries to be cleaned from its aqueous dirts and salines in the kidneys. During one blood circuit, a tenth of all the mass of blood is cleaned, as was shown. Although circulation repeats itself again and again, some parts of the blood, however, do not pass through the kidneys for as long as a day. During this interval of time the salts with the blood can exert their effects in the different parts of the animal. Many salts and serous juices remain in the blood. This is not because they are not separated by lack of exaltation nor because they are not liberated from their ties with the blood and thus cannot be absorbed by the renal tubes. This is rather because they do not pass through the kidneys. Once this is understood, all surprise disappears.

The different composition of the molecules has also something to do in this process. Many different particles can mix with others to compose considerably deformed molecules. We experience that with common aqua fortis which can enter and penetrate through the pores of silver but not through those of gold. But addition of ammonia salt alters so much the shapes of the molecules of this corrosive water that it can penetrate through the pores of gold and no longer through those of silver. But even without such addition, it is enough to change the arrangement and situation of the elements to obtain very different shapes of the molecules. This appears from many experiments. For example it is possible that particles of salt and of tartar not only mix with aqueous serum but also that they be included in and mixed with particles of blood. Hence it can happen that saline particles included in molecules of blood do not alter the external configuration of the blood, no more than they modify water, and then they cannot be absorbed by the urinary tubes, hampered by the coat of blood which covers them. But, on the other hand, they can be admitted by the veins.

Finally the famous author concludes: "that only those which, free and exalted more than normal by internal fermentation, are almost disjoined from the others and necessarily flow out, are separated in the kidneys and excreted as useless elements. Hence, besides the structure of the kidneys, steady fermentation of the blood contributes in separating urine. This fermentation dissolves and frees the salines and facilitates absorption of the particles of water by the existing structure and their separation from the rest".

Here the famous author clearly and self-assuredly expresses his opinion that urine is separated and excreted in the kidneys by fermentation and an adequate configuration of the canals. Since what he self-assuredly enounces is not very obscure, it is at least likely that the author considers this mechanism to be ignored or very obscure for others but not for himself.

But anyway I think to have proved clearly that neither continuous nor intermittent fermentation is required to separate urine from blood. However, I do not deny the motion which is carried out in all composite bodies by internal displacement of the elements. This alters neither the appearance nor the shape of the mixture.

But I consider as false this peculiar separating fermentation which can only dissolve the tight link and conglutination between urine and blood.

As was shown, fermentation in narrow tubes and during momentary passage is not only difficult and useless but even impossible. Such separation can be replaced mechanically by forcing the blood at high speed into the capillary tubes of the milking artery.

Chapter X

On the function of the liver.

Among the internal workshops of the animal you do not find any which is more famous than the liver. Ancient and modern authors wrote much on it and its function. As usual, I will briefly recall what others have reported, pointing out what I find questionable, and I will bring forward some points which were not discussed by others.

Proposition CXLIII

Recalling what famous authors wrote on the structure and function of the liver.

I think that the famous observations carried out as well by ancient anatomists as very accurately by the distinguished authors Glisson, Wharton, Willis and

Malpighi are to be highly praised. They discovered the structure of the liver. The vena porta ramifies into the liver and exerts the function of an artery expressing and eructing blood through countless capillary tubes as the aorta pours blood in its smaller branches through all the body of the animal. Like in the kidneys, there are two kinds of capillary vessels in the liver, the veins and the biliary pores. The famous Malpighi even adds the glands in which the vessels mentioned above are intertwined and ramify into capillary vessels of a third kind. The famous Willis adds a fourth kind of nervous vessels besides a branch of the hepatic artery. Moreover, the famous Bartholin adds lympathic vessels.

Some people believe that the function of this workshop is to make blood. Others think that this function is to purify the blood either because the ferment of the bile is added to it or because the bile is separated from it as a useless excrement.

Proposition CXLIV

Structure of the glands and their mechanism.

Nobody can deny that glands are workshops in which some ferments are elaborated and composed which afterwards are admirably used in animal economy to macerate food and make chyle, blood and spirits. But it is difficult to believe that all the glands are larders and workshops for ferment. The lobules of the glands appear to have the same function as lumps of earth in which the capillary roots of plants are ramified and to which they tightly adhere. Their gaping orifices are insinuated in the interstices of the grains of sand of the earth like leeches, to suck the juice which is there. The lobules of the glands of the animals, however, are different from lumps of earth and from the glands of plants because these have vessels of one kind alone, which can only suck nutritive juice. But the glands are clusters surrounding the extremities of three or four kinds of vessels: (1) those of arteries which carry and eruct blood like rain between all the lobules and diverticula of the glands; (2) those of nerves which pour there spirituous juice and sometimes absorb something from the gland itself; (3) those of excretory vessels which sort out and absorb from the eructed blood either some excremental juice to be eliminated in the same way as urine is separated in the kidneys, or useful ferment which is carried in other workshops in the animal; (4) those of veins which suck blood either cleaned of ordure or enriched by new proviants, and carry it to the main stream of the vena cava.

Nature is forced to disseminate the roots of the plants in different parts of the lumps of earth to collect nutritive juice which is not in one place but is scattered in different places and hidden in the interstices of the lumps. Besides this ramification, Nature pursues another advantage: these flexible and soft capillary roots are supported so that they do not fall or are not squeezed or closed or do not recede from their due places.

But many vessels had to be distributed and supported in the glands for greater and more precise efficiency. The capillary vessels which bring blood and other juices and irrigate all the recesses of the glands like rain could not be connected with the veins and excretory vessels by anastomoses. If the arterioles continued into the veins nothing could be mixed with or retrieved from the blood. Then the workshop of the gland would have been built to no avail. The arteries and veins would be ramified there into capillary tubes close to the nerves and excretory vessels in vain. On the other hand, if the arteries were linked with the excretory vessels by anastomoses, blood could not be absorbed by the veins. Therefore, to avoid these inconveniences, it is necessary that the arterial blood and other arriving juices which irrigate the glands be poured and left in some other place like a larder separated from the supplying and absorbing vessels. In such other place the arriving juices can mix in small quantities since everywhere in the different pores of the gland very narrow canalicules meet, bringing juices like rain. Then the excretory vessels can conveniently absorb from this mixture the juice which is appropriate for them and the veins remove pure blood as required by the similarity in shape of their pores and the extracted elements.

It is doubtful whether the lobules or the pores of the glands do nothing and behave purely passively like small jugs which only absorb the juices thus mentioned or whether there are storing ferments which modify the blood during its passage. Actually, one can believe that some glands can carry out the two functions. Such are those in which chyle, blood and spirits are elaborated and improved. I do not think, however, that the spongy substance of these glands is similar to a mass of fermentative flour or to the wood of a vinegar barrel which confers acidity to another mass or to wine poured in such a barrel. Such fermentation requires some delay and cannot be completed during the momentary passage of blood through the glands. The fermentative juice in the glands would be washed out and carried away by the very quick flow of blood. Therefore, its macerative virtue would be more and more weakened and finally would disappear completely as we see that a vinegar barrel washed profusely with the water of a fountain loses all its acrimony.

I rather suspect that ferments are steadily provided by nerves and other vessels in the glands, mixed like condiments with minute particles of blood and carried away with them. It seems likely that this is the reason why fermentations can occur in the glands of the animal.

As for the glands which must only separate the excremental juices from the blood, I do not think they need any ferment. Heterogeneous liquids can more easily be parted from each other mechanically by being wedged through very narrow orifices and, after separation, by being collected in two kinds of vessels provided with orifices corresponding to the liquids, as explained above.

Proposition CXLV

Bile can be separated from blood in the liver mechanically without the help of any ferment.

It can no longer be doubted that bile is separated in the liver from the blood passing through the vessels of the latter. Experiments by modern anatomists clearly suggest it. Only could it be questioned whether this separation occurs in the liver by fermentation or mechanically by wedging and filtering in the very small vessels ramified from the vena porta and in the pores of the lumps or glands of the liver in the same way as urine is parted from blood in the kidneys. There are two reasons which suggest fermentation. Firstly, the bile particles seem to be tightly encircled by the blood of the vena porta, and not united by simple contact. This appears from the fact that in oxymel the vinegar mixed with honey by simple contact stings the organ of taste by its bitterness whereas the very bitter bile is not tasted in the blood of the abdomen. Therefore, its corrosive bitterness must be cancelled, its stings being entangled among the glutinous elements of blood, in the same way as distilled vinegar is sweetened by lead and by chalk of corals.

Secondly, blood is not projected very quickly in the liver as it is from the arteries in the kidneys but it flows slowly from the vena porta. Therefore, if a strong link usually requires a long time before being dissolved by some ferment, it is not unlikely that Nature separates bile from blood by way of fermentation in the liver.

But after closer consideration, such fermentation seems to be very difficult and unlikely since it would require that ferment of the same nature, i.e. as bitter as bile, resides in the glands of the liver or is brought from somewhere else. Both alternatives are contradicted by experience. The glandular substance of the liver has a sweet taste and there is no appearance whatsoever of vessels carrying the bile into the liver from somewhere else. Moreover, it seems difficult that contortion and boiling of fermentation can take place in the vessels from the vena porta and in the very narrow pores of the glands of the liver during the passage of the blood, without some rest.

On the contrary, everything points to a mechanical action. A motive force is present. This is the impulse of the blood achieved by the compression of the abdomen and of the inspired air. A steady flow of blood is present. Very narrow canals ramifying from the vena porta and pores of the glands are present. There are also two kinds of vessels aimed at reception. Therefore, nobody will deny that a mechanical operation is possible and is even easier. It is thus likely that secretion of bile results from a mechanical operation rather than from fermentation.

To enable us to hypothesize whether the bile is separated from all the mass of blood which flows in the big artery and is collected by the vena cava, the following lemmas must be presented.

Proposition CXLVI

Of all the mass of blood, only a part, less than the twenty-fifth, passes through the liver during one circuit of the blood.

During one circuit of the blood, no other blood enters the vena porta and leaves it than what is sent into the mesenteric and coeliac arteries. The diameters of the mesenteric arteries are less than a fifth of the diameter of the aorta at its origin from the heart. The ratio of circular cross sections is equal to the square of the ratio of their diameters*. Therefore, the cross sections of the mesenteric arteries are approximately the twenty-fifth of the largest cross section of the aorta. Since the volumes of liquid flowing with the same velocity and during the same time are proportional to the cross sections of the pipes through which they flow, the volume of blood flowing from the mesenteric arteries and absorbed by the vena porta is one twenty-fifth of the volume of blood which is sent into the aorta during one circuit since blood flows in all the arteries with the same velocity. Actually, during one circuit of the blood, all the mass of blood flows through the aorta and afterwards through the vena cava. Therefore, during the same time the volume of blood which is absorbed by the vena porta and which leaves it is not more than one twenty-fifth of all the mass of blood which passes through the vena cava.

Proposition CXLVII

It does not seem that the bile which is collected by the bile vessels in the liver can be separated from the mass of blood which is outside the abdomen.

The total mass of blood is usually not more than 20 pounds in man. If the bile which is not more than 2 pounds was uniformly diluted in the blood, then only that part of blood which passes through the liver during one circuit of the mass of blood should be cleaned of its bile. Then during the first circuit, one twenty-fifth of all the bile which exists in the blood, i.e. one ounce, would be extracted. Afterwards the cleaned particles of blood would mix with the foul remainder of blood so far tainted by bile. Therefore, during each of the next circuits another twenty-fifth of the remaining bile must be extracted. Using the same calculation as for the elimination of urine, it appears that after six hours, during the hundredth circuit of all the blood, 1/250 of all the bile is extracted from the blood. This is less than a tenth of an ounce. After twelve hours, during the two hundredth circuit, only a thousandth of all the bile which is in man is retrieved. This is an eightieth of an ounce.

The bile is extracted from the mass of blood and no dirt is produced except from new food. Therefore, after one day of fasting, the liver would excrete bile very scarcely and eject a small and imperceptibe droplet of bile. This bile separated from the liver cannot stay or be retained in the bile duct or in the cystic duct since their orifices are open without any obstacle. Bile thus is always expressed from these vessels since they are compressed by the abdomen and the inspired air, like the veins. Consequently, after a fast of one day the bile vessels would remain completely empty. But this does not occur. After a fast of several days and even some months, the gall-bladder and the bile duct found in snakes

* The cross sections of circular tubes are proportional to the squares of their radii and not, as written by Borelli in his latin text, to the squares of their diameters.

are as turgid with bile as they were previously. Consequently, bile is not separated from the mass of blood which circulates through the arteries and vena cava.

Proposition CXLVIII

The bile which is discharged by the liver in the abdomen during one day in a man fasting is sixty times more than what exists in the blood.

The bile ducts are not similar to arteries or to the mesenteric veins. They do not end in capillary tubes. They are rather like the branches of the vena cava. This is like streams which form larger canals from the small rivulets collected in their course. Similarly, the vessels carrying bile discharge through large and open orifices in the duodenum. They always seem to be full of bile. They resemble the vena cava concerning the flow of liquids and the velocity at which the liquids move through both. These liquids are pushed by the same cause, the constriction of their circular fibres and the general compression of the abdomen. Liquid volumes flowing from inequal canals with the same velocity during the same time are proportional to the cross sections of the vessels. Therefore, the ratio of the amount of bile emitted by the bile canals to the volume of blood flowing through the vena cava in one day is equal to the ratio of the cross sections of these vessels. The diameter of the orifice of the bile duct is not more than a thirtieth of the diameter of the vena cava at its largest and the diameter of the orifice of the cystic duct is the same. Therefore, the sum of the circular cross sections of these two ducts is more than a two hundred twenty-fifth of the cross section of the vena cava. The liquids flowing from these orifices during the same time are in the same ratio.

During any blood circuit, 20 pounds of blood pass through the vena cava, i.e. 240 ounces. Therefore, during one circuit of all the mass of blood, 1/245 of this volume flows from the bile ducts. This is a little less than one ounce of bile. Since there are 16 circuits of blood in an hour, in 24 hours 7680 pounds of blood pass through the vena cava and a four hundredth of this quantity, i.e. 34 pounds of bile must flow through the bile ducts into the duodenum*. But all the mass of bile contained in the human blood does not seem to exceed two pounds. Consequently, the amount of bile which is sent from the liver into a human body fasting is seventeen times greater than the quantity contained in the mass of blood.

* Checking Borelli's calculations yielded different results. In 24 hours there are $24 \times 16 = 384$ circuits of the blood. Thus about 384 ounces of bile passes into the duodenum. This makes 32 pounds (384/12) and not 34. On the other hand, $7680/400 = 19.2$ and not 34. In the title of proposition 149 Borelli mentions 32 pounds.

Proposition CXLIX

An amount as great as 32 pounds of bile is not produced by way of fermentation in the abdomen of a man fasting.

I cannot imagine that, without food, bile can be produced again and again in the liver from the blood which would ferment there and be transformed as wine is changed into vinegar by an acetous barrel. What greater nonsense could be imagined than corrupting blood, the treasure of life, which Nature generates, maintains and increases with so much care and work? Would it not be surprising that Nature would corrupt such excellent substance and transform it into harmful excrement so that she would be forced to separate and reject it to avoid poisoning the remainder of the mass of blood by contagion and thus endangering the life of the animal?

If such transformation of blood into bile was carried out by a ferment outside the liver, then Nature would also act stupidly by providing so much ferment which would generate bile not only beyond need but which would have to be expelled by continuous work because of the danger of its noxiousness. Such inefficiency and inability cannot be attributed to wise Nature. Therefore, it must be concluded that the amount of bile thus mentioned is not generated in the animal, neither by fermentation nor by any other means.

Proposition CL

The same volume of bile expelled from the liver is relatively often brought back to its origin in a circular movement.

Circulation of the bile in the animal is suggested by the same reasons by which circulation of the blood was demonstrated by Harvey. The heart uninterruptedly pours blood into the arteries and the whole mass of blood (which does not exceed 20 pounds in men) is poured from the heart in 3 or 4 minutes. From this, one is forced to infer that it is the same mass of blood which, after having completed its itinerary through the whole body, returns to the heart to start again. Similarly, the whole mass of bile cannot be multiplied in the absence of new food and does not exceed 2 pounds. It is always expressed from the bile ducts and instilled into the duodenum. During one circuit of the blood, a little less than an ounce of bile flows from the bile ducts, i.e. 24/40 of an ounce. Thus during 40 circuits of the blood, in two and a half hours, all the mass of bile (which does not exceed two pounds) flows from the liver into the duodenum. Therefore, for this flowing from the liver to continue, the mass of bile, after completing its itinerary, necessarily must be carried back to the liver to renew its source and this circulation repeats itself again and again.

Proposition CLI

Circulation of bile seems possible with the bile being carried, not through the heart, mixed with arterial blood, but only through the abdomen.

Since we cannot know the route of the bile in the body of the animal, we must look into its source, the liver, and consider through which path and on which vehicle the bile proceeds and from there whereto it moves. It appears that bile is brought to the liver by the blood of the vena porta and hence flows into the duodenum. It is then mixed with nutritive juices macerated by the stomach and is carried along the long way of the intestines. The vena porta carries blood flowing from the mesenteric and coeliac arteries. This blood, before entering the vena porta, passes through the spleen and the mesentery where it is impregnated by different juices collected there. It is doubtful whether the bile which is separated by the liver was stored in these arteries mixed with arterial blood or rather whether pure blood, by cleaning the cloacum of the abdomen, collects the biliary ordures inside the mesenteric veins and transmits them to the vena porta.

Certainly, if bile was mixed with the arterial blood, expelling the two pounds of bile which are in the human blood would never be possible through the bile ducts or at least instillation of bile would be very scarce after 12 hours fasting, as shown above. But this is contradicted by experience: indeed cystic and bile ducts are always found turgid and pour bile profusely into the duodenum.

If bile was cleaned away by the lymphatic vessels and carried to the heart by the thoracic duct and hence circulated through the arteries back to the abdomen and liver, it would be mixed with the arterial blood. This would entail the same nonsense as described above. Moreover, if the bile is a useful juice improving blood, blood without bile would be imperfect. Then it would be stupid for Nature to separate the bile from the blood. In so doing she would be doing wrong to the animal. But if bile is a harmful excrement it would be stupid also to suck the biliary ordures from the cloacum of the intestines to infect and corrupt blood. Therefore, it does not seem likely that Nature would have created a circulation of the bile through the heart and arteries. Consequently, circulation of the bile can be achieved only through the abdomen since it can be assigned to no other path than the two thus mentioned.

Proposition CLII

How the bile circulates through the abdomen.

Certainly bile must flow in the duodenum continuously, forced by the compression of the abdomen, whether food macerated in the stomach is present or not. When there is food, after this has completed an additional fermentation in the intestines, a small amount of bile is excreted with the faeces. But most of it is mixed with the chyle and is collected again, sucked from the food by the pores of the intestinal walls. When fasting, bile or any other juice present there must also be absorbed mechanically by the pores in the walls of the intestines.

But I claim that the chyloferous vessels of Asellio cannot suck the bile. As a result of the shape of their orifices they can only absorb pure chyle or lymph. If they absorbed bile, they ought to carry it to the heart through the thoracic duct and mix it with the blood. This does not occur, as we showed. Consequently, bile in the intestines must be absorbed by other vessels. It does not seem possible that these other vessels originate from the mesenteric veins. Firstly, besides chyle and blood vessels, no third kind of vessels appears in the mesentery. Secondly, in all muscles, in the heart and the spleen, the cross section of the veins does not exceed that of the associate arteries by more than three or four times since the ratio of the diameters of their cavities is less than two. However, in the placenta the ramifications of the umbilical vein are more numerous and larger as compared to the associate arteries than in the biggest muscles, in the heart and the spleen of the same animal. The reason for this is that in muscles the veins must absorb about the same volume of blood as what flows from the associate arteries. It is thus enough that the cross section of the veins exceeds that of the arteries in inverse proportion to the velocities of blood in the arteries and veins.

But the function of the umbilical vein consists of absorbing new blood provided by the placenta. This blood is added to that transmitted by the umbilical arteries which are used as its vehicle. Therefore, the ratio must be increased. For example, if the umbilical veins must receive from the placenta a volume of liquid equal to half the volume which is transmitted from the umbilical arteries and if the velocity of the blood in the arteries is three times that in the umbilical veins, it is not enough that the veins have a capacity three times that of the arteries. They must be much larger, i.e. by four and a half times. Then they can absorb the arterial blood and the additional blood flowing from the placenta. It thus appears that veins can carry back another liquid besides the blood sent from the associate artery. This is also confirmed by experience. Indeed, those who remain in a bath for a long time urinate profusely. Water thus must be absorbed by the veins and be separated from the blood in the kidneys. Like in the placenta, in the mesentery the mesenteric artery has a diameter which is only a fourth or a fifth of the diameter of the vena porta which collects all the mesenteric veins. Therefore, the cross section of the artery, or its orifice, is not less than from 1/16 to 1/25 of the cross section of the vena porta*. Although the slowness of blood in these veins requires more width, such considerable and superfluous excess, however, seems to contradict the rules of Nature which uses to build proportionate vessels to carry and to absorb blood, according to the requirements of the flesh or viscera which this blood must feed.

Hence it can be surmised that such abundance of large mesenteric veins enables them to absorb another liquid besides the blood sent by the associate ar-

* The cross sections of circular tubes are proportional to the squares of their radii and not, as written by Borelli in his latin text, to the squares of their diameters.

tery. Therefore, they likely absorb from the intestines much chyle with which bile is mixed and which is used as a vehicle of bile as we shall demonstrate. This bile is collected in the vena porta and separated from the blood again in the liver. It is then sent into the bile vessels. It leaves these and repeats its circuit through the intestines and the mesenteric veins again and again.

Proposition CLIII

Necessity and function of the circulation of the bile.

Animals always need new food as long as they live. Food is prepared from beginning to end in the intestines where it is transformed into chyle by the fermenting action of bile and pancreatic juice. Bile is poured profusely into the duodenum and mixed with the food which was previously macerated in the stomach. Such abundance is necessary (since Nature hates superfluity). Food can ferment in the intestines only if mixed with much bile which is carried with the food through all the length of the intestines. The bile must not be expelled completely with the faeces. Firstly, the chyle with which the bile is mixed would also be expelled. Secondly, bile is useful and cannot be eliminated without considerable harm to the animal. It cannot be regenerated by the food abundantly and so quickly. Neither could so much bile remain in the intestines since the food by moving from the beginning to the end of the intestines would carry the bile together with the chyle, as if it were a sweeping broom. Therefore, much bile necessary for future use had to be removed from the intestines after having completed there its fermenting action. It had to be carried to some convenient place from where it could be sent again to the beginning of the intestines, in the duodenum. To carry out this circuit, Nature, according to her habit, did not create a new and unusual mechanism. She decided to use the simple and familiar movement of blood, through the mesenteric veins according to their character. They suck the bile like leeches and carry it together with blood, not to some remote and inconvenient place in the heart, but close to the beginning of the intestines where it is separated from the blood and instilled again into the duodenum. To this end, Nature provided the vena porta with the function and privilege of an artery and gave to the liver the function of a sieve. They transmit the juice thus excreted to the bile ducts and these pour it again into the duodenum.

As usual with Nature, one and the same operation results in several advantages. Firstly, the bile by its displacement in the intestines provokes expulsion of the faeces by its stimulating bitterness. It stimulates the circular fibres of the intestines to contract. Secondly, by provoking the peristaltism of the intestines, it causes the expression of juices and their suction into spongy pores.

Thirdly, unctuous and mobile particles of bile are mixed with the chyle in the intestines and with the blood in the abdomen. In these places, fermentation can conveniently occur as a result of the size of the vessels and of the very slow, nearly stagnant, movement of blood.

Fourthly, for fermentation of blood to take place easily in the mesenteric veins and in the vena porta, blood had to be mixed with bile by small quantities. This obviously can be done in these huge vessels by a beating movement urged by compression of the abdomen and of inspired air.

Fifthly, the same beating admirably helps to maintain fluidity of the blood in the abdomen. These are thus the functions which are known to me so far. I leave other investigations to more sagacious people.

Chapter XI

On the flowing of spirituous substance through the nerves.

In the animals, besides liquids such as blood, there is another extremely spirituous fluid substance which is the direct motive cause of the animal body. This appears from the effects of this substance. This spirituous humour is not wind or air but has a liquid consistency such as spirit of wine. It is generated from blood in the brain and diffused by the nerves. All modern authors admit this point. The exact structure and composition of the nervous juice, although unknown, can be surmised somewhat from its motions through the nerves.

Proposition CLIV

Nervous juice contributes with blood to the feeding of the animal.

Modern authors have shown with many experiments and arguments that feeding is possible only by a juice originating in the brain and transmitted to all elements through the nerves. These arguments can be found in Willis and others. Personally I am impressed by the huge profusion of nerves which are sent to and ramify in the abdomen, viscera and glands. In such places they are used neither for sensations nor displacements. Since they transmit some juice, they must not be there purposelessly. Chyle is elaborated in the mesentery for the nutrition of the animal. Thus, nervous juice contributes to the making and confection of chyle and, consequently, to the nutrition of the elements of the body. Moreover, from the extremities of all the other nerves, some juices are always instilled into the muscles and other elements even during sleep when neither sensations nor displacements are carried out but only nutrition can take place. Therefore, at such time this nervous juice either completes or helps nutrition of the flesh and other parts.

But it seems difficult to believe that all the elements of the animals can be repaired and fed by the nervous juice alone in so small quantity, without blood. Therefore, I think that the opinion of Willis is true that the material of nutrition is the blood but the vivifying and plastic mould is provided by the nervous juice.

Proposition CLV

The spirits are activated by opposite movements through the nerves.

The will from which displacements originate in the animal is exerted in the brain. Its orders are obeyed by the muscles which powerfully swell, contract and raise considerable weights. The paths through which these orders are transmitted are the nerves. When these are divided or tightly ligated, movement and sensation stop. No physical action can be conceived at a distance without some displacement of some body. Therefore, this implies some displacement or impulse by spirituous juices from the brain to the muscles through the nerves.

Inversely, in sensations, motions coming from external objects are transmitted through the nerves to the brain, more exactly to the sensitive area of the brain. Spirituous juices thus must be carried through the nerves in opposite movements, from the origins of the nerves in the brain to their extremities when the limbs are moved and from the extremities of the nerves to their origins to transmit sensations.

Moreover, experiments by modern authors show that nutritive juice flows through the nerves. This nutritive juice together with blood repairs lost elements of the body. This displacement is carried out from the brain to the extremities. Contrariwise, drugs and poisons are transmitted from the ends of the nerves to the brain. Consequently, two movements outwards and two opposite movements inwards must take place in the nerves.

Proposition CLVI

The opposite actions inwards and outwards through the nerves result neither from an immaterial faculty nor from shaking of tense fibres.

I will not spend time refuting these immaterial faculties about which almost everybody laughs. It is sufficient to recall that an immaterial force does not need paths or the canals of the nerves to be transmitted. Thus tight ligature of the nerve could not impede the passage of an immaterial sensitive or motive faculty.

Others think that sensations and orders of the will are not transmitted by movement of the nervous juice inside the tubes of the nervous fibres but by the tense fibres themselves and membranes. In the same way, if the tense strings of a zither are touched at one extremity, the vibrations immediately diffuse and are transmitted to the opposite extremity.

Actually, this comparison cannot apply to the nerves of the animal because the nerves in the animal body are neither pulled directly nor tense nor hard and rigid like iron rods. They are soft, incurved, loose and wrapped about muscles, vessels and viscera which are soft and yielding. They are not attached to fixed extremities like nails but rather to soft ones, the brain and flesh. Consequently, it is impossible that shaking, tension, striking or trembling diffuse or are transmitted either to the brain from some external sensitive object or to the muscles from the will. Similarly, when an extremity of a soft, loose and folded thread of cotton is stricken, the motion yielding to the blow is not transmitted to the remainder of the thread. Soft and loose bodies indeed differ from hard and tense bodies in that one part of them can be moved whereas the others remain immobile. It is impossible to move one part of a hard and tense body without activating and shaking the whole body, i.e. all the parts of the body together at the same moment and with the same velocity.

Proposition CLVII

Mechanism by which the spirituous juice carried through the nerves can be activated outwards and inwards.

We said that the spirits are activated through the nerves by two kinds of opposite movements. The first is that by which the limbs move and sensations are transmitted to the brain. We shall first deal with it. Before analysing these movements, it must be repeated that the nervous fibres neither are completely solid, full and waterproof, nor are they hollow and empty tubules like reeds. They are canals full of some spongy substance similar to the marrow of elder. But this spongy marrow of the fibres can be easily and continuously moistened and soaked up to turgescence by spirituous juice of the brain with which it is connected, as we see that sponges and filters are soaked by adjacent water.

A bowel of sheep would be made turgid if filled with water alone as if its cavity contained a big spongy cord and this was impregnated by water. In both instances there would be turgescence. Whatever slight compression or blow would be delivered at an extremity of the turgid intestine, the shaking would be transmitted to the other extremity immediately. In the same way the tubules of the nervous fibres provided with spongy marrow ean be soaked up to turgescence by the spirituous juice of the brain. Then, if one extremity is compressed, pushed, stricken, shaken or excited, the commotion and shaking, or the wave, must be transmitted immediately to the other extremity because, as a result of their contiguity, the successive parts in serial order press the next ones until the striking and impulse are transmitted to the last one.

Consequently, this slight motion of the spirits provoked by the will in the brain can shake or excite the fibres or spongy ducts of some nerves turgid with spirituous juice. As a result of this convulsive irritation which shakes all the length of the nerves, some spirituous droplets can be expressed and spilled from the orifices of their extremities into the corresponding muscles. This results in the boiling and bursting by which muscle is contracted.

Inversely, when the extremities of the sensitive nerves which end in the skin, tongue, nostrils, ears or eyes are slightly compressed, stricken or tickled, the shaking, waving and tickling of the spirituous juice contained inside the tubules must be immediately transmitted over the length of the nerve to the well-determined area of the brain to which the nervous fibres are attached. There the sensitive faculty of the mind can judge the cause of the movement from the place attained in the brain, from the violence of the blow and the characeristics of the motion.

This irritation at an extremity of a nerve can diffuse and be transmitted instantly to remote places without that the extremities be materially touched, stung or irritated. This occurs because nerves are of a nature such that if one of their extremities is pricked, pain is perceived in another place at a distance. This appears in many experiments. Slight tickling of the nostrils or of the ears shakes and irritates the nerves of the chest, diaphragm and abdomen and provokes violent convulsive movements with sneezing and coughing. Sometimes when the spinal cord is injured pain is perceived in the knee and the leg.

Consequently, the movements which were thought to be opposite movements in the same nervous canals actually are no progressive displacements with passage from one extemity to the other so that, when moving a limb, the spirituous juice instantaneously travels all the way from the brain through the length of the nerve to the muscles, and inversely, when a sensation occurs, the spirit is carried from the teguments of the animal to the brain. Some shaking and waving of the juice contained in the turgid tubes are enough. We showed, using the example of a turgid intestine, that such shaking can easily occur. Consequently, this operation is not only possible but is even very likely.

A problem remains to be solved. How can two opposite shaking movements occur together at the same time in the same nerve? The tongue can move and feel at the same time. For example, a piece of rhubarb is rejected from the mouth and at the same time its bitterness is perceived. All muscles perceive the pain due to the sting of a sword while they contract vigorously.

I think that this difficulty can be explained in two ways. Firstly, the two opposite shakings do not occur through the same fibrous canals but through distinct canals so that the fibres which transmit the movement of the order of the will do not accept the movement of pain which is transmitted through other canals.

Secondly and more likely, a wave of the juice contained in the tubes cannot be conceived without reciprocal movement, pushing to and fro towards the two opposite extremities of the nerve alternately as happens in shivering. Actually, these opposite shakings do not occur simultaneously but at different times which cannot be distinguished and are thus obscured by their brevity and frequency so that we are sometimes deceived by many interruptions of the sensations and motions.

I pass to another opposite movement which occurs through the nerves.

Proposition CLVIII

The nutritive nervous juice is different from the spirits which exert the motive and sensitive functions in the animal.

The nutritive nervous juice is different from that which conveys sensations and voluntary movements. This is surmised from the fact that these animal operations induce consumption, exhaustion and diminution of the spirits which give impetus to all of the body of the animal rather than that they induce their repair. Therefore, sensations and movements not only impede nutrition but they contribute to the loss and consumption of the parts of the animal. On the contrary, we observe that the lost animal spirits and solid elements are repaired and restored and that moreover the animal grows and develops only when it neither carries out voluntary movements nor perceives sensations, during sleep when the animal rests as if it was dead. This is confirmed by the long deep sleep of the embryo in the womb of the mother and by the frequent and deep sleep of infants. During such sleep embryo and infants are fed and grow considerably and quickly. On the contrary, during the short and interrupted half-sleep of elderly people, not all but only some of the lost parts are repaired slowly and uneasily because some parts are dried, consumed by decline and unfit for nutrition.

It thus can be surmised that the motive and sensitive spirits are different in nature from the juice which is used for nutrition. If they were one and the same, the nervous juice would diffuse from the brain to wet all the muscles when we are fed at a maximum. Then the muscles ought to contract together at the same time and produce some general convulsion. This certainly does not occur. Moreover, during nutrition sensations from external objects could not be reported to the brain because the rapidity of the flow of nutritive juice descending from the brain not only would impede but would drive back the impulses and sensitive waves tending to go upwards.

Nutritive nervous juice thus seems to be different from the motive and sensitive spirits. As far as character and energy for action are concerned, the latter are most noble, bitter, sulphurous, saline and very active, like spirit of wine. The former are very sweet, inducing quiet sleep rather than dissolution and exhaustion of strength.

Proposition CLIX

Paths in the nerves through which nervous juice is transmitted from the brain to the elements of the animal to be repaired.

The nutritive nervous juice thus appears to be different from the motive and sensitive spirits. It thus cannot be carried together with these through the same canals at the same time to achieve its functions, for the reasons given above and because their different natures and characters would be confused, weakened and altered and they would hamper each other. Hence the actions which they would have to carry out would be reduced or twisted which would be against the laws of Nature. It must thus be admitted that the paths through which the nutritive juice is transmitted from the brain through the nerves to the damaged elements to be repaired are different from the tubules of the nervous fibres through which the motive and sensitive spirits are transmitted.

The possibility of such different paths in the same nerves does not seem illogical. Nerves are bundles of fibres or small canals contained in purses or membranous sheaths. Like in comparable bundles made of very thin glass threads, the interstices between the threads play the role of glass tubules. These interstices absorb liquids and raise them like a sponge or the tubules of siphons.

It is thus possible that in the nerves, the interstices between the fibres contained inside membranous sheaths play the role of canals of another kind. Through the interstices of this type another juice can filter different from the one which fills the cancellous tubules of the fibres. This may be the nutritive juice which is sent from the brain to the extremities of the animal for their nutrition together with nutritive juice contained in the blood.

It remains to discover how the juices of different kinds present in the brain can enter the distinct canals of the nerves without confusion. This will be explained later in due place.

Proposition CLX

Opposite movement of the nervous juice towards the brain.

The theory explained above seems to apply more easily to the nerves which are almost not used for sensations and which are not aimed at moving the muscles of the limbs. Such are the nerves which end and spread in the glands, in the viscera and in the abdomen. These also consist of bundles of fibres enclosed in a membranous sheath. In these nerves, however, it seems certain that opposite motions occur. Nutritive juice is transmitted from the brain through the nerves to the parts of the animal. Inversely, spirituous juices are transmitted from the glands, viscera and all the skin to the brain. These juices are either purified or soiled by contact, mixing, separation or alteration of useful or noxious elements as appears with drugs and poisons which cannot be brought to the brain through other paths than the nerves. This is deduced from their very quick and instantaneous action. If drugs or poisons were transmitted to the brain through the veins or through the lymphatic vessels, firstly they would be weakened considerably by their mixing with the mass of blood and with other humours. Their energy would be dissipated. Secondly, a small droplet of viper poison or of diabolical oil could not be mixed and diffused uniformly in all the blood occupying all the vena cava during its swirling course. It could happen by chance and seldom that a part of the blood infected by a droplet of poison be driven into the ascending branch of the aorta rather than in the descending branch which is twice larger and from which it could be expelled through the kidneys or through other excretory vessels of the viscera. In such instance the poison would provoke no harm. If it actually reached the brain, it would have a dizzying effect fairly delayed by such a long and rough route. But do we not observe that the animal strength is broken immediately by the bite of a viper? Thus, the droplet of poison is not carried slowly to the brain by the blood through the long and complicated path of the veins, but through the nerves themselves. This is confirmed on the other hand by the instantaneous deficits produced by squeezing or sting of nerves. Therefore, drugs and poisons undoubtedly are carried to the brain through the nerves.

It remains to discover whether these juices are moved together with the nervous juice from the extremities of the nerves to the brain or whether only their influence or their motion is transmitted.

It seems to me that the second term of the alternative is impossible. Indeed juices, however spirituous and active they are, are material and thus cannot act at a distance. Without physical contact they cannot increase, vivify or weaken the animal spirits. They cannot increase, expel or transform the animal spirits which are also material without being present and mixing with them. Therefore, this action does not seem conceivable without displacement of the nervous juice travelling all the length of the nerve up to the brain.

Secondly, it must be found out whether such retrograde movement occurs through the cavities of the nervous tubes or through the interstices between these fibres.

The paths through which drugs and poisons are brought to the brain do not seem to be the tubules of the fibres which are full of the spongy substance mentioned above and are always turgid with spirituous juice. Therefore, drugs and poisons could uneasily penetrate these encumbered and rough paths and travel to the brain.

The paths through which the nutritive juice is difffused seem to be more expedient, i.e. through the interstices of the nervous fibres which constitute bundles. I claim that the displacements of the juices can be carried out up to the brain through these paths, not simultaneously but at another moment than the nutritive juice is carried from the brain. There is nothing new in the fact that two distinct liquids travel through the same canals in animals in opposite directions, as long as they carry out their displacements at different times so that they cannot impede each other mutually. This occurs in the salivary ducts through which saliva is ejected from the glands into the mouth. In an opposite displacement wine and clear soup beaten in the mouth are absorbed by these same ducts and carried to the glands and from there they are transmitted immediately through the nerves to the brain and restore the strength of the animal.

Proposition CLXI

Mechanism of the very quick movement of the nervous juice descending from the brain and returning to it.

From the experiments reported above on the quick repair induced by drugs and the quick harm provoked by poison in the brain, it is deduced that these juices travel their long way along the length of the nerves very quickly, almost instantaneously. Hence it can be assumed that flowing of the nutritive juice from the brain also occurs with the same velocity.

The mechanism and cause of this velocity must be explained.

The nervous bundles consist of threads or fibres as long as the nerves themselves. Therefore, the interstices between these threads have the same length, extending over the whole nerve from the brain to the extremity of the nerve.

These interstices play the role of many very narrow small canals. But in very narrow glass tubes or in the interstices between these tubes collected in a bundle, water is sucked very quickly pushed by its own weight and rises the higher the narrower the cavities, like in sponges. Water is impelled by no other cause than its weight as we showed in our book on natural movements resulting from gravity. Consequently, the nervous juices must be sucked very quickly by the very narrow interstices by the same mechanism, either from the brain towards the extremities or from the extremities towards the brain since the small canals thus mentioned are very narrow, always wet and the same impulsive cause of pressure is present, i.e. the weight of these fluids.

Proposition CLXII

Flowing of nervous juice from the brain for nutrition of the animal occurs during sleep and is the cause of sleep.

The paths through which the nervous juice flows from the brain for the nutrition of the animal are known. It remains to discover whether such flowing occurs during sleep and why it must occur during sleep.

Nobody doubts that nutrition actually takes place during sleep, either completely or for its most part. Indeed we observe that we are repaired and nourished by sleep whereas if continuously awake, even if well fed, we weaken, soften and our face becomes gaunt. It may be argued that, just as fortune mostly results from low expenditure of the goods which one has, similarly during sleep the spirituous and organic substance of the animal is not dissipated in hard exercises, displacements and sensations. Therefore, we experience beneficial and pleasant rest from work since we feel comfortable every time we are freed from hard work. This is partly true but it does not appear sufficient: sitting or complete rest of our body or prolonged inactivity not only do not repair our strength and are not pleasant but, on the contrary, induce weakness and painful numbness the treatment of which consists of moving the body. It is thus not saving but spending the strength which induces repair.

Hence we surmise that moderate movement and moderate exercise of the muscles and senses do not destroy but rather repair the strength and substance of the animal. If restoration of the animal and repair of the lost elements occur during sleep as taught by experience, this is due to another and completely different necessity. If I am right, it results from the fact that flowing of spirituous nutritive juice cannot occur without some boiling of this juice taking place in the brain. This effervescence, rarefaction and contortion result in an obstruction of the orifices of almost all the tubules of the nervous fibres gaping in the brain. When these are obstructed, it is not surprising that neither oozing is transmitted to the muscles to move, by order of the will, nor notice of sensations perceived by the extremities can be reported to the brain.

Sleep results from this boiling and effervescence occurring in the brain. This is suggested by the fact that wine, drunkenness and soporifics provoke in the head surprising disturbances, heat, dizziness, pain, swelling and reddening of the eyes and face. All these symptoms are provoked usually by boiling and heat in the brain. Since these soporifics produce deep sleep, it is likely that normal sleep cannot occur without such boiling taking place in the brain.

The tubules of the nervous fibres are the paths through which the motive force is diffused to the limbs and the sensations are reported to the brain. On the other hand, their origins or orifices can easily be choked by juice boiling in the brain. Therefore, it is likely that the nutritive juice boiling in the brain produces sleep by choking the orifices of the tubules of the nervous fibres. As a result, feeling and movement stop in the animal.

This theory is confirmed by the lack of strength, weakness and numbness with yawning and desire of lying down which we experience in spite of ourselves immediately after awakening. This is not due to actual weakness since strength has rather been restored and increased by sleep. In the numbness which follows awakening, voluntary motions, although freely stimulated in the brain, cannot transmit the spirituous juice through the nerves to the muscles to be moved perhaps because the orifices of the nervous fibres are not yet completely cleared.

Only two questions remain to be solved. The first is why during sleep irrigation of the openings of the nervous fibres by the nutritive juice and its passage through the interstices of their small canals do not push nor express spirit or motive juice. Where these ducts are turgid, all the muscles should shake and contract frantically during sleep as a result of their expression.

It can be answered to this question that some convulsive shaking carried out by an act of will is required to carry out a movement. But this irrigation of nutritive juice, soft and not pricking, is unable to provoke such convulsive irritation. Therefore, the juice contained in the spongy tubules is not expressed. Moreover, the flow of nutritive juice through the interstices between the fibres must not necessarily express the juice contained in the spongy tubules: firstly because the interstices are not obstructed; secondly because the juices are not pushed violently through them as if it were by a piston. Therefore, the nutritive juice can flow freely and conveniently.

The other problem is that nutritive juice flowing profusely into the muscles ought to produce strong and violent motion and swelling similar to the contraction which is usually provoked by motive spirit.

My answer is that sweet nutritive juice is unsuited to stimulate contraction in the muscles because it lacks the bitter and fermentative character which is necessary to provoke such contraction. Actually, the aqueous and sweet nutritive juice can rather quiet and is apt at inducing numbness. It thus damps the fermentative force and rather produces tranquillity which rests the muscles.

All this could probably not happen if the paths of the nervous tubes were obstructed or rather if the orifices of the spongy tubules in the brain itself were choked by boiling of the nutritive juice which irrigates these orifices. The convulsive excitation which seems necessary to the flowing of motive juice from the tubules of the nervous fibres into the muscles can be easily impeded as a result of this irrigation.

Proposition CLXIII

Why pulsation of the heart is not impeded during sleep.

The theory presented above concerns the movement of the nutritive juice through the interstices between the nervous fibres. Although it explains the voluntary movements of the limbs, this theory cannot apply to the movements of the heart, diaphragm and intercostal muscles which move during sleep as well as when awake. Thus, the juice boiling in the brain does not impede flowing of spirits to these muscles. Seemingly there can be no doubt that nutritive juice boils in the cerebellum and medulla oblongata. Therefore, the two vagus nerves which reach the heart and thorax should be obstructed there as occurs for the other muscles of the limbs since the muscles of the heart and thorax are repaired by nutritive juice during sleep as are the muscles of the limbs. Thus it may be suspected that either the vagus nerves which ramify in the thorax are not obstructed in the brain or the movement of the heart obeys another law independent of will, like an automaton, as well during sleep as when awake.

After further thought it appears that the difference of action can be explained easily in two ways while retaining the same hypothesis. Firstly, it can be assumed that the nerves of the heart do not have their origins in the area of the brain where the nutritive juice is boiling, but further. Their roots would prolong higher to the apex of the brain and thus would not be moistened by the nutritive juice. Therefore, the orifices of their canals could not be obstructed during sleep and the movement of the spirit which provokes pulsation of the heart could not be impeded. On the other hand, nutrition of the heart could be carried out because the nutritive juice boiling round the nerves of the heart far from their origin could easily enter their lateral pores (as happens for a rope immersed in water) and flow through the interstices between the fibres to feed the heart. The motive juice for the pulsation of the heart could, however, pass through the unobstructed tubules of the fibres.

Secondly, the same phenomenon can be explained by assuming that the orifices of the tubules of the fibres of the nerve of the heart or the pores of the spongy substance which they contain have a shape such that they cannot admit nutritive juice but are pervious only to motive spirituous juice. Therefore, they could not be choked by the nutritive juice mixed with the motive spirits. Similarly, in the kidneys the orifices of the urinary tubes are not obstructed by blood. They obviously select and absorb aqueous serum and discard the blood which was mixed with this serum. The bile and gall canals also retrieve the bile mixed with blood and are not obstructed by blood. The same is verified in other viscera, in glands and in our sieves and riddles. The nerves of the heart thus are different from the other nerves of the limbs because the orifices of the tubules of the latter are large, like funnels, so that they can be easily choked by nutritive juice. It is different for the nerves of the heart. The orifices of their tubules are narrow. Their perforated shape is impenetrable to the elements of the nutritive juice which are shaped differently, and can admit the motive juice only.

Thus, two different juices, spirituous and nutritive, are mixed in the brain during sleep. They flow in two different types of small canals in the nerves of the heart, the tubules of the spongy fibres and the interstices between these fibres. The tubules of the fibres have a shape suited to absorb the spirituous motive juice only but not the nutritive juice. The interstices are suited to admit the nutritive juice but not the motive juice. The motive juice necessary for the pulsation of the heart and the nutritive juice necessary to repair the substance of the heart thus can be carried simultaneously without that the displacement of one impedes that of the other. This otherwise general law that during sleep the voluntary movements of all the muscles stop to allow for repair and feeding thus suffers an exception.

Proposition CLXIV

Why breathing is not impeded during sleep.

The above theory seems to be contradicted by the nerves of the thorax and diaphragm which are involved in breathing and are branches of the two vagus nerves and of the intercostal nerves which also innervate the heart, since respiration is not voluntary and instinctively occurs when we are asleep or when we are awake but do not notice it. It is also voluntary since we can, however, hold, increase, decrease or accelerate our breath voluntarily. This does not force us to change the hypothesis presented above. It is sufficient to notice that nerve branches ramify in the thorax and in the diaphragm not only from the two vagus nerves and the intercostal nerves but some also derive from the spinal cord. Hence it may be that when awake normal breathing occurs independent of the will, through the nerves originating in the cord, and during sleep these spinal nerves or their roots in the brain can be choked by boiling nutritive juice. The other branches of the two vagus and intercostal nerves which innervate the thorax and the diaphragm have the same shape and structure as the branches which innervate the heart. Their nervous roots can absorb motive juice during sleep as well as the heart continuously absorbs it during sleep. Thus breathing can continue during sleep although when awake we can modify it at will but not impede it completely.

Proposition CLXV

Why the limbs sometimes move during sleep.

During sleep the nerves are obstructed: neither can sensations be transmitted from external objects to the brain nor movements be ordered. Why then do sleeping people turn over, sometimes move the arms and kick? Sleep-walkers walk and carry out other motions. Although their internal faculty is not asleep but is alert, activating and unrolling pictures previously imprinted in the mind, it is, however, difficult to understand how displacements can be carried out while the paths of the nerves are obstructed. I suspect that these movements are sometimes convulsive because the nerves are irritated by some fermentation of the juices which they contain in their tubules. This can occur mostly during sleep.

The problem of the sleep-walkers is certainly more difficult to solve. Sensitive perceptions of external objects are completely impeded and motive juice to contract the muscles cannot at all be instilled freely. Perhaps the boiling nutritive juices are too narcotic in sleep-walkers. They numb sensations but cannot really refrain the motive force because the imagination instead of the will orders movements which are carried out normally by acquired habit.

Chapter XII

On semen, its genesis, movement and nature.

After discussing the animal spirits which originate in the brain and their motions we must deal with spirits of another kind, the very lively and unctuous semen which is made in the testicles. We shall discuss its origin, movements and nature.

Proposition CLXVI

Structure of the testicles as discovered recently.

The ancient anatomists investigated the structure and function of the testicles rather carelessly. Some said that they are masses of flesh and were made by Nature for nothing else than to dilate vessels by their weight. Others believed that they are workshops for semen. They thought either that serum is sucked from semen by the spongy structure of their pores (Galen bitterly opposed this view) or that blood or lymph is changed into semen by their medullary and glandulous substance.

Recent and accurate anatomical observations clearly refuted all these views of the imagination. The famous Auberio when he taught anatomy in Pisa, on the ides of May 1657, showed to myself, the famous Malpighi and others, that all the mass of a boiled testicle of a sheep consists of countless white columnar filaments, like worms, extending from both nervous envelopes to the median axis of the testicle. These fibres were full of white semen. Slightly squeezed they secreted semen drop by drop. This appeared even more clearly in the testicles of a wild boar killed by order of Ferdinand the second, Grand Duke of Tuscany, at the time when these animals use to mate. Then their testicles are considerably swollen by semen and the vessels which are as large as the quills of chicken feathers are obvious without any preparation as a result of blood caught between the white fibres.

At my suggestion Auberio published the same year this beautiful observation which he illustrated by an elegant copper-plate of Florence.

Actually, since nobody must be deprived of the glory which he deserves, the famous R. de Graaf ulteriorly in 1668 published the same structure of the testicles and discovered many remarkable facts which had not been noticed by Auberio. He observed that part of the spermatic arteries penetrate the substance itself of the testicles and part of these arteries irrigate the epididymis and the tunica albuginea. This appears after injection of coloured water with a syringe. He unfolded the duct of the epididymis into a greater length than Auberio did. He also saw the foramen epididymis. He made visible its tortuous and serpentine prolongation by injecting water with a syringe. He, moreover, made other beautiful observations.

From the observations of these famous authors we know that arteries carry blood to the testicles, that proper white nervous vessels constitute the substance of the testicles. These vessels prolong into a nervous duct extending over the axis and then into the single very long canal of the epididymis. They end in the ductus deferens which discharges semen in the seminal vesicles. From there the semen is spilled into the urethra together with the liquid secreted by the prostate.

Proposition CLXVII

Structure of the seminal ducts.

The naked eye cannot see the composition and intimate structure of the wormy small columns which compose the mass of the testicle. It is surmised, however, that these columns are not simply perforated tubes like reeds so that semen just flows through them as blood flows through the arteries and veins. They seem to be filled by some spongy substance. Hence it is conjectured that the milky juice which they contain does not flow spontaneously from their ducts but requires some squeezing. Similarly, we express water from a wet cloth by squeezing it with our hands.

This has been confirmed. A testicle divided into two, even when compressed and squeezed considerably, retains, nevertheless, a considerable volume. Therefore, the columns were not filled by liquid alone but were made turgid by a thick mass of spongy substance.

If their internal cavities were hollow and smooth like those of the arteries and veins, blood could not be transformed into spermatic liquid. It seems more likely that these spongy areas are similar to honeycombs full of some liquid ferments. The prolonged action of these ferments transforms the blood into very unctuous and lively semen. The origin and continuous progression of these small columns cannot be seen but it is likely that they are rooted in the tunica albuginea. There they suck from the blood a glutinous white juice together with its albugineous liquid. These bloody liquids are also sucked from the arterial capillaries through their lateral pores. Therefore, many small veins are disseminated throughout the internal substance of the testicle to retrieve the other elements of the blood which are not suited to make semen. The small columns then end in the nervous bundle lying in the axis of the testicle. They spill the semen in the very long and serpentine epididymis duct and finally discharge it through the spermatic duct into the seminal vesicles and the urethra.

Proposition CLXVIII

Elaboration of the semen.

The spermatic arteries are a little anterior and, after entering the testicles, their course is not straight nor very short as is usual in all the parts of the body. They are tortuous, winding sideways to and fro, forwards and backwards, proceeding like in a labyrinth and they surround the substance of the testicle. Moreover, all the vessels together with the extremity of the epididymis are covered by the tunica vaginalis. Hence surged the opinion that the prolongations of the artery would become white and would be transformed into the epididymis so that the seminal ducts would be nothing else than the prolongations of these arteries.

But the famous de Graaf injected coloured water with a syringe and saw that the arteries ramify into the substance of the testicle, below the epididymis and surround both. In summary, the arteries are distinct and different from the epididymis.

However, semen is elaborated from the blood which irrigates the testicles. It does not comprise all the blood but its purest and selected glutinous, white, spirituous and lively elements which can be extracted as a result of the similarity of their shape with that of the pores which perforate the seminal ducts. This is the same mechanism by which Nature separates bile from blood in the liver and urine in the kidneys. This selected white element of the blood alone is absorbed by the substance of the testicle, i.e. by its spongy small columns. This is surmised from the fact that many small veins spread in the depths of the testicle evacuate the other thicker, impure and red elements of the blood, and lymphatic ducts also retrieve aqueous serum with other dirts.

Then not only the most noble spirituous juice from the nerves is added to this selected part of the blood which remains in the testicles but it is also improved in long chemical reactions by the ferment contained in the spongy small columns of the testicles: ties are released and the spirituous and mobile elements are liberated and can exert the most active and vivifying energy.

Such fermentation can occur in the testicles. This is suggested by the long delay and slow movement which retain the semen in the cavities of the spongy columns and in the long and intricated epididymis duct. Similarly, chyle is elaborated by fermentation in the long and tortuous intestines of the animals.

Moreover, during their passage, a wedging and almost a drilling, through the labyrinthine ducts of the testicles, the spirituous juices and their elements acquire remarkable and different situations, positions, shapes and animal motive impetuses similar to and of the same character as those by which the different parts of the animal which must be generated are suited to move. There is general agreement on this point. However, this motive force given by the semen is so strong and efficient that it can penetrate immediately through the pores of the womb like magnetism or poison or plague. With unutterable energy it can move, develop and give life to the embryo formed in the egg and to its organic elements so that they are able to start motions in the way and at the time required by the vital and animal operation.

Proposition CLXIX

The substance of the testicles and their organic structure are in some ways similar to those of the brain.

We use to distinguish the similarity of viscera or their diversity from what appears to our senses such as their consistency, colour, taste, configuration and structure on one hand, and from their functions and operations on the other hand. Thus we are convinced that the brain is different from the heart, liver, spleen, kidneys and stomach because the brain has a soft consistency, intermediate between glandulous and medullary substances. It is white, sweet and milky, made of tortuous tubes. The other viscera have a fleshy consistency. They are red and their shape and structure are different from those of the brain. Moreover, the brain is the workshop of spirits. The other viscera make, purify or complete and improve either the chyle or the blood.

The thymus, pancreas, parotid and other glands have a soft and white substance, sweet and milky to the taste. They are made of lobules in which arteries, nerves, veins and lymphatic vessels ramify. The arteries and nerves bring liquids. The veins and lymphatic vessels suck them away as do the roots of a tree in the earth. Moreover, they have excretory ducts. Hence we conclude that all of them are glands and have similar nature and function.

In the same way, we see that the substance of the testicles is similar, not to the glands or viscera, but to the brain (their substance indeed is soft, white and has the same taste as the brain). The testicles have the same configuration as the brain since they consist of small spongy columns like the nervous fibres, filled by sweet and milky juice. Blood is sent into them by ramified arteries and the remainder is evacuated by veins and lymphatic vessels like in the brain. In the testicles spirituous and vivifying semen is elaborated from the best elements of the blood in the same way as spirits or vital and animal juices are made in the brain. Finally, nervous prolongations of the epididymis and vessels carrying the semen leave the testicles as ramifications of the nerves which carry the animal spirits to vivify and move all the body spread from the brain.

What thus opposes considering the testicles as another brain on its own? Their anatomy seems to have been destined by Nature not to the benefit of the individual but to the preservation of the species so that their vital and animal spirits do not deal with sensations and movements of the individual but procreate another individual similar to the first.

This may be why so few and such small nerves ramify from the brain in the testicles that some anatomists even denied their presence. The testicles are abundantly provided with their proper nerves ramifying on their own from the brainy substance of the testicles which by its nature is not subjected to the brain but to the royal animal faculty.

The conformation of the brain is achieved immediately from the beginning in the embryo, when the animal life, i.e. nutrition and sensation, starts. But the conformation of the testicles is achieved only at puberty when the emancipated animal is able to procreate.

Proposition CLXX

In the act of generation animal spirits are transfused from the brain to the testicles profusely and with great impetus.

The brain and the testicles are distinct workshops in which different functions are carried out and spirits different in character are elaborated. Since they are contained in the same body it is not surprising, however, that they influence each other. But the admirable harmony between the testicles and the brain seems to result from a completely foreign cause. We indeed observe that during coition the animals are disturbed considerably, they are affected by some convulsive passion like a pleasant epileptic fainting. This could not occur without energetic activation of the animal spirits and shaking of the brain and all the nervous system.

Such rapid movement of the animal spirits occurs from the brain through the nerves to the testicles and the spermatic ducts. This is suggested by the fact that not only the well-known external colours and features of the face, eyes, fingers, contortions and deformities are imprinted in the embryo by the father semen but also the good or bad dispositions of the mind, viscera and limbs are inherited. Since no physical action can be conceived without material contact it must be concluded that some particles from all the body of the father are transmitted with the semen carrying characters, deformities, diseases, etc.

Particles of all the material and spiritual parts of the father do not pre-exist in the semen present in the testicles but are added at the moment of coition. This is proved by the fact that semen is made from the blood. Blood, however, is not made from particles of all the animal but from new chyle. Therefore, particles which were in the chyle will also be present in the semen which will lack particles retrieved from all the body.

Proposition CLXXI

The spirituous juice elaborated in the testicles is a very powerful elixir or balsam which not only is prolific but also develops and improves the mind and body of the begetter.

As far as voice, strength and manners are concerned, male children look rather like women than like grown-up men. They change during puberty widening suddenly their trachea, deepening their voice, growing beard, acquiring strength in their whole body and the appearance of a man. All these changes always occur when semen begins to be elaborated in the testicles. Every time making of the semen is impeded or prevented by disease or by some external cause, the signs of virility are missing: voice does not become deep, beard does not grow, mind and body do not strengthen. Thus, semen elaborated in the testicles is the actual cause of these effects. It is so powerful that it can confer virility. Since semen is considered as the cause of any of these effects, when present the effect follows, when absent the effect is missing.

This is confirmed by observation. At certain periods when they strive to generation, i.e. when their testicles are swollen by semen, animals are stronger, bolder, irascible, fearless and bellicose whereas in other times, when their testicles are empty of semen, they are peaceful, frigid and shy. Thus semen elaborated in the testicles improves and strengthens the mind and body of the animal.

Castrated animals lack their previous strength of mind, quality, boldness, intelligence and body strength and courage. They are weak, frigid, shy and effeminate in temper and body. This appears not only in men but also in untamed horses.

Spirituous and most noble seminal substance is retrieved from animals during coition. This thus should rather debilitate them. On the contrary, in castrated animals this spirituous substance is not dissipated and their strength thus should be increased. But this does not occur. Production of semen in itself is the most important cause of strength and perfection of the animal since the most noble juice and vital elixir or balsam which is sufficient to propagate the species and remains in the begetter, is elaborated in the brainy substance of the testicles as a result of chemical fermentation of the best of blood, and mechanical sieving. This semen diffuses irresistibly, like poisons or the germs of plague, through all the limbs of the animal, increasing its mind and body strength.

Some modern authors concede that some gas rises from semen during reproduction. This alters the body and flesh as appears in rams who owe their stench to the semen. The meat of boars when they mate spreads a foul smell whereas the meat of castrated swine smells good.

In the same way, receiving masculine semen changes the body of females as taught by experience. The famous de Graaf observed a she-dog which suckled a cat. When the dog mated, the cat felt repugnance for her milk.

Such alteration cannot be understood without physical contact. Therefore, it is necessary that the spirituous corpuscules of semen diffuse from the testicles to the brain, heart and all the parts of the animal by its very powerful motive faculty and high activity. Similarly, poisons and the germs of plague move and corrupt the spirits and all the parts of the animal body.

Proposition CLXXII

The spirituous juice elaborated in the testicles is spread to all the parts of the body through the nerves alone.

Any physical action needs material contact. Therefore, if the spirituous corpuscules elaborated in the testicles are not carried to the brain, nerves and muscles, they seem to be unable to develop, strengthen and vivify the spiritual and organic elements of the animal. The elaboration takes place in the testicles which are remote and almost separated from the body of the animal. There seems to be no exchange through apparent paths besides veins, lymphatic vessels and nerves. It must be sought by which of these ways diffusion occurs. Firstly, spirituous semen is not transmitted to the brain and the other parts of the animal through the blood vessels. This is suggested by the fact that blood arrives quickly through the arteries and is left in the testicles. It is thus impossible for semen to be carried to the heart against the torrentuous flow of blood since it would be repulsed by the blood flowing from the heart. Neither can it be carried through the veins because the soiled, impure and red part of the blood must firstly be separated from its nobler part and then the best of the blood must be elaborated, fermented and improved by a long process in the very long proper ducts of the testicles.

Immediately after entering the testicles, the arterial blood whitens. This does not happen because the red is destroyed in all the blood which has entered. Indeed if this was the case, the small veins would remove the white remnants of the blood. This does not occur and those veins appear red. Consequently, blood whitens when entering because its impure red part is separated and sucked by the capillary veins.

It thus seems impossible that the most noble semen be absorbed by the veins and carried to the heart. Firstly, these veins absorb the impure ordure of the blood. Secondly, they suck blood immediately after its entry in the testicles, thus from an area very remote from where elaboration of semen is completed. They absorb the remnants of blood not from the spermatic ducts but from the body itself of the testicles. Thirdly, semen would be corrupted and become spoiled by mixing with the dirt of the blood which was at first retrieved.

It does not seem possible that semen be carried to the heart and brain by the lymphatic vessels for the same reason. Lymphatic vessels are excretory vessels for serum which contaminated the most noble substance of semen. This serum would corrupt again and spoil semen, thus making useless the previous work of Nature.

It remains for semen to be carried through the small nerves of the sixth vagus pair, through the spinal nerves and through those which communicate with the seminal vesicles, the urethra and the penis. Similarly, we observe that malignant juices of venereal disease are sucked from an infected womb through the pores of the nervous cavernous bodies of the penis and immediately transmitted to the whole body. The occurrence of opposite movements in the nerves is not an obstacle since such opposite movements can take place at different times. Juices and dirts enter and leave the long caecal intestine of birds. Similarly, spirituous semen can enter through the nerves and be carried in a direction opposite to that in which the animal and nutritive juices are diffused from the brain through the same paths but at different times.

I do not deny, however, that semen can be spread through the veins to the spermatic ducts and the seminal vesicles with the blood flowing back to the heart, then to the brain and hence to the nerves to be strengthened. But I think it very likely that most of it is carried directly through the nerves because of the shortness of the route and its facility. This is confirmed by the convulsive fainting which uses to affect the spinal cord, brain and nervous system during the venereal act. There is agreement that this is transmitted directly through the nerves and not through the veins. If indeed this path from the spermatic ducts to the brain through the nerves exists and if semen is frequently moved through it, why should we doubt that semen is also diffused through the same path to strenghten the brain and the other elements of the animal?

Proposition CLXXIII

In some way semen circulates in the animal like the blood.

It was shown that there are several opposite movements in the nerves.

The first is that of the animal spirits from the brain to the limbs to be moved and that transmitting the sensations from the extremities to the brain.

The second is that which carries drugs and poisons through the nerves to the brain and that which oppositely spreads the nutritive juices from the brain to all the parts.

The third is that by which animal spirits rush to vivify semen in the testicles and that by which oppositely semen affects the brain and the nervous system during the act of coition and by which it strengthens the spirits and the other parts of the body.

All these movements, although they occur at different times through the same paths of the nerves, suggest that there must be a circulation of spirits. The fact that the same mass of blood leaves the heart and returns to it in a continuous flow suggests blood circulation. Similarly, from the continuous flow of spirits from the brain and their return to the same area of the soul, a circulation of the spirits can be inferred. There is general agreement on this point. The same total volume of blood which left the heart returns to it. All the mass of blood is expelled by the heart in 4 minutes. This mass cannot be dissipated in such a short time nor can an equal mass be regenerated. Therefore, the flow of blood from the heart would stop if the same mass of blood did not return to fill again the fountain of the heart after having circulated. The animal spirits also must be spread abundantly to vivify and move all the body and its parts. They cannot be repaired and restored so quickly. Therefore, it can be suspected that the same spirits (which on the other hand we know are transmitted to the brain), after circulating, return to repair and replenish the brain, the fountain of spirits.

This suspicion is confirmed by the fact that the flow of spirituous juice from the brain must be as much quicker than the flow of blood from the heart, as the mobility and violence of the spirits are greater than the impulse of the inert blood given by the muscle of the heart.

You will say that the paths are not clearly visible. I answer: if blood circulation has remained hidden so long, who knows whether in the future the paths of the circulation of the spirits will not be discovered?

Since life consists of continuous movement and since the movement of the spirits must be very quick, if the spirits were dissipated and destroyed with the same velocity, they could certainly not be regenerated from the blood so quickly. Therefore, we can suspect that they return to the brain after their circuit.

Moreover, if this did not occur, not only the spirits which leave the brain would have to regenerate with the same velocity but also those which flow to the brain to carry poisons, drugs and strength from the semen. This even increases the difficulty.

But why look for conjectures since it appears that spirits move from the parts to the brain? It is thus reasonable that vital and animal spirituous juices carry out some circulation during the whole life.

Chapter XIII

On the generation and development of plants.

It is normal to proceed from easier and better-known facts to less-known ones. Plants differ from animals in that plants have an easier structure with a smaller system of organs and simpler ways of acting and faculties than animals. Nature operates using the same laws and mechanisms in all beings to which these apply. It is thus very useful to know the generation of the plants to understand the genesis of the animals.

Proposition CLXXIV

In the seed of any plant there is a small plant of the same species perfectly designed.

The Ancients knew that in the seeds or in the eggs of the plants a nobler particle is present which, like a bud, would be some rough beginning of the future plant. But later it appeared that this bud is actually the whole plant with shaft, branches, leaves and roots, comprised in two thick leaves or white and fat follicles from which the small plant can feed from the beginning.

Finally, the famous Malpighi in his excellent work on the anatomy of plants increased and improved this knowledge using the microscope. Therefore, I will only present the theory which I learned from B. Castelli and which I have then diffused.

Proposition CLXXV

Building and using a Santorio thermometer
as an introduction to the life of plants.
Tab. XVIII, Fig. 13.

Santorio was the first who designed an instrument to measure the temperature of the air. This was later called a thermometer. Its structure is as follows. A glass phial AB is provided with a neck which is a long glass tube BC the orifice C of which is open. The vessel AB is warmed with a hot cloth, the hands or some other means until the air which it contains is made rarer than the ambient air. The aperture C of the tube is immersed in the water contained in a bowl DEF. Cooling of the ambient air brings the enclosed air to its initial state of condensation. This enclosed air thus occupies less space in the tube in which water rises up to H. Thus prepared this machine shows any new change of the air. When the air ABH is warmed more than usual and rarefies, the level of the water H is lowered. If the air is cooled and condenses, the level of the water rises from H to I.

This is the Santorio thermometer. To adapt this instrument to our purpose, we assume that the phial ABC is soft and extensible like warm wax or something similar. We also suppose that all the cavity and above all the tube BC are spongy. Then when the air contained in the spongy pores ABI is rarefied by sun rays, the water which attained the level I cannot retrocede because the internal spongy structures act as valvules and thus impede descent of the water once this has been taken in. But the air warmed in AB requires more space. Therefore, by swelling the soft porous vesicles it increases the whole mass AB. When cooling occurs at night or as a result of wind the air contained in the spongy space AB condenses. Hence water CI rises further up to B. This phenomenon repeats itself again and again. By sucking more water, all the mass of the vessel ABC swells and increases.

Proposition CLXXVI

The shaft of a plant divided and immersed in water
seems to be fed and to grow by the same mechanism
as the thermometer described above absorbs water.
Tab. XVIII, Fig. 14.

ABC is a twig of any delicate plant or flower. Its divided shaft C is immersed in the water of a bowl DEF. It appears that the rod BC is filled by spongy fibres. The upper extremity covered by leaves or flowers BA is soft and spongy too. The very abundant pores are filled by air mixed with aqueous juices.

If the sun shines, the air enclosed in these pores must rarefy and thus needs more space. It cannot find it by expelling the aqueous juice through the lower orifice C of the rod BC because the valvules formed by the compression of the spongy structure prevent the exit of water. Therefore, the upper walls of the small pores which are soft and extensible must expand and in this way the volume of the bud of leaves and flower increases and grows. When shadow or night cold or cooling by wind occurs, the air which was expanded and rarefied in the pores, condenses. Therefore, additional water rises from the bowl DEF through the spongy tubes CB. This water must fill and saturate the empty spaces in AB. By the repetition of this phenomenon the leaves and flowers AB can be fed and grow.

Proposition CLXXVII

Beginning and generation of a plant from its seed.
Tab. XVIII, Fig. 12.

Germination of the seeds of plants must now be explained. Firstly, the spongy pyramid of roots sucks adjacent water either contained in a vessel or exhaled in vapours in the earth or mixed with air. Then the roots, acting like a sponge or the tube of Santorio described above, swell and lengthen because of their pores extensible longitudinally and not sideways as a result of the narrowness of the fasciae which oppose dilatation. This growth of the roots can be seen when seeds of plants are moistened. The buds or small plants at the tip of the pyramid of the roots swell at first by sucking water like a sponge and lengthen little by little. Then the small plant subdivides in several twigs. The movement of growth of the roots is outwards by diffusion of aqueous nutritive juice towards their extremity. Water is sucked through the lateral pores of the roots.

When the roots are sufficiently lengthened to go deep into the wet earth, an opposite movement begins. The spongy extremities of the roots suck water and carry it to the ball or egg of the seed. In the meantime growth of the roots continues. These opposite movements may not be simultaneous but take place at different times so that they do not impede mutually.

After the preparation of the seed, two hard, thick and turgid follicles AB and CB begin their action and are used like two phials of a Santorio thermometer. The air contained in their pores warmed by the sun and rarefied during the day is cooled by the cold of the night. Water is sucked by the roots D, rises through the spongy canals DB and progressively penetrates the pores of the leaves AB, CB. This water is impeded to retrocede by the valvules or by closing of the spongy membranes. When air is rarefied again, the pores of the follicles swell and the leaves AB, CB become turgescent and grow. After several repetitions of rarefaction and condensation of air, the leaves AB, CB not only are swollen and turgid but the external envelopes of the follicles harden and resist further stretching. Then the growth of the follicles stops. But in the meantime the air contained in the expanded and multiplied pores of the follicles is rarefied again by a new episode of warmth. It exerts a force to dilate the narrow spaces of the pores. Such expansion is impeded by the resistant cortices of the follicles. The air is thus forced to dilate the pores of the bud or small plant IH comprised between the follicles. When cold occurs again the absorbed water is no longer driven into the pores of the follicles but rather into those of the bud which is soft and extensible. Development and growth thus proceed in the small plant IH while it is fed by the suction carried out by the teats of the follicles AB, CB. After the small plant IH has developed sufficiently to be able to replace the follicles, its auxiliary thermometers become useless and progressively dry up. Then the plant continues by itself its development and growth.

If I am right, such is the possible and likely period of germination and feeding of the plants from their seed, which I discovered, guided by my eminent professor B. Castelli.

Proposition CLXXVIII

Water is the nutritive juice of the plants.

Seeds of plants are compared to the eggs of oviparous animals in that in the seeds there exists a complete small plant and in the eggs a small animal with teguments and organs which help in their growth and development. However, they are different above all in what concerns the substance on which they must feed and grow from the beginning. Indeed eggs contain nutritive juices for feeding, unlike the seeds of plants. The eggs of animals have no need for external juice but they carry out growth and development needing only moderate warmth. The eggs of plants become rather dry and die when exposed to external warmth alone. They begin to develop only if moistened by water. This water is not used only as a vehicle carrying another nutritive substance from outside. Actually, seeds and plants irrigated by distilled water alone develop and grow without addition of any other substance. Neither is water used to dilute the farinaceous substance of the follicles of the seed and to make it slippery like gruel from which the small plant would be fed. The scarce substance of the follicles would be consumed quickly. Therefore, neither the follicles from the beginning nor the plant could grow to its huge size without addition of another material substance from outside which is nothing else than water. Water without which plants cannot grow thus is their actual and proper food.

Two principal circumstances must be noted in the development of plants. Firstly, while the root in the seed lengthens, the two follicles and all the other substance of the seed do not seem to be altered because they are covered by a complete, hard and dry envelope. Therefore, growth of the root does not result from some internal juice in the seed, liquefied or diluted by external moisture introduced inside. I observed that in bay berries which were found above fallen leaves in the forest. These grew long black roots through the earth. These roots were ligneous and like threads. Some of them were half a foot long. But meanwhile the cortex of the berries was intact and dry. The substance inside the seed was white and hard with the same taste, the same shape and size as those of the other berries without roots. Secondly, I observed that this considerable growth of the root resulted, not from outflow of the internal substance of the seed but from an addition from outside. The volume and the weight of this long root were about the same as the volume of the berry itself which was not reduced at all. Consequently, the substance of the root was undoubtedly created by the external moisture introduced into the pores of the pyramid of the bud of the seed.

This is confirmed even more clearly by the experiments of van Helmont. He observed that a branch of willow increased by five pounds in five years, to weigh 169 pounds. This increment was due to water alone since the earth in the vessel in which the willow grew was almost not reduced. The famous Boyle and others observed the same fact in garden plants. Personally I saw a poplar grow to considerable height, fed only by water through the roots which were floating in a water-main, like an abundant hair. I also know that oysters do not eat anything and absorb only water through their gaping mouth. From this aqueous food they increase not only their internal meat but also their marble-like shell.

From all this it does not seem unlikely that plants are fed by and grow from water alone as if it was blood.

I dare not deny that, besides water, plants absorb salts and sulphurous particles from the earth. But these appear as condiments of the main nutritive aqueous juice. They help considerably the germination and fructification of the plant. Indeed seeds and plants are fed by and grow from tasteless and distilled water alone but not as well as when provided also with salts.

Proposition CLXXIX

It does not seem possible that water is transformed into a plant by ferment present in it or in its seed.

It remains to examine how plain water is transformed into plant by acquiring hard consistency, sweet or bitter taste, and countless other qualities.

The problem is difficult. The juice contained in the two follicles of the seeds of the plants seems to constitute the best spirituous and nutritive substance since it is abundantly provided with oily fat and saline particles and is similar to the white and yolk of the egg of animals. The liquids of an egg are like milk with which the embryo is fed initially and grows until birth. There seems to be agreement that a plant also is fed initially and grows from the milky, spirituous, fat and saline liquid of the follicles of the seed. In a fertilized egg these spirituous juices activated by the external warmth begin to feed according to an organic need and expand the umbilical roots. The spirituous milky juices of the follicles of the seeds of a plant, activated by emanations of water, also grow roots and afterwards feed the plant itself.

From this likeliness some people conclude that the water which is sucked by the roots of the plants actually is rough and unsuited by itself for feeding and that it must be fermented and digested by the very powerful juice which is kept in the pores of the follicles and in the plant itself.

The duty of a philosopher is to avoid carefully being impressed by these deceiving metaphoric words of fermentation, digestion, maturation, exaltation and plastic virtue. It is thus wise to find out whether these are suited to our subject.

Thus, it must be noted that all the operations which are ncessary for the nutrition of animals are not required for the nutrition of plants. Food is digested by ferments in the stomach of the animals. It must not be deduced that sap ought to be prepared in the same way in plants. Animals need more organs and workshops which are missing in plants.

Firstly, animals have teeth with which they divide and grind solid food. Plants need no teeth since they do not absorb big and hard pieces of meat and fruit. They suck through the narrow orifices of their roots liquids and minute particles which need no further division.

Secondly, animals have a stomach and intestines in which food macerates, is cooked and ferments, i.e. the conglutinated elements of food are separated and dissolved so that the saline and spirituous particles and other mobile elements being freed can exert their motive quality. They can induce either bitterness or sweetness by the unveiled edges of the salts and opened clusters. For this reason all the activated mass can be warmed up and boil. This is called fermentation. Ferment thus is like a rasp of Nature which resolves an aggregated body into minute particles.

In plants food can be divided into the smallest particles through the very narrow capillary roots which can absorb water. Afterwards a series of vessels can receive fat particles, others saline solutions in water, others air particles, others different liquids, depending on the different shapes of the orifices of these vessels. Consequently, fermentation is useless.

Thirdly, animals have excretory vessels to expel solid and liquid excrements. Plants do not need them since their roots do not absorb dirts but only useful and clarified liquids.

Finally, animals have lymphatic vessels to sort out chyle with their multiple tubes, leaving impurities in the intestines. They collect and carry chyle into the mesentery. There finally, only after so much work, do the animals use the same numerous and capillary roots through which plants begin their nutrition, sorting out from the earth and sucking the juices suited to feed them. Plants thus need no fermentative preparation.

Ferment would be useless in plants. This can be demonstrated as follows. Simple water by nature is tasteless and inert. It does not seem to have any active force. Therefore, when water is mixed with this ferment which some people think to exist in the pores of the plants, water is certainly not transformed by this contact so as to become active whereas it was inert, i.e. becoming sweet whereas it was tasteless, or bitter, or smelling, etc. Water only dilutes and is used as a vehicle to carry the particles of ferment with which it is mixed, like a stream flowing through salt mines absorbs some particles of salt.

In the shaft of the willow described by van Helmont, which was fed with distilled water for five years, all the ferment present in the twig would have been consumed necessarily. The continuous flow of water passing and transpiring was due to wash out of all this ferment which on the other hand could not be repaired and restored by pure water. The plant thus would have perished soon. This did not occur. The plant lived and grew for five years and would have survived and grown longer if it had not been extirped. I saw a poplar which grew tall for a long time fed only with water, as I mentioned above. The water was not so impure as to have to be digested by a ferment, soon dissipated and consumed, which would have been in the poplar initially.

It must also be noted that the molecules of water are not primary simple bodies. They are composite, made of primary elementary atoms. The link and structure of these molecules can be dissolved by incision with acute harder bodies, moved violently and delivering blows.

I certainly do not deny that some very powerful ferments sometimes can break the links of molecules of water. But I do not admit such ferment in plants for two reasons. Firstly, from an incision made at the extremity of a twig of a tree liquid water seeps of the same nature as the water which irrigated the roots, mixed, however, with other saline, milky and fat juices which are additional condiments. Similarly, clear soup retains its nature of water even if muddied and tainted by particles of meat, vegetables and aromatics.

Consequently, if these ferments imagined in the plants cannot alter by their contact the most part of the water absorbed by the roots, there is no reason that they dissolve and change the other water which is transformed into nutritive juice.

Secondly, even if these ferments are able to dissolve the structure of the molecules of water, another considerable and very difficult operation remains which seemingly cannot possibly be completed by the same ferment. Indeed the particles thus separated would have to be united again and conglutinated in a new and artificial combination to result in the structure of the nutritive juice of a quality so different from plain water. It seems impossible that this new combination be carried out by a ferment since a ferment characteristically acts with some blind and incidental violence like a flame. It can induce destruction rather than a combination. Therefore, it seems impossible that the nutritive juice of the plants be composed from pure water by fermentation.

If such a digestion of water cannot occur in plants by a ferment it must be concluded that such ferment does not exist in plants since Nature does nothing in vain. Consequently, the fat milk which is contained in the follicles of the seeds of the plants is no ferment but nutritive juice provided by Nature to feed the delicate plant.

Proposition CLXXX

Making of the nutritive juice of plants must be completed by the sieve-like structure of their vessels.

Plants thus are fed by water alone, pure or modified by saline and sulphurous condiments. It was shown also that this water cannot be transformed into food for the plant by the action of a ferment present in its pores. It must be found how and by what mechanism plain water can acquire faculties and qualities so numerous and diverse as to result in the constitutions, colours, tastes and active virtues which we observe in so many plants.

Such mixing and composition cannot occur without breaking the previous structure of water. Afterwards a new form is induced by a hidden and active plastic faculty. Otherwise the particles could not be collected in their due places, be disposed, placed, linked and shaped regularly to achieve the perfection and faculties designed by Nature.

All this no doubt must occur immediately, not by way of some prudent and intelligent process but by a natural mechanism designed by the Divine Architect. Actually, as mentioned when we dealt with the blood composition, we think that this necessary, natural and mechanical combination cannot occur without absorption of different and well-determined juices by the very narrow tubes of the vessels of the plant acting as a sieve. A juice can enter any tube the orifice of which has the shape corresponding to this juice. Then it is necessary that these small vessels, curved and intertwined like a network, end in appropriate places. There, by instilling the droplets which they contain they can complete a structure similar to a mosaic work. They thus make the proper mixing and composition of sap which is the blood of plants.

I think that plants can be fed and grow from water by this mechanism. Transformation of water into the hardness and consistency of ligneous plants is no problem. When frozen by intense cold, the same water is changed into hard ice by modifying only the position of its elements. Similarly, the stony shells of oysters are made from water alone since oysters do not absorb any other food than water.

Proposition CLXXXI

The main cause of development of the plants is air.

Nobody questions that plants contain countless particles of air inside their pores. It is commonly thought that this is the cause of their lightness which appears in water. They lose it and descend to the bottom of the water when soaked after some time or when the vesicles of air have been expressed from straw by compression with the hands. Then bubbles of air are seen rising in the water.

Plants could not be fed and grow if they were deprived completely of air because aqueous juices, as a result of their weight, cannot rise by themselves from the roots to the shaft and branches. They must be pushed by an external force. This external force is the atmospheric pressure which drives water in the roots of the plant and pushes it through the vessels up to the last twigs because the pressure of the air contained in the pores of the plant is less. Similarly, water is sucked in the Santorio thermometer or rather is raised in the plant by the weight of the external water which is driven through the very narrow pores or small canals. This also occurs in sponges and capillary glass tubes as explained in my book on natural movements resulting from gravity, chapter 8.

The action of a sponge or a filter is not sufficient to suck completely the water up to the extreme twigs and leaves of a tree. This is demonstrated by the following experiment. A twig is cut from the top of a vine or from another tree.

Sap drops from the cut for some time. A small tube or a filter is set perpendicular to the water of a bowl. Water does not flow from the top of the tube or filter. The tube or filter or the upper edge of the filter must be bent so that they are lower than the surface of the water in the bowl, as illustrated in Fig. 15, Tab. XVIII.

Thus rising of water in the divided vine does not occur by the action of a sponge. Hence, this must occur by an operation similar to that by which water is sucked by the Santorio thermometer, as described above.

But one can question whether the action of the divided vine and that of the Santorio thermometer are the same and result from the same cause. Actually, if the top of the tube of Santorio is cut or perforated, suction of water stops completely. The opposite occurs in the divided vine in which not only suction from the roots continues but also a significant quantity of sap flows out.

This discrepancy in the effects seems to result from the difference between the structure of the common glass of the thermometer and that of the plants. Firstly, in the plants the spongy valvules of the vessels impede regression of water. The droplets of water pushed into the pores of the twigs replace the lower bowl of the instrument of Santorio. Secondly, in plants there is not only one phial full of air like in the glass instrument. There are as many vesicles as there are pores full of air. These exist not only in the hard part of the plant but are also disseminated in the sap of the plant. Thirdly, these vesicles not only are expandable as a result of the softness of their skin or of the slipperiness of the liquids but also they can transpire vapour from their small pores. These can be closed at once. They are similar to the pores of our skin which emit sweat and vapour and close immediately by contraction of the soft skin. In wax also or in boiling water the small pores from which vapour transpires close spontaneously.

Fourthly, all the twigs of plants have the power to retract laterally under tension and the power to contract the fibres and fasciae surrounding the twigs circularly. This appears after the cortex has been incised longitudinally. The edges of the cortex gape in the same way as scars widen in animals.

Aqueous juice is sucked by the small thermometers of the plant and is carried everywhere up to the extremities of the twigs by the mechanism described above. This suction is not stopped or interrupted by excising a twig because many vesicles of air remain intact. They suffice to carry out the operation of the thermometer. Meanwhile the raised water cannot retrocede, as a result of the closure of the spongy valvules. Therefore, after incision of a twig, sap is expressed by the constriction of the fasciae and can seep drop by drop.

I now return to the growth of the plant in which two different actions must be considered: one is sucking of water from the roots of the plant, the other is growth itself.

Concerning the first action, in the Santorio thermometer and in those of the plant the cause of the ascension of water undoubtedly is not the difference of specific weights between the ambient air and the air contained in the vesicles. Actually, internal and external airs are equally rarefied by the rays of the sun and condensed by the cold of the night. It would be impossible, because of this permanent equilibrium, for external air to raise water by its pressure. Therefore, in this instance the specific weight of the air contained in the tubes must be made lighter than that of the ambient air by another circumstance.

This is as follows. In any closed hard phial containing air, even when air is condensed by cold, the phial retains the same specific and absolute weight as previously. The volume of its cavity is not decreased and thus is filled partly by condensed air, partly by vacuum. Contrariwise the ambient air which is not contained in a vessel occupies less volume when it condenses as a result of cold. It thus becomes heavier than the air enclosed in the tubes. It can, therefore, by its pressure, push water upwards with a force of several pounds.

Growth of the plant does not result from cold condensing air but from warmth of the sun which rarefies the air enclosed in the vesicles. Because of the rarefaction the soft and extensible vesicles must dilate and grow. Growth, however, is unequal. It is more pronounced towards the top of the plant. In the part of the shaft towards the roots, expansion is considerably hampered by the hardness of the cortex and sideways by transverse fasciae. The soft part towards the top gives way easily and lets itself be pushed and stretched.

The direction in which a plant grows is remarkable. It always tends vertically upwards when its progression is not impeded or deviated by excessive weight of the branches or other obstacles. This appears in the twigs which despite being below or at the side of a branch and looking to the ground initially, afterwards bend their course and continue to grow vertically. This certainly follows Archimedes' laws of statics. The effort by which the lighter ambient air compresses the aerial phials of the plants can occur only vertically.

Such compression, however, could not occur without the external air communicating with the internal air contained in the vesicles of the plants so that it can freely pass through the roots and pores of the cortices. This is how nutrition and development are carried out in plants.

Proposition CLXXXII

Plants also breathe although imperfectly. Their life depends on and is maintained by this respiration.

So that nobody refutes as a paradox that plants breathe in some rough manner, like animals, I will present publicly the reasons which move me to this opinion.

Plants certainly suck air mixed with water continuously from their roots. This can be compared with the inspiration in animals. In plants, although there are no muscles such as those which dilate the thorax of the animals and send air into their lungs, there are however tense and rigid ligneous fibres which can be separated from each other by wedging in of droplets of water, and brought back together after water has left. This appears from my experiments in the Accademia Del Cimento. The vessels and the rings of boxwood, although they are very hard, obviously swell when soaked as ropes thicken when wet. Moreover, the ligneous fibres of the plants which develop are dilated by warmth and the air which they contain in their vesicles is rarefied. They logically would contract as a result of expression and transpiration of molecules of water if there was no subsequent new sucking of water from the roots, not at the same time but later when the warmth abates and the air which was rarefied in the vesicles condenses again.

Hence, the alternate changes force the ligneous fibres of the plants to contract and relax not only at different times but also at the same time in different parts of the same plant since, when vesicles are expanded by rarefaction of the air, the adjacent ligneous fibres necessarily are compressed and squeezed. Water contained is thus expressed. On the contrary, when the interstices between the ligneous fibres are swollen and expanded by arrival of water, the vesicles of air are compressed.

These alternate lateral dilatation and constriction of the fibres of the plant obviously compare whith the expansion and constriction of the thorax in animals. These movements must be carried out by muscles in animals. They are carried out by the ligneous fibres in plants. In both instances they result in entry of new air. Therefore, plants also have their respiration, although less complicated. By this mechanism air mixed with water is continuously driven through the roots and carried to all the parts of the plant. There an oscillatory movement like that of a pendulum occurs as a result of the elasticity of the small machines of air and of the alternate constriction and dilatation provoked by external causes, the rays of the sun, the shadow and above all the winds. This oscillatory movement is the principal and most important cause of the life and development of the plants. By the same mechanism, the vital movement is maintained in animals by the oscillatory movement of air.

Chapter XIV

On animal generation.

The Wisdom of the Divine Architect shines the most in animal generation. The famous authors Fabrizio d'Acquapendente, Harvey, Malpighi, de Graaf and others discussed this subject and observed many facts which had not been noted by the Ancients. We, therefore, gratefully like to mention their works and to praise them. I will add a few speculations which, if they do not discover the whole truth, will at least stimulate more perspicacious people to find better.

Proposition CLXXXIII

Both male semen and an egg formed by the female contribute to the generation.

Male semen is required for generation. We knew this by experience. But hitherto it was questioned whether females also make their semen and whether the foetus is procreated by both. Finally, research and sagacity of modern authors discovered that viviparous as well as oviparous female animals do not produce a genital juice but eggs, i.e. they elaborate a complete body with all the organs and elements necessary to the development of the animal. They also observed that in viviparous animals the small membranous purses which the Ancients called the female testicles are ovaries. Finally, modern authors discovered and demonstrated definitely that all generation consists of the formation of the egg by the female, and its fertilization by the male semen.

Proposition CLXXXIV

Male semen does not fertilize the egg by some immaterial force or by a gas or breath but by contact and mixing of its corpuscules.

The famous Harvey did not find male semen in the womb of does after coition. He thought that its liquid had been filtered by the uterus and that the inside of the uterus had been fertilized only by an immaterial breath or by some spiritual irradiation. Afterwards R. de Graaf believed that the eggs contained in the ovaries were fertilized not by an immaterial irradiation but by a gas or a wind sent into the Fallopian tubes. Then others said that all the semen or its more delicate part is transmitted to the blood through the veins and so fertilizes the egg.

The opinion of Harvey was rejected by everybody as unbelievable. Who will accept that male semen can activate the liquids of the womb and form the organic and vital embryo by an immaterial force, indivisible, imperceptible like intelligence? And how, after its long and complicated elaboration in the testicles, can semen carry out an immaterial action?

Even if nothing is found in the womb after coition, this does not mean that all or the principal part of the semen was not absorbed by the uterus. It is possible that actual paths and canals through which the semen is transmitted do not appear. It can be sucked by blind ducts. Similarly, water transudes through the inconspicuous pores of a baked clay vessel; water, oil and even much more rapidly spirit of wine penetrate through the pores of the skin and through those of meat. These liquids are material. They do not change their nature but penetrate intact. Therefore, it is not necessary to resort to irradiation or to a vaporous breath since semen can most easily be sucked through the pores of the spongy uterus and be carried to the eggs to be fertilized.

Transmission of semen inside the blood through the veins does not seem less absurd. Firstly, the orifices of the veins do not gape inside the cavity of the uterus so that they could absorb the liquids contained in the uterus like leeches. They are rather disseminated inside the parenchyma of the uterus where they can absorb the blood flowing from the arteries.

Secondly, the veins of the uterus called hypogastric do not go to the ovary but to two crural branches of the vena cava. Some drops of semen can occupy only a very small part of the vena cava descendens and taint that small portion of blood which they touch and with which they mix. They cannot touch the blood which flows to the heart from other veins and from the vena cava ascendens. Consequently, the portion of blood tainted by semen is a minimal part of the blood which passes through the heart. Therefore, it can easily occur that this tainted portion of blood is driven into the ascendens artery rather than into the descendens or, if by chance it enters the descendens artery, that it does not reach the recess of the ovary arteries since it rather circulates through the largest arteries to the head, to the kidneys and through the whole body. Therefore, it can happen that semen is never carried with the blood to the ovary. Then most of the eggs would remain unfertilized, frustrating the desire of Nature, for lack of a direct and shorter route through which semen should be carried to the ovary. Who will believe that Nature organized so inappropriately her ways of generation?

Some people may say that all the mass of blood is contaminated by some drops of semen, the contamination propagating successively from one part of blood to the next until no particle of blood remains intact. But it is easy to enumerate the absurdities which would result.

Firstly, the power and spirituous energy of semen would be made inefficient as a result of its useless dispersion in such an abundant mass of blood. Similarly, a few drops of wine dispersed in an amphora of water lose their virtue. Secondly, all the mass of blood infuriated by the very powerful ferment of semen and its very spirituous warmth would provoke fevers, dizziness and other symptoms. Thirdly, Nature would stupidly achieve with such complication, difficulty and risk what it could obtain simply through a very short route, by introducing semen into the ovary through the pores of the womb.

There may even be a shorter way for semen to fertilize eggs which have parted from the ovary and are in the vicinity of the uterus itself. We see indeed that the eggs of chickens do not wait for the arrival of the semen of the cock in the ovary but, although unfertilized, they are carried to the uterus and grow on their way. Therefore, as semen of a cock can fertilize growing eggs close to the uterus, the women's eggs parted from the ovary and close to the uterus can be fertilized by male semen by contact with this semen sucked by the spongy substance of the uterus which semen swells and vivifies wonderfully by its fermentative faculty, as observed by Harvey.

It appears that famous authors were deceived by valvules and obstacles which were found in the Fallopian tubes and in the cornua of the uterus. These valvules allow easy descent of the egg to the uterus but not its regression. Consequently, the semen which is fluid cannot be carried to the ovary. But it may be able to penetrate through blind ducts, or through spongy pores of the womb, the Fallopian tubes and their thin valvules. The course of a stream seems to be stopped by a transverse heap of sand but filtering through the interstices of the sand is not impeded. The same phenomenon clearly can occur more easily in the uterus as suggested by the very experiments brought forwards to confirm the opposite opinion. If drugs, by rubbing penetrate the skin and flesh and if the same authors concede that the spirituous semen is carried to the internal veins through pores or blind ducts of the womb, why do they deny that it can a little later penetrate the adjacent ovary?

The eggs of insects are fertilized by contact and sprinkling with semen as observed by the famous Malpighi. This suggests that the same occurs in the other animals. This is also in agreement with the fact that Nature always operates economically and abhors complications. But this will appear better in what follows.

Proposition CLXXXV

All the organic construction of the animal is generated in the egg not by the mother alone but mainly by the semen of the father.

If we were to trust our senses, we should say that fertilized eggs are not at all different from unfertilized ones. Their liquids are the same in quality and quantity and both seem to contain the same membranes. Hence the conviction prevailed that fertilization of the egg concerns only the mother and that the regular arrangement of the substance and the organic conformation of the embryo are achieved and completed by the female alone. Thus, as in plants the seminal eggs contain the bud or the small plant formed regularly with its membranes, vessels and juices necessary for its growth, similarly membranes, vessels, humours are present in the eggs together with the organic embryo. The eggs of plants are different from the eggs of animals in that all of them are fertilized whereas in animals only the eggs which absorb a motive force from the male semen are fertilized, i.e. those on which the vital virtue has blown. The others die. The fertilized eggs can be compared with a clock made of cogwheels with an engine or a hanging weight. The unfertilized ones are similar to the same clock without engine.

This opinion is partially true. But I cannot agree that the male semen, besides its motive force, does not provide the egg with anything else nor that all the organization depends on the female. It seems to me that the organic construction of the foetus is completed and shaped by the male semen.

It seems unlikely that nothing else occurs in the brainy substance of the testicles than selection of mobile particles from the best of the blood and their liberation from ties with rougher particles as occurs in distillation of alcohol from wine. I think that in such complicated chemical process and filtering through the labyrinthine canals of the testicles, the elements of semen are placed, arranged and displaced in many surprising ways. This results in some general aspect composed of many vital elements shaped in different manners. In a word, the vital organism is created. This organism is added to the substance regularly arranged in the egg which comprises the organic beginnings of the embryo with all the organs, liquids and workshops necessary for the feeding and growth of the animal. This addition results in a perfect and fertilized egg.

My conjecture is confirmed by the observation of the famous Malpighi who in all the eggs sprinkled by the semen of a cock found the frame or embryo of a chick similar to the bud of the eggs of plants. But he found no traces of an embryo of a chick in the cicatricule of unfertilized eggs. In these the liquid was a shapeless substance. Therefore, the frame of a chick which is found in fertilized eggs results from the addition of male semen which is mixed and conglutinated with the substance of the egg. This frame is the embryo which thus has been composed and shaped by the male semen.

Neither can it be said that single male semen can transform the liquid of the egg into an organic embryo by its plastic virtue since a homogeneous body provided with one simple motive force such as flowing water can carry out different operations in different instruments, like grinding corn in a mill or indicating the time in a clock. However, it seems impossible that the same flow of water builds the mill from wood or the clock from metal. Similarly, if semen was a homogeneous body provided with one single motive force it could not shape the embryo in the cicatricule of an egg from the simple substance of the latter. It appears as if semen exerts some plastic power and is an organic body. How can it apply itself to elaborate the different parts of the cicatricule of the egg and, by displacing them in different ways, shape all the organs which constitute the embryo without elements of different qualities and without different instruments and movements?

During ejaculation, all the parts of the animal body shake somewhat convulsively and with something like slight epileptic fainting. Such excitement indicates that spirituous juices rapidly rush from the brain and all parts of the animal to vivify the semen, mix with it and be driven into the womb of the female. We are thus convinced that semen impregnated by these spirits is not a substance completely homogeneous provided with one single motive force but that it has a somewhat complicated structure and some organization. Either the elements of semen itself or the added spirits must have different tempers, shapes and motive qualities. They thus build a general structure which is not simple and homogeneous but very composite and almost organized.

This organic seminal substance poured into the female egg can move, twist and shape the material liquid contained in the bud of the cicatricule and arranged at best until this acquires the shape of an embryo. Afterwards, animating this embryo by vital spirits makes the egg perfect and fertilized.

Spirits which rush to vivify semen during coition carry and mix different natures and characters with the semen. This is suggested by the nerves and signs very often imprinted in the foetus, the heredity of the conformation of the limbs of the father, of his temperament or of rheumatic or nephritic diseases, etc. of the viscera and limbs which affected the father. This clearly could not occur if all the embryo was completed by the mother in the organization of the egg and if male semen provided one single motive force alone. Therefore, it seems very likely that, in their quick flowing during coition, the spirits of the father scrape some particles from his organs and perhaps carry hereditary diseases of these parts to the foetus.

Proposition CLXXXVI

Hypothesis on the mechanism of the fertilization of the egg and why the egg is activated only if stimulated by the foetus or by an external impulse.

Among the technical achievements, we do not see anything more comparable to animals and plants than an automaton or a clock made of cog-wheels. In such devices we must consider three points above all:

(1) the shape of the wheels which can carry out different movements;
(2) the addition of a motive force such as a hanging weight or the force of wind or a flow of water;
(3) the way of action of the automaton.

In the preceding proposition we attempted to explain how the male semen imprints the configuration and shape of the embryo in the substance contained in the cicatricule of the egg. It must be explained how the animal motive force is imprinted in the embryo. Since semen is a very powerful ferment vivified by its own spirits and by the added spirits of the father, gifted for considerable activity and mobility, we presume that its motive faculty infused with its liquid through all the limbs of the embryo already imprinted and shaped in the egg by the same male semen provides this embryo with a vital motive faculty. This may occur in the same way as magnetic spirits or particles, by moving and activating a whole mass of iron, organize its elements properly and imprint a motive force in the iron, according to the laws of magnetism.

We can compare an egg fertilized by male semen with a clock made of cog-wheels and which would be not only ideally built but even provided with a motive faculty, i.e. with an engine or a hanging weight.

Actually, if an embryo vivified by animal motive force is present in a fertilized egg, why is such vital force dormant and immobile, like in a corpse, in eggs of birds and reptiles and why do vital movements begin only when stimulated by external warmth?

Although we ignore the precise mechanism used by Nature in this piece of work, we perceive however its possibility from some analogy with comparable achievements by Nature and technique. In dry hay, warming maceration is stimulated only if the hay is moistened by water vapours or by sprinkling with water. It is obvious that water in itself has no motive and warming force but can only soften the dry parts of hay by mixing with them. Afterwards the active spirits retained in recesses of these parts not only can exert their motive character but they are pushed and shaken by the movement of water when this water was sucked.

We observe the same in the seeds of plants which germinate only if stimulated by moistening.

In other vegetables and composite bodies sufficiently moistened, fermentation starts only when they are slightly activated by moderate warmth. Too much warmth dries them and too little is unable to induce fermentation.

Similarly, the eggs of birds need external warmth, not too much since then the white would disappear but moderate so that slow movement activates the spirits quiescent in the embryo of the cicatricule. Then vital fermentation can proceed.

But the question still remains why the seminal spirits present in the egg cannot start vital movements from the beginning, without an external instigator.

These spirits in the egg may be impeded and the obstacles removed by the ferment of warmth. But since this answer is not satisfactory, we shall attempt to find something more precise from an artefact, from a clock composed of cogwheels with an oscillatory pendulum. In such a clock, although the motive force of a hanging weight is present, the machine remains immobile and inactive if the first oscillations are not provoked by somebody.

The mechanism of this achievement results from the nature of the pendulum which is moved from a vertical position until the forces are balanced and then becomes immobile. The hanging weight always exerts the same force. When the pendulum is raised by the hand higher than required by its equilibrium, it falls and exerts its oscillatory power. Since this is always revivified and renewed by the impulses of a horizontal wheel, the oscillations continue and with them the movement of all the wheels of the automaton.

It results from what was said above that there are oscillatory pendulums in the animal machine. Countless particles of air are included in the blood and liquids of the animal. They are oscillatory machines. They are the main and most powerful causes of the vital movement of the animals. Being successively compressed and rebounding they exert a continuous oscillation in the animal. Shaking and activating the liquids and spirits of the animal they act like the wheels of a clock and maintain the vital movement of the machine by their pendulum oscillation.

In the organic animal machine, the movement of a clock which maintains life also starts it. Thus, when the first oscillatory movement begins in the egg, then the vital movement starts. Hence I presume that the moderate warmth of the chicken gives the first impulses to the small oscillatory machines of the embryo in the egg. These impulses set in continuous movement all the organic parts of the embryo according to the requirements of the animal machine. Some particles orient their course to shape rudimentary vessels for nutrition, others to prepare the food, others to expel the excrements. Other particles distend the membranes and envelopes of the egg. Others improve the fire and stimulate in the heart the perpetual movement necessary to life and transmit blood to the heart. Others achieve tasks according to laws prescribed by the Divine Architect.

I could add here many other achievements which I deliberately omit not to remain too long on this matter.

Chapter XV

On insensible perspiration.

The animal is ruined and destroyed according to the same natural law by which it is born, grows and feeds. Nutrition and destruction combine in the animal so that one cannot be without the other. They stimulate and help one another. Didactically loss of the parts must be considered before their repair.

Proposition CLXXXVII

Famous theory of Santorio, confirmed by experiments, on insensible perspiration.

In his Static Medicine, the famous Santorio observed that if the food of man amounts to eight pounds a day, about five pounds perspire insensibly through the pores of the body and through the skin and about half a pound through the mouth. This perspiration is twice more abundant during sleep than when man is awake. We perspire the more the more we are fed.

Moreover, when vital movements are disturbed by disease or any other cause, insensible perspiration is impeded proportionally to the disturbance. Therefore, if perspiration is not by itself the absolute cause of health both are at least connected tightly so that good or bad vital movements are always combined with normal or abnormal perspiration of elements of the substance of the animal.

Proposition CLXXXVIII

Insensible perspiration is necessary to maintain the life of the animal.

The life of the animal consists of continuous movement of its particles, as suggested above. The degrees of life and perfection of animals can be distinguished by no other sign than the number and velocity of the movements which they carry out. The more perfect animals such as dogs, horses, etc. are activated by both more numerous and more energetic movements than are less perfect animals such as oysters and the like, and these by more than plants since the first carry out more and nobler sensitive and motive vital actions than the second and the second more than the plants. These motions are carried out by liquid and solid particles of the animal, either inside the animal by fermentation or circulation or rather by diffusion, by transmitting food to particular elements and by expelling harmful and useless excrements. The excrements which are expelled through the pores of the skin are said to perspire. Thus sweating is nothing else than expulsion of the mentioned particles through the pores.

Such perspiration is absolutely necessary to maintain life. The building of the animal is not the same as an inert and dead structure or as a mosaic work in which stones laid to represent different shapes are sealed and retained in their due places. The elements of the animal are rather comparable with a stream or a flame or a marching legion of soldiers the components of which are arranged in a certain order but are not quiet and move continuously. Such movement is possible in the animal only if some elements successively part and leave, i.e. if they perspire, and if others arrive to replace them, filling the spaces left empty. Perspiration thus is necessary for two reasons: to eliminate the useless elements so that they do not corrupt the organic structure by their foulness and to make possible the subsequent arrival of nutritive juice to restore the lost parts.

This arrival could not succeed if the blood was not pushed in a violent course by the heart. Such impulse in turn would not be possible without circulation of the blood. This could not occur if impulses of spirits from the brain did not produce heart pulsation. Consequently, perspiration appears to be necessary to maintain and prolong the life of the animal.

This is confirmed by the fact that, when perspiration is impeded or altered in its periods of occurrence, diseases result as everybody knows. Perspiration can be impeded not only by reducing the pores, as the ancients thought. It can also result from many other causes, about which famous authors wrote.

Chapter XVI

On nutrition of the animals.

Since there are beautiful observations and speculations about the feeding of animals, there is little which I can add to what others have reported.

Proposition CLXXXIX

Preparation by which food is transformed into chyle by Nature.

Composite bodies cannot change their form if their component particles are not arranged in an order, place, position and configuration different from what they were previously. This is not possible without all the composite body being divided into minute particles. Such a division, if the composite body is hard, cannot be achieved without harder tools such as hammers, knives, rasps or millstones. Thus hard food to be transformed into chyle must first be divided and ground. In most animals Nature achieves this by using incisor and molar teeth to chew food. But in birds (with few exceptions) food is shaken, corroded and triturated by a muscular stomach. Its callous internal walls compress each other and by means of hard and sharp stones which are contained in it they replace the teeth for dividing and squeezing the food. The internal walls of this stomach rub against each other like millstones. They thus corrode and grind the food, reducing it into the finest flour. In Pisa I experimented to find that this is true by order of Sereniss. Grand Duke Ferdinand the second. I introduced glass spheres, empty phials, lead cubes also excavated, wooden pyramids and many other objects through the mouth into the gizzard of Indian chickens. The next day I found the masses of lead squeezed and eroded, the glass reduced to powder and the same for the other ingested objects.

Hard food is triturated in a different manner by birds of prey and fishes which are deprived of teeth and have not a muscular but a membranous stomach similar to that of the quadrupeds.

These animals digest meat and bones with some very powerful ferment as corrosive water corrodes and dissolves metals. This corrosive juice is instilled by glands with which the membranous substance of the stomach is stuffed. I observed this most evidently in the stomach of a dolphin in which the glands are particularly large and prominent.

Proposition CXC

Structure and functioning of the muscular stomach of the birds.

The muscles of the stomach of birds are somewhat comparable to the heart since in both the fibres are convoluted spherically and by their contraction they constrict and reduce the cavity which they delineate. There is, however, a difference: all the fibres of the stomach do not act simultaneously but successively so that they seem to consist of several muscles. There are four main groups, two in each face connected by a strong tendon and each of them radiates muscular fibres spherically like rays. Beside, there are other transverse muscles. All seem to consist of layers of fibres which are interrupted and separated by membranous expansions. This appears clearly in the huge stomach of the swan.

Their arrangement and mode of action are shown in Tab. XVIII, Fig. 15.

The section of the stomach ABDE is made along the directions of the opposite tendons AF and CD. The opposite muscles AB and DE contract simultaneously whereas the other two FE and CB are relaxed. For the connection E to remain immobile, it must be pulled in two opposite directions, from E towards D by the contraction of the muscle DE and from E towards A by the contraction of the muscle BA. The connection B remains at the same place for the same reason. Since the extremity A is pulled towards B by shortening of the muscle AB, the muscular wall EFA is stretched and elongated. It is forced to a transverse movement from the fixed extremity E via A towards B. Therefore, it rubs and erodes the solid food R with the internal rough stones, like an upper millstone. By the same mechanism the opposite lower millstone BCD is forced transversely from B towards E via D. The food R thus is carved by the opposite corroding movements of the muscles EA and BD and then by the muscles BF and EC, like by rasps. Meanwhile the walls of the stomach constrict spherically towards the centre like a winepress so that the contained food is squeezed and triturated.

Proposition CXCI

Conjecture on the considerable force exerted
by the muscular stomach of birds.

Besides the mentioned difference between the stomach and heart of birds, there is also a difference in the shape and length of the fibres. The fibres of the heart are convoluted in a particular way whereas in the stomach they expand radially. They are also different in strength and consistency: the fibres of the stomach are compacter and harder than those of the heart. Their effects are also different: the heart expresses blood from its cavities, the stomach grinds and softens hard food contained in its cavity.

But it is equally difficult to measure the motive faculty of both viscera. We surmised the force exerted by the heart by comparison with the force of the masticator muscles. We are also compelled to conjecture the force exerted by the muscle of the stomach of birds by comparison with the same human muscles, since no other and more reliable method of investigation can be found.

Some hazel-nuts have a shell so hard that it is difficult to break between the molar teeth. I introduced some in the stomach of Indian chickens through their mouth and found them broken and ground the next day. It could be questioned whether these ligneous shells may have been macerated and softened by some ferment. Again I introduced in the stomach of other chickens glass balls so hard that they could hardly be broken with the teeth. These balls were found reduced to powder in the excrements the next day.

The function of both, the teeth and the muscular stomach, is the same. They act by compression like a wine-press and overcome the same resistance, i.e. the hardness of the glass balls. Consequently, we can presume that the forces which they develop are equal. It was shown[1] that the force of the muscles closing the human jaw is more than 1350 pounds. Therefore, the force of the stomach of the Indian chicken is not less than 1350 pounds.

Proposition CXCII

Some animals seem to feed on sandy earth alone.

It can be suspected that some aquatic animals such as shellfish, mussels and other small testaceans discovered beneath the sand of the sea, feed on sand alone since nothing else is found in their venter than the simple sand in which they are hidden. They cannot move to look for pastures. Remaining in the same place they are able only to open the mouth and to absorb ambient wet sand which perhaps they can resolve into minute particles by way of salt water or ferments with which they are abundantly provided. Similarly, corrosive waters corrode the hardest stones and metals and dissolve them into liquids. Then these particles of sand may be the food of the shellfish and of its parts. The

[1] Comment on Prop. 88 Part I

same apparently can be said about these water-caltrops or worms which are generated in very hard rocks and fill cells excavated exactly to the size of the small animal. These cells and the enclosed animals, however, have different sizes. The animals thus grow there and progressively enlarge the cells since one does not find a minute animal in a large cell which it does not fill. Therefore, it can be assumed that the cells increase by internal corrosion of their walls and that the small animal feeds on the scrapings.

I do not think that shellfishes feed only on salt water. To what end would they ravenously eat so much sand and absorb it into their viscera? And why would water-caltrops corrode the walls of their cells if they could feed well on plain water?

Such transformation is not impossible. This is demonstrated by an opposite operation. Blood generates, feeds and increases bones, teeth and stones in viscera of animals. All have the nature and consistency of stone. Consequently, in a reversed way, stony sands can be dissolved into blood from which the meat and the shells of these animals can be fed and grow.

Proposition CXCIII

Some animals and plants may be able to feed on water alone.

Oysters fastened to rocks or superimposed upon one another eat neither sand from which they are remote, nor grass by which they are often not surrounded, nor small fishes, fragments of which are never found in their stomachs, as I learned from oyster farmers. They seem to be able to absorb water alone with their open mouth. The same is observed in limpets or barnacles. These feed and grow with the mouth downwards and always attached tightly to rocks like surgical leeks applied on the flesh. They do not corrode sand or rock and cannot eat grass or small fishes since they are always glued to the rocks and remain as if they were welded. Therefore, they can suck nothing but water and, however, they are fed. It seems also that plants can feed on water alone as appears in herbs developing over water (seaweed). Even those which send their roots into the ground do not seem to feed on earth but on water alone since the earth on which they grow is reduced neither in volume nor in weight although a huge solid mass of tree is sucked and retrieved from it. This was observed by van Helmont and others[2]. Irrigating a willow with plain water, even distilled water, they say that the tree grew without any sensible reduction in the dry weight of the earth which they had poured in a huge vessel.

Ficus trees rooted in clefts in buildings or in marble rocks are fed and grow considerably without eroding these hard stones. Their swollen roots rather cleave the rocks. The vines which grow and bear fruit in the marble sand close to the beach of Catania can feed only on rain water. I also observed a big poplar which grew, sucking water through its roots which were floating in a

[2] Prop. 175

water-main like a large and abundant hair. One can thus rightly suspect that these plants and the animals mentioned above develop and grow feeding on water alone.

Proposition CXCIV

It can be suspected that, with their food, birds take stones which they ravenously devour.

It thus does not seem incredible that some animals can feed on sandy earth alone or on water alone. It must be determined whether birds which have a muscular stomach and so ravenously eat stones retrieve some nutrient from these stones.

This hypothesis is suggested by repeated inspection of stomachs of swans, which are huge and stronger than what I have ever seen in any other bird. The swans spend their life flying or swimming or devouring sand on the shores of stagnant waters. Their stomach always seems to be stuffed with abundant sand. Small terrestrial or aquatic animals or their fragments or plants are never found in them except for a few thin herbs which are eaten together with sand. Ser. Grand Duke Ferdinand the second declared that he experienced this fact repeatedly over 40 years.

Such a huge bird continuously needs abundant food. Since this food must be digested continuously, some of it, entire or squeezed and broken, would certainly be found in its stomach sometimes. The stomachs and gizzards of all birds before their night rest are stuffed with small worms, insects, plants and vegetables together with stones. But in the stomachs of swans only sand is found so stuffed that there is no place for any other food. Therefore, it does not seem impossible that this sand squeezed and dissolved into liquid becomes the food of this animal as the sea shellfishes appear to feed on sand.

After I had written this, I went to the city of Burgesià where a swan is fed among other birds. The fowler told me that, when brought with another which soon died, this swan did not eat anything for some days. Then the fowler drove fragments of apple together with bran into its throat and finally he accustomed the bird together with some ducks to a food made of wet bran. I do not know whether this results from a new habit. As Gassendi indeed mentions, it happened that a sheep which had fed on bread, fish, cheese and meat while in a ship, disliked very much graminaceae when back on land.

However, domestic chickens and doves which feed only on wet bread, bran and flour mixed with water need no stones to crush and grind their food which is minced. Therefore, they do not work in vain for a natural need by collecting stones almost the whole day. Similarly, we do not use our teeth when we eat gruel. Therefore, it is likely that stones are eaten by the birds thus mentioned to triturate and dissolve their food and not to injure their stomachs at the cost of useless work. Some parts of these animals are fed by juice extracted from the stones. These parts perhaps are the feathers the abundance of which in the swan may be fed by this juice.

This is not impossible when we see that everything is fed and grows from earth and water. Plants and animals feed on them. There is a double process: earth and water are transformed directly into plants and then into animal substance. Possibly this double process is carried out simultaneously in birds. I think that formation of composite bodies must not be sought for otherwise than from a difference in place, position and configuration of particles of earth and water, as mentioned above[3].

Proposition CXCV

Why perfect animals feed on herbs, fruit and animal meat.

Nobody doubts that similar composite bodies can be transformed one into another more easily than bodies of different natures since a rough body acquires a nobler form only after several successive and long preparations and elaborations. Hence, it is easier to make chyle from fruit and meat which are more similar to perfect animals than from dissimilar elements such as simple earth and plain water. The reason why more perfect animals must be fed with shorter and easier elaboration seems to be the necessity of later and nobler transformations. Indeed organs and spirits of high quality and perfection must be formed, able to carry out sensitive operations. Therefore, the animal faculty ought to be less occupied at servile and very laborious works.

Before investigating the route of chyle, the next lemma must be presented.

Proposition CXCVI

Two volumes of liquid flow during the same time from two soft tubes of inequal cross sections, equally turgid and compressed by the same force. These volumes of liquid are unequal and proportional to the orifices. Tab. XVIII, Fig. 16.

Of two soft and equally compressible tubes AB and DE one has a diameter AC larger than the other DF. Both are filled by the same liquid and compressed with the same force. I claim that during the same time, two inequal volumes of liquid flow from their orifices. The ratio of the volume flowing from AB to that flowing from DE is equal to the ratio of the size of the orifice AL to that of the orifice DF.

Two tubes full of liquid are compressed during the same time by the same force, i.e. by the same impulsive force. The liquids thus are expressed and leave the orifices AC, DF with the same impetus and the same velocity. But the volumes of liquid flowing out with the same velocity and during the same time are

[3] Chap. 13 Prop. 179

in the same ratio as the orifices. This was demonstrated in our book on natural movements resulting from gravity[4]. Therefore, the ratio of the volume of liquid leaving the tube AB to that leaving the tube DE is equal to the ratio of the orifices AC to DF.

Proposition CXCVII

The volume of blood leaving the splenic vein is about a quarter of that which flows from the mesenteric vein during one circuit of the blood. Tab. XVIII, Fig. 17.

AB represents the mesenteric vein and GH the mesenteric artery, DE the splenic or coeliac vein and LK the splenic or coeliac artery. The coeliac vein DE and the mesenteric vein or vena porta AB are two tubes equally soft and compressible always full of blood sent into them by the corresponding arteries. They are compressed by the same force exerted by the viscera squeezed by the abdominal muscles, the diaphragm and inspired air. The length of the coeliac vein is DE. Its orifice DF is about a quarter of the size AC of the equally open and unimpeded orifice of the mesenteric vein AB (since the diameter DF of the former is half the diameter AC of the latter). Therefore, during one circuit of the blood the volume of blood flowing from the orifice of the coeliac vein DE is about a quarter of the blood flowing from the mesenteric vein AB.

Proposition CXCVIII

The quantity of liquid flowing through the mesenteric vein is three or four times larger than the quantity of blood which was sent from the mesenteric artery during the same time. Tab. XVIII, Fig. 17.

In the same figure, the diameter L of the splenic artery LK is about equal to that G of the mesenteric artery GH. Blood, pushed by the heart, flows in both with the same velocity. Therefore, the volumes of blood flowing from them during the same time are about equal or at least the volume from the splenic artery is three-quarters of that from the mesenteric artery. The splenic vein DE actually retrieves from the spleen the same volume of blood as that sent by the splenic artery LK, except for a small portion used for the nutrition of the spleen. Thus the volume of blood flowing from the splenic vein DE, if it is not equal, is at least three-quarters of the blood which has flown during the same time from the mesenteric artery GH.

But is was shown that the volume of blood ejected from the splenic vein DE during the same time must be about a quarter of the liquid which flows from

[4] Chap. II Prop. 224

the mesenteric vein AB. Consequently, the volume of blood which flows from the splenic artery LK or from the mesenteric artery GH which is equal, is about a quarter of the blood which flows from the mesenteric vein AB during the same time.

Proposition CXCIX

Some of the chyle can be carried to the liver through the mesenteric veins.

Ground and macerated by the ferment of the stomach and fermented in the intestines by the bile and pancreatic juice as well as by condiments, the food acquires the consistency and character of the chyle. It then meets two kinds of vessels which perforate the walls of the intestines like a sieve: milky vessels which have orifices suited to absorb pure chyle as a result of their shape, and mesenteric veins which have orifices appropriate to absorb the blood flowing from the arteries mixed with that portion of the chyle which has been soiled by bile.

Two reasons suggest that some part of the chyle soiled as a result of its mixing with bile is absorbed by the mesenteric veins with the blood. First, the liquid which can be brought back through the huge mesenteric veins and the large vena porta from the intestines to the liver is three or four times more than the blood brought by the mesenteric artery and much more than what remains of this blood after the portion to feed the intestines has been removed. Therefore, it is more than likely that the remaining space in the mesenteric veins is filled by some body and that some other copious liquid besides blood flows through them to the liver. The quantity of additional liquid is by far more than the small amount of bile which is present in the mesenteric veins (since the two pounds of bile are distributed in the liver, in the gall-bladder and, mixed with the food, in all the intestines). Consequently, some other liquid than bile must be carried through the remaining space of the mesenteric veins. It seems that this other liquid cannot be anything else than chyle.

The second reason results from the great number of mesenteric veins which originate everywhere round the jejunum and duodenuum. Nature would not have stupidly put there so many veins to absorb the small quantity of blood flowing from the artery for the feeding of the duodenum and jejunum. It also seems nonsensical that so many veins would be set there to remove immediately after its arrival the bile just infused in the duodenum and jejunum: what fermentation could indeed take place without some delay, before all the food going through could be mixed or even be in contact with the bile? Therefore, it seems that this large and abundant network of veins suck from the duodenum and jejunum some other liquid, besides blood and bile. This other liquid can only be chyle considerably diluted by drink.

These reasons seem to suggest that some portion of the chyle is transmitted to the liver not to make blood there but only to purify the mixture of chyle and blood and to free it from its soiling by the bile before carrying it to the heart. Meanwhile the remainder of the chyle purified through the vessels of Pecquet is carried through another route to the vena cava and the heart, the source of blood. It does not seem unusual or difficult that the chyle be transmitted to the heart through two routes equally convenient.

The red colour of the mesenteric veins is not difficult to explain. New colours indeed can change very easily. I think that the chyle can easily become red by an alteration of its structure or by mixing with some other substance.

Afterwards, chyle acquires the consistency and form of blood after several transformations in the heart, lungs, viscera and the remainder of the body. Finally all the parts of the body feed on it.

Proposition CC

Blood with nervous juice is the direct nutritive substance of the animals.

Blood is the actual and main substance on which all the elements of the animal feed. This seemingly cannot be denied since blood is composed of liquids similar to those which feed, very rapidly increase and improve the embryo in eggs. Blood indeed has its albumen which corresponds to the albumen of the egg in consistency, taste, colour, coagulation in fire, and nutritive quality. Moreover, blood comprises a glutinous substance similar to the brain and nerves. This glutinous substance is discovered in the red part of the blood after repeated washings, solidified in white fibres and membranes. All these important substances are sent to all the parts and recesses of the animal. We see that all these parts are irrigated by the blood. On the contrary, all the parts which dry and become sclerotic are deprived of blood. Therefore, either blood restores the lost elements or acts as a vehicle carrying food. Blood contains parts of the best substance and in the animal no other element is found sufficiently abundant to compensate such large consumption. Consequently, the elements of the animal are fed by the blood.

Two objections can be raised against this claim. Firstly, in animals which have died of starvation, all the blood ought to be consumed since death resulted from a lack of nutritive liquid, i.e. of blood. However, not only are the cadavers of these animals not bloodless but the volume of blood which they previously had is retained, hardly diminished. Secondly, severe haemorrhage ought to weaken the animal proportionally to the loss of blood, i.e. of nutrient. This, however, does not occur. Significant weakness is perceived only after most of the blood has been lost.

Some people answer that the blood left in the cadavers of starved people lacks its glutinous elements. It is a stale liquid deprived of its nutritive elements.

It is deduced from this not that the animal is not fed by blood but that nutrition is not completed without spirituous nervous juice, as modern authors have remarked. Since the small amount of nervous juice is insufficient to restore so many lost parts, it must be admitted that nervous juice induces shape and vital animal faculty in the elements repaired by blood. Hence the animal is hardly weakened after severe haemorrhage because it retains the spirituous nervous juice on which the strength depends. In people who have died of starvation the nervous juice is consumed completely.

Proposition CCI

Explaining the mechanism of nutrition.

It remains to explain the mechanism of nutrition of animals. Since some particles continuously are dissolved, perspire and leave as gas from the liquids of the animal and also from its flesh, nerves, membranes, tendons and bones, empty cavities must remain in the places from which these particles parted. Violent impulses of the blood due to the heart and contraction of the arteries can wedge in elements in these cavities with the same impetus as that with which blood is pushed. Actually, any particle of blood cannot be driven indiscriminately into any cavity. Well-determined elements are forced to enter the appropriate cavities with a corresponding shape. The elements for bones have the same shape as the pores of the bone, those for flesh the same shape as the cavities of flesh. This is the mechanism of a sieve. Hence, every element absorbs the nutrient that is similar and conforms to this element. Repair and nutrition thus can be completed.

External air also is involved in nutrition. Absorbed by breathing and penetration through the pores of the skin and flesh or pushing other aerial particles contained in the liquids and flesh of the animal (as occurs in plants), external air helps and promotes now perspiration, then wedging in of nutritive particles. It also maintains and increases the vital movement by its elasticity and impetus.

Finally, nutrition occurs mostly during sleep, as mentioned above. Then nervous juice can flow from the brain through all parts. It provides the particles repaired by the blood with their vital shape and mobility by its plastic and active faculty, sprinkling and soaking them in the same way as magnetism vivifies iron.

Chapter XVII

On hunger and thirst.

Nature remarkably took care of the survival of the animals relying on two powerful stimuli, pain and pleasure, by which they are violently driven to avoid destructive harm and to pursue salutary and life-preserving activities. Since the elements lost by continuous perspiration had to be restored by food and drink, Nature set up a painful feeling from hunger and thirst so that animals seek food to maintain life. She relied not only on this mechanism to satisfy such an important need but also lured us in making the food pleasant. To be able to consider this admirable cleverness of Nature we must first see:

Proposition CCII

Causes of hunger and thirst.

Hunger and thirst result from a lack of food. This lack as such has no name. It seems that the cause of this feeling is vacuum itself, i.e. nothing. But this is absurd since painful feeling is some operation. Since an operation must be carried out, not by nothing, but by an actual cause which operates, it is deduced that this pain and thirst must be produced by some actual cause. To be able to analyse this cause, I observe that hunger does not originate from pure weakness and lack of whole parts of the animal. Most often sick people have not the stimulus of hunger although their spirits, blood and solid parts are considerably diminished. Some healthy people are weak after a long fast but do not endure hunger which they felt some time previously. Consequently, the actual cause of hunger is a prickle of the walls and small nerves of the stomach which is stimulated only when completely empty. Then the corrosive juices from the excretory glands with which the substance of the stomach is abundantly provided are spilled into the cavity. They sting and prick the nervous envelopes with their darts. Not only do they pucker them but they also induce this painful feeling which we call hunger. This is suggested by the following fact. After the stomach is full, hunger disappears. But acid or bitter drinks provoke hunger again. Hunger in dogs cannot be appeased by any food because the membranes and nerves of their stomach are always pricked by corrosive juice as a result either of its considerable bitterness or of its abundance, flowing profusely from the excretory vessels of the glands.

When hunger combines with deprivation and considerable exhaustion of the strength, then it can be somewhat appeased or at least moderated by smells and drugs because they restore the spirits, or by movement which anaesthetizes somewhat the nervous fibres and gives them some rest.

Proposition CCIII

The cause of thirst.

Drinking is necessary to liquefy and macerate the solid food in the stomach and to make the chyle fluid so that it can more easily be absorbed by and leave the narrow milky vessels, and finally to wet the fibres and wall of the stomach. If they were too dry they would be unable to move and to express the food.

Thirst results either from dryness of the mouth, throat and stomach or from the bitterness of the salts with which the walls of these organs are stuffed. In both instances the cause of thirst is the irritation of nerves either because they are compressed as a result of the dryness of the walls and their fibres are stretched by the movement of these walls, or because they are stung by saline darts.

Proposition CCIV

How hunger and thirst are appeased and cause
of the following pleasant feeling.

Nature itself teaches that relief of pain is highly desirable and it is followed by some pleasant feeling. Hunger is a condition which originates in pricking of the membranes and nerves of the stomach by corrosive juices. Therefore, relief and cure of this painful condition must induce pleasure and joy.

We must analyse how this cure is achieved. Hunger is not relieved by removing and expelling the corrosive juice from the stomach. This juice is rather useful. Removing it would be harmful in impeding nutrition. Relief of pain is obtained by adding some substance, food which absorbs the corrosive juice like a sponge. Herewith Nature achieves two advantages. Firsty, the membranes of the stomach are freed from corrosive inconvenience. Secondly, bitterness and corrosive force are used beneficially in macerating and fermenting ingested food. Hence, making of chyle begins.

Actually, food cannot clean the internal wall of the stomach completely of this corrosive juice. Henceforth gnawing pain of hunger would not disappear. Thus Nature needs the additional help of drinking by which she achieves several advantages which we discussed at length in proposition 134 concerning the kidneys. Firstly, drinking washes the wall of the stomach, removes the saline fermentative juice and thus appeases hunger. Secondly, wetting the throat, the oesophagus and the stomach walls makes them softer and more flexible. Thirdly, saline particles dissolved by drink more easily promote maceration of food by entering into all its recesses. Fourthly, drink is used as a vehicle of food which, made more slippery, can easily flow though the intestines. Moreover, chyle becomes more fluid and, consequently, can be carried to the heart through the very narrow canals of the milky and mesenteric vessels. Fifthly, after blood has been made, water impregnated by alkalified salts is expelled through the kidneys and leaves blood pure and clean.

The pleasant feeling which follows quenching thirst is perceived mostly by the tongue, the throat and oesophagus because dryness is suppressed as is the unpleasant roughness of these organs which impeded motions. Saline particles retained in these organs sting and corrode their membranes. They thus dry further and in the absence of liquid are neither washed nor carried away. Therefore, removing the cause of this inconvenience must be followed by a pleasant feeling, according to the laws of Nature.

Chapter XVIII

On pain.

Discussion on the movements of the animals would not be complete if only the causes and mechanisms of movements which are carried out normally and in good health were considered. Disturbed, unpleasant and pathological movements should also be examined. Since this is beyond my competence, I am forced to suggest a few concepts rather than to describe them precisely. Firstly I consider pain in general.

Proposition CCV

The principal cause of pain is not a solution of continuity.

If the common definition of pain as a solution of continuity was true, undoubtedly no solution of continuity could occur in the live animal without pain. And conversely pain would never occur without a solution of continuity. Very often this does not seem to be verified. Not only bones can be corroded and divided but even portions of brain, viscera and flesh can sometimes be excised without any pain. Moreover, if solution of continuity as such produced pain, pain should be perceived all the time during which parts are divided, for example as long as a wound is not healed. This does not occur. We perceive pain only at the time of a sting or a cut and soon afterwards pain disappears.

Only nerves, membranes and tendons are sensitive in animals. The other parts deprived of nerves lack sensitivity and, therefore, do not perceive pain. However, acute and intolerable pain does not occur when membranes or nerves are divided but rather when they remain intact. Finally, if nerves are stung by a needle or pricked by saline juice or squeezed by a hammer, then they experience the most severe pain. Consequently, pain is not elicited by solution of the continuity of their fibres. It does not seem possible for the sharpest point of a needle or of salts to cut the thinnest fibrils which compose the nervous bundles and membranes. Similarly, the point of a nail does not cut the fibres of wood or of a cloth. It only stretches them somewhat and separates them from each other so that they yield to the introduction of the nail. The fibres are divided only by the sharp edge of a knife. Finally, even if it is conceded that some nervous fibres are cut by the point of a needle and that this induces so severe pain, obviously division of a whole nerve or even of a part of it, composed of thousands of fibrils, would result in much more severe pain. But this does not occur since, after complete division of a nerve, pain disappears. Consequently, pain is produced by something very different from a solution of continuity.

Proposition CCVI

The main cause of pain is not a dissolution of the composition and texture of nervous fibres.

In the preceding proposition it was shown that simple division of fibres, nerves and membranes is not the main cause of pain. It remains to see whether dissolution of the composition and texture of these fibres is likely to be the cause of pain. Actually, if contusion of nerves resulting from a blow with a hard body such as a hammer, caused pain, it could be thought that pain results from a dissolution of the fabric of the fibres. A contusion indeed could actually produce such dissolution. But more often acute pain is provoked by a single sting by a needle or by the darts of salts. Their sharpest points are unable to dissolve the texture of nerves because, even if the fibrils or parts of the fibres are a little stretched and separated from each other by the width of the point of the needle, it does not seem possible that their texture be dissolved. That is why they are pulled back immediately by their glutinous consistency, glue together and restore their previous connections, as the strings of a zither spontaneously contract immediately after being pulled.

Such minimal distension of the fibres is not painful by itself. Indeed when the limbs are grasped and the muscles are compressed the fibres of the nerves ought to be more stretched and elongated than by the introduction of the point of a needle and, however, this compression does not induce pain.

Even without a sting, pain is provoked when for example the nerves or the membranes of the eyes are touched and rubbed by a straw. Then the nervous fibres are not stretched nor is their fabric dissolved. Consequently, the main and most likely cause of pain is not a dissolution of the composition of the fabric of the nerves.

Proposition CCVII

The main and immediate cause of pain seems to be irritation of the nerves. Explanation of its mechanism.

Solution of continuity and dissolution of the fabric of the nervous fibres being excluded, only irritation and corrosion of these fibres remain as the possible cause of pain. The very sensitive nervous fibres and the spirituous juice present in their spongy marrow seem to be easily irritable. When activated by an irregular and disturbed movement they may transmit to the brain disproportionate shakings which induce this convulsive, unpleasant and tormenting feeling which we call pain.

This feeling can be induced in the sensitive centre of the animal, as suggested by the tormenting inconvenience produced by the acute noise made when a craftsman sharpens a knife or files a sheet of metal. This dissonant noise unpleasantly irritates the drum so that there are people who support more willingly a blow delivered with a fist or a foot than to hear this disturbing whistling. And who understands that this noise which is an acute vibration of air provokes a tear, rupture or destruction of the drum? Nothing more than irregular and dissonant irritation seems able to produce disturbance of the harmonic rhythm at which the animal spirits are normally suited to move.

This is confirmed by the convulsive horror of the viscera which is provoked by remembering the nausea which was once induced by an emetic. Obviously, this painful shaking of the viscera cannot be produced by any solution of continuity but by a movement of spirits irritating and shaking normal nerves.

Pain results from such irritations. This is confirmed by the way it is cured in many instances. I often experienced the painful inconvenience provoked by slight contact or rubbing with a straw or by a fly in the nostrils, in the eyebrows and on the forehead. This feeling is so unpleasant that to stop it some people slap themselves in the face, strike the walls with their head and scratch their skin with their nails until they bleed. This irritation thus needs corrosion and division of the skin with the nails, a remedy as severe as the painful corrosion itself. Consequently, irritation by a straw or by contact of a fly is much more painful than corrosion and laceration of the cutaneous nerves. Therefore, pain consists in this irritation and itching of the nerves because the animal spirits are shaken in the brain irregularly against their natural tendency.

That is why pain is easily sedated by narcotics which induce sleep. Toothache is sometimes sedated for a short time by stupid talk whispered by mountebanks. A firm belief by the patients that they can be cured by these words induces opposite movement of the animal spirits which interrrupts the irritation and sometimes stops it completely. This is clearly confirmed by the way this contracture of the lower leg which is called cramp is treated. We free ourselves of this excruciating pain in the twinkling of an eye if we shake the foot and energetically move the affected leg. It is thus necessary to disturb and impede pain by another irritation.

Chapter XIX

On fatigue.

The feeling of weakness is very different from tiredness. This is obvious. Weakness results from simple lack of strength or impetus to act. Tiredness is some uneasiness perceived after exhausting work. Children who cannot raise heavy weights because of a lack of strength, joyfully dance and trample tiredlessly. But strong men after violent motions of the body perceive this unpleasant and painful feeling of the limbs and spirits with anxiety and deprivation which is called fatigue. This feeling has several causes.

Proposition CCVIII

How fatigue can be produced by dissipation of the spirits.

Sometimes people in good health experience an unpleasant feeling of tiredness as a result of some activity of the body after a long fast. The locomotor apparatus then is neither worn nor injured. Consequently, this fatigue results from an inability of exerting an effort. Spirituous juices must be instilled by the nerves in the muscles together with blood to stimulate the capacity of the animal to move. In the instance mentioned above fatigue thus originates in a lack of such spirituous juice. We shall attempt to find out the mechanism of this fatigue. When several men raise and pull a light weight which can be moved by one man, they become slightly tired or not at all because they can carry out the work one after another each during a short time and then rest. If the same weight is pulled all the time by the same man, not surprisingly he becomes tired and finally exhausted. Spirituous juices and blood provided abundantly can be imagined as several stevedores carrying a load. When spirituous juices are abundant a part of them easily succeed in supporting the weight during determined periods of time. It is different when spirits are scarce: forced to work continuously without any rest, they yield to the load.

To apply this example to our problem, it must be considered that spirituous juices are instilled by the orifices of the nerves inside the muscles. When the spongy tubules of the nervous fibres are abundantly filled by such spirituous juice, this can be provided continuously without any interruption. But when, after a fast or a long secretion, the tubules are impoverished and almost dry or, as a result of wasting, the spirits are almost unable to carry out contraction, they must then be transmitted from the brain at the cost of a great effort. Such dearth of spirits cannot occur without a feeling of tiredness and exhaustion. This is how fatigue can be explained.

Proposition CCIX

How painful tiredness can result from injured organs.

Fatigue can result from injured organs in three ways. Firstly, spirituous juices cannot be carried to the muscles to move because their paths are obstructed. Secondly, stretched or lacerated muscular and tendinous fibres cannot be moved without pain. Thirdly, dirt stuffed between the fibres impedes the movement, as the cog-wheels of a clock can uneasily turn in powder or rust.

Concerning the first cause, it is clear that motive spirituous juices, however abundant they are in the brain, cannot move the limbs if they are not carried from the brain to the muscles through the nerves. Thus, when the paths, i.e. the orifices of the nerves, are obstructed, the spirits cannot contract the muscles as if the animal was deprived of spirits. This is the cause of the inertia, apathy and tiredness which we experience immediately after awakening which induces the same fatigue as hard work. Then the spirits have been abundantly restored. But so far the paths of the nerves are obstructed by nutritive juices and these cannot be pushed by the will to move the muscles, as mentioned above.

The same seems to occur during some acute diseases in which strength is suddenly decreased not because of a lack of spirits but because the spirits cannot be carried to the muscles to be moved. This is suggested by the fact that after the paroxysm of the condition, strength comes back without being restored by more food. If delirium occurs it even moves the muscles very powerfully.

Secondly, the same tiredness is produced by an injury of the locomotor apparatus without any lack of spirits. Muscles can contract powerfully only if their fibres are widened by wedges of rarefied substance driven into their pores. They cannot move and raise heavy weights without suffering some dilaceration of their muscular or tendinous fibres by exaggerated stretching. Joint cartilage and ligaments caught between the limbs must also be injured by the incumbent weight of the body of the animal. Such distensions, lacerations and squeezing never occur without the position, order and connection of the particles constituting the fibrous conformation being dissolved and disturbed. This, when carried out beyond normal, results in irritation and corrosion which are painful and unpleasant. It is thus not surprising that they entail anxiety.

Thirdly, the movement of muscles is impeded by the stinging of bitter juices scattered between the muscular fibres. The darts of these juices induce pain when the muscles contract. Therefore, painful weariness appears in these soiled bodies.

Fourthly, fatigue results from the presence of excremental viscous liquids which fill the pores of the muscles and thus impede swelling necessary for muscular action. Similarly, the wheels of a clock, if rusted, cannot rotate without difficulty. I think that fatigue affects the animals mostly as a consequence of such difficulty.

Tiredness is cured by rest for two reasons. Firstly, injury and irritation of the muscular fibres stop. Secondly, the injured and lost particles are restored progressively and are replaced in their due positions from which they had been removed. The divided elements are glued by the nutritive juice which arrives. All this happens with some pleasant feeling of repair.

Proposition CCX

Why standing is more tiring than walking.

When standing we maintain the body machine erect. Then the muscles extending the legs and the spine do not work alone since the greatest part of the load is supported by the bones of the legs and spine as if it were by columns. Contrariwise, when walking the same extensor muscles, besides maintaining the body erect, must also carry and push forwards all the weight of the animal body since walking cannot occur without flexion of the lower legs and when the lower legs are flexed all the body which is bent is no longer supported by the columns of the legs. It thus must be raised with strength by the extensor muscles. Consequently, walking should be much more tiring than standing.

This, however, is false. The cause of the tiredness which we undergo when standing thus must be sought for. If I am right, this results from the fact that the continuous action of a muscle in supporting the same weight, although small, is very unpleasant and induces more fatigue than if the muscle carried a weight ten times heavier for a much longer time in an intermittent action interrupted by frequent pauses. The reason is that the action of a muscle can be continued without any rest only if the muscle remains tense and contracted. Therefore, it is necessary that spirituous juices be expressed from the nerves oozing continuously. Otherwise new explosions would not occur in the muscles without rest. Actually such continuous outflow of spirits empties the nervous canals and dries them out because more leave them than what can be introduced and transmitted from the brain. Consequently, the outflow of spirituous juice into the muscle stops or at least is reduced to some droplets expressed from the brain with great effort. Hence come the tiredness and exhaustion which are experienced when standing. During walking the muscles rest repeatedly. Thus the nervous tubules which end in them can be refitted and refilled successively by spirituous juice. In this way, acting with interruptions, the muscles can carry and move the animal body easily without significant fatigue.

Besides a lack of spirits, I consider that during a moderate and alternate action of the muscles such as walking it is necessary that the muscular and tendinous fibres be moderately stretched repeatedly and recover during rests in between. In standing the extensor muscles of the limbs not only work without any interruption but they exert a powerful effort not because they carry a heavy weight but because they are considerably loosened beyond normal as a consequence of the extension of the limbs, knees and hips. Therefore, normal contraction of the fibres is made ineffective. Similarly, when the string of a bow is loose, even if it is pulled it is not powerful enough to tighten the bow. The fibres of the muscles thus must be pulled more than normal. Since they are deprived of repeated recovery, they are lacerated in some way. This injury is painful by itself so that, when standing, the sensitive faculty in the brain perceives this aching fatigue. On the contrary, during walking the frequent stretchings of the muscular fibres are quickly cured. Therefore, they are not painful and cannot provoke so much fatigue.

Moreover, in standing the skin of the plantar soles, the membranes, cartilages and ligaments which are placed between the bones in the joints of the legs and spine, like cushions, are squeezed considerably by the continuous pressure of the load of the whole incumbent body. This squeezing entails painful laceration of some elements. This is not cured as long as the standing position does not allow for repair and rest. But the sensitive faculty of the animal is more affected by pain in any part of the body than by the usual work of all the muscles of the limbs and by the interrupted squeezings and recoveries which occur without pain during walking.

This is demonstrated admirably by experience. Prolonged sitting, even comfortably, produces notable tiredness and exhaustion as is experienced by people who are carried in a small boat or in a litter. And, surprisingly, the remedy of this inconvenience is to walk. That is how these people recover with much pleasure after a long journey. Not only does prolonged sitting compress the glutei muscles and all the limbs and thus lacerate their fibres but it also impedes blood circulation and nutrition of the compressed parts. The consequence is aching tiredness and exhaustion which afterwards are relieved and cured not by rest but by laborious contraction of the muscles which must occur in walking.

During prolonged standing, besides painful laceration of the extensor muscles, tendons and cartilages, we also experience the inconvenience that blood circulation through the muscles is impeded by their continuous tension and swelling. The appropriate remedy for all these inconveniences is not rest but walking by which blood circulation is restored, diffusion of the spirits through the nerves is ensured and lacerations are repaired by nutritive juices.

Proposition CCXI

Conjecturing the cause of the aching in limbs and joints induced by energetic and unusual movements.

Experience shows that energetic and unusual motions carried out with great effort in a ball game or by gladiators in training provoke aching not immediately but the next day after a night rest and recovery. Then the joints and muscles of the arms and legs cannot carry out the usual movements without aching. No doubt, this pain is not produced by laceration and rupture of fibres, muscles and tendons. Firstly, such injury ought to be felt immediately during the violent exercise when laceration and rupture occur and not the next day when the injury has had time to heal and repair. Secondly, this aching is not cured by rest but by repetition of the same violent exercise. It seems thus that this pain may be provoked by a profusion of serous and viscous humour which fills the pores of the flesh exaggeratedly dilated like a sponge by the previous explosions. As a consequence of this swelling and stuffing the fibres can no longer move without painful stings. This swelling and stuffing can resolve only by insensible perspiration through the pores of the skin or after the glutinous serosities have been clarified and evacuated through the veins. These operations are helped and amazingly enhanced by subsequent hard exercise until these stagnant humours which constituted something like an abscess are diluted, clarified and pushed away. They can then be evacuated easily and the cause of pain thus disappears.

Repetition of such exercise achieves another advantage. This is an ability to carry out energetic motions in the future because blood can flow more easily through dilated and open new paths in the muscles and feed the fibres better. These fibres, stronger and more mobile, can then move more quickly.

Chapter XX

On convulsions.

Convulsion is defined as an involuntary movement of extension or flexion of some limb due to an abnormal cause. We shall express some remarks on this movement.

Proposition CCXII

Convulsion does not result from contraction or loosening of nerves.

It is commonly thought that convulsion or spasm is a disease of the nerves which shorten as a consequence of too much dryness or exaggerated moistness. Then the muscles in which they ramify contract and flex or extend violently the limbs to which they are attached. This is an obsolete opinion which is refuted by experience and reason.

Nerves are never observed to dry and dessicate. They thus cannot contract for this reason as hair contracts and twists when burned. Moistness which suits them naturally never contracts them, no more than does too much warmth or frost.

Even if nerves actually contracted they could not exert a force sufficient to flex or extend limbs because they are not attached to two hard and fixed extremities like nails, as muscles are linked with bones or fixed tendons. The origins of the nerves are rather connected loosely with the soft and yielding medullary substance of the brain or of the spinal cord and their extremities are tied to soft muscular tissue. In between they are not straight but wind loosely. Therefore, nerves cannot flex a limb like a tight rope attached to a nail.

Proposition CCXIII

Spasm results from an involuntary contraction of muscles provoked by pathological pricking of nerves.

When limbs are flexed or extended by convulsion, it is obvious that the same muscle which was destined to some voluntary contraction in good health, when in spasm remains contracted, tense and hard. Thus convulsion is not different from a voluntary movement in its action. They are different in their cause: the latter results from an order of the will, not the former. If what was achieved by a voluntary movement in good health is achieved in the same manner by a pathological cause, then both movements are carried out by the same mechanism. Actually, during a voluntary movement animal spirits irritate the origins of the nerves in the brain, and by shaking them express droplets of nervous juice into the blood contained in the muscles. This provokes a contraction. Consequently, in spasm, the same irritation in the nerves and outflow of nervous juice must occur as a result of a pathological cause. Such irritation is achieved in the nerves of the legs in eviscerated frogs every time they are stung by a needle or touched with some saline. We experience on ourselves that convulsive sneezing or coughing is provoked by irritation of the nerves of the nostrils or of the trachea. Similarly, stinging or gnawing of nerves of the limbs by some bitter or saline juice can produce convulsive movements. These movements are transitory if the irritation immediately stops. They are prolonged and continuous if the irritation persists. This is confirmed by the fact that in those who perish in convulsions, the malposition stops after death and the limbs which were flexed become straight. This occurs because once sensation has disappeared in the nerves, there is no longer unpleasant feeling by which they were shaken. Consequently, they no longer secrete spirituous juice, the cause of contraction, into the muscles. Then the violent contraction of the muscles ceases.

It remains to explain how enough spirituous juice can flow out of the nerves to achieve continuous contraction during several days in the convulsed muscles. The solution of this problem is found in what was said above where it was shown that abundant outflow of spirituous juice from the nerves is not required but some droplets are enough. They act as odoriferous particles of civet perfume or as purgative particles which are diffused in wine from glass of antimony.

Chapter XXI

On shivering.

We consider in general the shivering of bodies in the book on the force of percussion[1]. Only the causes and mechanisms of shivering in animals, either voluntary or involuntary and convulsive, must be presented.

Proposition CCXIV

How voluntary shivering occurs in animals.

Shivering is nothing else than frequently repeated agitation in opposite directions. The cause and the organs of voluntary shivering must firstly be found. When we wish to shake the hand in a trembling movement, antagonistic muscles are contracted rapidly and frequently, successively one after the other at will. Similarly, in scales balanced with two hands, we depress now one scale and then the other. In the same manner, when speaking if we wish to pronounce the letter R, we contract antagonistic muscles which produce frequent

[1] Chap. 30

vibrations. In these shivering movements, the hand or the tongue is supposed to be balanced by equal, proper and normal contractions of the small machines which constitute the antagonistic muscles. These are activated successively by voluntary tensions resulting from the contractions described above and produce shivering.

Proposition CCXV

How involuntary shivering can result from loosening and contraction of the small machines which constitute the fibres of the muscles.
Tab. XVIII, Fig. 18.

It was said above that, as oscillation of a pendulum can result from its natural arrangement and from its weight, a spring or a similarly flexible and resilient machine can vibrate as a result of the material structure of the machine alone. Muscular fibres are chains of small machines linked to each other. We thus understand that they can relax and contract alternately in a frequent and regular vibration. To discover the mechanism of shivering let us consider the following example. The column or bone AB is articulated with the subjacent bone CD about the joint B. Two muscular ropes AC and AD are attached to the top A and end at C, D on the opposite sides of the subjacent bone. Obviously, if the opposite ropes were pulled with the greatest force without loosening, no doubt the column AB fixed by a tonic pull would remain immobile and would not undergo any shaking. But the small machines which constitute the muscular chains AC, AD are compressible and can resile. Therefore, the ropes are never pulled with so much force that they cannot loosen somewhat. In such instance, if the column AB tilts towards D as a result of a strong impulse, the rope AC must be pulled and the opposite rope AD be loosened that much. The rope AC exerts a greater force as a consequence of the violent stretching of its small machines than the rope AD the small machines of which are loosened. Therefore, after this impulse is extinct, the greater force exerted by the rope AC must necessarily bring back the column AB to a vertical position. In the return movement new impetus is acquired, opposite to the first so that the column tilts towards C, stretching the rope AD and loosening the rope AC. The column comes back for the same reason and tilts again towards D. Thus the movement is frequently repeated alternately to and fro. Shivering thus occurs and persists as long as the external impulse is renewed. A flag streams waving as long as wind blows which pushes alternately the two faces of the flag and acts like the two ropes AC and AD. This must certainly be verified even if the wind is continuous and uniform. If initially the right face of the flag a little inclined is stricken by the wind, it is straightened and then moves to the opposite side as a consequence of the acquired impetus. The left face thus is exposed to the same wind. Streaming results from such repeated oscillations.

In the same way, the limbs of animals can be pushed by external causes in a thousand ways. Thus, antagonistic muscles alternately loosen and contract, resulting in shivering.

Proposition CCXVI

How a lack of strength induces shivering.

People exhausted as a consequence of fast, poor health, senility or after hard work cannot carry a small weight in their hands without trembling. This results from a lack of the faculty which gives the impetus, i.e. a lack of the spirituous juice which is transmitted through the nerves. This is suggested by the immediate recovery achieved by a drink of wine or of cardiac liquor which restores the strength so that the weight can then be carried without trembling.

This is how this shivering occurs. As a consequence of a dearth of spirituous juice, droplets necessary for steady contraction of the muscles can no longer flow continuously from the nerves. The outflow is interrupted by frequent delays during which the muscles relax and loosen. Carrying the weight thus is interrupted frequently and repeated. Such is the origin of trembling.

This is confirmed by the fact that men otherwise strong cannot support the weight of their own arm held in extension, continuously during a whole hour without trembling. As a consequence of continuous outflow from the nerves the spirituous juices are dissipated and missing. Hence they can be expressed with interruptions and by minute seeping.

Proposition CCXVII

Why fear and cold induce shivering.

Steady action of muscles requires continuous seeping of spirituous juice. This can stop either by dearth of juice or because its arrival and diffusion are impeded. Arrival of spirituous juice to the muscles to contract can be impeded in three ways: firstly by inducing rest in the spirituous juices themselves; secondly by activating them by an opposite movement, i.e. back towards the brain; thirdly by obstructing the canals of the nerves or their orifices. We all know that cold by its nature provokes numbness and immobility. Cold seems to be nothing else than deprivation and lack of warmth. Cold thus induces dearth and lack of warmth which is the main motive cause in animals. It is thus not surprising that when cold unexpectedly occurs the movement of spirits into muscles is stopped or at least delayed. Contraction of the muscles is repeatedly interrupted which gives rise to shivering.

Fear consists of a kind of flight of the spirits inwards. This is suggested by the paleness of the face, the darkening of the eyes, the diminution and interruption of the voice, the lack of strength, the trembling of the limbs, the acceleration of the pulse. All these symptoms occur in animals in good health in which the spirits are abundant. They are thus the consequences of a retrieving and flight of the spirits produced by the fear of danger and of imminent calamity. Therefore, not surprisingly, the spirits retrogress during such disturbance. They are scarcely instilled into the muscles and contraction is interrrupted again and again. This is the origin of trembling.

Proposition CCXVIII

Cause of the pathological tremor which occurs in palsy, fevers and senility.

Besides the shivers mentioned above which result from a lack of spirits or from some obstacle to their involvement, there are other shivers which are pathological. This is obvious. Injury of the extremities of nerves or tendons in the limbs entails some paralytic tremor. In fits of epilepsy and in bouts of fever the limbs shake violently. Fits of anger provoke less violent trembling. All these shivers no doubt must be produced by some pathological irritation since they occur involuntarily and cannot be impeded or refrained by the motive force of animal spirits.

These violent shakings of the limbs occurring in spite of the animal faculty use to be called convulsions. They are provoked by irritation and pricking of nerves and membranes. Because of the resulting pain, spirituous juices are expressed from the nerves into the muscles which they continuously contract. Therefore, the limbs remain either flexed or extended.

In paralytic tremor this convulsive pricking of nerves also occurs. However, it is not continuous but interrupted by pauses. This results in frequent oscillations like those of a pendulum. The cause may be either that the saline and pricking juices are driven into the nerves not in a continuous flow like from small fountains but droplet by droplet, or rather that the nerves are compressed and the saline darts driven into them can sting the nervous fibres and membranes interruptedly by some peristaltic movement. This is confirmed by the fact that these irritations are provoked or increased by disturbed activation of the spirits as happens in paralytics who tremble more violently when they are angry or sad. In healthy people a fit of anger induces trembling of the limbs because violent stings can be provoked during turbulent movement of the spirits.

The most violent shivering, however, is that which occurs in bouts of quartan fever and which induces aching in the spinal cord where saline juices separated by previous fermentation create this painful pricking as we shall say in due place.

Finally, senile trembling, although comparable to that of palsy, is enhanced considerably by a lack of strength and spirits.

Actually, in any shivering the movements to and fro must be of about the same duration according to the law of oscillation, because the antagonistic muscles must contract alternately with the same frequency, like a pendulum.

Proposition CCXIX

Numbing force of the torpedo (numb-fish) and shiver of the porcupine.

Two fabulous stories are fairly spread. One concerns the poison of the electric ray-fish which numbs the hand of the fisher from a distance with a dart.

The other concerns the porcupine which by tightening its skin propels at a distance the long spines which cover its back. About both animals I will tell what I saw with my own eyes.

The torpedo is a flat fish like a ray-fish but covered by a soft skin. This skin when pressed by the fingers gives a violent electric discharge so that the hand of the individual who touches the fish experiences a painful numbness about the same as that which is produced by striking the elbow on the edge of a table. Such numbing energy does not originate from a poisonous faculty or some poisonous gas since: (1) if a torpedo is touched or palpated when at rest it does not injure the hand; (2) at the time that a fish shakes, if its lateral edge is grasped by opposite fingers, the hand is not injured. This shows that this fish does not emit any poisonous gas. It is true, however, that, when its thorax is compressed by the fingers close to the spine where most of the nerves and muscles are, then a violent shaking numbs the touching hand and provokes a kind of spasm. A famous English anatomist, however, exaggerated this pain. He had been shaken by paralytic tremor. In the presence of Serenissime Ferdinand Grand Duke of Tuscany he asserted that touching a torpedo gave him pain in the arm for two days. I do not know whether his opinion increased the harm. I can speak on good grounds after having experimented. Every time I squeezed slightly the thorax of a torpedo with extended fingers, I tolerated the trembling of the fish without any inconvenience. But when my fingers were flexed, above all the thumb, I received the trembling percussions of the fish perpendicularly and then experienced a spasm. I thus learned that in this instance the tendons and nervous ligaments of our fingers which are very sensitive are strongly shaken by repeated blows provoked by the shakings of the fish which thus can produce a spasm similar to that provoked by a blow on the elbow.

The porcupine actually does not throw its long spines but it keeps them erect, moves and vibrates them. This is done by the muscles of its skin and by semilunar muscles which contract the skin. These muscles erect and shake the roots of the spines.

The theory of countless other pathological motions which medical authors have discussed at length ought also to be considered. But since I cannot present anything better I will submit only a few ideas.

Chapter XXII

On fevers.

I will consider fevers. These appear to our senses and we unfortunately experience them. However, their causes and mechanism escape human sagacity so far, despite much study and research. The most learned of the physicians do not hesitate sincerely to recognize that they have not understood the nature of fevers. Thus, if in such a difficult and mysterious matter I do not bring anything new, the erudite Readers will not make me more responsible of this defect than all the others. But they will appreciate that I discussed and presented what is taught commonly, following new lines and according to other principles.

Proposition CCXX

Common definition of fever and suggestion of its shortcomings.

The name fever was given by people to some transport of the movements of the heart and arteries (not as a result of anger or some violent movement of the body but spontaneous and permanent) usually associated with burning and heat of all the body and followed by other symptoms: exhaustion, weakness, aching and several others. But fever uses to be defined more briefly as heat ignited in the heart. Fevers are subdivided into intermittent and continuous, periodical and irregular, beginning with stiffness and cold or immediately with heat, very hot or minimally warm. These are thought to be very pernicious sometimes.

Concerning the causes of a bout of fever, people think that it is due to the innate heat of the heart. The fuel is the blood which is abnormal either in quantity or in quality. The blood would act as the wood which enables a pasture to burn. Most often fevers present with periods of duration and intermission so that they recur at determined intervals more precisely than a clock. Therefore, something must originate in the blood or be transmitted to the blood from somewhere else, which can re-ignite the fever at certain periods in a well-determined measure, in equal proportion and time. According to these people, this something is some excremental and corrupted humour hidden in some recess, increasing significantly at a well-determined time so that the blood is affected and ferments. This recess or focus is commonly supposed to be in the mesenteric veins or somewhere else. The substances are the four cardinal humours. This opinion was rejected deservedly after blood circulation was discovered since blood stagnates neither in the mesenteric veins nor anywhere else. Recesses or dens cannot be imagined in the animal body without considerable tumour or some abscess. Consequently, all this fable fades away.

Proposition CCXXI

Fever is not ignited by a fire or a ferment present in the heart.

Other modern authors say, as does Descartes, that blood in fever boils from the heat contracted in the heart, that this appears from experience, i.e. from violent and increased pulse, from intolerable heat in the precordial area and from scorching thirst. They assume that the main ferment of blood is in the heart itself which is the fire-place of heat.

Facts themselves protest loudly against this imaginary explanation of Descartes. This boiling heat which has been so exaggerated is not observed in the heart of live animals. On the contrary, the temperature in the heart of the animal is found to be moderate, equal to that of its other viscera and the other internal parts of its body, even when it is irritated by acute and excruciating pain while being cut and thus very furious (which by itself increases the heat in the heart). This temperature does not exceed 40° and is thus equal to the temperature in the sun in the summer.

Moreover, no vessels containing ferment are found in the heart. There is nothing else to see there than fleshy cavities, hard and clean, which are continuously washed by the passage of blood. Nobody has found there any taste, bitter, acid or of any kind other than that of blood.

This and other arguments which are usually raised against Descartes by others, clearly show that heat or a fire-place in the heart or ferment is not the cause of the bout of fever.

Proposition CCXXII

Fever is not ignited by alteration and fermentation of blood, and pricking of the heart does not provoke a bout of fever.

There are famous authors who, influenced by the prejudices of the ancients, even now think that blood plays the principal role in this picture. They indeed say that the blood is altered, fermented and can be easily corrupted as a result of its abundance and turgescence or of its mixing with some ferment. Hence, blood acquires movement, effervescence and heat from this fermentation. Similarly, squeezed and macerated hay and young wine are effervescent, boil and are activated violently with much heat. After the blood thus fermented has become somewhat bitter and stinging it must, they say, prick and sting the fibres of the heart during its passage. Similarly, in a heart retrieved from the body, contact of stinging juice such as aqua fortis starts again and increases the pulsation which had stopped. Because of this sting the heart must move and pulse very energetically with considerable ardour and heat. This is produced by the heart itself or rather by the blood as a result of its own inborn heat or stimulated by a ferment added to the blood.

These authors then attempt to confirm that fever results from the corruption of the blood by the fact that urine of febrile patients is sometimes altered and corrupted. Since urine is separated from blood in the kidneys it is forced to carry with it the particles of impurities which soil the blood. As long as the urine remains altered, fever persists. This is an unquestionable sign of corruption of the blood.

There seem to be many objections to this plausible opinion which was accepted by many famous men. It is worth subjecting these objections to an examination by more learned people.

Firstly, the fact that the urine of febrile patients is altered in its colour and consistency does not mean that the blood with which urine was mixed is also corrupted. Alteration of urine may result not from the blood but from another cause. When pure and tasteless oil is separated from water with which it was mixed, it is possible that the sweet or bitter taste which water presents was acquired from contact with sugar or rhubarb and not from the tasteless oil. Even if altered urine is separated from blood in febrile patients it thus cannot be concluded that the blood is affected by the same corruption and was the cause of the alteration of the urine unless it is certain that urine could not be contaminated by another cause. Who indeed knows whether in the stomach, intestines or mesenteric glands one could not find some humour, salt or ferment which could have achieved this disturbance of the urine?

This opinion is also contradicted otherwise. If indeed altered and corrupted urine indicates a corruption in the mass of blood which generated fever, then as long as the alteration of urine persists, the disease of the blood also persists and, henceforth, its effect, the fever. But alteration of urine persists after the bout of fever not only days and hours but even during all the quiescence and intermission between bouts in tertian and quartan fevers. The disease of the blood thus persists during the days without fever. Then circulating blood continues to pass through the heart. It thus would stimulate ignition in the heart and the heart should pulse violently as a result of its bitterness. Therefore, fever should be ignited during the intermissions. As a consequence, tertian and quartan fevers should be continuous without intermission. This does not occur. Consequently, the hypothesis is false.

Secondly, if a disease as pernicious as is fever is ignited only by an alteration of the blood which has become rotten and thus is far from normal, then no doubt the blood of febrile patients would be very different from that of healthy people. Indeed rotten vegetables or animals appear to be very different from fresh ones in colour, smell, taste, consistency and activities. But blood obtained from healthy people by blood-letting is not at all different from that of febrile patients even during a bout of fever. In both a solid red part separates from a liquid white part similar to white of egg, which coagulates in the fire. The taste and smell are the same in healthy and ill patients. The colour may differ somewhat between different febrile patients but the same differences are observed in the blood of perfectly healthy people. Consequently, the blood of febrile patients is neither rotten nor corrupted. Therefore, fever is not ignited and maintained by corrupted blood.

Thirdly, an abscess in the extremity of the foot provokes fever. The same occurs in inflammatory arthritis. Then, however, there is nothing wrong with the blood.

Fourthly, in dyspnoeic catarrh and epilepsy, fever is not stimulated by the blood but by inflammation of the brain.

Fifthly, if corruption of the blood is supposed to be the cause of fever one cannot explain the bouts of the perodic fevers such as daily, tertian and quartan fevers, which recur at well-determined intervals and are separated by intermissions. They pass from not being to being. It is not possible to explain these effects even by assuming different degrees of alteration and corruption of the blood which would produce the bouts of fever. The authors are forced to invoke the mysteries of the movements of the moon, the qualities of numbers and similar things. Moreover, if rotten blood provokes fever when passing through the heart by irritation or poisoning of the latter, how can a bout of fever stop? They answer that the foul humour boils in a paroxysm and is expelled by Nature through the pores or ejected from the main places to the intestines or some other place as if they were cesspools. If it is so, why does fever recur after a certain time of intermission? They say that ferment remains which can corrupt blood again after a well-determined time and thus provoke a new bout of fever. Actually, after a paroxysm, febrile ferment immediately acts beginning anew the alteration of the mass of blood, increasing corruption gradually in a steady way and not with intermissions, as occurs with fermentation of vegetables and humours. Since blood beginning to be corrupted continuously circulates through the heart, it will affect the heart according to its own degree of alteration and decay. Therefore, fever will always continue and there would be no rest and complete intermission. There would be a period of continuous increase up to a paroxysm. Hence fever would decrease down to a minimum where it would not stop but would start increasing again. This is what occurs with the oscillations of a pendulum and with the ebb and flow of the sea because of the permanence of the cause of the movement which increases when its cause increases and decreases when its cause decreases. This is not what is observed with fever. Absence of fever persists several hours or whole days. Consequently, corruption generated and increased in the mass of blood cannot provoke morbid fever.

Some authors assume that fermentation occurs and is kept in some cavity of the animal body. This is contradicted by experience in anatomy. In the bodies of febrile patients such cavities or cesspools full of putrid humours are never found. Blood and humours flow to and fro continuously and uninterruptedly in all viscera, vessels and flesh. There are sometimes internal abscesses but this is very rare and almost always fever occurs without such abscess. But periodic fevers, daily or tertian, sometimes are provoked by some bad weather, hot or cold, by lack of sleep, anger or sadness and no abscess is found. Thus, fevers are not always provoked by blood diseased and mixed with some ferment, kept in some recess of the animal body.

Proposition CCXXIII

Fever is not ignited by altered chyle from the blood, which induces fermentation and boiling of the blood.

But, before proceeding further, it is worth-while to examine the theory of a very learned modern author on the same subject. He supposes that corrupted blood can provoke a bout of fever by itself every time the spirits and the sulphurous component of the blood are exalted and overcome the other elements of the blood. Since they cannot combine with these, there is some turmoil which creates boiling and fever. He adds that ferment which alters the blood comes from outside and is a nutritive juice different from blood. Therefore, it can uneasily be assimilated and, when filling vessels, it can induce fermentation and boiling. He foresaw that the biggest problem would be to explain the recurrences of intermittent and periodic fevers and attempted to meet this difficulty with the following argument. Although, after a paroxysm, the nutritive juice different from blood is immediately mixed little by little with the blood, fever is not ignited immediately. One must wait until the vessels are filled and turgescent for violent febrile ebullition to be able to occur. This is not different from beer or young wine poured and hermetically sealed in goatskin bottles to prevent evaporation. Although they ferment, they are contained without ebullition or explosion. But with multiplication of the vapours thus enclosed all the mass swells, boils and considerably ferments so as to break the bottle. Such operation is carried out and completed in a well-determined time like in intermittent fevers.

I have some objections to this theory. Firstly, the fact that in fever the blood is activated and heats considerably does not mean that this heating results from fermentation and ebullition of the blood since such activation of the blood can be provoked by any other cause. On the other hand, people concede that pricking and tickling of the heart can produce violent shaking of the latter and thus shaking of the blood which resembles boiling. Therefore, fermented blood must not necessarily be the primary cause of the bout of fever.

Secondly, a fit of anger is exactly similar to fever: violent and almost convulsive pulsation of the heart and arteries, precordial heat, hot blood circulating through all the body, thirst and dry mouth, shaking of viscera and trembling of the extremities. However, blood retains its normal constitution, is sweet without any bitterness and does not present signs of abundance of salts or of being exalted by sulphurs. A bout of fever thus can result from a cause very different from blood fermentation.

Thirdly, it is asserted that in fever spirit and sulphur of the blood are excited and exalted so that they overcome all the other components of the blood and, since they cannot unite with them, they induce turmoil and ebullition. If I may tell what I think, these are metaphoric words said self-assuredly without any ground. They neither describe the cause of fermentation of the blood nor demonstrate its existence.

I should rather say that there are countless particles of fire in the mixtures and also others which are mobile spontaneously. These are diversely linked and held together by ties. Similarly, small particles of fire are retained in lime. If it happens that moisture or usual motion, either internal or external, dissolves their ties, then the mobile particles having acquired their freedom exert their mobile character. The other particles of the mixture are activated by them, turned over diversely and mixed. This can occur in wine, lime and other similar substances as a result either of easy dissolution or of abundance of spirits or small particles of fire or something analogous. In blood, however, such effervescence does not seem possible since there is no abundance of oil or spirits as there is in wine. It is doubtful whether these bitter spirits which are extracted from blood by distillation originate from the action of and mixture with fire (as they are mixed in lime). The very learned Boyle noticed that in many instances.

Fourthly, concerning the way the same author attempts to explain the recurrence of paroxysm in intermittent fevers, I note that the example of must enclosed inside goatskin bottles, of its boiling, and of its explosion as a consequence of its exaggerated swelling poorly applies to the fermentation of blood inside the veins which would result in recurrences of the paroxysm in intermittent fevers. Must enclosed inside goatskin bottles always ferments and boils at the same pace: it warms up evenly and generates bubbles at regular intervals in equal numbers at the same frequency whether the bottle swells or not. The vessel explodes when swelling is exaggerated not because the enclosed wine ferments with more violence or heats more and generates bubbles in greater proportion and more quickly. But then filling of the bottle is completed and the mass of air evaporated from the must in greater and greater abundance can no longer be contained in the space of the dilated goatskin bottle. This mass of air exerts a force and seeks a larger space equal to its own volume, thus disrupting the bottle. But, if after the first paroxysm the blood inside the veins ferments like must and boils without swelling like must, I do not see how turgescence of the veins and arteries must be expected to ignite the next bout of fever. Rupture of the blood vessels must not be feared since the blood vessels at the beginning of a new paroxysm neither swell nor contain boiling blood. On the contrary the vessels are empty, small and cold so that the pulse is short and small. Very often the patients are stiff and shivering at the same time. Ebullition and heat of the blood are not produced by swelling but by the same fermentation of the blood as that which occurs from the beginning (as conceded by this author himself) and could not be compared with fermenting must. The origin of this error consists of comparing the intermission of intermittent fever which is an absence of blood fermentation with the state of a goatskin bottle remaining full of must before its turgescence, without explosion and rupture. In this state must ferments and boils, unlike blood. Indeed, if blood contained in the vessels fermented and boiled throughout the intermission, like must in a non-turgid goatskin bottle, during its passage through the heart it necessarily would provoke heat and a bout of fever depending on its own degree of heat. Therefore, the bout of fever would persist throughout the intermission although less intensely. Consequently, there would be no intermittent fever but all fevers would be continuous, steadily increasing and decreasing as said above. But this does not occur. Thus, the bout of fever is not created by fermentation of the blood.

Proposition CCXXIV

Experiments demonstrating that blood altered by heat, decay or mixing with salts or sulphurs does not produce fever.

All philosophers agree that one must proceed according to the rule and principle established by Nature whenever there is doubt whether something is the certain and unquestionable cause of some effect appearing in some subject. If the suspected cause is always followed by the same effect in the absence of anything else being involved and if it is verified that when this suspected cause is removed or absent the effect always stops, we shall assert without risk of error that this is the actual cause of the effect thus mentioned. If it must be discovered whether something is not the cause of some effect, this rule is infallible. If the suspected cause is absent and the effect is produced once, we must say that it was not the cause of the effect thus mentioned. But if the suspected cause is applied to the due subject without any obstacle to the effect and nothing happens, then it can be concluded that there is no cause of an effect not produced and absent in Nature. We wonder whether actual heat of the blood entails fever. We observe that heat of the blood resulting from the high temperature of summer air or from a Turkish bath or from drinking abundantly hot clear water, although it is applied intimately to the blood, diffused through all the blood particles and thus not impeded, does not produce fever at all. We thus can conclude without hesitation that alteration of the blood by actual heat is not the cause of fever. That does not impede fevers to be provoked sometimes by sun rays in dog-days. This occurs because of some damage to the brain and nerves and not because of the heat as such. This is suggested by the fact that these fevers do not start during the act, while the blood is hot, i.e. when the sun rays are heating the head, but after some time when this heat has disappeared.

Contrariwise, there may be absence and deprivation of actual heat in the blood. Despite coolness of the blood in the winter and stiff cold in quartan fevers, the effect, i.e. a bout of fever, follows. Consequently, actual heat of the blood is not the cause of fever.

On the other hand, alteration of the blood by addition of spirits and exalted sulphurs or salts, is not the cause of fever. This is suggested by the following fact. Sulphurous and saline thermal waters drunk copiously mix intimately with the blood and are spread throughout the body. They should above all irritate and prick the heart which they irrigate like a torrent, and thus provoke considerable bouts of fever. But this does not occur. As a consequence of urine retention in patients with calculi, very bitter urinary ammonia salts contaminate the entire mass of blood. They should prick the heart itself continuously and thus provoke severe and continuous fever. But this does not occur. When I lived in Pisa the excellent anatomists Carlo Fracassati and Silvestro Bonfiglioli, in my presence, injected one or two drachms of sulphurous oil into the jugular vein of a live dog. The vein was ligated and after some howling the dog was released. It remained alive and well without fever. It then gnawed bones ravenously and in due time ate and slept without presenting any sign of being unwell. It remained so until the eighth day when it fled from the dogkennel in which it was being detained. If thus blood altered by so much salts and bitter sulphurous oil could not give fever it must be concluded that alteration of the blood by salts and sulphurous juices and spirits is not the cause of fever. This is confirmed by the reverse: blood of febrile patients at moderate temperature has a sweet taste neither pricking nor bitter. Thus there is no bitter, saline and sulphurous alteration of the blood and such alteration cannot be the cause of fever. Alteration of the blood thus is not the cause of fever. However, in drunken people the pulse is accelerated and the body hot: but the primary lesion is in the brain and nerves as suggested by headache, dizziness, sleepiness, convulsions, delirium and similar symptoms. One thus must rather think that acceleration and heat of the heart are a consequence resulting from a lesion of the brain and nerves. This will be confirmed below.

Concerning the decay of blood, what can be imagined as more soiled and putrid than pus in a pleuresy? And, however, pus when eliminated with the urine must have been carried through the lungs to the left ventricle of the heart and hence flushed into the aorta so that about half of it spreads in the head and arms through the ascending branch of the aorta and is carried back to the heart through the jugular and axillary veins. The remainder of the pus flushed through the descending branch of the aorta is not discharged by the kidneys all at once but only a third or a quarter of it depending on the size and capacity of the renal artery compared to that of the descending aorta, as we showed. Therefore, the remaining pus is spread through the whole body, collected by the veins and brought back through the vena cava to the right ventricle of the heart together with the pus descending back from the head and arms. From the right ventricle it spreads in the lungs a second time and is carried back to the left ventricle of the heart with an additional amount of pus which was stagnant in the lungs. Hence circulating again another small part of the pus is left in the kidneys. Thus, circulation must be repeated several times before all the pus is excreted through the urinary vessels. Meanwhile, not only the whole volume of blood must remain infected and soiled during hours or days but the heart must be affected and pricked by the soiling pus at each diastole and pulsation. Consequently, such soiling of the blood by the decay of pus ought to produce high fevers. Since this does not occur often, it must be concluded that fever is not produced by decay of blood but by a cause completely different.

I do not deny that blood may be affected by such vicious and malignant alteration which can produce in the animal considerable lesions and finally death. This has been the subject of experiments by my friends praised above, in my presence. Aqua fortis was injected with a syringe in the jugular vein of a live dog. If this aqua fortis was not diluted and was injected copiously, the blood coagulated acquiring an ugly consistency almost like stone. Filling all the vein it impeded further passage of blood so that aqua fortis could not be carried to the heart. If diluted aqua fortis or some other liquid was injected in the vein little by little some of it reached the heart. Then the dog suffered convulsions and spasms. Writhing with pain and howling it soon died. After opening the thorax curdled blood was found in the heart and vessels. Oil of tartar provokes the same convulsions, pain and death by deprivation but the blood does not coagulate and remains fluid. I observed also that some acid mixed with the serum or albumen of the blood and heated impedes coagulation. Hence I note incidentally that, when clots are found in the heart and veins, blood can in no way have been infected by some acid.

These experiments, however, do not demonstrate that alterations of blood by spirits and bitter salts produce fevers. If it happens that fever sometimes is accompanied by these symptoms I will always deny that the bout of fever was caused directly by such saline soiling of the blood. This results from the fact that all the other countless fevers are not produced by alteration, fermentation or corruption of the blood, as was shown.

Proposition CCXXV

Spirits or nervous juices made more bitter than normal and irritating the nerves and heart are the primary and direct cause of fever.

After having refuted the causes of fever reported by others and having indicated their unreliability, it is worth finding out whether we can conjecture something more likely concerning the nature of fever. The most characteristic feature of fever is the pulse and violent commotion of the heart and arteries which are followed by heat and other symptoms. Not only does almost everybody agree on that point but it is logical and suggested by experiments. Shortly after violent motion and shaking of the heart, some new heat appears in the entire body of the animal which was not there previously, as occurs in anger and in energetic physical exercise. This motion of the heart, after it has decreased and slowed, is followed by coolness, stiffness and cold which did not exist previously and pervade the body of the animal as occurs in fear and in the shiver of quartan fever. Conversely, exaggerated heat of the body and viscera does not always entail violent reaction of the heart as we experience in summer heat, in a Turkish bath, and after drinking hot thermal water abundantly. Consequently, violent commotion of the heart is the cause of heat in the body but the reverse is not true: heat of the body is not the cause of violent motion of the heart.

It must now be seen whether the principal and direct cause of violent motion of the heart may be also the principal cause of fever.

This cause is no doubt the same which in good health moves the muscles of the heart quietly and regularly and which, when increased by some forces, moves them more quickly and energetically. Similarly, the same fire which produces moderate heat is felt to give more heat if one comes closer to the fire or if its flame is increased. In the same way, hands, feet and the other muscles move slowly and softly but also energetically and quickly, by order of the will through the intermediary of spirit or nervous juice. The heart, like an automaton, can move and pulsate repeatedly with short pauses independently of the will. Its pulsations must be produced by the same spirit or juice instilled by the nerves into the heart[1], slowly in good health, violently in fever. Therefore, to move the heart more quickly and energetically nothing else is required than this spirituous juice made more bitter and instilled more frequently. Thus fever is different from good health only in the velocity and violence of the movement of the heart. Consequently, in fever the spirit or nervous juice is instilled into the heart more quickly and more frequently as a result of its bitterness.

This shows the possibility of our position on the nature of fever and the simplicity of its apparition. We now shall bring forward arguments and experiments to confirm this assertion.

Firstly, I observe that the simple smell of wine or any spirituous and nutritive vapour amazingly restores the strength of a weakened and staggering animal and immediately vivifies and increases the weakened pulsation of the heart. No doubt the odoriferous vapours are not transmitted through the veins

[1] Chap. 6 Prop. 77

or the bronchi to the heart. Indeed there is no open passage through the lungs. The next experiment is convincing. The breath is held and odoriferous vapours creep in through the nose. After the subject has breathed out he breathes in external air with his mouth applied to a hole in the wall of the room. He keeps breathing through the hole. Certainly the odoriferous vapours enclosed in the room do not reach the heart through the lungs but, creeping in through the nose, they reach the brain by tickling the mammilar nerves. Hence the motion of the odour and its smell can be transmitted to the heart through the nerves very quickly, almost instantaneously. Therefore, it seems not only possible but even necessary that the heart thus restored moves and pulsates more strongly.

Secondly, this is confirmed by the reverse operation of foul smell which is carried to the brain through the nose and produces a weakening of the pulsation of the heart.

Thirdly, when wine is retained in the mouth and agitated, spirit of wine transmitted to the glands through the salivary ducts and hence through the nerves to the brain and the heart seems very quickly to restore and vivify the strength and pulsations of the heart.

Fourthly, this is even more obviously confirmed by the sufferings of the soul. As a consequence of simple apprehension of some horrible object the heart immediately slows down and reduces its pulsations. Conversely, expectation of some sweet and long-desired object increases and accelerates the pulsations of the heart. Nobody in his senses will deny that such feelings of fear or joy are transmitted through the nerves from the brain, seat of the imagination, to the heart. Consequently, if spirit, i.e. nervous juice, alters the movement of the heart so easily and almost instantaneously, it is not difficult to believe and it is very likely that the same spirit or nervous juice which has acquired some bitterness in febrile patients entails such considerable, violent and disordered activation of the heart.

As a consequence of the violent pulsation of the heart, the blood where it was at some rest, flows from the vena cava and must be pushed into the arteries to the extremities of the capillaries by the same powerful movement of the heart. This results from the laws of blood circulation. Obviously violent and powerful circulation of the blood throughout the entire body of the animal provokes further heat which did not exist previously, as we suggested. I think that the origin of this heat must not at all be sought for in the movement as such but rather in the nature of the blood itself. The blood indeed contains spirit or oil or rather sleeping particles of fire which, if freed from the ties which bound them can exert their mobile character and then develop heat. If it does not result from squashing of heterogeneous elements of blood, heat and fervour can result at least from mixing with salts produced from lye, spread through the viscera and stored in other parts of the body which the blood reaches as a consequence of the violence with which it is pushed by the heart. The same happens in a mixture of acid of vitriol and salt of tartar. Willis demonstrated this experimentally. He mixed with hot blood spirit of wine, or of stag horn, or of soot, or of vitriol or of another spirituous liquor or saline and saw amazing ebullition and effervescence provoked in the blood.

Proposition CCXXVI

How nervous juice fermented and bitter can be transmitted to the heart to provoke fever.

It is not enough to know that nervous juice instilled in the heart can produce violent fever by its bitterness. It is necessary to describe how this juice is carried to the heart from the most remote parts of the body, to be able to complete there its febrile operation. But first of all, not to appear to work on an imaginary and fantastic hypothesis, it is worth showing that such operation occurs in animals: noxious juice can be transmitted from the extremity of a foot through the nerves to the brain and hence to the heart which it activates violently and shakes. This occurs with venoms which are harmless if they only touch the skin. But if applied on wounds or ulcers they are carried to the brain very quickly provoking dizziness and numbness. Then they are transmitted to the heart, modify its rhythm and induce palpitations. The venom is not carried to the heart through the veins but through the nerves: the brain is affected and disturbed by the bite of a viper or by oil of tobacco before any lesion of the heart is produced. The small veinous orifices in ulcers are obstructed tightly enough by some gluten since this same juice also closes the orifices of the arteries adjacent to the small veins so strongly that blood which otherwise uses to leave the arteries violently cannot flow through them. Neither can venom in fresh wounds be transmitted through the veins. Its entry is impeded by the bleeding of the small veins. We have drawn attention to this fact in the Accademia Del Cimento when we experimented on the highly poisonous effects of the oil of tobacco. This, when poured copiously into a fresh wound, did not do any harm whereas, one or two minutes after it was rubbed into the flesh with a wet cloth, the animal became numb and shortly afterwards died.

This very quick transmission of venom thus must necessarily occur through the small apertures of the nerves which gape in ulcers and wounds. They immediately spread up to the brain where they produce many severe symptoms and are transmitted as quickly to the heart which they shake and where they provoke fever.

This is confirmed by the fever which occurs when there is pus in ulcers and pustules. In these the corrupted liquids which make and ferment pus are not transmitted to the heart through the veins but to the brain through the nerves. This is suggested by the fact that the heart is not affected by contact with similar fermented liquids during the passage of pleural pus through the heart. Therefore, the fermented liquid of the pustules which can prick considerably the nerves ending there can easily transmit its venom to the brain and hence provoke commotion in the heart, alter its rhythm and produce fever.

The same can be confirmed by the fever occurring in arthritis. It is obvious that the acute articular pain results from pricking of the nerves of the membranes and tendons. No doubt such pricking is carried out by saline and tartarous liquids which meet there, ferment and are activated. These pricking juices either irritate the nerves up to the brain or are transmitted to the brain and hence to the heart where continuous fever can originate.

The same must be said of any symptomatic fever in all of which it is verified that either irritation or pricking juice is transmitted to the brain and hence to the heart. This increases and accelerates pulsations.

The existence and truth of these symptoms were shown. It remains to find out why irritation in the nerves must be transmitted from the brain rather to the muscle of the heart than to the other muscles of the animal, to which nerves ramify from the brain as they do to the heart. There are ducts carrying strength and motive faculty to these muscles as well as to the heart. Therefore, as the heart is violently activated and shaken following this diffusion from the brain, all the other muscles should also be shaken and present some spastic paralysis and trembling.

Actually this does not occur although the existence of the symptoms thus described cannot be denied. If we refuse to resort to words such as sympathy, agreement and convenience which have no meaning and hide ignorance, we must admit that the organic structure of the heart is considerably different from that of the other muscles. The same motive faculty indeed carries out different operations only if being used by different organs. For example, the flow of the same stream either crushes and grinds wheat or saws beams or beats metals or plays symphonies using pneumatic flutes and organs.

Let us find out how the organic structure of the heart is different from that of the other muscles. Although not obvious this difference can be surmised from the difference in their actions. All the muscles of the limbs, although they do not lack motive force, or spirits or juices in their nerves, remain quiet and do not move without an order of the will. On the contrary, the heart is a particular organ which carries out its pulsations always in the same way and at the same rhythm, willy-nilly. Hence I wonder whether there is no other cause of the movement than the spirit or juice flowing from the nerves into the muscles. In the muscles of the limbs this outflow can take place only if the origins of the nerves are violently shaken by an order of the will so that they suck spirituous juice as leeches would do. In the heart, however, this outflow must always occur, droplet by droplet, without any command because the narrow orifices of the nerves remain always open and gaping. And so our problem is easily solved, if I am right. Indeed if the activation or effervescence of the spirits or juice is supposed to be little and slight in the brain, the muscles of the limbs must not or even cannot move without being ordered to do so by the will. It is different for the heart because the outflow of spirituous juice into the heart cannot be impeded, as a consequence of the permanent opening of the orifices of the nerves. Therefore, the heart must necessarily be shaken and, by carrying out more violently its usual pulsations, produce this movement which we call fever.

We said that this can occur if some slight commotion of the spirits in the nerves is assumed. But if this activation of the juice inside the nerves and the brain is more pronounced as a consequence of too much mixing of bitter salts, then not only the heart is shaken very violently and gives more severe and higher fever but all the muscles or the main muscles of the whole body are activated and suffer convulsions. Then the bitterness and activation of the nervous juice are so violent that they can force the orifices of the nerves despite the resistance of the will. That is what is observed in pernicious fevers.

Proposition CCXXVII

Where and how nervous juices are altered so as to produce periodic fevers.

To describe exactly the nature of fever, it is not sufficient to have indicated that fermented and bitter nervous juice can be transmitted to the heart and produce a bout of fever by activating it violently. The direct causes of the alterations of this juice must also be described as well as the places where the alterations take place. To facilitate this it is necessary to observe the effects of fever in cadavers. In these no significant change is found in the blood but very often the lungs appear to be inflamed, ulcerated and sometimes necrotic or sprinkled with abscesses. Sometimes, the spleen and liver are found to be injured. But more frequently the spleen is hardened and scirrhous. Always glands are observed either hard and scirrhous or yellow and blackish, above all in the mesentery.

Red hepatization of the lungs is certainly not the cause of fever. It must not be considered as an inflammation or as a lesion of the lungs since it is found also in all the cadavers of healthy animals. It occurs in moribunds because breathing stops before heart pulsation. The blood which is carried to the lungs at this extremity of life remains there and swells the lungs since it cannot be discharged, breathing having stopped.

Tuberculous lesions or abscesses of the lungs are seldom the cause of fever and then indirectly. They are more often the effects of fever since the lungs are not found to be corrupted in this way in all cadavers of febrile people. Conversely, in asthmatic patients the lungs often present lesions and are even full of them without fever. Either this corruption of the lungs is completed at death's door or, if earlier, it is not the direct cause of fever. Only lesions of the glands remain to be considered. They are always found in the cadavers of febrile people.

From the observations of the very learned and very intelligent Wharton and of others, it appears that nerves, arteries, veins and lymphatic vessels ramify from the countless glands of the animal, sending countless roots like those of trees. In some there are excretory ducts and canals such as the salivary and pancreatic ducts.

This structure implies that glands are as many workshops in which some important work is carried out. But what is it? Perhaps arterial blood is purged in the glands of some impurities and serous humours which are absorbed by the excretory vessels such as the salivary ducts. But if this was true what would be the use of the nerve which ramifies in the gland since glands do not move and have no feeling? The nervous roots in the glands thus must either absorb something from the arteries or spill something in the collecting vessels such as the veins, or in the excretory vessels such as the salivary ducts. It seems unlikely that the nerves in the glands receive blood from the arteries since arteries ending in abundance in the brain can provide blood copiously there very simply whereas it would be very inconvenient for the nerves to be forced to beg blood from the arteries in the glands. It seems more likely that some juice is discharged from the nerves into the glands. The nature of this juice is well-determined and depends on the various shape and capacity of the orifices of the nerves. Mixed with particles emitted by arterial blood, this juice becomes able to carry out some subsequent action in different parts of the animal, in the mouth, oesophagus, stomach, intestines and anywhere else where it is involved in fermentation and digestion of food and ulterior operations, feeding and vivifying parts of the animal.

These juices destined to be expelled from the nerves and let in the glands occasionally may be retained in the nerves if the orifices of the latter in the glands are obturated as a consequence either of too much juice or of some gluten contained in the nerves. These juices thus retained in the nerves can easily degenerate by some fermentation and take on an altered nature harmful to the animal. This is nothing new: indeed semen in the animal in perfect condition, destined to be expelled to the end of generation, if retained for a long time, becomes corrupt and poisonous. Even more likely it can happen that, not only the outflow from the nerves is prevented but irritating, harmful and malignant particles are transmitted by these nerves, as a consequence of obstruction and scirrhosity of the glands or of corrupt juices collected in them or transmitted from elsewhere, hardened and fermented. This occurs in venereal disease. From all this, it is obvious that the brain can be affected by some irritation or transmission and that the consequent commotion of the spirits or nervous juice is propagated to the heart through the nerves.

Consequently, the places in which these nervous juices ferment are above all the glands and the roots of their nerves which are obstructed, and pathologically affected and irritated. The cause of fermentation actually is the retention by force of elements which ought to be excreted from the nerves or an infection transmitted to these glands and able to alter the spirits by retention.

Proposition CCXXVIII

The reason why a bout of fever produces either heat all the time or at first a shiver and heat afterwards.

We have shown that in general nervous juice can ferment because of an obstruction or lesion of the glands and of their nerves. The heart can be irritated by such fermentation, and produce a bout of fever. This general theory must be adapted to the particular instances so as to deduce the causes and mechanisms of all the phenomena of fevers.

But firstly, to avoid any ambiguity and uncertainty, the following must be noted. It seems likely that complete obstruction of the glands and of the nerves which end in them must be progressive rather than quick and almost instantaneous. Therefore, as a consequence of the deprivation, the bout of fever starting from normal should increase over some time, passing through all the intermediate stages to reach its maximum. It does not jump from a minimum to a considerable degree of commotion and heat or of painful stiffness which cripples ill people, as shown by experience. Although disturbed sleep, troubles, dizziness, rings under the eyes and nystagmus precede the paroxysm, all these slight symptoms are rather signs than parts of the fever.

The problem can be solved, I think, by analysing the experiment mentioned above[2] in which oil of tobacco and other poisonous juices applied on the orifices of divided veins were harmless although the veins communicate with the brain or the heart, because the outflow of blood impedes entrance of the poisonous juices into the veins. We speculate that several glands are infected by some corrupt and excremental juice which obstructs most of their excretory ducts so that fermentation can proceed in the glands. Then the nervous canals and blood vessels can also be obstructed. Nervous juice continuously flows out of the nerves and blood from the arteries. Both can be carried away through the veins. Obviously, as long as juices flow from the nerves, bitter, glutinous and fermented particles from the glands are mostly impeded entering the nervous canals. Therefore, they can prick neither the nerves nor the brain nor the heart. Consequently, then the bout of fever is not provoked. But afterwards when the glands are full as a consequence of addition and fermentation of glutinous and viscous excrements, finally not only the pores of the nerves are obstructed but they are even forced to admit some bitter and fermented particles which overflow from the full glands. This can occur rapidly without long delay as soon as the glands are full of glutinous juices. The other effects mentioned above follow this filling. The cause of the paroxysm of fever thus is the obstruction and pricking of the nerves with consequent transmission to the brain and heart. This explains the rapid and almost immediate change from normal to fever. This is, however, preceded by the slight symptoms of disturbed sleep, troubles, dizziness, etc. which are produced by some benign irritation by fermented juices of the glands occurring in a few capillary nerves already obstructed. After the glands are full of gluten, the orifices of the nerves are immediately closed and the nervous juices thus retained and soiled ferment. Then the heart is irritated and the bout of fever develops with considerable and violent heat.

If there is no peculiar cause of stiffness and shiver the bout of fever is always accompanied by heat from the beginning, by its nature. Indeed blood circulates very violently through the body and provokes heat as a result of the violence of the pulsation of the heart, irritated by spirits or fermented nervous juices[3].

Finally, the cause of the stiffness and cold with which some bouts of fever begin must be explained. Perhaps the blood is never completely deprived of urine since animals urinate even after abstinence of drink for two or three days. It is thus possible that in the obstructed glands arteries spill nitrous particles and other salts which copiously contaminate the urine present in the blood. These particles acquire a vitriolic and freezing nature as a consequence of a peculiar ferment originating in the diseased glands. Afterwards, when filling and obstruction are completed, they are carried from the glands through the nerves to the brain and spinal cord. This indeed seems to be the quickest and simplest route from the nerves of the abdominal plexus which are linked to the lumbar roots of the spinal cord. Since the nervous fibres are very easily excited and irritated they can induce a feeling of cold and shiver at the beginning of

[2] Prop. 226
[3] Prop. 225

the bout of fever as a result of the peculiar nature of nitre. As a consequence of this contraction and shiver the spirits are numbed and hence the pulse becomes slow and small. This happens when the commotion of the nervous juice is weak.

Feeling of cold at the beginning of a bout of fever can occur in another way. From the glands obstructed and altered by ferment, some juice not unlike spirit of vitriol is transmitted through the nerves, as we suggested above. We also supposed that urinary salts are left by the arteries in the spinal cord. Since they retain their ammoniacal nature (ammonia salts use to be extracted from urine), when mixed with freezing spirits they can produce cold boiling. We indeed found by experiment that such mixture boils with considerable cold. Stiffness, cold and shiver thus can occur in both ways, mainly in the area of the spinal cord with numbness of the spirits and small pulse because the heart is slowly stung. This cold boiling does not disappear immediately but can persist one or two hours because the juices thus mentioned can be instilled and mixed little by little through the canals of the vessels from distant places. In the meantime the spirits or nervous juices set in movement can irritate the heart by their bitterness with a quick impulse as a result of the aperture of the orifices of the nerves which end there. Therefore, the pulsations by increasing continuously must create hot fever for two reasons. The first reason is that the nerves disseminate the nitrous salts and expel them partly into the veins. The second is that the blood diffused throughout the body with violence by the strong pulsation of the heart can provoke heat and a bout of fever, as results from what was said.

Proposition CCXXIX

On the causes of the symptoms accompanying a bout of fever.

Not only the beginnings of the bouts of fever but also their progress and all the accidents which accompany fever are worth our attention, either for themselves or to support our theory by different evidence and arguments.

Firstly, it is worth giving the reasons for the signs or feelings preceding the paroxysms, such as disturbed sleep, troubles, dizziness, etc. Just before the paroxysm the glands, although they are not turgid, are however abundantly provided with ferments which up to a certain point can affect the brain by the intermediation of the nerves. This disturbs the sleep. The troubles occur for the same reason because the spirits or nervous juices thus infected cannot carry out motions or transmit usual sensations normally and quietly. Dizziness and rings under the eyes have the same origin: they result from the beginning of disturbances of the brain and spirits. The nystagmus which is a benign convulsion may be provoked by slight pricking of the nerves. Consequently, it appears that one cause alone, i.e. the direct lesion of the nervous juice and brain, can produce these accidents and, considerably increased later, give the bout of fever.

When the bout of fever has begun, there is often bilious vomiting because, during the febrile stiffness and shivering, the membranes of all the body are shaken by the same nervous juices soiled by bitterness, mainly the stomach, the pylorus and the bile duct which are abundantly provided with nerves. As a consequence of the spasm, the bile is discharged from the gall-bladder thus compressed, into the duodenum and from there into the stomach, and finally vomited through the mouth.

Almost intolerable precordial heat results directly from the very violent displacement of the blood. Its indirect cause, however, is the pricking of the muscle of the heart by bitter nervous juices which provoke such energetic pulsations.

Unquenchable thirst is due not only to the heat but above all to the bitter and saline particles which are stuffed abundantly in the tongue, throat, oesophagus and cardia as a consequence of the obstruction and fermentation. These particles corrode and tear the nerves which end there and the nervous papillae of the tongue and they make them pustulous, black and dirty above all in acute cases. These reactions, however, are very different from those which pepper and bitter salts on the tongue and in the throat produce. These also excite much thirst but without dryness of the mouth. They provoke profuse salivation contrary to fever. This discrepancy clearly confirms our theory. If the excretory vessels, i.e. the salivary ducts of the glands of the mouth, were not obstructed in febrile patients, the salts pricking the nerves which end there should provoke an outflow of saliva as occurs in good health. Therefore, this considerable dryness of the mouth in febrile patients suggests that the excretory ducts of the glands are actually obstructed as supposed in our theory. By the way, it is commonly thought that blackness of the tongue results from vapours of soot ascending from the stomach. Nothing more stupid can be said or imagined.

It is very easy to explain aches and principally headache in febrile patients, by our theory. Bitter and stinging saline salts flowing through the nerves can prick membranes and nerves, above all in the brain where the most exquisite feelings are. They thus can entail ache and spasms.

Weakness and tiredness are experienced by febrile patients. This is explained easily by our theory. Since the lesion is in the nerves the spirits or nervous juices which are substances of small volume but producing impetus in the animals are dissipated and altered. They are not in the blood since blood-letting even when abundant hardly weakens the patient unless most of the blood is lost.

Sleepiness and delirium are usually associated with the paroxysms. They are also explained easily by our theory. Both occur in the brain as a consequence of bitter and stinging juices irritating the brain and disturbing the order of the movements of the spirits.

Faintings occur in many different ways. Their direct causes are the same. But in fevers which we discuss now they result from pricking of the cardia as a consequence of the abundance of nerves from the sixth pair which end there. Therefore, the brain and heart are affected. Such pricking can be produced by any bitter juice, venoms, earthworms and others of the same kind. Nausea has the same cause.

Convulsions during the paroxysms can uneasily be explained outside our theory. Indeed, muscular contractures against the will can occur only as a consequence of the bitter and biting nature of the spirits and juices flowing out of the nerves into the muscles and producing these involuntary movements by irritation.

Finally, in acute cases, abscesses in the largest glands and rash seem to confirm our opinion wonderfully. Nervous juices can be so malignant and poisonous that, excreted in the largest glands or in the papillae of the skin, they can produce abscesses and necrosis. This usually does not occur at the beginning of the fever but after some paroxysms, after the orifices of the nerves ending in the glands have been de-obstructed violently by a flow of circulating blood, as we will show.

Proposition CCXXX

Cause of the diminution and end of fever and of the alteration of the urine.

We must also explain why and how the bout of fever, after it has reached its paroxysm, regularly decreases until it stops completely. Because of the turgescence of the glands or of the viscosity of their contents the orifices of the nerves which end in them are obstructed. As a consequence the nervous juices thus retained and diseased ferment and sting the small nerves. They produce fever by irritating the brain and afterwards the heart. Consequently, as long as fever persists, the blood pushed violently from the arteries into the glands must wash, scrape and remove the corrosive obstacles which obturated the orifices of the nerves in the glands by means of its boiling fluidity and rapid rasping. It must even open some excretory ducts. As a consequence of this opening, bitter and fermented nervous juices, no longer impeded, can flow out and be excreted from the nerves into the glands. Hence they partly ènter the veins ending in these glands (according to the laws of Nature which destined these veins to absorb juices) and can partly be excreted through the excretory vessels some of which were recently de-obstructed. As long as these bitter and fermented juices which were the cause of fever by their irritation are expelled little by little from the nerves, the bout of fever must necessarily keep softening and decreasing. Finally, when the nerves are completely discharged and purged of their load of bitter juices, irritation and violent commotion of the muscle of the heart necessarily stops. The heart thus quiets down and the bout of fever ends.

But the nerves can get rid completely of the fermented and bitter juices by evacuating these into the glands without all the spirits having flowed out so that the nerves are empty and dry (this could not occur without death of the animal). One must rather think that not all the mass of nervous juices must have been contaminated equally and uniformly like a barrel full of vinegar from which bitterness cannot be removed without throwing away all the vinegar. It is rather like lees of wine separated from sulphurous and oily elements and remaining at the bottom of the barrel. The lees can conveniently be retrieved from the wine and oil through a hole opened in the bottom of the barrel. Similarly, the bitter and fermented juices can persist in the distal parts of the nerves close to the obstructed glands and flow from their orifices while the other part of the nervous spirits which is clean and not soiled by salts remains.

It is easy to understand that the heart can be irritated otherwise although not touched directly by bitter salts stored in remote parts of the nerves. Nerves are of a nature such that, if one extremity is pricked, pain is felt at a distance. For instance when there is injury to the spinal cord by contusion, sometimes pain is felt in the knee and in the leg. Similarly, if the inside of the ear is irritated, lesion and commotion are provoked in the diaphragm and in the intercostal muscles which results in violent coughing. There are many other examples to confirm this point.

The diminution and end of the bout of fever thus can be explained adequately by our theory.

This other problem can also be solved. At the beginning of the disease, urine is not altered but only well after the first or second paroxysm. The possible reason seems to be that, before the first paroxysm, the saline juices which are the ferment of fever were not retained in the veins but inside the obstructed glands and afterwards inside the nerves. Therefore, it is not surprising that urine separated from the blood is immune of impurities which are not yet in the blood. After the paroxysm, the saline juices are discharged from the nerves mostly into the veins. These salts thus must contaminate and impregnate the serum of the blood, as required by the nature of water. As a consequence, urine presents with the foulness and corruption which result from these salts.

Proposition CCXXXI

How a bout of fever can recur after a well-determined interval of time.

We said in the preceding proposition how a bout of fever quiets down and finally can stop completely during the intermissions. But the most difficult problem remains to be solved: how and why can a bout of fever recur after a well-determined interval of time or more exactly what clock determines the times? The intimate, proper and direct causes of this difficult phenomenon are not known hitherto. Therefore, we shall have to content ourselves with general conjectures which can be suspected. After the end of a bout of fever, recurrence is possible only if the remnants of ferment and the viscous humour increased and multiplied obstruct again the excretory vessels for a second time and fill and sting the orifices of the nerves of the glands. After the first bout of fever, the violent outflow of blood, which maintained the orifices of the glands open, stops. On the other hand, the remnants of ferment in the glands ferment the viscous and bitter juices which arrive. These thus can stuff, obstruct and sting for the second time the excretory ducts and afterwards the orifices of the nerves. Obviously all this fermentation and obstruction of the vessels in the glands must occur and be completed in a well-determined time and not instantaneously. If the ferment and the viscous substance were the same in all fevers, with the same bitterness and consistency, and if the vessels were arranged in the same way in the glands, this operation no doubt would be completed in the same time. Thus in all fevers the intermissions would have the same duration. But the intermissions between two bouts of fever are considerably different in duration. The interval of time between the beginnings of two successive bouts of fever is twenty-four hours in daily fevers, forty-eight hours in tertian fevers and seventy-two hours in quartan fevers. It must be concluded that the strength of the ferment and the viscosity of the substance are different and proportional to the times required to complete these bouts of fever. It is commonly believed that this difference results from the alteration and fermentation of the four humours. The long intermission in quartan fever would result from the inertia and numbness of the melancholic, cold and earthy humour. The shorter intermission in tertian fever would result from the mobility of the yellow, warm and igneous bile and the higher frequency of the paroxysms in daily fever would be produced by the very noble, aerial and more lively nature of the blood. But all this theory of the four humours with their properties and actions is obsolete and was never supported by solid arguments. It cannot explain the phenomena of fevers satisfactorily. That is why we shall attempt to solve the problem using another hypothesis.

The inequality of the intermissions can be explained in different ways. The one to be chosen, in my opinion, should not only explain all the phenomena but also explain them in the same easy and simple way.

Such is my hypothesis. The bout of fever cannot be provoked without obstruction and stinging of the orifices of the nerves. If a second obstruction and irritation could not occur before exactly 24, 48 or 72 hours, the second bout of fever should occur after the same precise intervals of time. The causes of obstruction and irritation of the orifices of the nerves ending in the glands are the viscous, slow and bitter juices collected in the glands. Therefore, if these juices had a consistency such as to be able to achieve obstruction in the inequal times thus mentioned, the effects of fever should also occur after the same unequal times.

How can the consistencies of viscous juices be different in a given proportion? This does not seem to be difficult. In three cups we pour equal quantities of water, for example a pound in each cup. Six ounces of flour or of another powder forming gluten are mixed in the first. Three ounces are mixed in the second cup and two ounces in the third. In water the atoms of flour occupy a space equal to their volume. They cannot be at equal distances from each other since six ounces of flour contain twice more granules of flour than three ounces. In the cup with three ounces of flour the atoms are at a double distance from each other as compared with those in the cup with six ounces. During the descent of the flour to the bottom of the vessels gluten of the same consistency is formed in unequal times. It is formed twice more quickly in the cup in which flour was twice more abundant. Let us assume that the gluten of the first cup arrives in some canal in which wine flows very slowly and lazily, and obstructs this canal after one hour by forming a clot of particles of gluten. Obviously the quantity of gluten in the second cup being half that in the first cup is also able to obstruct the canal but more slowly, in a time twice as long. Finally, the gluten of the third cup is a third of that of the first cup and two thirds of that of the second cup. It will take longer to obstruct the canal, three times as long as that of the first cup and one and a half times as long as that of the second cup. The same can occur in our problem of the bouts of fever and we can surmise that it is likely that it occurs. At rest, nervous juices flow lazily from the orifices of the nerves into the glands. Therefore, the viscous and bitter substance which is in contact with these orifices, when swelling as a result of fermentation, can easily enter, obstruct and sting these orifices. This should occur in times inversely proportional to the consistency of the gluten, i.e. to the quantities of substance thus mixed and forming gluten. If, for example, a scruple of this substance is mixed and can complete the obstruction after 24 hours in daily fever, then half this quantity of the same substance can complete the obstruction after 48 hours in tertian fever and a third of the first quantity of the substance will complete the obstruction after 72 hours in quartan fever. The intermissions or intervals of time between the first and the second bout of fever thus will last 24, 48 and 72 hours respectively. I think that these generalities can be stated on this problem as a hypothesis.

We now proceed to continuous fevers which seem to differ from the intermittent ones only in the persistence of uninterrupted fever with ups or downs during the same one-day or three-day periods. In these continuous fevers two opposite actions must be assumed to take place together. One is the same fermentation of gluten which, as a consequence of its abundance or malignity, can continuously produce the obstruction, irritation and stinging of the nerves in the glands with more or less velocity depending on its strength and abundance. The other is the activation of the blood by the commotion of the heart due to fever. This continuously scrapes and de-obstructs the orifices of the nerves and of the excretory ducts ending in the glands.

It seems that the two opposite actions can be carried out according to the same theory. When the considerable turgescence of the glands and the quantity of bitter juices (which must be assumed to be always present in continuous fevers) are reduced to a minimum, the fever decreases to a minimum. When the glutinous and bitter substance increases, the glands swell and fever increases to a maximum. Hence the violent energy with which blood pulsates and flows can scrape and remove a great part of the obstructions of the nerves. Therefore, bitter juice begins to flow and be excreted from the nerves. The more of it which is expelled from the nerves the more the bout of fever diminishes.

But, while the paroxysm decreases as a consequence of the expulsion of bitter ferment from the nerves, the arrival of new viscous, glutinous and bitter substance sent into the glands by the arteries and lymphatic vessels does not stop nor does fermentation in the glands cease. Therefore, before the decreasing fever has completely disappeared and temperature is back to normal, the glutinous and malignant juices collected in the glands and fermented perhaps can achieve a new obstruction by filling the excretory ducts and by stinging and pricking the orifices of the nerves. When the paroxysm has diminished, blood is pushed more slowly. Because of this weakness of the impetus not only is it unable to de-obstruct completely the orifices of the nerves but also it cannot impede the arrival of new substance and a new obstruction of the vessels. Consequently, before the first bout of fever is completely finished a second must begin for the same reasons. I think that this is the likely cause of continuation of the fever: one bout is not yet finished when another follows.

It is commonly said that in continuous fevers periodic fevers are exactly intricated, for instance two tertian, or two or three quartan. I must concede that if this was true it could not be explained so easily by our theory. But they are perhaps mistaken those who think that bouts of inequal periods are intricated, as the very learned Willis remarked. The error can originate in that the days are numbered and not the intermissions in hours between two successive bouts. If, for example, the intermission is of twelve hours one bout of fever will occur during the night, the next one during the day. These are not two daily fevers but rather one one-half-day fever. The same would be true if the intermissions did not stretch considerably on either side beyond these limits. If the intermission is of 16 hours, three bouts of fever will be completed in two days. This will not make exactly two tertian fevers but one period of two-thirds of a day. In the same way if the intermission is of 18 hours, five bouts of fever will occur in three days. If the intermission is 40 hours, three bouts will take place in five days. We can argue similarly on intermediate intermissions. Therefore, it may be untrue that in continuous fevers, several fevers are intricated. It seems that continuous fever can be better explained by our hypothesis. Some physicians say they have observed that complicated fevers must be distinguished not only by the periods of their paroxysms but also by the variety of their symptoms. We observe that fever is provoked by an abscess at the extremity of a foot. It can also occur that there are two distinct obstructions in two important glands. These glands then achieve the maximum of swelling and fermentation at different times and thus irritate the nerves and the heart at different times also. So it seems possible to explain that two bouts of fever result from two foci which ferment and blaze up at different times.

Proposition CCXXXII

On the causes of other bouts of fever and symptomatic fevers.

Besides periodic fevers there are also bouts of fever which we will show can all be produced by the same cause described above. It is worth mentioning catarrhal fever which usually is produced by unexpected cold: very often sweating is provoked by exercise or staying in a heated place, and then a draught occurs. Sometimes progressive cooling of air provokes a catarrh. To understand its pathogenesis it must be admitted that not only considerable and countless glands are present in the skin of our face, neck and cheeks but all the surface of the body is provided abundantly with small glands with their own nerves and sweat ducts. This was observed by the famous Malpighi and other modern anatomists. The outflow from the excretory ducts of the sweat-glands aimed at insensible perspiration thus must be impeded to continue either by unexpected cold or draught or wind. The same happens when the flow of water from the pipes of small fountains is impeded when beaten by contrary winds (by the percussion of these winds). Therefore, the juices and particles which had to leave the skin are kept in check. They are thus retained and stopped by force. This delay in outflowing and retention are followed by obstruction of the pores. Once these are obstructed, the other nervous juices and sweat or urine on their way to get out are prevented from doing so. Therefore, they accumulate in the glands, and swell them as appears in swollen tonsils. Afterwards they retrogress into the nerves which they obstruct. They thus must ferment. Irritation can be transmitted from the pricked nerves to the brain and then to the heart. The pulse accelerates and fever can begin as stated above. But the truth of this theory is better confirmed by the cure of the catarrh as explained below.

There is another very pernicious disease which is commonly called latent and traitor fever from which the patient dies despite a normal pulse and normal urine. I would deny that this disease is a fever. It is rather an illness similar to that produced by venoms which kill without fever. They disturb, alter and fix the spirit or nervous juice so quickly that they break and destroy the strength before they can provoke fever. In patients presenting with this malignant disease and in poisoned patients, the nervous juices are pathologically altered and the blood does not coagulate (as some imagined). This appears in experiments which we carried out at the Accademia Del Cimento on animals who were killed by a viper bite or by oil of tobacco. In these animals the blood was always found fluid and clean so that these animals killed by venom were eaten without any harm to those who ate them. Similarly, in the cadavers of patients killed by malignant diseases the blood was also found fluid and generally similar to normal blood. But nervous juice is considerably affected as appears from the numbness, lethargy, paralysis, convulsions, delirium and abscesses in the emunctory organs.

There is a fever usually called hectic, which is not periodic but slow and continuous. However, it blazes up after a meal because new chyle in the stomach, pylorus and intestines is infected by acid salts and then mixed with salts produced from lye which are abundant in the glands of the mesentery. These salts can cause some heat and boiling (as required by their nature). As a consequence, the heart pricked through the nerves can provoke commotion and fever.

One can also mention here the fits of gout and other syndromes with which we dealt above. We showed that all are produced by the same direct cause, an irritation of the heart due to bitter nervous juices.

Proposition CCXXXIII

How fever is extinguished and cured so that the animal is returned to normal.

Intermission in periodic fevers is a temporary cure. The cause of recurrence is renewed fermentation originating in the remnants of ferment staying in the glands. Consequently, complete and absolute cure of fever cannot be achieved before all the ferment of fever is removed and eliminated completely. It is useful to indicate how this can be obtained by using the example of the cure of the familiar catarrhal fever. In this fever, the heart violently activated pushes the blood through the arteries with great impetus. As a result of its impetus the blood can scrape the glutinous obstructions in the nerves and excretory ducts of the glands in all the body and above all near the brain where the lesion mainly exists. Once the obturations have been removed from the sweat-glands and from the large salivary ducts in the throat, nose, mouth and chest, the altered and fermented nervous juices can be cleaned and the impurities expelled or neutralized. Most of the salts are usually sent into the veins and hence excreted with the urine. Their remainder is emitted by the sweat and saliva. In this way, once the nervous juices are neutralized so that they regain their normal sweet taste similar to that of the brain and spinal cord, fever stops completely as does the catarrh.

This is clearly confirmed by the excretions which are expelled at the end of the condition. Their taste is sweet and pleasant, similar to that of the substance of the brain and spinal cord whereas in the beginning of the disease they were bitter, salty and corrosive.

The other fevers are cured in exactly the same way when the excretory ducts of the glands in the viscera and skin are de-obstructed and the ferment present in the glands is partly rejected in the sweat or insensible perspiration, partly neutralized and partly removed by the veins and hence evacuated little by little in the urine or neutralized by the arrival of normal chyle well constituted.

It thus seems possible to deduce that fevers can never or seldom be cured by copious purgations and evacuations of humours since the ferment of fever is usually in small quantity. This is demonstrated by the cure of fevers occurring with cold, by the febrifuge root recently imported from India and which eliminates fever without any excretion either through stools or sweat or urine*. It is sufficient that this little ferment finally be removed from the nerves and expelled or be mixed with and diluted in other humours or that its fermentation be stopped, numbed or modified. This is clearly suggested by the fact that the most tenacious quartan fever sometimes is cured by a single fit of anger, deep anxiety or fear. The explanation seems to be that very powerful and violent movement of the spirits or nervous juices can disturb, stop or modify the fermentation of these juices. Such introduction of a new movement can eliminate fever completely.

This cure of fevers is achieved by Nature spontaneously in most instances. But the endeavour of Nature can be helped by the talent of an expert physician working according to the rules of his art. Everybody agrees on that. Actually, the usual medicines are unreliable and questionable as the most able and learned physicians sincerely acknowledge. Although the results sometimes seem to justify the use of specific medicines, this is, however, accidental and misleading because in most instances fevers are of good prognosis. In these,

* Probably Ipecacuanha: root of a South American shrub used as an emetic and purgative.

the physician acts either well and according to his art, or poorly and harmfully, or not at all and, however, the patients recover perfectly well. In such instances, different and even opposite operations and medicines seem to help. This shows that we may be relying on bases completely misleading and unstable.

But sometimes fevers are so malignant that any medicines neither help nor harm. As a conclusion it is safer to abstain from any artificial medicine except if urgently necessary.

But I do not at all condemn Medicine as useless or harmful and I do not want to proscribe it. I know that among such unreliability and as long as prejudices are discarded, several medicines have been tested by long and wise observation so that they always or very often help. This is obvious in daily experience and we can use them with due caution.

Actually, as results from what was said, the principal purpose of a physician in the cure of fevers must be to remove the obstructions of the excretory vessels and to neutralize the fermenting salts by using light and liquid food and drink. It is indeed useless to hope for separation of the salts from the humours and their elimination because salts once imbibed in and incorporated with the juices of the glands cannot be separated easily from them and expelled. Salts can be separated from sea water, and even not completely, only by distillation with evaporation of water. It is possible rather to neutralize and sweeten the mentioned salts either by adding appropriate fluid solvents or by using opposite salts. This we experimented with aqua fortis which is made of the saline bitterness of vitriol and alum. It is so powerful that it can corrode silver and dissolve it into minute atoms. If ammonia salts, much more bitter than vitriol and alum, are added to this aqua fortis, not only does it not become more corrosive but, on the contrary, it loses its power and becomes so blunted and weakened that it can no longer corrode silver. Similarly, a mixture of salt of nitre and sulphur is inflammable. After addition of ammonia salt it can no longer be ignited. On the other hand, salt of sloe is composed of the same burned nitre and sulphur. It suppresses inflammation and promotes melting of coagulated juices. Salt and powder of hartshorn and crabs blunt and remove the most acid bitterness. Vinegar itself, when distilled, after addition of lead, acquires a sweetness similar to that of sugar. These examples and others which can be mentioned show that the saline bitterness of the ferment of fever can be blunted, sweetened and destroyed completely by addition of other salts of an opposite nature.

Finally, the usage of phlebotomy must be mentioned. Some people praise it highly. Others condemn it as pernicious. Personally I think that it can hardly help or harm. If it was so beneficial no doubt more patients proportionally would recover in France and Spain where blood is let from all febrile patients than in Italy and in other places where blood is never let. Conversely, if it was so harmful more patients proportionally would die in Spain and France than in Italy. Since neither happens, it must be concluded that blood-letting is not worth mentioning either for its help or harm.

Moreover, phlebotomy does not weaken the strength by removing spirit and vital balsam. This appears from the experience in haemorrhages and profuse blood-letting in healthy and ill people in whom strength is not diminished unless the body remains almost bloodless.

Experiments carried out by some people to test the advantage or the absence of advantage of phlebotomy seem not to be able to demonstrate anything. These experiments were carried out in favourable fevers in which small errors do not do any harm. Blood-letting and no blood-letting are opposite. Both have been found beneficial. This indicates clearly that the experiments are misleading and that they involve a fallacy in which something which is not the cause is considered as being the cause.

It cannot be denied, however, that phlebotomy sometimes helps and sometimes is harmful. During some epidemics all patients from whom blood is retrieved die. But in other conditions phlebotomies are beneficial and helpful. Therefore, there may be instances in which they are harmful and others in which they are beneficial. If this was not true, it would not be verified always or in most cases.

The cause why phlebotomy sometimes can help may be that, at the end of any bout of fever, the bitter and malignant juices flow from the nerves of the glands into the veins. This contaminates all the mass of blood which cannot be cleaned completely in the kidneys. Therefore, phlebotomy eliminates these bitter juices with the blood and those which remain in the blood can be neutralized by the arrival of chyle. In this way blood is easily and rapidly restored and renewed.

Although altered blood is not the primary and direct cause of fever, I think, however, that it is not harmless. It can indeed disturb the economy of the animal in many ways. Therefore, its return to normal can promote healing considerably.

Sometimes phlebotomy can be useful. When the normal movement of the blood is altered, it happens that a new movement is introduced in the brain and nerves. Then the fermentative and morbid activation is disturbed and may be improved. This can be confirmed by the fact that often a haemorrhage is stopped by blood-letting from a vein. Similarly, suffocating asthmatic coughing or epilepsy is cured by phlebotomy.

The alteration of the movement in the blood which can result from phlebotomy consists of an acceleration of the circulation in the arteries and a slowing down in the veins. As long as blood leaves an incised vein, arrival and circulation of blood in the adjacent artery are not impeded and this blood can circulate more quickly. Previously this volume of blood moved slowly through obstructed paths and impeded and forbid the exit of the arriving blood. On the contrary, as long as blood flows from an incised vein no blood is provided by the wound to the proximal aspect of the vena cava. Therefore, the blood remaining in the vena cava is carried to the right ventricle of the heart more slowly and with less impetus not only as a consequence of the reduction of its volume but also because it is no longer pushed as it was previously. This alteration can result in some disturbance of the movement of the blood in the brain and in the nerves. The spirits or nervous juices can be affected and irritated by this slight disturbance and be able to move at a new rhythm.

This disturbance of the movement of the spirits must not necessarily result in an ordered, normal and favourable motion since very often it can be changed for the worse. Sometimes thus blood-letting can be so harmful as to kill.

Purgative and emetic medicines also can sometimes help and sometimes harm. But the abundant and corrupted humours are not the causes of fevers. The cause is rather the little amount of ferment transmitted in nerves and glands. I do not see how this ferment could be excreted by purgative medicines. It seems like a dream to think that a drug, by some attractive force of selection like a magnet, sorts out in the confused mass of humours, collects and carries away some corrupted juices, leaving all the other favourable juices. Purgative drugs, by their bitterness and the poisonous force which make their efficacy, rather act by pricking the membranes of the stomach and intestines. These by their contractions express juices contained in the excretory ducts of the glands and in blood vessels and pour them in the cavity of the intestines. In about the same manner, pepper and some corrosive salts when on the tongue and in the mouth elicit profuse salivation. It thus can be concluded that purgations expel all the humours which are in the mentioned vessels whether good and useful or corrupted. That is why purgation can be useful and harmful.

The violent commotion produced by purgative drugs, however, can alter the fermentation of fever and change it for the best or for the worst. So, depending on each case, the drug can be sometimes beneficial or harmful. Nevertheless, in very severe diseases, it matters to take some risk rather than to perish surely.

Purgative medicines can be beneficial for another reason. Their saline poison can attack, neutralize and sweeten the salts of the ferment of fever (as results from what was said above). Similarly, the poisonous character of cantharides in vesicatories sometimes can be beneficial for the same reason. But all these drugs are unreliable. It can often happen that they do great harm.

If what we described of the theory of fevers is not erroneous, i.e. if we did not roam too far away from the truth, we hope that, with time, very learned and wiser physicians will discover some more reliable medicines against fever.

End

F I N I S

References

Accademia del Cimento (Academia Experimentalis Medica) "One year after Borelli arrived in Tuscany the Accademia del Cimento held its first session: the year Borelli left, the Cimento quietly died. Indeed, Borelli seems to have been the principal animus of the academy, but lest he appear the sole mover, we should recall the documentation, especially for the extensive experimental work performed during this Galilean epoch, in Giovanni Targioni Tozzetti's "Atti e memorie inedite dell'Accademia del Cimento e notizie aneddote dei progresse delle scienze in Toscana". In fact the Tuscan court had been thoroughly infected by Galileo's ideas and those of his pupils. Grand Duke Ferdinand II, from the time of his accession to power in 1628 until his death in 1670, maintained a personal laboratory as did Prince Leopold. From the time of the death of the Master, Galileo, informal gatherings met at the court and presented and discussed experiments. At first Torricelli was the most prominent figure; after his death in 1647 Viviani presided over the activities.

"Then, possibly under the crystallizing influence of Borelli, Leopold asked for and received permission from Ferdinand to organize formally an academy for purely experimental research. Under Leopold's aegis it met for the first time in June of 1657. Among its more distinguished members, besides Borelli and Viviani, were Antonio Oliva (d. 1668), Carlo Rinaldini (d. 1698) and Francesco Redi (d. 1697). Nicholas Steno arrived in Florence in 1666 and soon thereafter joined the group. Lorenzo Magalotti, after attending the University of Pisa as a student, was appointed secretary in 1660. The Cimento had adopted a policy of submerging the identities of its members and presenting itself as a group. Accordingly, when Magalotti brought out the "Saggi di naturali esperienzi fatte nell'Accademia del Cimento" in 1666-1667, it appeared anonymously and refrained from identifying the individual contributions of the members. Actually the Saggi presented only part of the work performed; it tended to emphasize the identification and description of physical phenomena and the perfecting of measuring techniques. It failed to present other interesting investigations, including some potentially controversial observations and discussions of comets.

"During the life of the Cimento dissension appeared among the membership; Borelli may have originated some of it. He seems to have chafed under the requirement of anonymity, and by all accounts he was a touchy person to get along with under any circumstances. In any case, toward the end of 1666 and just after the publication of his important work on the theory of the motions of the moons of Jupiter, Borelli made his decision to leave Tuscany and return to Messina. In 1667 Leopold was created a cardinal and thus had some of his energies diverted. Rinaldini moved on to the University of Padua and Antonio Oliva went to Rome where he came under the suspicion of the Inquisition and died by throwing himself from a window of one of its prisons. In December of 1667 Steno converted to Catholicism and shortly thereafter set out on a series of journeys. How or whether any of these events may have been connected is not known with any degree of certainty. But at this point the Cimento effectively ceased to function, even though it apparently was not formally dissolved, and even though Prince, now Cardinal, Leopold continued to direct some experimental work until he died in 1675. As far as Borelli was concerned, he had been, and afterward remained, on excellent terms with Leopold; and Leopold maintained his high regard for Borelli." (ThB Settle in Gillispie Dictionary of Scientific Biographies (DSB), vol 2, article on Borelli. New York, 1970, pp 309–310)

Apollonius of Perga (Apollonius), 262–180 B.C. His treatise on conical sections (circle, ellipse, parabola, hyperbola) is one of the great works of Greek mathematics. (GJ Toomer in Gillispie DSB, vol 1. New York, 1970, pp 179–193)

Aristotle (Aristoteles), about 384–322 B.C. Greek philosopher. His observations and experiments on different animal species enabled him to draw the rough lines of a classification. To describe their structure and functioning he was led to distinguish in every being a "substance" (being in power) and a "shape" (entelechy, being in act). His concept of Nature is finalist: each being is organized and tends to perfection; there is a hierarchical order of the species up to man provided with reasoning. (GEL Owen in Gillispie DSB, vol 1. New York, 1970, pp 250–258; L Minio-Paluelo, idem., pp 250–281)

Aselli or *Asellio* Gaspare (Asellius), 1581–1626. Italian physician, he discovered the chyloferous vessels in 1622. (L Premuda in Gillispie DSB, vol 1. New York, 1970, pp 315–316)

Auberio (Auberius), taught anatomy in Pisa in 1657. No other information was found concerning this gentleman.

Bartholin Thomas (Bartholinus), 1616–1680. Danish physician, he was the pupil of Vesling in Padua. He also studied philosophy, jurisprudence, philology, Arab, archeology, etc. He did research in anatomy. He wrote "Anatomia nova" (Leiden, 1641). He investigated particularly the lymph vessels and studied the Harvey's theory on blood circulation. (CD O'Malley in Gillispie DSB, vol 1. New York, 1970, pp 482–483)

Bellini Lorenzio (Laurentius Bellinus), 1643–1704. Italian physician. He attended the courses of Borelli. When 19 years old, he published "Exercitatio de structura et usu renum" (Florence, 1662; Leiden, 1726) and gave his name to small canals in the kidneys. When 20 years old, he became professor of philosophy and theoretical medicine in Pisa and then of anatomy. He also published "Gustus organum" (1665) and "De urinis et pulsibus et missione sanguinis" (1683). (TM Brown in Gillispie DSB, vol 1. New York, 1970, pp 592–594)

Bonfiglioli Silvestro (Silvester Bonfigliolus). No information was found concerning this gentleman.

Boyle Sir Robert (Boylius), 1627–1691. Irish physicist and chemist, he improved the thermometer of Galileo and the pneumatic pump of Otto von Guericke. Although his first scientific interest was chemistry his first published scientific book was on pneumatics: "New Experiments Physico-Mechanicall, Touching the Spring of the Air and its effects" (1660). In an appendix to the second edition of "New Experiments" (1662) he developed his last discovery into a quantitative relationship (that volume of gases varies inversely with pressure) rightly called Boyle's law. His principal written work was "The Sceptical Chemist" (1691). He was a very active member of the Royal Society of London. (M Boas Hall in Gillispie DSB, vol 2. New York, 1970, pp 377–382; T Birch, The history of the Royal Society of London for improving of natural knowledge, vols 1–4. 1776–1777, London: A. Millar)

Casserio or *Casseri* Giulio (Casserius Placentinus), 1552–1616, of Piacenza. He was at first a servant of Fabrizio d'Acquapendente, became his pupil and his successor in teaching anatomy and surgery. He described the sense organs in "Penthaestheseion, hoc est De quinque sensibus liber" (Venice, 1609). He also published "Tabulae anatomicae LXXIIX, omnes novae nec antehac visae" (Venice, 1627). He studied the development of the embryo and wrote "Trattato iconografico, Tabulae de formato foetu" (Amsterdam 1645). (L Premuda in Gillispie DSB, vol 3. New York, 1971, pp 98–100)

Castelli Benedetto (Castellus), 1578–1643. Italian mathematician and Benedictine monk. He was a friend of Galileo and became famous for his work on hydraulics. He taught in Pisa. His "Della misura dell'acque correnti" (1628) represents the beginning of modern hydraulics. (E Carruccio in Gillispie DSB, vol 3. New York, 1971, pp 115–116)

Cesalpino Andrea (Cesalpinus), 1519–1603. Italian physician, director of the botanical garden of Pisa. His works on blood circulation make him a precursor of Harvey. (K Mägdefrau in Gillispie DSB, vol 15 [Suppl]. New York, 1978, pp 80–81)

Christina (Christina), 1626-1689. Queen of Sweden (1632-1654), of the house of Vasa. She displayed huge intellectual curiosity and extensive cultural knowledge. She corresponded with all the learned in Europe and invited Descartes to her court. Converted to Catholicism, she abdicated, travelled through Europe and settled in Rome where she launched her Accademia Reale in 1674.

Croone (or *Croune*) William C. (Croon), 1633-1684. English physician, he taught anatomy in London. He was a very active member of the Royal Society of London. He wrote "De ovo" and "De ratione motus musculorum" (Amsterdam, 1676), in collaboration mostly with Th. Willis. (LM Payne in Gillispie DSB, vol 3. New York, 1971, pp 482-483; T. Birch, The history of the Royal Society of London, vols 1-4, 1776-1777, London: A. Millar)

De Graef Reinier (de Graaf), 1641-1673. Dutch physician and physiologist. He is known for his studies on the pancreatic juice and his description of the ovarian follicles. (M Klein in Gillispie DSB, vol 5. New York, 1972, pp 484-485)

Descartes René (Cartesius), 1596-1650. French scientist and philosopher, he admitted only reasoning in sciences and thus opposed the scholastic philosophy. He published in 1637 his "Discours de la Méthode" and three essays which are its application: "Dioptrique, Météores, Géométrie", followed by "Méditations métaphysiques" (1641), "Principes de philosophie" (1644), "Passions de l'âme" (1649). The mathematical application of his method was the creation of analytical geometry. He enunciated the laws of refraction in optics. He also applied his method, with mistakes, to other sciences such as biology, medicine, psychophysiology and wrote "Traité de l'homme et de la formation du foetus". (TM Brown in Gillispie DSB, vol 4. New York, 1971, pp 61-65)

Dumel D. (Dumel, Professor Patavinus). No information was found concerning this gentleman.

Euclid (Euclidius), about 300 B.C. Greek mathematician, he was the founder of the school of mathematics of Alexandria. (I Bulmer-Thomas in Gillispie DSB, vol 4. New York, 1971, pp 414-437; J Murdoch, idem., pp 437-459)

Fabrizio d'Acquapendente Geromino, Fabrici Girolamo (Fabricius ab Acquapendente), 1533-1619. Pupil of Fallopius and teacher of Harvey, he carried out many anatomical observations and embryological researches on chickens. He published: "De visione, voce, auditu" (1600), "De locutione et ejus instrumentis liber" (Venice, 1601), "De brutorum loquela" (1603), "De venarum ostiolis" (Padua, 1603), "De formato foetu" (1604), "De musculi artificio, ossium de articulationibus" (1614), "De respiratione et ejus instrumentis, libri duo..." (1615), "De motu animalium secundum totum" (1618), "Hieronymi Senis de totius animalis integumentis opusculum" (1618), "De formatione ovi et pulli" (1621). (B Zanobio in Gillispie DSB, vol 4. New York, 1971, pp 507-512)

Falconieri Alexandro (Alexander Falconerius). The following was found in the proceedings of the Royal Society of London for the meeting of March 12, 1668: "There was presented to the society by Signor *Magalotti* and Signor *Falconieri* in the name of prince *Leopold* of Tuscany, and as a mark of his highness's esteem of the society, the book of experiments of the academy del Cimento at Florence, intitled, Saggi di Naturali Esperienze fatte nell'Academia del Cimento in Firenze, printed in that city in 1667 in fol." (T Birch, The history of the Royal Society of London, vol 2, 1776, p 256, London: A. Millar)

Fallopius, Fallopio Gabriele (Fallopius), 1523-1562. Italian surgeon and anatomist, pupil of Vesalius and teacher of F. d'Acquapendente. He taught anatomy in Ferrara and then in Pisa, anatomy and botany in Padua. He described the passage of part of the facial nerve through the petrus bone (in the aqueduct of Fallopius), the ligament inserted in the iliac spine and in the pubic spine, the Fallopian tube which connects the ovary and the uterus. He wrote "Observationes anatomicae" (Venice 1561), "Lectiones de partibus similaribus - et selectorum explicationes" (Nürnberg 1575), "Opera omnia" (Venice 1584). (CD O'Malley in Gillispie DSB, vol 4. New York, 1971, pp 519-520)

Ferdinand II de Medici (Magnus dux Hetruriae, Ferdinandus), 1610-1670. Grand Duke of Tuscany (1621-1670), son of Cosme II. He protected Galileo and Torricelli, and founded the Accademia del Cimento, the first academy of natural sciences in Europe.

Fracassati Carlo F. (Carolus Fracassatus) was professor of medicine in his native city of Bologna, then in Pisa. He engaged in anatomy and wrote "Dissertatio epistolica responsaria de cerebro ad Malpighium" and "Exercitatio epistolica de lingua ad A. Borellium" which were published together with letters from Malpighi (Bologna 1665). He also published "Praelectio medica in aphorismos". (A Wernich in A Hirsch (ed) Biographisches Lexikon der hervorragenden Ärzte aller Zeiten und Völker, vol 2, 1930, p 589, Berlin, Wien: Urban and Schwarzenberg)

Galen (Galenus), about 131-201. Greek physician. By dissecting animals he made important discoveries on the nervous system and the heart. His physiology was based on a theory of the humours like that of Hippocrates. As a physician, Galen was a Hippocratic and, as a scientist, an Aristotelian. He composed several works while still a student in Pergamon: "On the Anatomy of the Uterus", "Diagnosis of Diseases of the Eye", "On Medical Experience", probably "On the Best Sect, for Thrasybulos". His "On Pleuritis, for Patrophilus" was incorporated later in "On the Constitution of Medical Art". He also wrote: "On the Doctrines of Hippocrates and Plato", "On Genuine and Ungenuine Hippocratic Writings", "On Critical days", "On Crises", "Therapeutic Method", "On Scientific Proof and Introduction to Logic". His most interesting and exciting work seems to be: "Hygieinia". (F Kudlien in Gillispie DSB, vol 5. New York, 1972, pp 227-233; LG Wilson, idem., pp 233-235)

Galileo Gallilei (Galileus), 1564-1642. Italian mathematician, physicist and astronomer, he discovered the laws of the pendulum. He enunciated the principle of inertia and studied the laws of gravity. He built the Galilean telescope. Supporting the Copernican theory, he was condemned by the tribunal of the Inquisition. He wrote his "Dialogi", "Il saggiatore". His final work was "Discorsi e dimonstrazioni matematiche intorno a due nuove scienze attenenti alla Mecanica e i Movimenti Locali" (S Drake in Gillispie DSB, vol 5. New York, 1972, pp 237-249)

Gassendi Pierre, called *Gassend,* (Gassendus), 1592-1655. French philosopher and researcher, he was a supporter of the system of Copernicus and an admirer of Galileo. He carried out many observations in astronomy and physics. He was an opponent of Aristotle and Descartes. (B Rochot in Gillispie DSB, vol 5. New York, 1972, pp 284-290)

Glisson Francis (Glissonius), 1596-1677. English physician and philosopher. He described the fibrous envelope of the liver. His theory of substance makes him, in philosophy, a precursor of Leibnitz. He published: "De rachitide" (1650), "Anatomia hepatitis" (1654), "Tractatus de ventriculo et intestinis" (1677), "Tractatus de natura substantiae energetica" (1672). He was a member of the Royal Society of London. (O Temkin in Gillispie DSB, vol 5. New York, 1972, pp 425-427; T Birch, The history of the Royal Society of London, vol 1, 1776, London: A. Millar)

Harvey William (Harveus), 1578-1657. Physician of James 1st and Charles 1st, he discovered blood circulation which he described in "Exercitatio anatomica de motu cordis et sanguinis in animalibus" (1628). He also carried out research on the development of the embryo of the chicken and the formation of the foetus of mammals: "Exercitationes de generatione animalium" (1651). (JJ Bylebyl in Gillispie DSB, vol 6. New York, 1972, pp 150-162)

Heraclites (Heraclitus), about 576-480 B.C. Greek philosopher of the Ionian school.

Hérigone Pierre (Herrigonius), died about 1643. French mathematician. Hérigone's only published work of any consequence is the "Cursus mathematicus", a six-volume compendium of elementary and intermediate mathematics in French and Latin. (P Strømholm in Gillispie DSB, vol 6. New York, 1972, p 299)

Hippocrates of Cos (Hippocrates), about 460-377 B.C. Greek physician, he relied on clinical observation and advised simple treatments. He practiced surgery. His physiology is based on the theory of the humours (blood, lymph, choler, melancholy or black choler) from which the theory of the temperaments derives. Their equilibrium means good health; excess or lack of one of them entails disease. (R Joly in Gillispie DSB, vol 6. New York, 1972, pp 418-431)

Lorrerius (Lorrerius): no information was found concerning this gentleman. Was his non-latinized name Lorrer? Lorrerio?

Lower Richard L. (Lower, Lowerus), 1631–1691. British physician, friend of Willis. He made one of the most significant contributions of his time to the anatomy and physiology of the heart: "Tractatus de corde, item de motu et colore sanguinis et chyli in eum transitu". He was a supporter of the theory of Willis on fevers and wrote on this subject: "Diatribae Thom. Willisii de febribus vindicatio, etc." (1665) and "Dissertatio de origine catarrhi etc." (1671). He was a very active member of the Royal Society of London until he was expelled for not performing his obligations to the Society. (TM Brown in Gillispie DSB, vol 8. New York, 1973, pp 523–527; T Birch, The history of the Royal Society of London, vol 3, 1777, p 233, London: A. Millar)

Lucretius, Titus Lucretius Carus (Lucretius), about 98–55 B.C. Latin poet, he was the author of "De natura rerum" in which he expounds the Epicurian physics with a moral purpose. (DJ Furley in Gillispie DSB, vol 8. New York, 1973, pp 536–539)

Malpighi Marcello (Malpighius), 1628–1694. Physician of Pope Innocent XII, he can be considered as the founder of microscopical anatomy. He discovered the glomeruli in the kidney. Working on blood circulation in the frog he confirmed the works of Harvey. His research on the embryology of the chicken led him to admit the theory of preformation. His first and fundamental work is "De pulmonibus" (Bologna, 1665). He also wrote: "De lingua" (Bologna, 1665), "De externo tactus organo" (Naples, 1665), "De cerebro" (1665), "De cerebri cortice" (1666), "De viscerum structura exercitatio anatomica" (Bologna, 1666), "De bombyce" (London, 1669), "De formatione pulli in ovo" (1673), "Anatome plantarum" (London, 1675 and 1679). He was elected honorary member of the Royal Society of London on March 4, 1669. (L Belloni in Gillispie DSB, vol 9. New York, 1974, pp 62–66; T Birch, The history of the Royal Society of London, vol 2, 1776, London: A. Millar)

Michelini Famiano (Famianus Michelinus), 1604–1665. Italian mathematician. In 1635 he taught mathematics at the Florentine court and in 1648 obtained the chair of mathematics at Pisa. He declared that all knowledge is derived from the exact sciences. Following this principle he applied the experimental method even to medicine. He is credited for paving the way for Borelli's theories. He published "Della direzione de' fiumi". (G Tabarroni in Gillispie DSB, vol 9. New York, 1974, pp 369–370)

Needham Walter N. (Needham), 1631–1691. English physician, member of the Royal College of Physicians. He wrote on the development of the foetus: "Disquisitio anatomica de formato foetu" (London, 1667; Amsterdam, 1678). He was a member of the Royal Society of London. (Pagel in A Hirsch, Biographisches Lexikon der hervorragenden Ärzte aller Zeiten und Völker, vol 4, 1932, p 335, Berlin, Wien: Urban and Schwarzenberg; T Birch, The history of the Royal Society of London, vols 1–4, 1776–1777, London: A. Millar)

Pecquet Jean (Pequetus), 1622–1674. French physician and anatomist, famous for his discovery of the chyle vessels. The receptaculum chyli in the lower aspect of the thoracic duct (ductus Pequetianus) where these vessels arrive is designated in French as citerne de Pecquet. (P Huard, MJ Imbault-Huard in Gillispie DSB, vol 10. New York, 1974, pp 478–481)

Philippe III (Philippus III), 1578–1621. King of Spain (1598–1621), son of Philippe II.

Plato (Plato), about 428–348 B.C. Greek philosopher, disciple of Socrates. In his Dialogues he dealt with the most important philosophical and metaphysical problems combining reasoning and poetic language. By his dialectic method the soul progressively rises from multiple and varying appearances to ideas (essences). (DJ Allan in Gillispie DSB, vol 11, New York, 1975, pp 22–31)

Ricci Giacomo (Jacobus Riccius). Dominican monk, secretary of the Congregation of the Sacred Index, he was the brother of Michelangelo Ricci.

Ricci Michelangelo (Michael Angelus Riccius), 1619–1682. Italian mathematician. He was with his friend Torricelli a pupil of Castelli. He published "Geometrica exercitatio" more usually called by a subtitle "De maximis et minimis" (Rome, 1666). His other mathematical contributions include his study of spirals (1644), his investigation of a family of curves more general than ordinary cycloids (1674) and the method by which he recognized fairly explicitly that the treatment of tangents is an operation inverse to that of the calculation of areas (1668). (L Campedelli in Gillispie DSB, vol 11. New York, 1975, pp 404–405)

Santorio Santorio (Sanctorius), 1561–1636. Italian physician, he taught in Padua and Venice. In "Methodi vitandorum errorum omnium qui in arte medica contingunt" (1602) he expressed ideas that prefigured the mechanistic explanations of the iatrophysical school. He also wrote "Commentaria in artem medicinalem Galeni" (1612). He is known for his experiments on the variations of weight of the human as a result of ingestion and excretion and for his measurement of temperature of the body in "De statica medicina" (1614). (MD Grmek in Gillispie DSB, vol 12. New York, 1975, pp 101–104)

Socrates (Socrates), about 470–399 B.C. Greek philosopher. The essence of his philosophy consisted of a belief in human reasoning by which man attains self-knowledge and happiness: "Know yourself" and "Nobody is naughty willingly".

Steno Nicolas, Niels Stensen, (Steno) 1638–1686. Danish anatomist and geologist. He taught anatomy in Copenhagen. Converted to Catholicism while on a journey in Italy, he became a cleric and a bishop. He discovered glands in the cheeks, beneath the tongue and in the palate, whose structure of veins, arteries, nerves and lymph vessels he also described. The excretory duct of the parotid salivatory gland is called canal of Steno. He published: "Adenographia" (1656), "Apologiae prodromus" (1663), "De musculis et glandulis" (1664), "Discours... sur l'anatomie du cerveau" (1669). (G Scherz in Gillispie DSB, vol 13. New York, 1976, pp 30–35)

Thruston (Thruston). Member of the Royal Society in London. We found his name mentioned only once in the proceedings of this society. (T Birch, The history of the Royal Society of London, vol 2, 1776, p 95, London: A. Millar)

Torricelli Evangelista (Torricellus), 1608–1647. Italian physicist, disciple of Galileo, he demonstrated atmospheric pressure by an experience from which the mercury barometer derives. He also demonstrated that two weighing bodies attached together can start moving spontaneously only if their common centre of gravity moves down (principle of Torricelli). Founder of researches on hydro-dynamics, he enunciated the first quantitative law of discharge of a liquid through a narrow orifice at the bottom of a vessel (law of Torricelli, 1644). He published "Opera geometrica" (1644). (M Gliozzi in Gillispie DSB, vol 13. New York, 1976, pp 433–439)

Van Helmont Jan-Baptist (Helmontius), 1577–1644. Flemish physician and chemist. He discovered the carbonic gas and distinguished different gases in air. He pointed out the role of gastric juice in digestion. He was a supporter of iatrochemistry. His "De re magnetica vulnerum... curatione" (Paris, 1621) was published against his will. His "propositions" were condemned by the General Inquisition of Spain and he had to acknowledge his "errors". (W Pagel in Gillispie DSB, vol 6. New York, 1972, pp 253–259)

Vesalius Andreas, André van Wesele (Vesalius), 1514–1564. Flemish anatomist and surgeon regarded as father of modern anatomy. He wrote a treatise of anatomy "De corporis humani fabrica libri septem" in which he criticizes the medical theory of the Ancients, particularly Galen, and recommends the experimental method. Accused of having dissected a man still alive, he may have been sentenced to undertake a pilgrimage to the Holy Land. On the way back, thrown by a storm onto the shores of the Island of Zante, he died in Zakinthos. (CD O'Malley in Gillispie DSB, vol 14. New York, 1976, pp 3–12)

Vesling Johann V. (Veslingius), 1598–1649. German anatomist. After a study journey in Egypt and Jerusalem, he taught anatomy and botany in Venice and then in Padua. He was famous for having written "Syntagma anatomicum, publicis dissectionibus in auditorum usum aptatum" (Padua, 1641). He also wrote "De pullitione Aegyptiorum et aliae observationes anatomicae et epistolae medicae posthumae" (Copenhagen, 1664). (E Hintzoche in Gillispie DSB, vol 14. New York, 1976, pp 12–13)

Wharton Thomas (Whartonius), 1614–1673. English anatomist, he wrote a treatise on glands: "Adenographia" (1656). He gave his name to the excretory duct of a salivatory gland (Wharton's duct). (W LeFanu in Gillispie DSB, vol 14. New York, 1976, pp 286–287)

Willis Thomas (Willisius), 1621-1675. English anatomist and physiologist, he particularly studied the brain and cranial nerves. By analyzing reflexes he expressed the hypothesis of two types of answers, spontaneous or voluntary movements ordered by the brain and natural or involuntary movements ordered by the cerebellum. He wrote "De febribus" in the 1650s and "Dissertatio epistolica de urinis" which were published together as "Diatribae duae medico-philosophicae" (1659). His "Cerebri anatome" (1664) is the foundation document of the anatomy of the central and autonomic nervous system. (RG Frank in Gillispie DSB, vol 14. New York, 1976, pp 404–408; T Birch, The history of the Royal Society of London, vols 1–4, 1776–1777, London: A. Millar)

Glossary

Agent (agens): efficient cause. See patient.
Agnani, lake of (lacus Agnani Puteolis): there are several places near Naples which are known for their emanations of sulphurous gases.
Air (aer): substance composed of small spiral machines like springs.
Alum (alumen): a double sulphate of aluminium and potassium, $KAl(SO_4)_2 \cdot 12\,H_2O$. (Schneider)*
Ammonia (ammoniacum): colourless gas with pungent smell and strong alkaline reaction, NH_3, spirit of hartshorn.
Ammonia salt (sal ammoniacum): ammonium chloride, NH_4Cl; eventually also a kind of rock salt. (Schneider)
Animal (animalis): organized being endowed with life; pertaining to the functions of animals; pertaining to animals as opposed to vegetables (anima = breath, wind, principle of life, soul; animus = soul, vital principle, life, mind, courage).
Antimony (antimonium): antimony (III) sulphide, Sb_2S_3. (Schneider)
  Metallic antimony (regulum antimonium): antimony, Sb. (Schneider)
  Glass of antimony (vitrum antimonii): antimony-oxide and sulphide, Sb_2O_3 and Sb_2S_3. (Schneider)
Apple of Marseille (malum Massilicum): the correspondence with a contemporary kind of apples could not be determined.
Aqua fortis (aqua fortis): nitric acid, HNO_3, about 50%. (Schneider)
Astrolabe (astrolabium): instrument formerly used for taking altitudes, etc.
Attraction (facultas attractiva): for Borelli there is no attraction between two bodies in nature. According to him everybody finds such attraction by some magnetic virtue ridiculous. This was some years before the work of Newton was published.

Bow (arcus): weapon for shooting arrows; contrivance made of two or several straight segments connected by ropes and forming one or several angles. See contrivance and spring.

Cantharides (cantharis): dried Spanish fly. The Spanish fly is a bright green insect dried and used for raising blisters, as aphrodisiac, etc.
Catania (Catana): city in Sicily, in the bay of the same name, at the foot of mount Etna.
Chalk of corals (corallium calx): burned coral.
Civet perfume (zibethum): strong musky perfume got from anal glands of civet.
Colocynth (colocynthis): bitter-apple, gourd plant with bitter-pulped fruit used as a purgative drug.
Contrivance (machina): mechanical device.
Contrivance (arcus): mechanical device made of articulated rods connected by ropes forming a shape different from a bow, for example a Z or several Z's. See bow and spring.
Corrosive water (aqua corrosiva): acid or alkaline distillate.
Cosine (sinus secundus): see sine.
Cosine's complement (sinus versus): see sine.

Delos (Delos): Greek island, the smallest of the Cyclades, was a religious center.
Diabolical oil (oleum zabui): the composition of this substance could not be determined.

* Schneider W (1968) Pharmaceutische Chemicalien und Mineralien. Frankfurt a.M. Govi-Verlag, Pharmaceutischer Verlag

Displacement, movement (motus localis): Borelli means movement of a limb or of a body from one place to another or displacement, by contrast with the internal movements of viscera, blood, etc. which constitute life. He also designates fever, pain as movements.

Drachm (dragma): 60 grains, 1/8 ounce.

Earthy elements (partes terreae): the four elements were the earth, water, air and fire.

Egeria (Aegeria): nymph of ancient Italy. Her cult was bound to that of Diana. She was supposed to have been the adviser of the Sabine king Numa Pompilius.

Entelechy (elechia): the becoming or being actual of what was potential, developed perfection.

Eolypile (eolypila): vapour-blowing machine. Small receptacle containing water and provided with a hole, destined to be placed near a fire. When heated the water is changed into vapour which is blown out and activates the fire as if it were wind.

Etna, mount (mons Aetneus): volcano in the north-east of Sicily, overlooking the plain of Catania, still active.

Fermentation (fermentatio): internal movement of the elements of a composite body which are activated by their own motive force or by that of another additional body (Borelli, *De Motu Animalium,* part II, prop. 135).

Finger breadth (digitus): 1.85 cm = 1/16 of a foot.

Fixed salts (salia fixa): salts prepared in such a way as to be no longer volatilizable, resisting fire.

Foot (pes): probably 29.57 cm, the attic foot comprising 16 finger breadths. But the author seems to equal a foot with twelve finger breadths in part I, prop. 175 and with eighteen finger breadths in part II, prop. 69!

Glass of antimony (vitrum antimonii): antimony-oxide and sulphide, Sb_2O_3 and Sb_2S_3. (Schneider)

Gluten (gluten): all kinds of glue.

Gunpowder (pulvis pyritium, pulvis pyrius, pulvis nitratus): mixture of carbon, sulphur and saltpetre.

Humours, four cardinal (quatuor humores vulgares): the four chief fluids of the body: blood, phlegm, choler, and melancholy (or black choler). According to an old theory, health consisted of the proper balance of these four cardinal humours and disease came from an excess or an insufficiency of one or another of them. Each person's "temperament" was his unique balance of the four cardinal humours. (Boorstin, *The Discoverers,* Random House, New York, 1983, p 341)

Icarus: son of Dedalus, was put with him in the Labyrinth by Minos. They escaped by using wings which they had made. Icarus flew so close to the sun that the wax which fixed his wings melted. He fell in the sea.

Line of gravity (linea propensionis): vertical line passing through the centre of gravity of a weighing body.

Line of support (linea innixionis): straight line joining the centre of gravity of a weighing body and its support.

Lye salt (sal lixivius): yellow potassium ferrocyanate. Lixivium: generally alkaline fluid produced from ashes. (Schneider)

Machine (machina): instrument that transmits force or directs its application.

Mass (mollis): mass or volume. In the text the distinction between mass and volume is not clear.

Metallic antimony (regulum antimonium): antimony, Sb.

Moment (momentum): means the actual action of a force and corresponds to what is now called either the magnitude of the force, or its moment (product of the force and its lever arm), or the product of the force and the velocity of the object moved by the force, or the importance of the force, depending on the context.

Moon, in the sphere of the (sublunaris): the bodies which are in the sphere thought to surround the moon were considered corruptible, those outside this sphere incorruptible.

Mortar (bombardum): short piece of ordnance for throwing shells at high angles.

Musk (moschum): odoriferous reddish-brown substance secreted in gland by male musk-deer, used for perfume and as stimulant.

Nitre (nitrum): saltpetre, potassium nitrate, KNO_3. (Schneider)

Numa Pompilius (Numa): second legendary king of Rome (about 715 to about 672 B.C.). Pious and pacific Sabine, he claimed to be inspired by the nymph Egeria. He supposedly organized religious life in Roma.

Oil of sulphur (oleum sulphuri): sulphuric acid, H_2SO_4. (Schneider)

Oil of tartar (oleum tartari): solution of potassium carbonate, K_2CO_3. (Schneider)

Oil of tobacco (oleum tabaci): product of the distillation of tobacco.

Oil of vitriol (oleum vitrioli): sulphuric acid, H_2SO_4, about 75%. (Schneider)

Oxymel (oxymeli): beverage made of sea water, honey and vinegar.

Patient (patiens): enduring subject. See agent.

Plaster (gipsum): calcium sulphate, $CaSO_4 \cdot 2\,H_2O$. (Schneider)

Powder of crabs (pulvis crancorum): was not found mentioned in Schneider.

Powder of pyrite (pulvis pyritium): see gunpowder.

Powder of hartshorn (pulvis cornu cervi): ammonium carbonicum pyro-oleosum. See spirit of hartshorn. (Schneider)

Pozzuoli (Puteoli): small city in the bay of the same name, 18 km west of Naples, was a very busy harbour in Roman antiquity.

Quicklime (calx viva): white caustic alkaline earth (calcium oxide, CaO_2) obtained by burning limestone.

Quicksilver (hydrargirum): mercury, Hg.

Rare (rarus): of loose structure.

Rarefaction (rarefactio): the author means a diminution of the number of particles of a substance in a given volume.

Rarefy (rarefacio): loosen the structure.

Resistance (resistantia): weight, counterforce, counterweight.

Rhomb (rhombus): oblique equilateral parallelogram, diamond or lozenge.

Salts of lye, salts produced from lye (sales lixivii, sales lixiviales): see lye salt.

Salt of nitre (sal nitri): mainly potassium nitrate, KNO_3. (Schneider)

Salt of sloe (sal prunellae): potassium nitrate, KNO_3, in varying proportion; potassium sulphate, K_2SO_4, and potassium nitrite, KNO_2. (Schneider)

Scruple (scrupulum) = 1/24 ounce = 20 grains = 1/288 jugerum

Sine (sinus): Let us consider a circle the centre of which is O and the radius r of which is equal to the unit. In the right upper quadrant of the circle the perpendicular BA drawn from the extremity B of a radius OB = r to the horizontal radius OD is the sine of the angle BOA. The segment OA on the horizontal radius OD is the cosine of the angle BOA. The segment AD on the horizontal radius OD is equal to 1 − cos

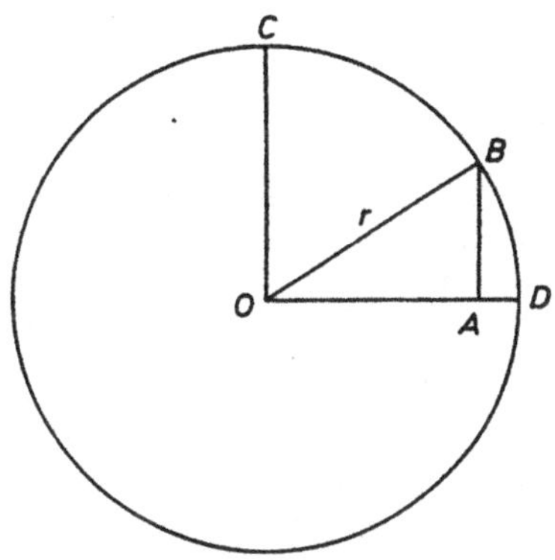

Thus: OB = r = 1

sin = AB

cos = OA

1 − cos = AD

For Borelli the radius of the trigonometrical circle is not equal to the unit and is called sinus totus. His sinus versus is equal to the difference between the trigonometrical radius and his cosine which he calls sinus secundus. We translated sinus versus into cosine's complement. Thus:

For Borelli	To-day
Trigonometrical radius (sinus totus)	= r
Sine (sinus)	= r sin
Cosine (sinus secundus)	= r cos
Cosine's complement (sinus versus)	= r (1 − cos)

Soul (anima): animating part of animals.
Spirit (spiritus): alcoholic extract. (Schneider)
Spirit of soot (spiritus fuliginis): product of distillation of soot. (Schneider)
Spirit of hartshorn (spiritus cornu cervi): diluted ammonia containing carbonates, soiled by other substances. (Schneider)
Spirit of vitriol (spiritus vitrioli, spiritus calchantinus): acids in variable proportions, SO_2 and H_2SO_4. (Schneider)
Spirit of wine (spiritus vini): alcohol of wine, ethanol, 40% vol. C_2H_5OH. (Schneider)
Spring (arcus): depending on the context arcus was translated as bow, contrivance or spring.
Sulphur (sulphur): sulphur obtained by melting or sublimation, S_8. (Schneider)
Systems (systemata): reference to the geocentric theory of the universe of Ptolemy (Ptolemaic system) and the Copernicus' heliocentric theory (Copernican system).

Tartar (tartarum): pink or red deposit from completely fermented wine, forming hard crust on side of cash; mostly potassium hydrogen tartrate, $KHC_4H_4O_6$. (Schneider)
Tartarous (tartareus): see tartar.
Temper (animus): habitual or temporary disposition of mind.
Temper (temperies): suitable combination of ingredients; resulting condition of consistency.
Trigonometrical radius (sinus totus): see sine.

Vital balsam (balsamum vitale): could not be determined precisely.
Balsams (balsamia): fragrant fluid preparations made from several drugs, fat or ethereal oils either by simple mixing or by recombination after extraction of elements. (Schneider)
Vitriol (vitriolum): sulphuric acid H_2SO_4 or any of its salts.
Vitriolic acid or acid of vitriol (acidus vitrioli): see vitriol.
Volume (molis): see mass.

Winding-drum (tympanum): drum moved by men walking inside, thus rotating a horizontal axle.

Name and Subject Index

Table I

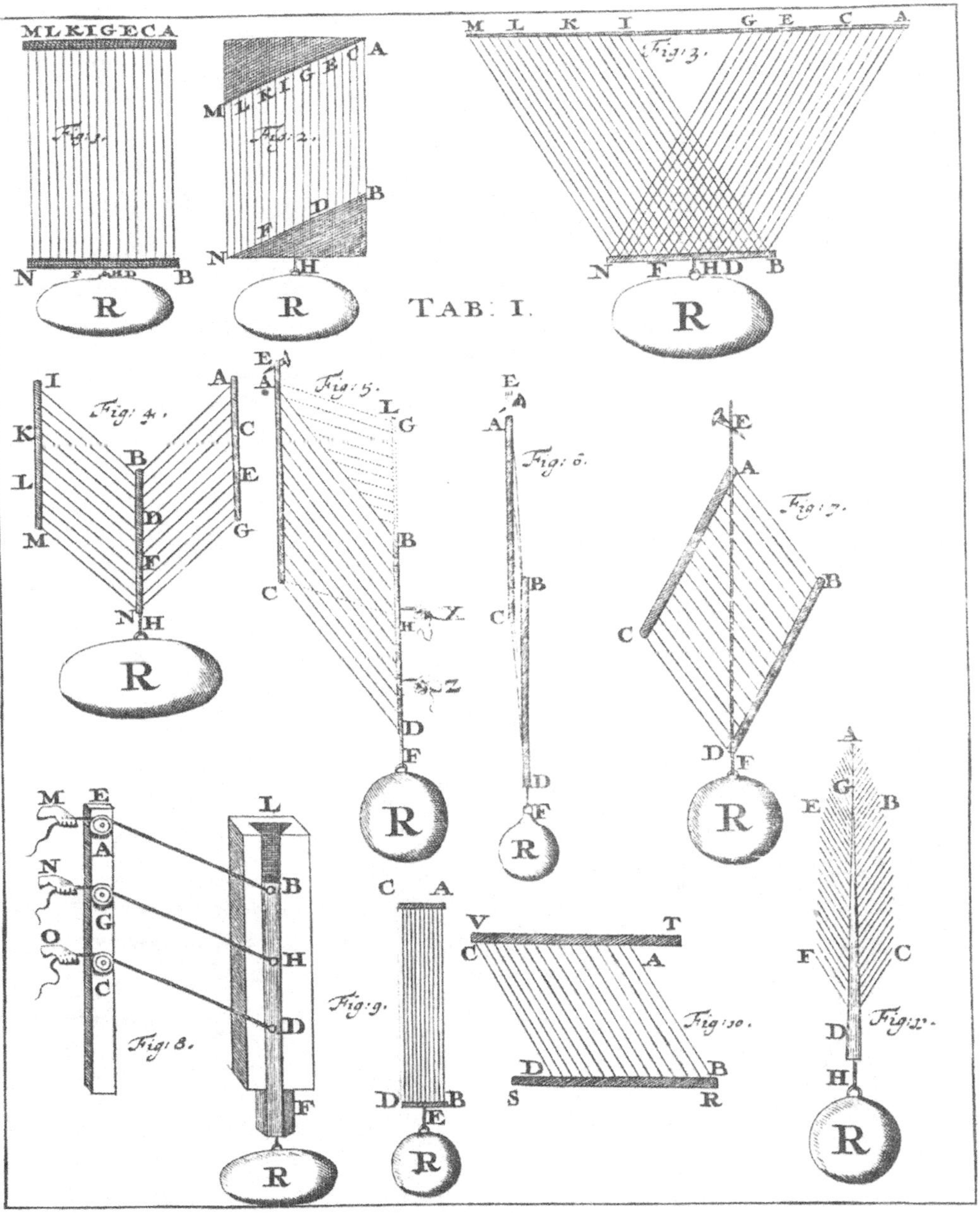

G. A. Borelli—On the Movement of Animals

Table II

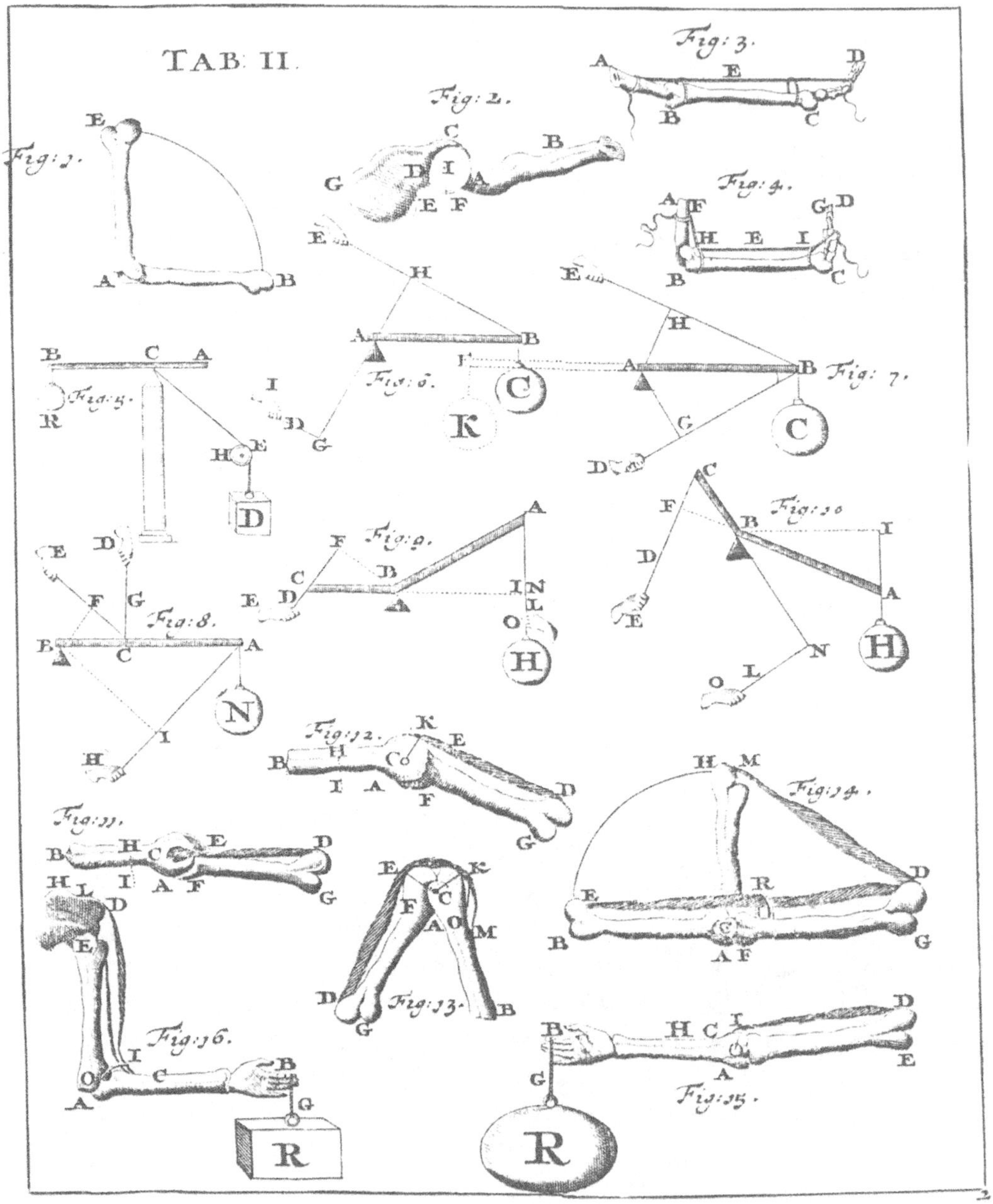
TAB. II.
Fig: 1.
Fig: 2.
Fig: 3.
Fig: 4.
Fig: 5.
Fig: 6.
Fig: 7.
Fig: 8.
Fig: 9.
Fig: 10
Fig: 11.
Fig: 12.
Fig: 13.
Fig: 14.
Fig: 15.
Fig: 16.
C
K
C
D
H
H
N
R
R

Table III

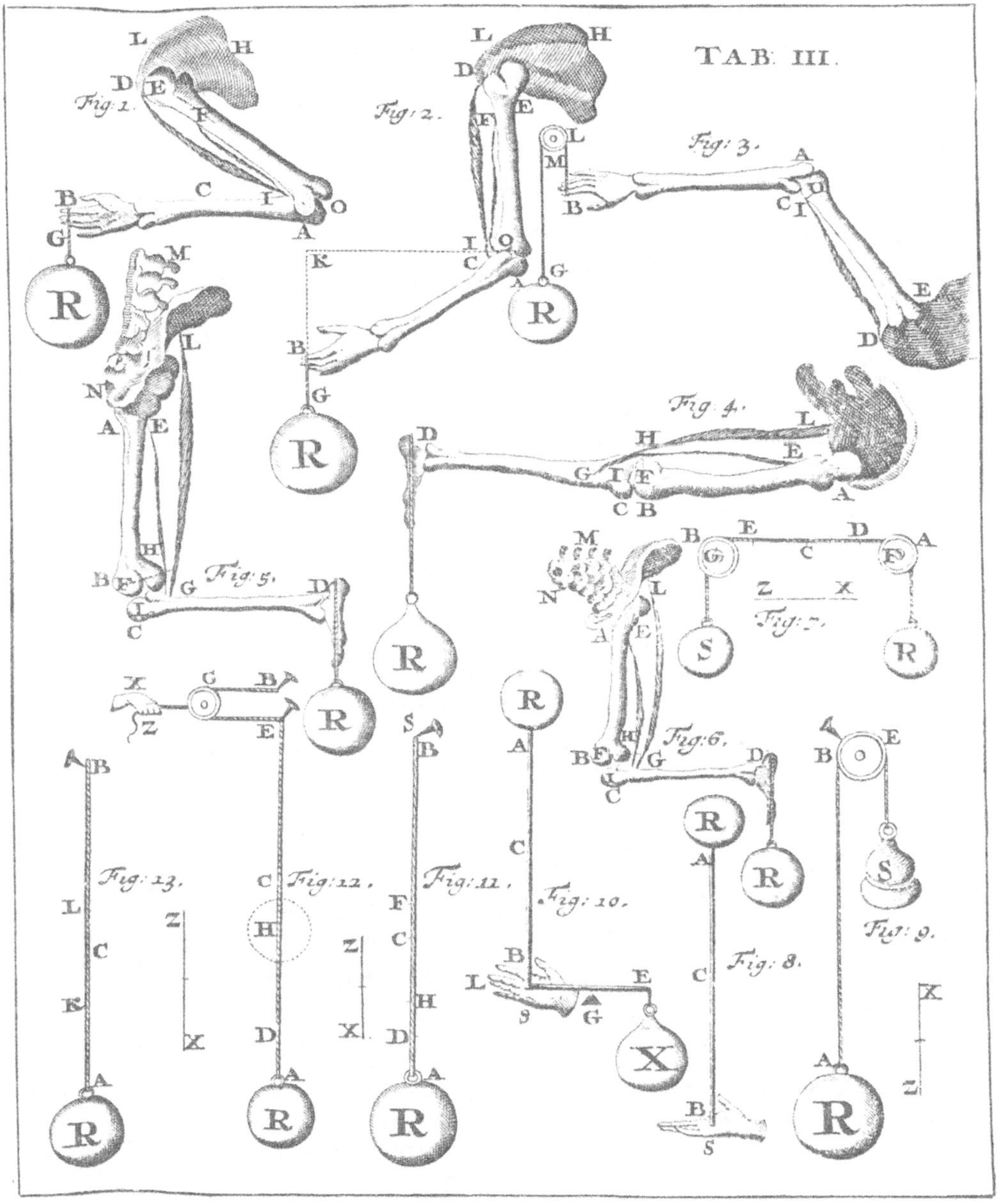

G. A. Borelli—On the Movement of Animals

Table IV

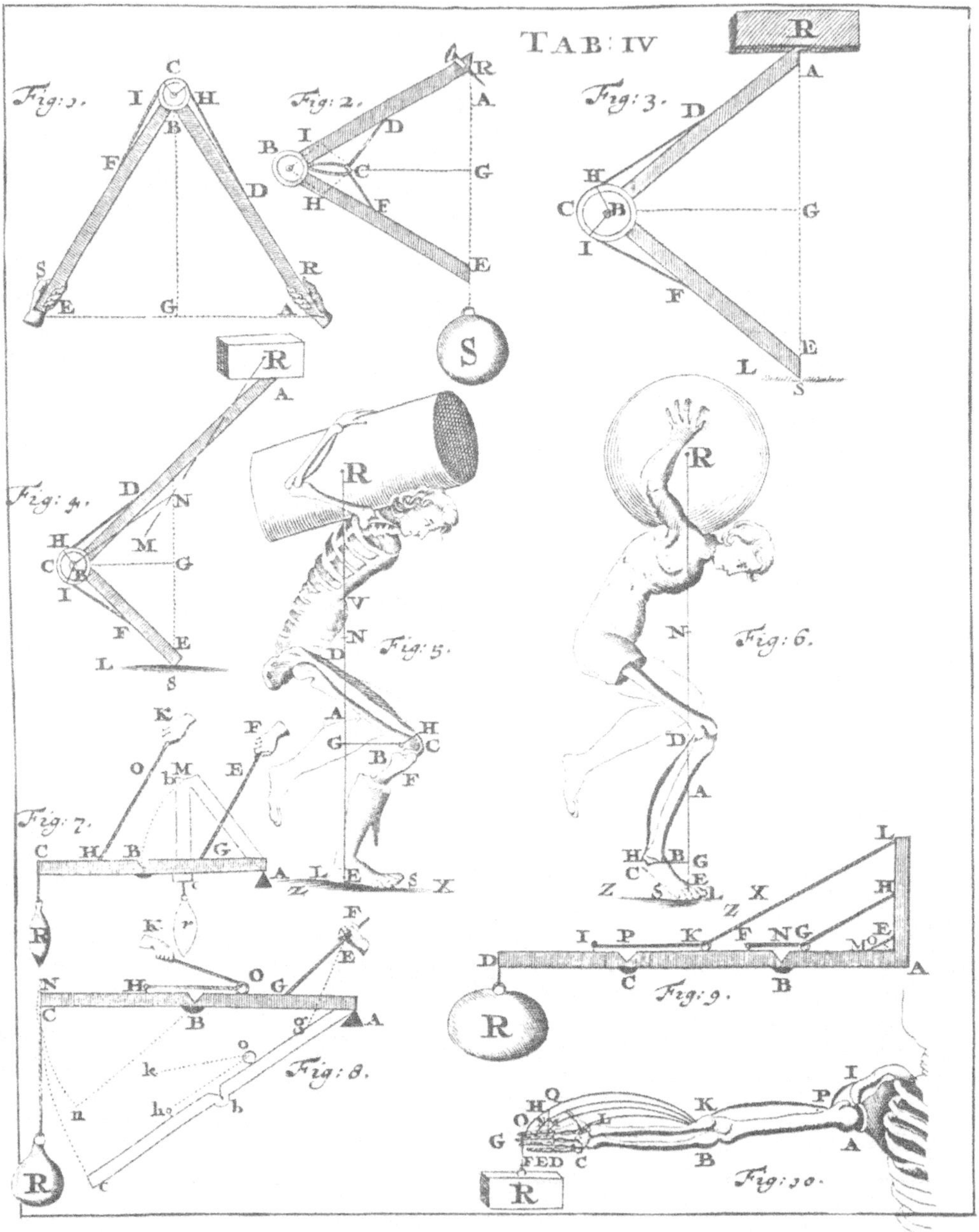

Table V

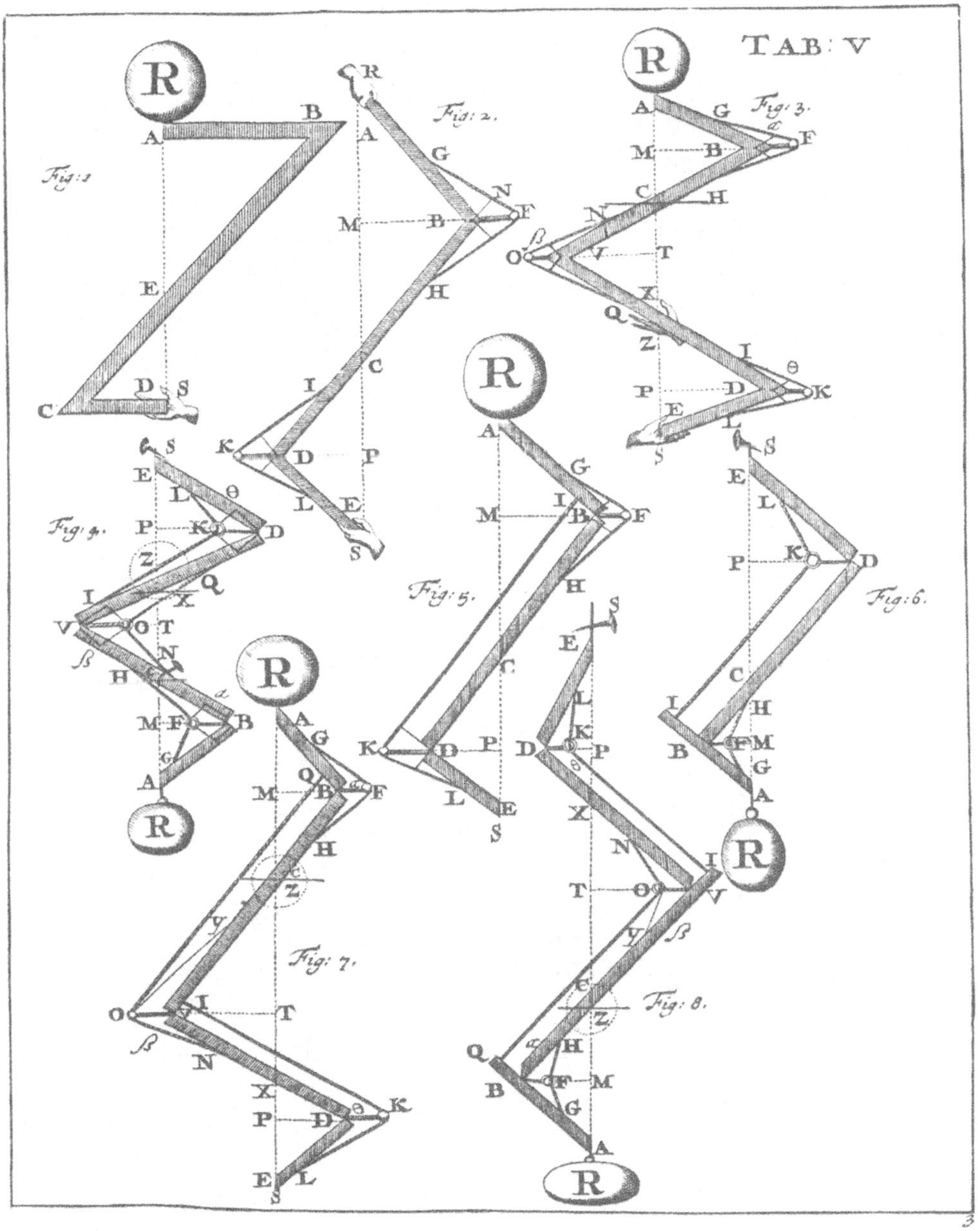

Table VI

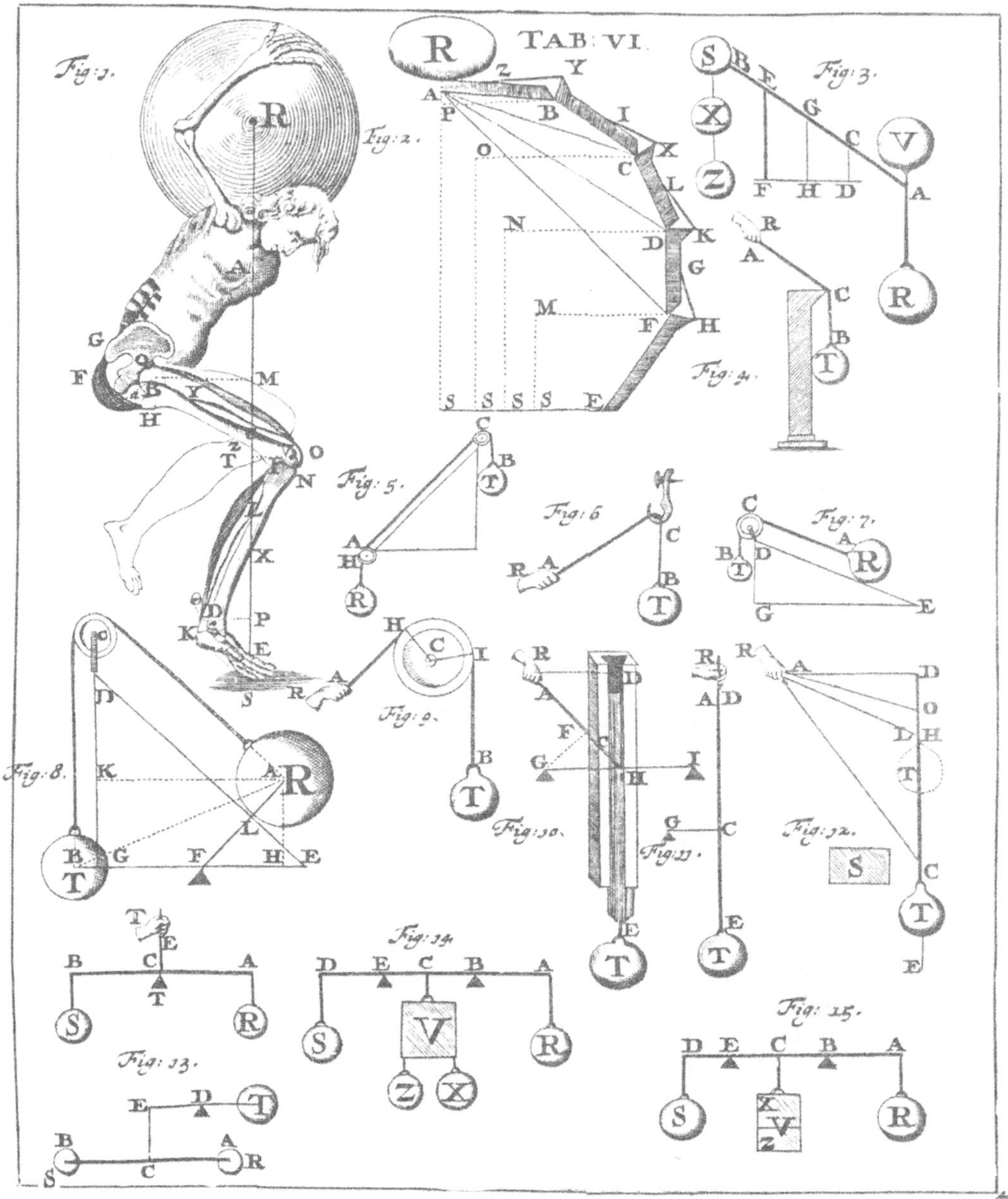
TAB: VI.
Fig: 1.
Fig: 2.
Fig: 3.
Fig: 4.
Fig: 5.
Fig: 6
Fig: 7.
Fig: 8.
Fig: 9.
Fig: 10.
Fig: 11.
Fig: 12.
Fig: 13.
Fig: 14.
Fig: 15.

Table VII

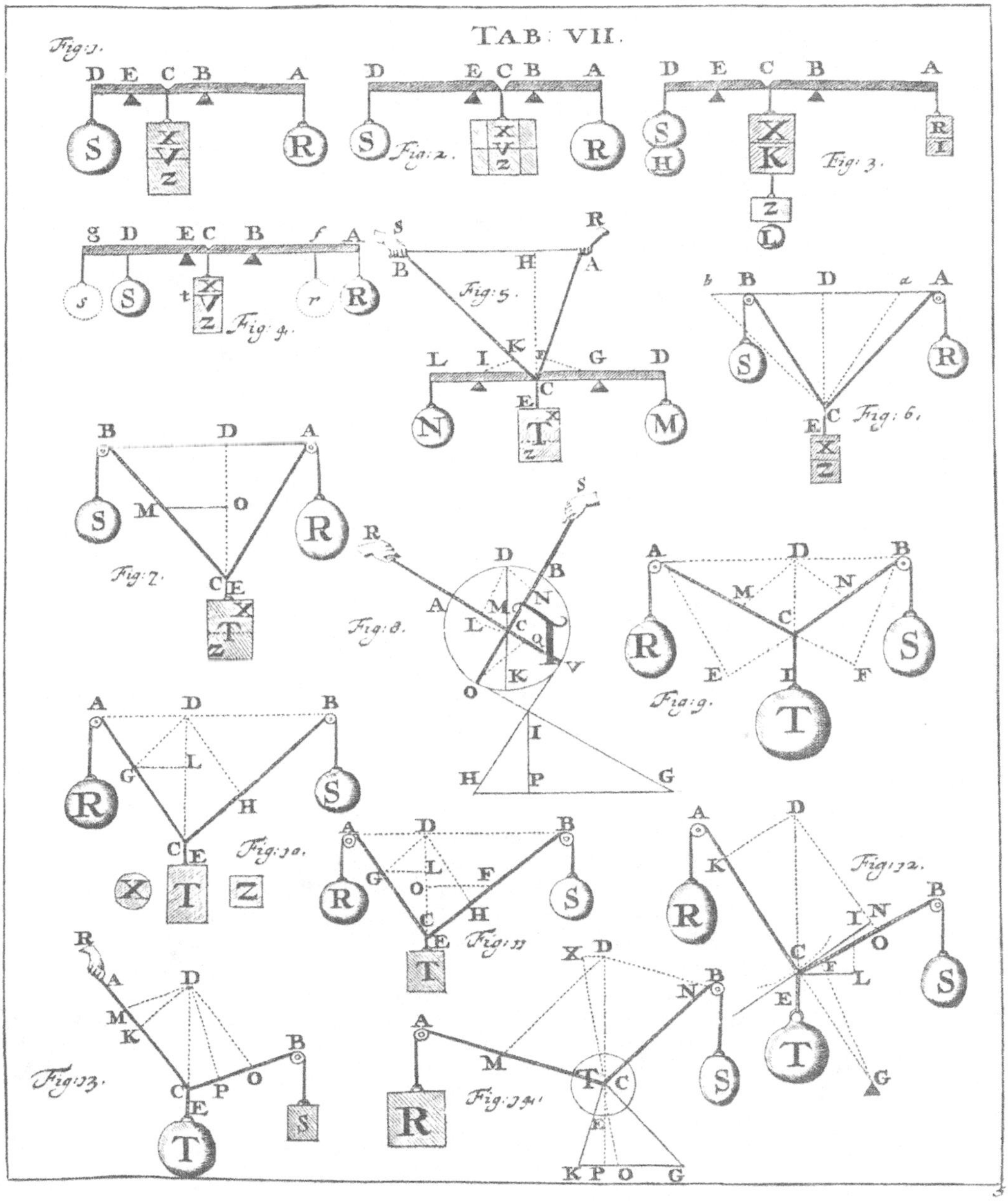

G. A. Borelli—On the Movement of Animals

Table VIII

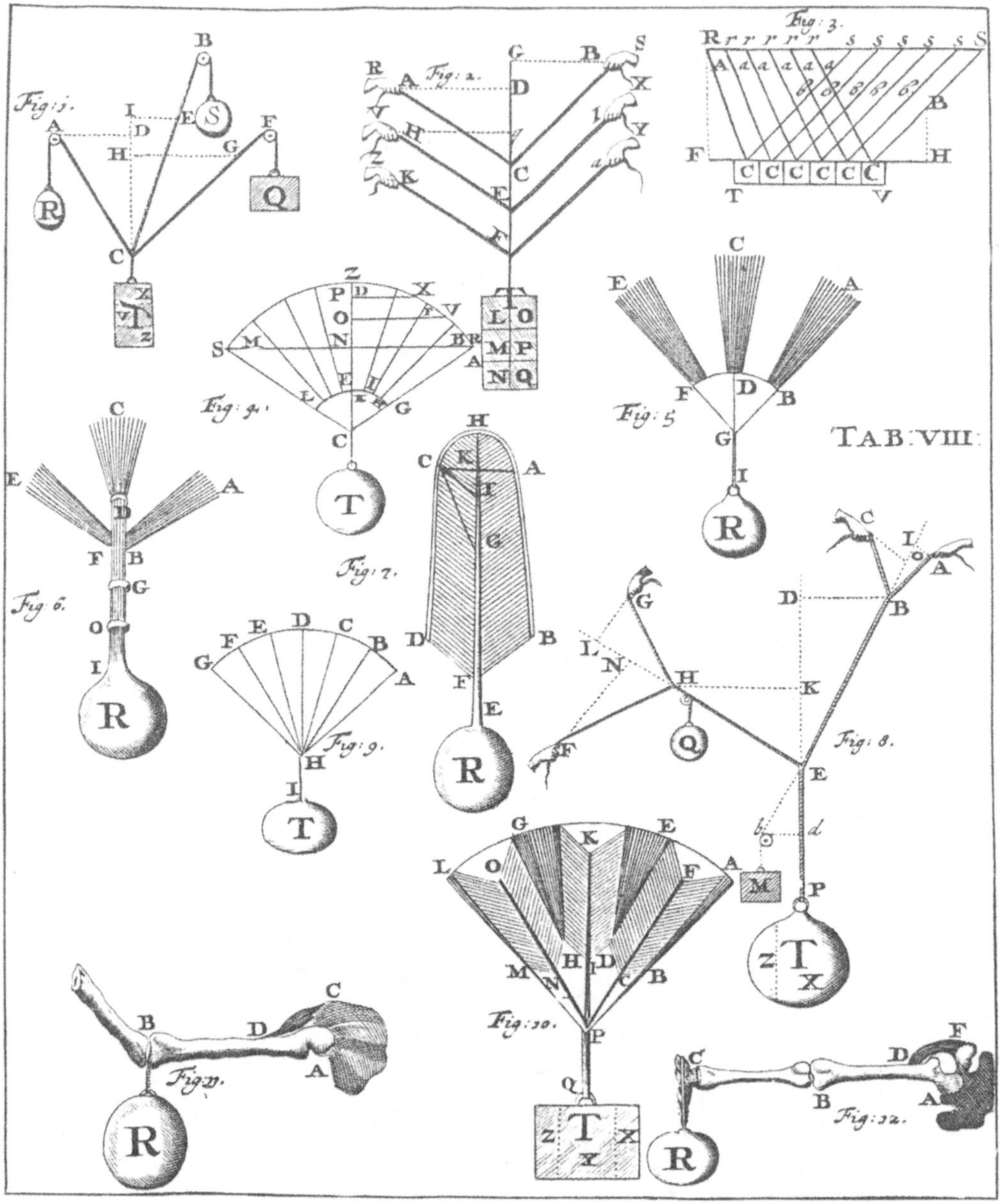
Fig: 1.
Fig: 2.
Fig: 3.
Fig: 4.
Fig: 5
TAB: VIII:
Fig: 6.
Fig: 7.
Fig: 8.
Fig: 9.
Fig: 10.
Fig: 11.
Fig: 12.

Table IX

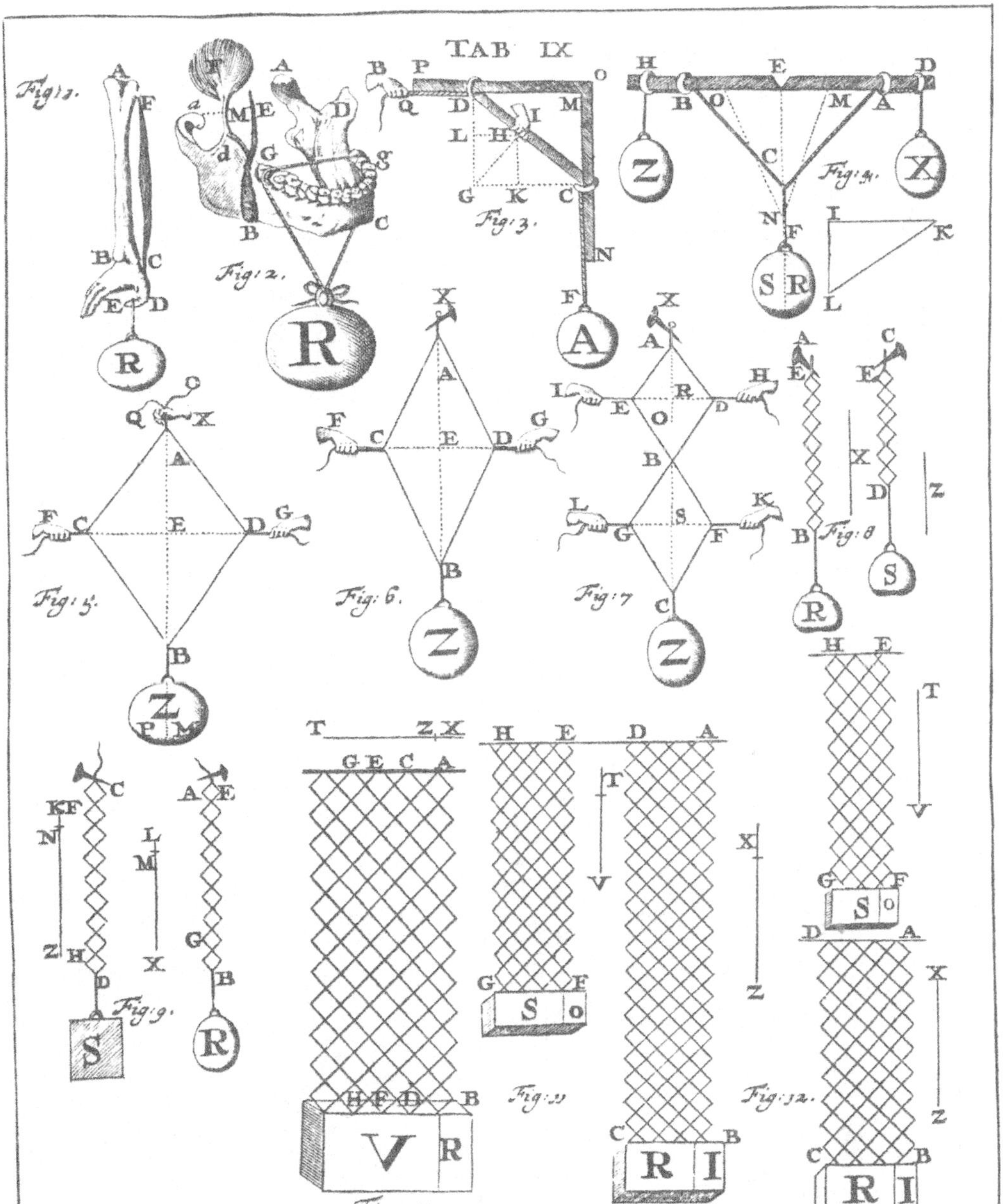

Table X

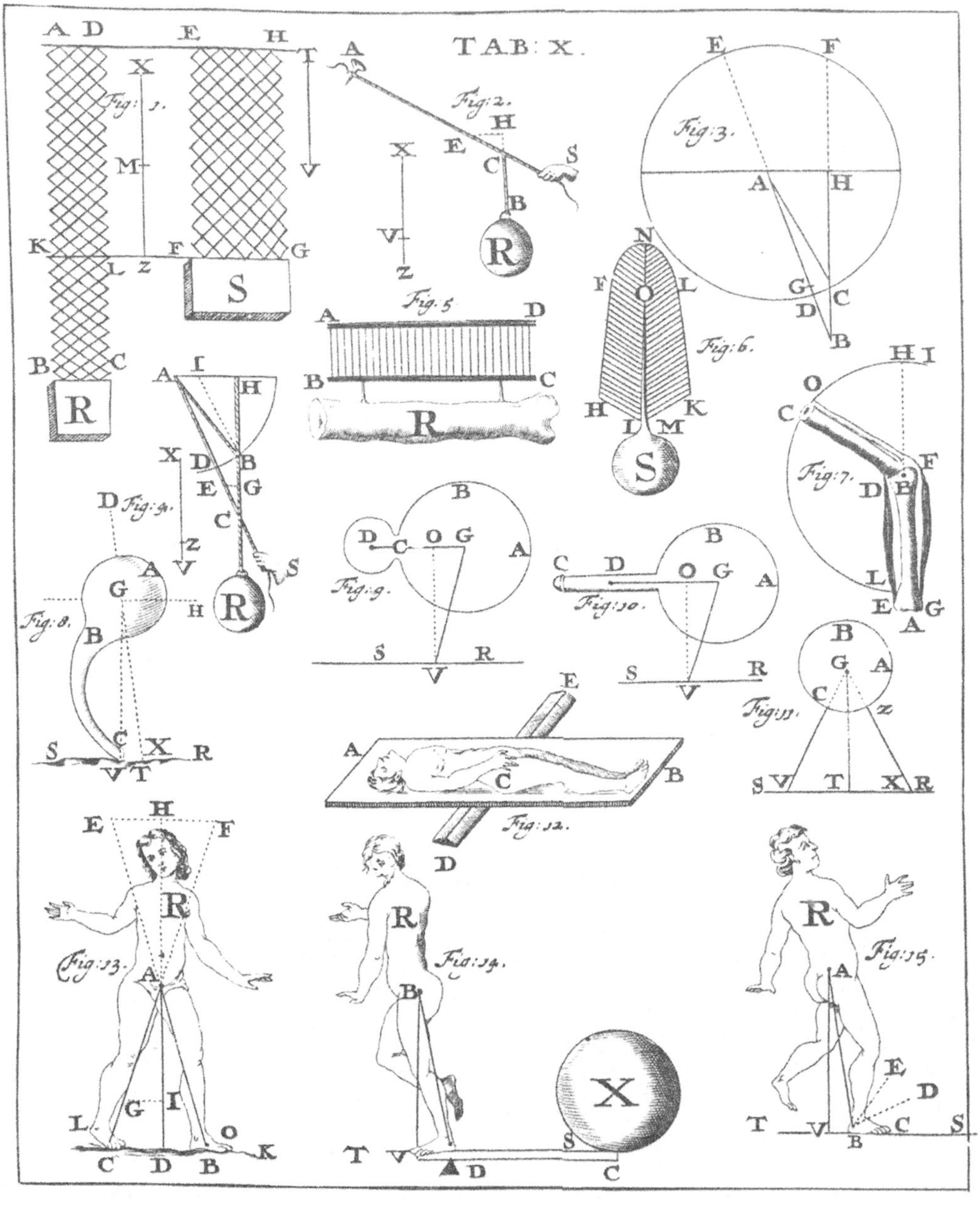
TAB: X.
Fig: 1.
Fig: 2.
Fig: 3.
Fig: 4.
Fig: 5
Fig: 6.
Fig: 7.
Fig: 8.
Fig: 9.
Fig: 10.
Fig: 11.
Fig: 12.
Fig: 13.
Fig: 14.
Fig: 15.

Table XI

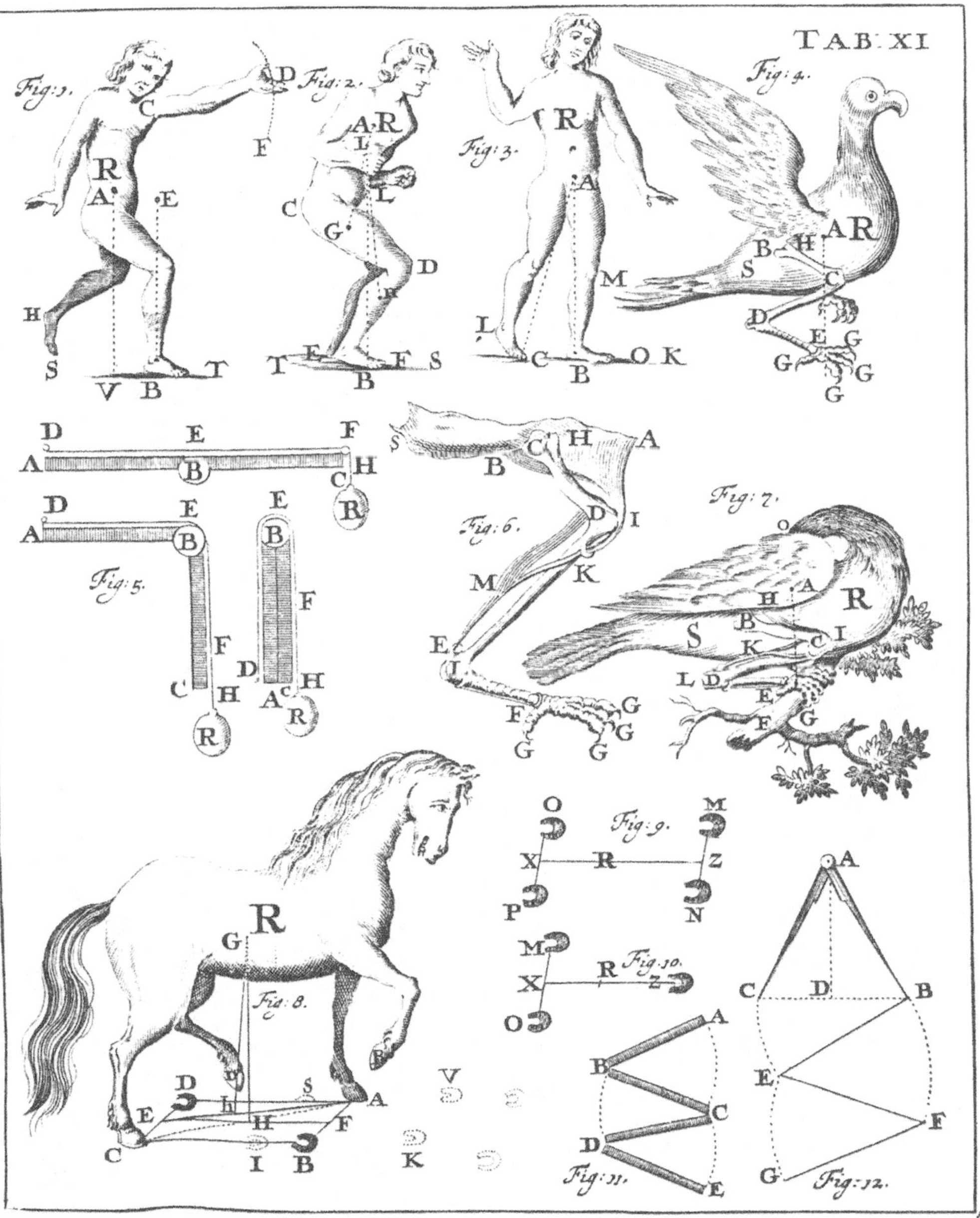
TAB: XI
Fig: 1.
Fig: 2.
Fig: 3.
Fig: 4.
Fig: 5.
Fig: 6.
Fig: 7.
Fig: 8.
Fig: 9.
Fig: 10.
Fig: 11.
Fig: 12.

Table XII

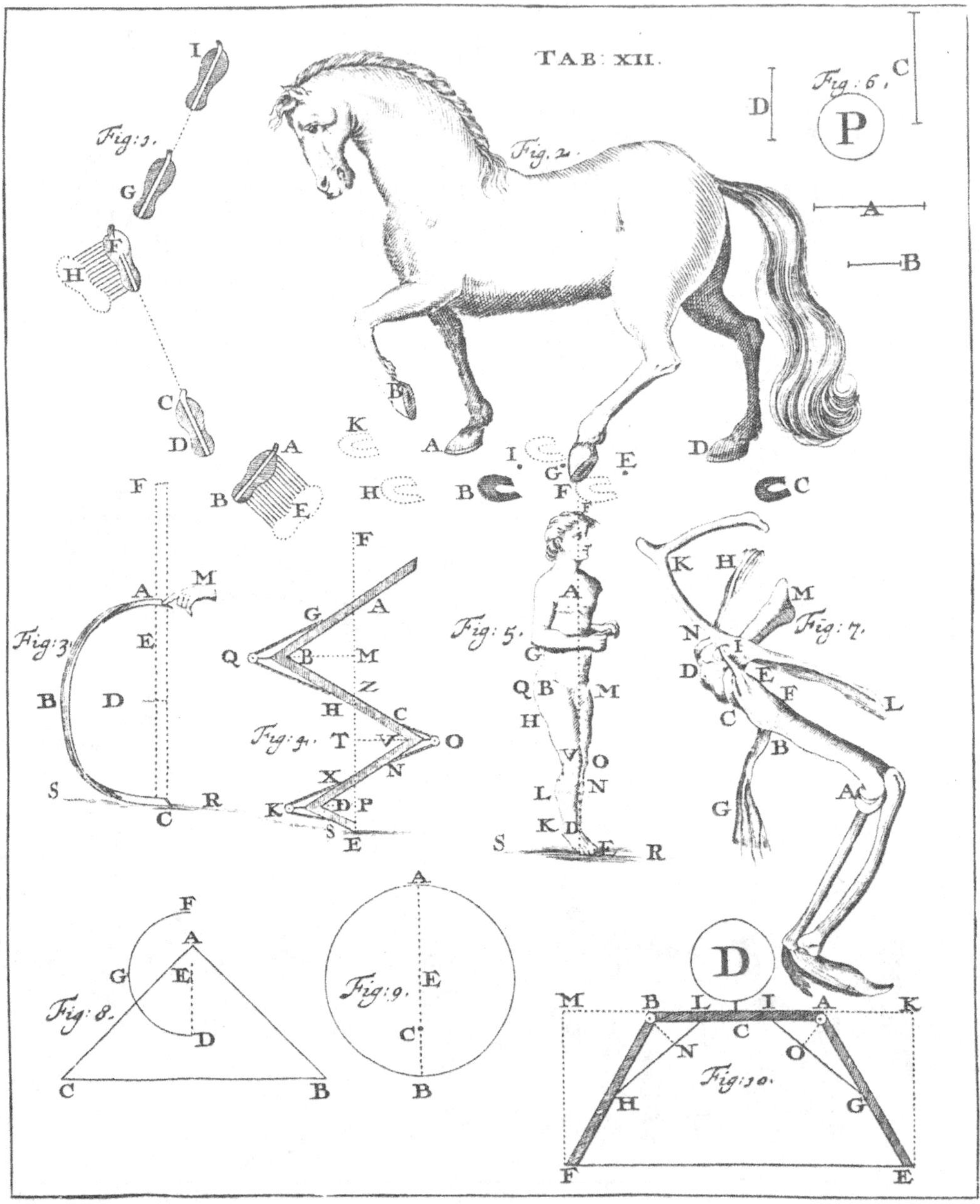

Table XIII

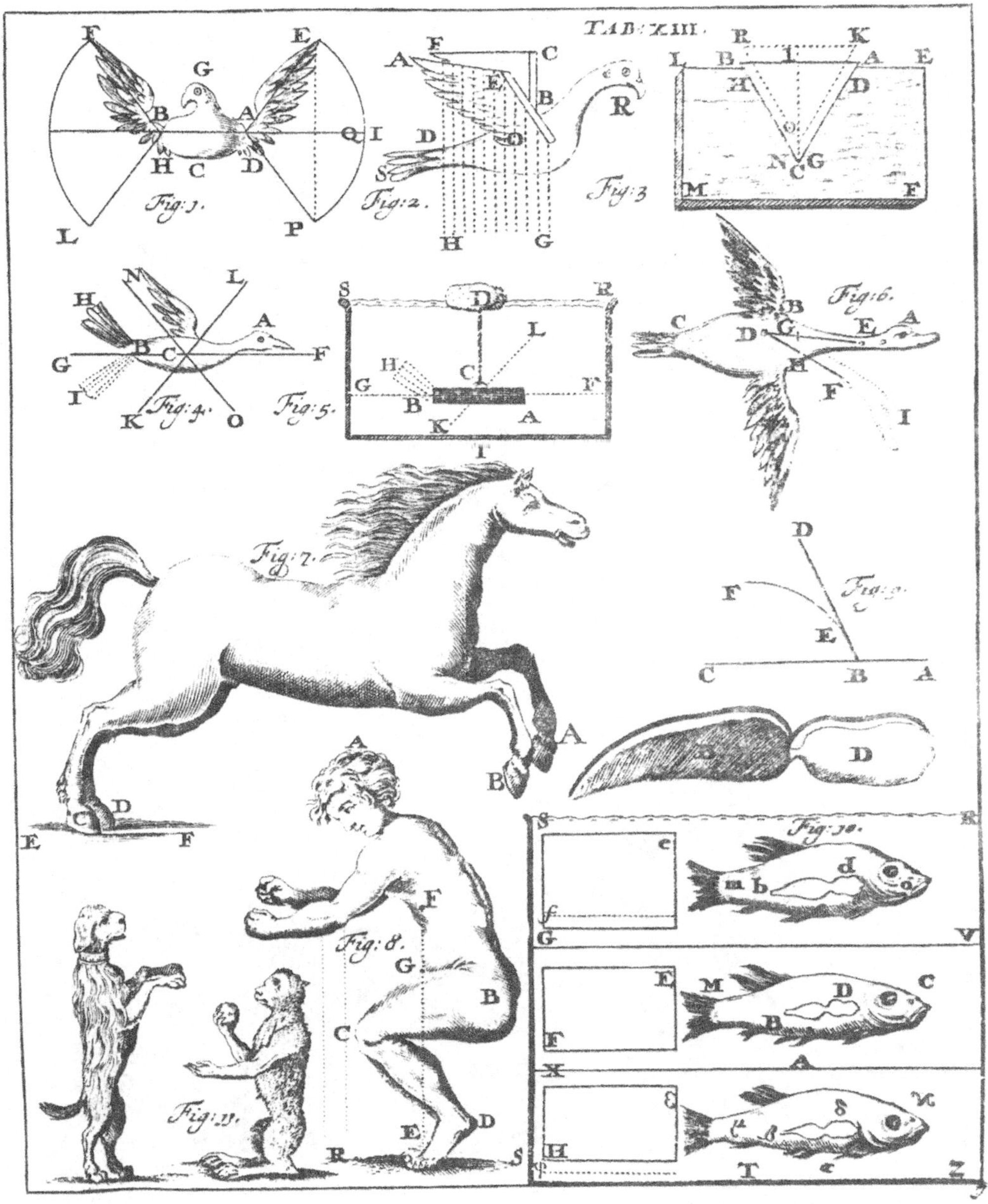

Table XIV

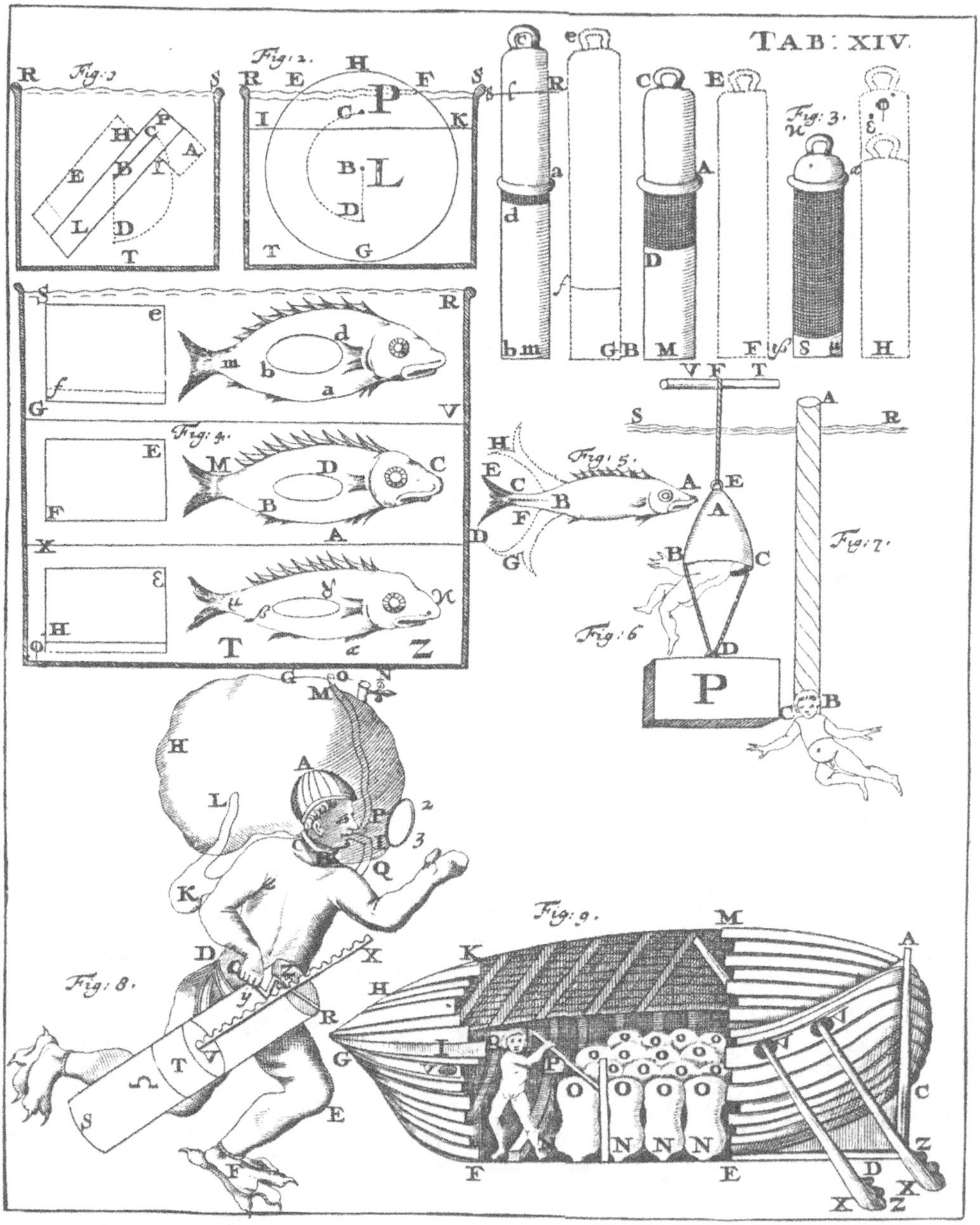

Table XV

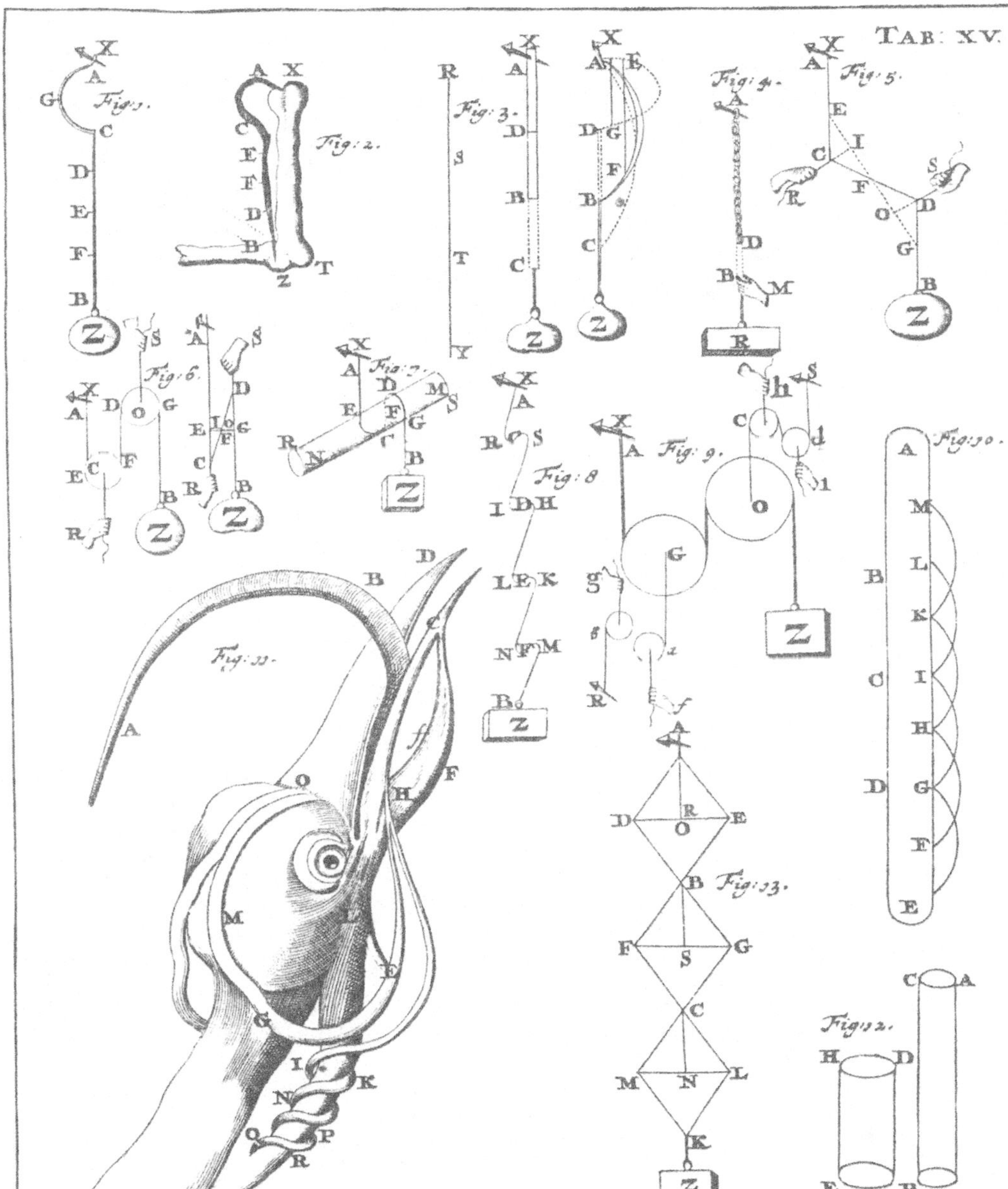

Table XVI

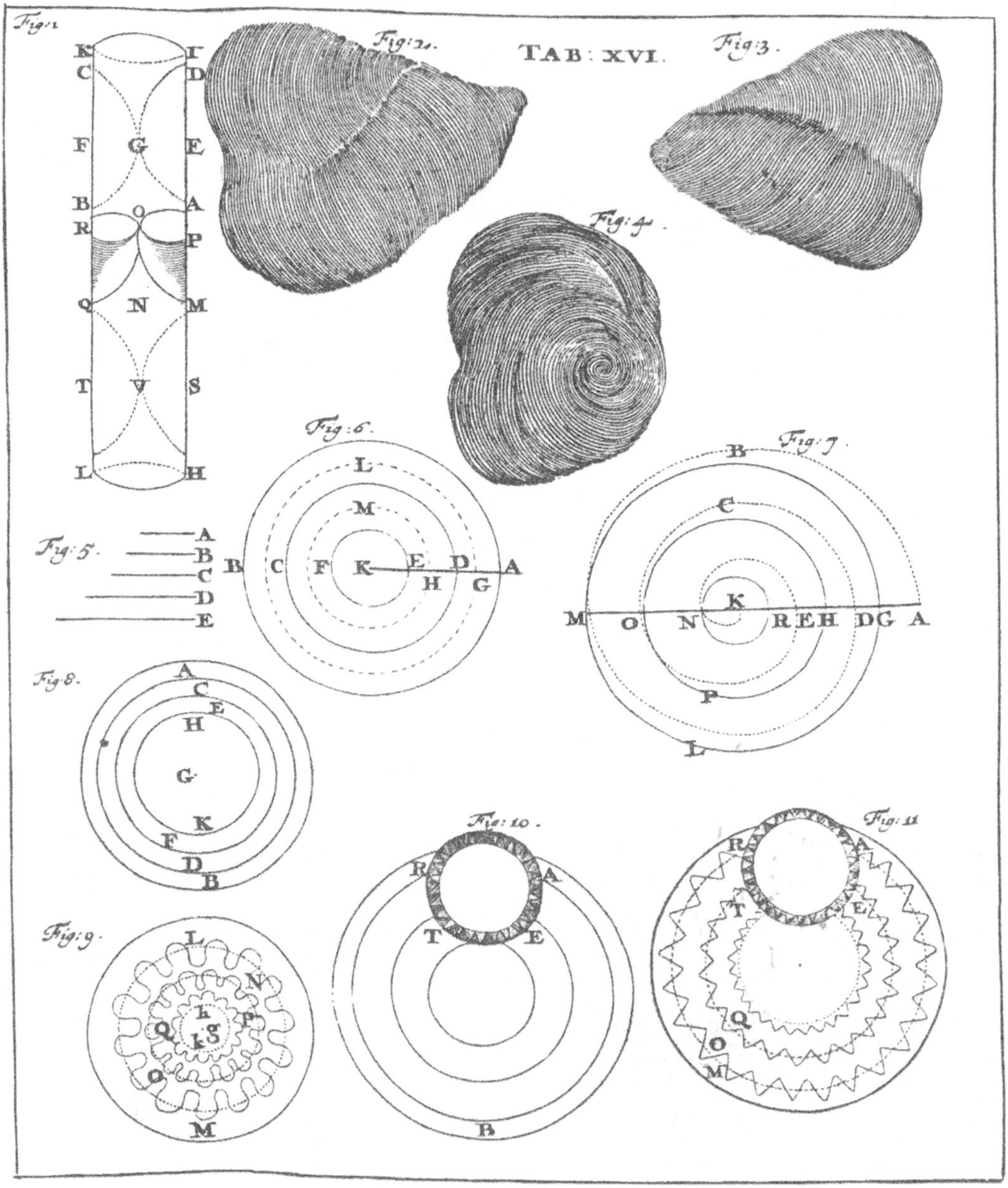
TAB: XVI.
Fig: 1
Fig: 2
Fig: 3
Fig: 4
Fig: 5
Fig: 6
Fig: 7
Fig: 8
Fig: 9
Fig: 10
Fig: 11

Table XVII

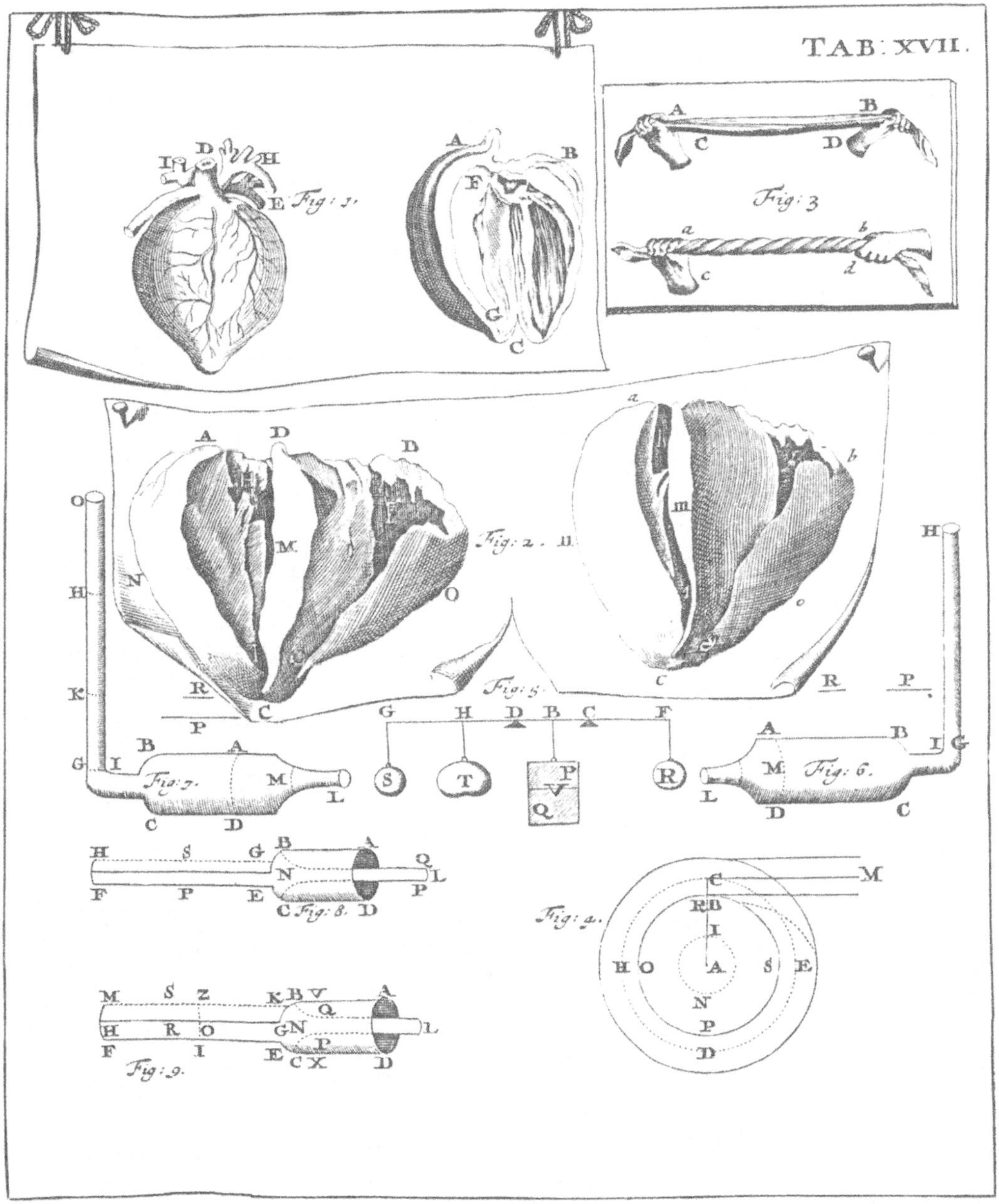

Table XVIII

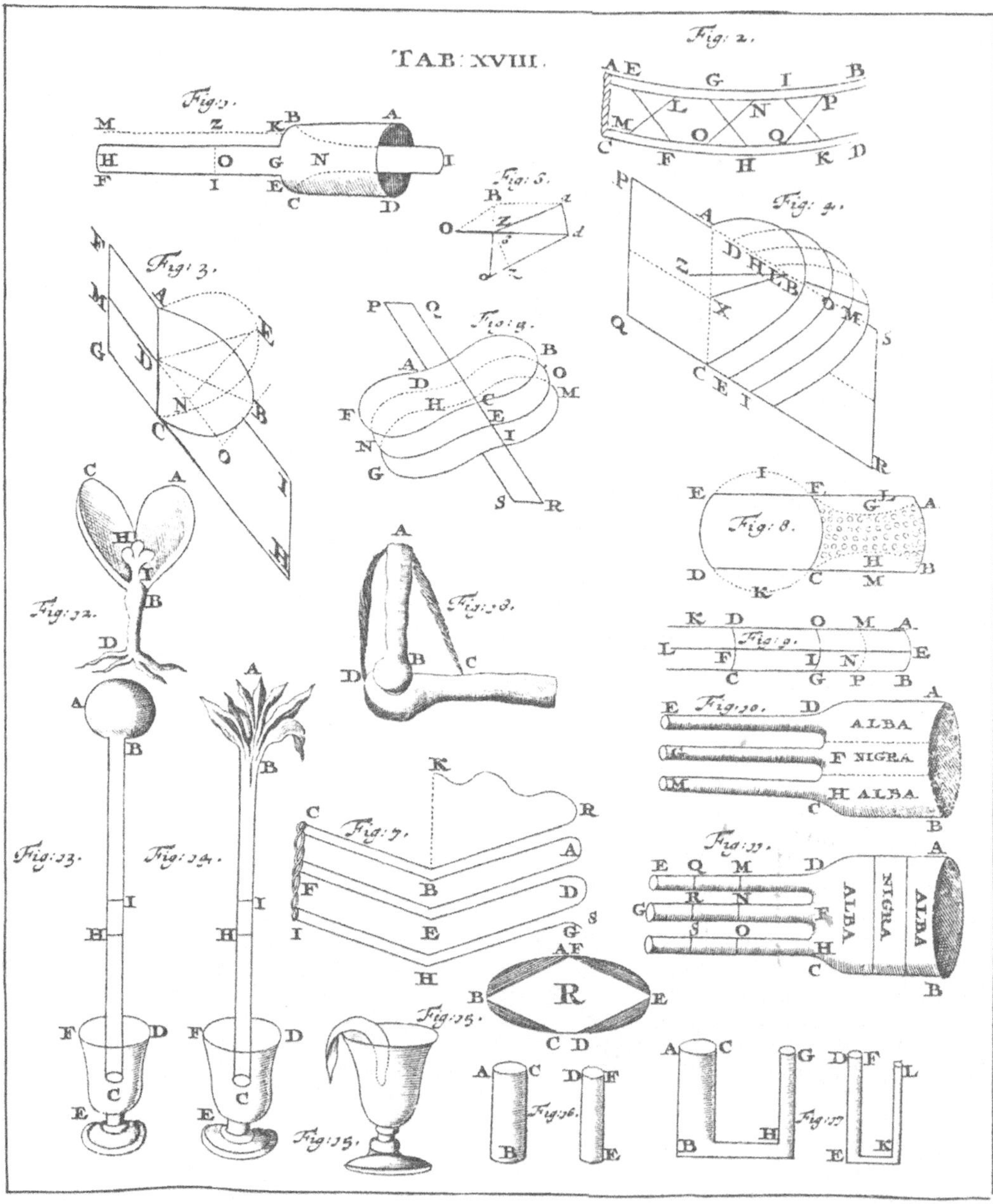

Made in the USA
Monee, IL
13 September 2026

62973776R10280